Leningrad Mathematical Olympiads (1961–1991)

Mathematical Olympiad Series

ISSN: 1793-8570

Series Editors: Lee Peng Yee *(Nanyang Technological University, Singapore)*
Xiong Bin *(East China Normal University, China)*

Published

The complete list of the published volumes in the series can be found at
http://www.worldscientific.com/series/mos

Vol. 24 | Mathematical Olympiad Series

Leningrad Mathematical Olympiads (1961–1991)

Dmitri Fomin

W♦ World Scientific

NEW JERSEY · LONDON · SINGAPORE · GENEVA · BEIJING · SHANGHAI · TAIPEI · CHENNAI

Published by

World Scienti ic Publishing Co. Pte. Ltd.

5 Toh Tuck Link, Singapore 596224

USA office: 27 Warren Street, Suite 401-402, Hackensack, NJ 07601

UK office: 57 Shelton Street, Covent Garden, London WC2H 9HE

Library of Congress Control Number: 2024061900

British Library Cataloguing-in-Publication Data
A catalogue record for this book is available from the British Library.

Mathematical Olympiad Series — Vol. 24
LENINGRAD MATHEMATICAL OLYMPIADS (1961–1991)

ISBN 978-981-12-5444-4 (hardcover)
ISBN 978-981-12-5497-0 (paperback)
ISBN 978-981-12-5445-1 (ebook for institutions)
ISBN 978-981-12-5446-8 (ebook for individuals)

For any available supplementary material, please visit
https://www.worldscienti ic.com/worldscibooks/10.1142/12782#t=suppl

To my parents,
who gave me plenty of opportunities to learn, to think, to explore,
to make my own mistakes and then make them again.
Thank you for being kind and patient.

A problem well put is half solved.

John Dewey

Solving a problem for which you know there's an answer is like climbing a mountain with a guide, along a trail someone else has laid. In mathematics, the truth is somewhere out there in a place no one knows, beyond all the beaten paths. And it's not always at the top of the mountain. It might be in a crack on the smoothest cliff or somewhere deep in the valley.

Yoko Ogawa

Contents

PART 1

Introduction

Preface to the English Language Edition

Mathematical competitions and problem solving in science have an ancient and rich history covering at least four thousand years. As far as we know, examinations in mathematics and logic began in Babylonia, and became common in Egypt and Greece. That tradition was followed in all universities and academies throughout the subsequent millennia, whether in imperial China or in republican France, ancient Byzantium or contemporary America.

In medieval Europe, a so-called *Quadrivium* curriculum, which covered arithmetic, geometry, music and astronomy, was considered a necessary step to the magister's degree—and of course, the students had to pass an exam. *Trivium*, which preceded it, was a simpler course but it included logic. In particular, the students had to solve simple logical problems similar to the famous syllogism:

Socrates is a man.
All men are mortal.
Thus Socrates is mortal.

Apparently, this is a later form of a similar syllogism from a book by Roman philosopher Sextus Empiricus (c.160–c.210 A.D.) [16], which read:

Socrates is a human.
All humans are animals.
Therefore, Socrates is an animal.

Trivium lectures on logic also included paradoxes, self-contradicting statements and exercises in what today we would call philosophical rhetoric.

With the advent of the Renaissance, interest in science and mathematics was renewed. Mathematics became so popular that local aristocracy and even monarchs held public contests in logic and arithmetic for the amusement of the courtiers and learned people. In 1535, in the course of such mathematical "duel", one of the most prominent Italian mathematicians of that era Niccolò Tartaglia famously defeated his lesser known colleague Antonio Fiore, demonstrating that he somehow learned the formula for the roots of cubic equations, which was supposed to have been a closely held secret known only to a couple of other contemporaries. The historians of science still debate whether Tartaglia deduced that formula on his own, bought or cajoled it from Bolognese mathematician Scipione del Ferro, or perhaps even stole it from the same del Ferro. In any case Tartaglia did not enjoy his fame for long, as his formula was shortly published by Girolamo Cardano who received that formula from Tartaglia himself after swearing not to reveal it to anyone else. Apparently, Cardano only could keep the secret for seven years after which he decided that the world deserved to know the truth, solemn promises be damned. Naturally, a huge scandal ensued—you can read all about it in [14], chapter 57.

About one hundred and fifty years later, discovery (or should we say, invention?) of calculus made mathematics widely popular again, even among the uninitiated. Those who could afford it hired local professors to teach them calculus, despite the fact that the theory at the time was extremely vague and even carried some air of magic, religion and general philosophy all mixed up together. After all, it was all about "infinitesimals", which, on one hand, were not really that much distinguishable from zero, but on the other hand, by adding them up you somehow managed to obtain non-zero magnitudes. However, we can be quite sure that the popular attitude among the aristocracy and the merchant class was not to really learn calculus but to familiarize themselves with a few buzzwords they could later use to wow the ladies at the next high society gathering. Nothing is new under the sun, indeed.

Of course, just as it happened with Tartaglia, Cardano and del Ferro, mathematicians themselves quite often joined the fray, becoming extremely competitive, especially when trying to claim solutions to some famous problems of the day. They still do that as the famous Poincaré Conjecture debacle, described in [22], demonstrated to our common dismay.

And for that, of course, there had to be a sort of understanding about which problems were the important ones, as well as about the degree of their difficulty.

During late antiquity and medieval times, when people talked about the important (and oh, so elusive) mathematical problems, they usually meant the most famous ones called "angle trisection", "squaring the circle", and "doubling the cube". These problems were geometric in form but their mathematical "core" was algebraic; on the surface, questions inquired about the constructibility of some geometric objects. In fact, however, they were about proving (or disproving) that some numbers could be represented as roots of certain polynomials with integer coefficients. Another world-famous question of that time had genuine geometrical (as well as logical) substance—we are talking here about "the problem of the fifth postulate", which meant deducing the fifth axiom (postulate) of Euclid's geometry from the first four. None other than the famous Persian mathematician, astronomer and poet Omar Khayyam thought that it would be the crowning achievement of his life were he to accomplish that feat.

Another much more formal example of how the most important problems were determined and "popularized" is the famous list of Hilbert's Problems. It consisted of 23 problems submitted by David Hilbert, one of the greatest mathematicians of the modern era, at the International Congress of Mathematicians in 1900. Actually, Hilbert only presented ten of the problems and then published the rest of them as a separate text. We can recommend book [18] as a good popular exposition of Hilbert's Problems and related developments.

One hundred years later, in 2000, to commemorate the centennial anniversary of Hilbert's Problems, the Clay Mathematics Institute published the Millenium Prize Problems list which comprised seven (only seven!) major mathematical questions. Quite differently from Hilbert's list, the Millenium problems came with a hefty monetary award—one million dollars for each problem solved. The oldest among them, the so-called Riemann's Hypothesis, was also included on the Hilbert's list. It remains unsolved after more than 150 years of constant attempts undertaken by the very best minds of humankind.

So this is what it has come to—from simple arithmetic questions offered to the lowly clerks vying for the position of a junior assistant priest somewhere in Abydos of the Old Kingdom epoch to the immensely complicated and extremely abstract problems of our time that carry with them promises of great wealth and fame.

$$* \quad * \quad *$$

Now, let us turn to another important aspect in the analysis of mathematical problem-related texts, namely their historical context and the issues of translation.

One of the oldest mathematical texts in human history, the so-called Rhind papyrus, is a problem book, a sort of a math teacher's helper text. Surely it was not used to run mathematical contests, although, who can really know?

One of the questions from the Rhind papyrus was this (presented here in a somewhat modernized version):

Question 79. *An estate has seven houses, every house has seven cats, every cat caught seven mice, every mice ate seven spelt[1] stalks, and every spelt stalk generates seven heqats.[2] Compute all the numbers for the full inventory in a table, as well as their total.*

Seems like there is not that much to solve. Compute a few powers of seven, and then add them up. So if we formulate the question as "*Compute* $7 + 7^2 + 7^3 + 7^4 + 7^5 + 7^6$", then there is really nothing else, and it would not look particularly interesting to historians besides the fact that Egyptian mathematicians of that era knew how to calculate geometric series. That looks okay for a compilation book of short mathematical problems of the ancient past, but it would certainly not do for a book on the history of mathematics or for any book, which intends to give the reader a bit more information than just a bunch of dry formulas.

The original question tells us about cats and mice, it shows us what units of volume were used, it hints at what kind of wheat was grown. It also (and this is my personal speculation) implies that the teachers in ancient Egypt tried to demonstrate to the students that the mathematics can be abstract. How else do you explain the last part of the question which asks us to calculate the sum of all the numbers? There is absolutely no practical justification for adding up numbers of cats, mice, volumes of wheat, houses, etc.

All this additional information enriches the document, and sometimes it might even help the reader in the process of solving the problem. At the very least, it certainly adds historical color.

Nowadays, almost every time a compilation problem book is published, the authors (and sometimes editors) are presented with the following question: should the text be updated to reflect the current terminology and social changes?

Imagine this scene in some future classroom circa 2319:

TEACHER: So here is this really old problem I would like you to solve. *In the country*

---

[1] A type of wheat.
[2] *Heqat* is a unit of volume equal to approximately 4.8 liters (of grain, we should assume).

of Eldorado its n cities are connected by the railroads so that each city can be reached from any other ...

STUDENT A: What is a *railroad*?

TEACHER: Oh, a very good question. You see, in those days...

STUDENT B: I am sorry, but what is a *city*?

TEACHER: Well, I guess you should have paid more attention during your lectures on ancient history...

STUDENT C: And what is a *country*?

TEACHER (giving up): Okay, listen. I will put it like this: *a finite connected graph with n vertices...*

STUDENTS: Ahh! You should have said it like that right from the start!

I am sure you see what I mean. But even if not so many years have passed, updating an older edition can still be quite tough. Add to it all sorts of complications one might run into when translating from one language to another, or trying to explain the historical concepts or social constructs of any age, culture and era to the readers from a different country and society.

Here is a nice short example from this very book.

Question 1979.05 (**original version**). *In the class each boy is friends with three girls and each girl is friends with two boys. The class has 19 desks and 31 pioneers. How many students are there?*

A contemporary Western reader of secondary school age would certainly have some difficulty even understanding the question. First of all, what do they mean by pioneers? What are the first settlers of the American Wild West doing there? And if we only have 19 desks, how can we have 31 people sitting there? Are these people even supposed to be sitting at the desks—after all, they are adults and not the schoolkids, right?

A Soviet reader from the 1970s, however, understood perfectly well that the *pioneers* were simply the members of the Soviet youth organization called Young Pioneers (almost every student in the middle school was a Young Pioneer; it was *that* sort of "voluntary" membership). So the line about 31 pioneers simply meant that there were at least 31 students in the class. Next, the desks. Again, at the time, virtually all the desks in the Soviet schools were the type which accommodated two students. Therefore that line can be simply restated as saying that there are no more than $2 \cdot 19 = 38$ students in the class. Now it all makes sense and we can go ahead and solve the problem.

So what should we do? Are we supposed to completely rewrite the problem so it is perfectly understood by the students of today, stripping away all color of the period? If we do so, then we must be very careful lest we erase some of the logical steps that the solver would have to make in order to translate the statement to the formulas, inequalities, etc. By no means should we reformulate the problem like this.

Question 1979.05 (**really bad reformulation**). *In the class each boy is friends with three girls and each girl is friends with two boys. We know that there are at least 31 and at most 38 students in the class. How many students are there?*

Of course, we can find some decent way to stay close to the original text while at the same time getting rid of all the outdated terminology and of the period-specific "hints", which back then allowed the students a relatively easy "translation" of the statement to the formulas.

I, however, prefer (at least in this book) to stay as close to the historical document and keep as much "old times" color and terminology as reasonably possible. My preferred solution to this problem is to enhance the text by adding footnotes and comments. Their purpose is to explain the specific terminology and provide clarifying details which used to be readily available and sometimes obvious to the local reader at the time when the book was published.

$$* \quad * \quad *$$

You will have noticed that some names and numbers occur rather frequently. Those often represent tongue-in-cheek references to the problem committee members, local landmarks, historical figures, cultural memes of the era, etc. For instance, the name **Theo** (**Fyodor** or diminutive **Fedya** in Russian) was used as a friendly nod to Fyodor Nazarov, a prolific contest problem author as well as a former Soviet IMO[3] team member. A few problems from the 1970s, which use the names **Nick** and **Vera** (**Kolya** and **Vasya** in Russian original), are a similar gesture towards Nikolay Vasiliev, a Moscow mathematician and olympiad enthusiast who was well known and admired by many colleagues around the country.

The numbers **30**, **45** or **239** are used quite often, and not because they possess some extraordinary arithmetic properties[4], but because they represent the three best sci-math public schools in 1960s–1980s Leningrad.

A few more informational tidbits:

- This collection spans 31 years of high level mathematical competitions for middle and high school students;
- It contains approximately 1150 unique problems from LMO, plus 120 additional problems from other math contests held in Leningrad;
- In addition to this compilation, problems and solutions for the last five Leningrad Math Olympiads (1987–1991) can also be found in book [12].

---

[3] International Mathematical Olympiad—the ultimate math competition for high school students; IMO has been held annually since 1959.

[4] **30** is a square pyramidal number, which also equals number of the edges in both icosahedron and dodecahedron; **45** is a triangular number and a Kaprekar number; finally, **239** is a twin prime, and also the only natural number which, when represented as sum of powers, requires maximum number of squares (four), positive cubes (nine), and fourth powers (nineteen).

For this translation, I have tried to fix all the typos and errors found in the previous editions. Several new solutions were added, some were updated or rewritten. With the only exception of "mysterious" Question **1962.51**, all problems are supplied with answers, hints, and solutions. More than twenty problems were added to the "Additional problems" section at the end of the book.

I have also substantially expanded some parts of the introductory sections "Brief Historical Overview" and "Olympiad: The Inside Look". Most of the additional information comes from [1], [26], [29] and [31]. Some photos came from personal archives, others—from public sources; namely, from Wikipedia, KVANT magazine, websites and archives of St. Petersburg State University's Faculty of Mathematics and Mechanics, St. Petersburg Mathematical Society, St. Petersburg Electrotechnical University, St. Petersburg ITMO University, Phys-Tech Institute and St. Petersburg State University of Architecture and Civil Engineering, university alumni websites such as Math-Mech.1967 and St. Petersburg State University Alumni Association. In addition, my thanks go to Alexey B. Aleksandrov, Daniil A. Aleksandrov, Sergey L. Berlov, Simon M. Bliudze, Anna A. Gaikovich, Leonid M. Koganov, Konstantin P. Kokhas, Boris A. Lifshits, Andrey A. Lodkin, Natanson family, Ilya Mironov, Galina E. Piolunkovskaya, Elena N. Sokiryanskaya, Nina N. Uraltseva, Sergey S. Vallander and Dmitry A. Volkov for their help with the image files (several photos came from their personal archives, others—from the web or hard-copy publications they have found per my request). The original of the A. S. Merkurjev's photo belongs to George's Bergman's image collection.[5]

Finally, in addition to all people already mentioned above I would like to express my sincere gratitude to Larisa A. Afinogenova, Evgeny V. Abakumov, Miron Ya. Amusia, Boris M. Bekker, Mikhail Yu. Bliudze, Sergey A. Bogomolov, Yuri D. Burago, Yuri A. Davydov, Sergey V. Fomin, Sergey A. Genkin, Vitaly E. Gol'din, Lev I. Gor'kov, Yuri I. Ionin, Vladimir S. Itenberg, Vsevolod V. Ivanov, Alexander P. Karp, Yuri F. Kazarinov, Piotr S. Lavrov, Boris B. Lurie, Vladimir G. Mazya, Alexander S. Merkurjev, Boris Z. Moroz, Aleksandr I. Nazarov, Vladimir V. Nikiforov, Vladimir P. Odinets, Ernest O. Rapoport, Leon A. Petrosyan, Iosif V. Romanovsky, Grigory V. Rozenblum, Marina V. Rozet, Viktor M. Ryabov, Oleg N. Serdobintsev, Nikolay A. Shirokov, Maxim V. Sorokin, Alexey L. Verner, Anatoly M. Vershik, and Anna B. Zhiglevich for sharing their archives and their memories, or helping me to find new information.

My special thanks to Inessa A. Blyumkina, Maria G. Bykova, Alexander D. Fomin, Chris Kischuk, Konstantin P. Kokhas, and Andy Liu for their assistance in making this book as error-free as possible. It goes without saying that the author is solely responsible for all the errors which, no doubt, still remain (hopefully, well-hidden!) in this text. If you find an error, omission or a typo, or you simply would like to send a comment, do not hesitate to contact me at ⟨ *author's last name* ⟩ *at hotmail dot com.*

DMITRI FOMIN
BOSTON, 2022

---

[5] Creative Commons license Attribution-ShareAlike 4.0 International.

Preface to the Original Russian Edition

This is a translation of the preface to the Russian editions of this book published in 1994 and 2021 ([11], [32]). It was updated for this publication.

In 1965, book [4], dedicated to the Moscow Mathematical Olympiads, was published in the USSR; its print run was $122,000$ copies.[6] Written by Andrey Leman, it contained the almost complete collection of MMO problems from 1935 through 1964, with solutions provided for every one of them. There was also an introductory section with training problems categorized by topic. Leman's book immediately became one of the main sources of math contest problems for numerous Soviet era enthusiasts of extracurricular mathematics. However, its publication also had an important historical aspect.

First, that book finalized the work of collecting and compiling all the previous math olympiads held in Moscow. That provided the future "olympians" with the first, most important, cornerstone in their work on the systematization of organized math contests in Russia.

Second, a wonderful article by Vladimir Boltyansky and Isaak Yaglom, which served as a preface to the book, contained a lot of interesting factual material and also expressed the true spirit of Moscow olympiads.

Naturally, any book like that is not just a compilation. It also represents something of a snapshot in time, a monument to the history of reason if you will. Perhaps our reader might reject the notion that such a book can be considered a cultural or historical object comparable to Michelangelo's David or to the first edition of "Hamlet". Still, I submit that a historical record of human ingenuity and intellect deserves at least as much respect as an artistic artifact.

Indeed, I truly hope that in a few centuries historians and intellectuals alike will be interested not only in material history but also in intellectual history (including mathematical and scientific history) as well. Just like nowadays the literary critics discuss and deconstruct the works of Joyce and Chekhov, so in the 23rd century the historians of the intellectual culture will surely engage in excited arguments over the International Math Olympiads of the late 20th century or the entrance exams to Leningrad State University (LSU) circa 1954.

---

[6] This is, actually, not a very high number for popular scientific literature published in the Soviet Union—the Russian original of [15] had the initial print run of $680,000$ copies. A year later it was nearly impossible to buy it in any large Soviet city.

Now, let us return to this collection of problems from Leningrad Math Olympiads. Alas, its fate is far more complicated and unfortunate than that of its Muscovite sister. There are several reasons for that, with the main two being the World War II blockade of the city and the shadow of provinciality that fell over Leningrad in the after-war years. The problems of the first Leningrad olympiads were not published back when they were still fresh—we should mention here that those olympiads started in 1934, one year before the Moscow contests. As a consequence, the archive of LMO completely lacks the pre-war problems sets (1934–1941), and, I am sad to say, there is no real hope of ever recovering them. The sets from the post-blockade years (1944–1960) are not complete, but at least we have considerable number of problems from those competitions. Finally, all the problems starting from 1961 have been found and the problem sets have been reconstructed (with various levels of confidence). Here we present the results of that endeavor.

Following the tradition established by the aforementioned preface article in [4], I wrote the foreword as both a formal historical survey and a collection of interesting anecdotes. Many facts came from the memoirs and reminiscences of mathematicians and teachers who were, or used to be, the members of the Leningrad olympiad committee. To all of them I give my sincere gratitude.

* * *

This collection contains all problems of the Leningrad City Mathematical Olympiads from 1961 through 1991; that is, from the 27th to the 57th olympiad.[7] The choice of the years was caused by the fact that the author was not able to recover or reconstruct full problem sets for the olympiads of 1934–1960.

For a similar reason we did not include the problems for the 5th grade olympiads from 1969–1978. That contest was—at least, in part—organized by a different olympiad committee and its records could not be found.

And, of course, while the problems from the later years' contests are fully available, in 1991 the city itself changed its name, returning to its roots circa 1703. Thus, the history of the Leningrad Mathematical Olympiads ended in 1991. Following that watershed moment, the city olympiad was accordingly renamed to the Saint Petersburg Mathematical Olympiad; hopefully, a similar volume of **SPbMO** problems and solutions will soon be published. Those who can read Russian are encouraged to peruse the books [17], [23], [24] or access the website of the SPb Math Olympiad.

We need to emphasize here that some of the problem sets were not reconstructed with absolute certainty. Many different, sometimes conflicting versions of those sets were uncovered in personal archives, which contained both preliminary and final contest printouts. Quite often some small changes were made immediately before the competition, and in some cases errors and typos were found and fixed when the olympiad was already in progress. As a result, in some printed copies those errors were fixed and in some they were not—this made it nearly impossible to determine the "real" problem set.

---

[7] Indexing of the LMOs presents a bit of a conundrum as it was understood that during the war at least two or three contests were not held. This, however, turned out to be incorrect.

As you will see, sometimes a few problems are included within a comments paragraph, attached at the end of the respective section. This is because in some years a few additional problems were provided for the especially successful competitors—those who managed to solve all the problems in their grade. They were subsequently either moved to the next grade's auditorium or given an additional problem. Their results on these questions were not counted for any official tally; this was done purely to provide a challenge. Due to small discrepancies in the olympiad committee records, it cannot be established for sure which of these problems actually made it into the final version of the contest list; in a situation like that, we provide both versions together with the relevant commentary.

All problems in this new edition are provided with solutions, hints or short answers.[8] I am very grateful to numerous enthusiasts—teachers, professors, students and math connoisseurs—who, over the years, have provided problems and solutions for these contests. Special thanks go to my friend and colleague Konstantin Kokhas, whose role in preparing the second edition of this book ([21]) cannot be overstated.

A few words about the authorship of the olympiad questions: for the vast majority of the problems from 1961 through 1979, the names of their authors were not recorded at the time, and so the chance of ever recovering them is extremely slim. On the bright side, virtually all the authors of the LMO problems starting at 1980 are known (the "authorship index" can be found at the end of this book).

I would like to express my deep gratitude to everyone who helped me in this endeavor by sharing their memories of those thirty years. The following people have graciously granted me access to their personal document and photo archives: L. P. Avotina, S. M. Ananievsky, M. I. Bashmakov, O. N. Bondareva, T. A. Bratus, Y. D. Burago, S. S. Vallander, I. Ya. Verebeichik, S. A. Gol'dberg, M. N. Gusarov, V. A. Zalgaller, O. A. Ivanov, Y. I. Ionin, K. P. Kokhas, L. D. Kurliandchik, B. A. Lifshits, A. R. Maizelis, A. S. Merkurjev, N. M. Mitrofanova, N. Yu. Netsvetaev, A. I. Plotkin, I. S. Rubanov, N. A. Sokolina, S. V. Fomin, V. M. Kharlamov.

Also, my thanks for all of the advice and support I received go to S. A. Genkin, I. V. Itenberg, A. L. Kirichenko, V. M. Golkhovoy, I. A. Chistyakov. Some of the solutions were communicated to me by K. P. Kokhas, F. L. Nazarov and M. N. Gusarov. Numerous solutions that have been added in the second and third Russian editions of the book were written and edited by F. L. Bakharev, S. L. Berlov, V. I. Frank, S. V. Ivanov, A. I. Khrabrov, D. V. Karpov, F. V. Petrov, D. A. Rostovskiy, O. Yu. Vanyushina.

DMITRI FOMIN

---

[8] There is only one problem, **Question 1962.51**, which is missing the solution. Repeated attempts to solve it were all, as far as we know, unsuccessful.

.

Brief Historical Overview

This is a translation of the corresponding chapter from the Russian edition; parts of it were updated for this publication.

The Leningrad Mathematical Olympiad is not the world's oldest official mathematical contest for secondary school students[9], although it is quite likely the oldest such contest held at the city level as a formal part of the national educational system. It was organized for the very first time in 1934, thanks to the groundbreaking efforts undertaken by the distinguished Soviet mathematicians Boris N. Delone[10] and Grigory M. Fichtenholz, with a lot of assistance provided also by Vladimir I. Smirnov, Onufriy K. Zhitomirsky, Vladimir A. Tartakovsky, and Dmitry K. Faddeev.

BORIS DELONE, GRIGORY FICHTENHOLZ, VLADIMIR SMIRNOV—THREE OF THE "FOUNDING FATHERS" OF THE LENINGRAD MATHEMATICAL OLYMPIAD

We must add here that the olympiad certainly did not appear out of thin air. In 1933, in a school on the embankment of the Fontanka River, a few enthusiasts established the first mathematical circle for students called *"Scientific Station for Gifted Schoolchildren"*, and its director, Olga A. Beloglavek, also played significant part in getting the first math olympiad off the ground.

---

[9] That honor goes to the Kürschák-Eötvös Competition held in Hungary since 1894.
[10] Another, better known outside Russia, spelling is *Delaunay*.

It should be also noted that the idea of mathematical competitions for Russian high school students was proposed way back in 1912, in imperial Russia, during the first Congress for Mathematical Education which was held, of course, in St. Petersburg.[11]

However, due to well known reasons, in the next twenty years neither the scientific community nor the secondary education system in Russia (or USSR) had time for anything like that. The idea was reinvented in 1930s, when both the Soviet science and college education were going through a very rough patch. In particular, several scandals in Leningrad scientific community, provoked by a few professors and agitators touting a new, "revolutionary" approach to mathematics resulted in the forced disbanding of Leningrad Society of Physics and Mathematics. In addition, in 1925 and 1929 new laws were put in place restricting access to the university and college education for anyone of questionable social background. If you were a high school graduate whose parent used to run his own little pharmacy, or had been a lawyer, or worked at some noticeable government post during czarist times, you were summarily banned from obtaining a college-level education.[12] Often even technical or vocational schools would not accept your application.

VLADIMIR TARTAKOVSKY, ONUFRIY ZHITOMIRSKIY AND VASILY KRECHMAR—PROFESSORS OF LENINGRAD STATE UNIVERSITY, ORGANIZERS OF THE FIRST MATHEMATICAL OLYMPIAD

It is quite likely—at least such were the rumors circulating around the Leningrad scientific community—that one of the main motivating factors in establishing the mathematical competitions for schoolkids was to create a small backdoor within this cruel and ineffective system. What if at least a few gifted kids could, by winning the olympiad, be granted an automatic enrollment into a college of their choice, without being vetted by ideological bureaucrats? Suggested by Boris Delone, this idea received wholehearted support and cooperation of his colleagues and then was successfully implemented for at least a few first pre-war contests.

* * *

---

[11] St. Petersburg (Sankt-Peterburg, later Petrograd) was Russia's capital until 1918.

[12] Also you would not be allowed to vote or be elected to the municipal or state-level organs of the government, receive welfare, serve in the army, occupy managerial positions and so on.

In that distant year of 1934, the olympiad was quite different from what we are used to nowadays. To begin with, only high school seniors[13] were invited to participate. Several introductory lectures by Leningrad University professors were organized for these contestants—they were given during the weeks between the second and the third rounds of the olympiad. At the time, quite a few students, boys and girls, had been getting their education at so-called *work faculties* (also known as *schools for the working youth*), and therefore, the first round of the olympiad was set up as follows: the problem set was mailed to all the high schools and work faculties of the city so that any eligible student of the proper age could take part. Their papers were sent back to the organizing committee which reviewed the solutions. Those who performed best at the first round were invited to the second round. It was still a written test but the participants sat at one common location. The winners of that stage were invited to the final third (all-city) round, which was organized in the style of an oral examination.

In her article published in collection [29], E. A. Belskaya tracked the winners of the LMO #1. Everyone of them received a university education, majoring in mathematics, physics and chemistry. Four of them (by an unhappy coincidence, all mathematicians) were killed on the front line of the Great Patriotic War, twelve received PhDs and doctorates in their respective fields. Several became renowned scientists, full professors, department chairs and deans at their universities and colleges.

Just recently we were able to unearth some additional information about the participants as well as the problems of the first Soviet mathematical olympiad. In particular, we tracked down the biographical information and photos for almost all the first prize winners. Finally, as a minor miracle, eighty-six years after that seminal contest, a nearly complete main problem set of LMO #1 was found, hidden in plain sight.

Currently the LMO-1934 is the only pre-war Leningrad olympiad for which any of the questions are known. At the end of this introductory chapter the reader will find section "The First Olympiad", which presents all this information.

∗   ∗   ∗

In the very first years of these competitions, all the problems of the main set were disclosed to the participants from the very beginning. First, solutions were checked by the "mere" supporting staff—assistant and associate professors of Leningrad universities and colleges—then the full professors joined the fray. Besides the ones we have already named above, among them were Vassily A. Krechmar, Yakov S. Bezikovich, Aron G. Pinsker, and Isidor P. Natanson. All these renowned scientists and educators (and many others) also previously spent a lot of time preparing and reading lectures in various fields of mathematics to high school students. Some of the talks were later edited and published as separate books.

The main set consisted of several versions (four to twelve). These versions were different but they complied with the same fixed pattern (e.g., one problem in algebra, one in geometry, and one—a so-called "word problem" or a question in combinatorics, etc.) Most likely this was done to discourage copying—each contestant was given one version of the main set, consisting of these so-called *introductory* questions.

---

[13] Until 1989, with some short-lived exceptions, the Soviet school system had 10 grades.

What exactly happened after the competitor solved all the introductory problems is not fully clear. However, based on the recollections of several "veterans" of the first after-war olympiads, we can make a good educated guess. Each such contestant was either transferred to a separate room, or was simply assigned to one of the "senior" solution checkers (as we already mentioned, at the very first olympiads those were the full university professors, the original *founding fathers*). Each one of these senior jury members had their own (!) list of additional problems (most likely, sorted by difficulty) which they offered to the students, one question after another.[14]

ISIDOR NATANSON, DMITRY FADDEEV, ARON PINSKER—THREE OTHER PROFESSORS, WHO TOOK PART IN DEVELOPING THE SYSTEM OF EXTRACURRICULAR MATH EDUCATION IN LENINGRAD

Since the introductory questions were similar in both their statements and in solutions to the regular school curriculum problems, the organizers chose the additional questions to encourage use of advanced logical skills, to award thinking outside of the box as well as the ability to use variety of conventional mathematical tools.

Nowadays, this way of conducting a competition would certainly strike us as illogical, not to mention blatantly unfair. To offer the contestants different problems based on which senior jury member they accidentally happen to be assigned to, is absolutely out of the question. However, back then the organizers did not feel the need to hold the olympiad as if it were a formal sporting event with fixed and rigid rules which absolutely had to be equally applied to everyone. To begin with, the participants did not even receive the same list of the introductory questions (although the different versions were supposed to be all of approximately equal difficulty). Similarly, at the first Moscow Mathematical Olympiad, each competitor was allowed to choose (!) one arbitrary geometry problem, one algebra problem, and one combinatorics problem from the common list. This was all a part of the grand experiment whose main objective was to search for and attract gifted schoolchildren with scientific inclinations. The organizers themselves were on a quest as well, looking for the most suitable and fruitful format of this new intellectual competition.[15]

---

[14] Obviously, this system was modeled after the conventional college-style oral examination.

[15] However, for the historians and olympiad enthusiasts, this means that the most interesting questions of the first twenty olympiads were, as far as we know, neither recorded in the jury archives nor published. Alas, we have to accept that, in all likelihood, the full problem sets for 1934–1956 LMOs will never be recovered.

* * *

It was inevitable that gradually everything—the rules, the character and the scale of the competition—changed. In 1936, the 6th grade olympiad in arithmetic was started, and the problems of that olympiad were then published by the city department of education. In 1939, the lineup was expanded with the olympiad for the 9th graders, and the next year, the Grade 8 contest was added as well. However, the Grade 7 olympiad had to wait until 1954, when it was organized by two young mathematicians Elena Sokiryanskaya and Alexey Verner (this is confirmed by the contents of the official booklets with the olympiad problems and training materials; we will talk about these later).

In 1969, the Leningrad Mathematical Olympiad grew to encompass the 5th grade. Since then, every year around ten thousand schoolkids from Grades 5 through 10 (in post-1989 notation, Grades 6 through 11) participated in the Leningrad All-City Mathematical Olympiad.

The changes affected not only the scope of the contest, they also touched on the organization, structure, and rules.

More changes came shortly after the Second World War. Just as before the war, each grade's problem set consisted of several versions, usually from three to eight. A participant was given one of these ostensibly equivalent versions, each of them following the same thematic pattern.[16] At the arithmetic olympiad of 1953, the number of those versions reached the record amount of twelve (!).

Since the olympiad now covered many more grades and contestants, the organizing committee decided to involve university students (math majors), with many among them being former olympiad participants and winners.

At the end of 1950s, the multi-version system was abandoned since the problem committee could no longer prepare so many nearly identical versions; especially because they now had to include a few non-standard questions. So starting around 1955–1957, the students in every participating grade were given one common list of six problems which were usually divided into two portions. The first (introductory) portion consisted of the first four questions, and the second (advanced) portion added two more questions which were generally more difficult and often required some knowledge which, while being formally within the margins of the school curriculum, used some extracurricular ideas and theories. So at the beginning of the olympiad, everyone was given the first four problems—they would be written on the blackboards for everyone to see. After a student solved n problems (usually, n was equal to 2, but sometimes 3 solutions were required) from that introductory portion, he or she was transferred to another room where two more (advanced) problems were added. Of course, those who made it to the second stage were given extra time.

By 1961, the system of mathematical olympiads was deemed a resounding success and so the first All-Russia Math Olympiad was organized. Finally, in 1967, this growing pyramid of competitions was topped by the All-Union Math Olympiad. Because of that, Leningrad was now required to select a team of students who would represent the city at these competitions. And so came to existence the Final round of LMO (sometimes also called Elimination or Selection Round). From 1962 to 1983, it was held as a competition external to the LMO

---

[16] For instance, one problem in algebra, one problem in geometry, one problem in "elementary" calculus.

proper, as it only served as a team selection tool. The winners (first and second diploma holders) of the LMO were invited to participate in it a few weeks after the final rounds of the All-City Olympiad, and the selection round's results determined the city team to the All-Union Olympiad.[17]

In 1984, that system was once again changed, and the elimination round was fully incorporated into LMO as its final, fourth stage.

The results of this expanded round (where every grade's problem set contained not six but eight questions) decided not just the distribution of awards, but were used to make a decision on the members of the Leningrad team as well.

*　*　*

Now, let us go back to the 1930s. We would like to mention here three peculiar rules introduced into the LMO system at that time.

First, the approximate number of the first diplomas was predetermined before the olympiad. It was ten (don't ask me why). This rule was quickly abandoned as it was, obviously, quite illogical and simply impossible to enforce.

Second, the winners of every year's contest were prohibited to take part in the following years' olympiads. The intent was to prevent the so-called *"professional problem solving"*, which apparently was foreseen by the organizers from the very beginning (oh, how right they were!). In 1935, two of the previous year's winners, Beniamin Minzberg and Ivan Sanov, took part in the Second LMO where they won yet again (both of them were high school juniors in 1934). We cannot be sure that it was this event that triggered the ban but it seems likely.

However, it did not take long for that rule to be discarded as well. After only three years, in 1937, the ban notwithstanding, Pyotr Kostelyanets, a member of the mathematical circle at Young Pioneers Palace and the winner of LMO-1936, simply walked in and took part in his second city olympiad. The organizing committee decided to overlook this, probably hoping it will not end up becoming a tradition. Then it happened again, in 1939, when a high school junior Georgiy Epifanov, one of the winners of LMO-1938, broke the "law" again. After that the committee concluded that it would be more honest to simply strike that rule from the regulations. As a result, Epifanov was able to compete in LMO-1940 as well—and he won the first prize once again, thus becoming the first ever triple LMO winner.

Third, there was a very peculiar rule which defined how the results were tallied. For a few years before the war, each problem was assigned a value (in points)—and the most amazing thing is that it was done after (!) the olympiad (presumably, based on the solution stats). To determine a competitor's result, the point total of the problems he or she solved was calculated. When the olympiads resumed after the war that approach was deemed unreliable, the points were abandoned, and from that moment on, a contestant's result simply equaled the number of problems solved.

---

[17] Two capitals, Moscow and Leningrad, were considered major educational centers and so despite them formally being cities within the margins of Russia they sent their teams directly to the All-Union stage, bypassing the All-Russia round.

PYOTR KOSTELYANETS   GEORGIY EPIFANOV

Overall, quite a few things have changed over the years—the rules, the structure, the results, the organizers, etc. But one thing was constant—the Leningrad Mathematical Olympiad final round was always held as an *oral* competition.

Tradition of the oral olympiad is a distinctly Leningrad phenomenon. We do not know of any other major city or other major scientific olympiad where the solutions to the contest questions have ever been delivered and verified orally, right there during the contest itself.

From the very beginning, the contests in Leningrad have inherited the oral traditions of mathematical circles, which were already rather firmly established by that time. The lectures and other sessions of math circles were, of course, oral; the proofs and solutions were given and discussed orally—this gave rise to the idea of oral olympiad. It must be said here that the system of school seminars and out-of-school mathematical circles played a huge role in shaping mathematical education in Leningrad—LMO is only one of the things that grew out of it.

As the experience accumulated, the olympiad movement gradually gathered strength and knowledge. The number and quality of olympiad-style mathematical problems reached a level at which it was not possible to ignore its role as an outstanding educational tool. The problem sets were published as small booklets—judging by the catalogs of Central Public Library and The Library of Academy of Sciences, those were the first books on that topic printed in the Soviet Union. All that already happened before World War II.

In 1946, the first comprehensive collection of math competition training problems was published, and almost every year afterwards the problem set of LMO was printed as a small brochure (number of copies was usually around several hundred). That became a tradition which, unfortunately, lasted only until 1960 (it is unclear why the publication was interrupted)—to be successfully revived in 1980.

It is not just a random coincidence that a major shift in the style of math olympiads happened at the end of the 1950s. Clearly, the groundbreaking changes in mathematics and sciences, as well as their standing in the society, were the most obvious underlying cause but we are concentrating here on the much narrower area of scientific education in high school and mathematical competitions. One of the reasons was that the old, school-style problem topics were virtually exhausted, and all previously used types of problems became exceedingly well known to the students who took part in math seminars and circles around the city. They were also helped by publication of numerous collections of contest problems. In contrast to the Moscow olympiads which do not seem to have had such a major turnaround, in Leningrad the style of the LMO problems has changed rather drastically just within three or four years: from 1958 to 1962.

Those same years saw several other events which changed the situation around LMO as well as mathematical and scientific extra-curricular education in the city.

First, the organizing committee has almost entirely changed its membership. Many of the veteran enthusiasts who actively worked on it before and immediately after the war have retired or moved.

Second, the establishment of the All-Russia Math Olympiad has necessitated the selection of the Leningrad team.

Third, in 1961–1963 the Leningrad City Department of Education was persuaded to set up several high schools with Science-and-Math specialization. These schools were also different in that their students would come from all over the city—in sharp contrast to all other public schools which took only the students from their local city district or even subdistrict.

P.S. 30, CIRCA 1990S

Three of the newly created specialized schools—P.S. 30, P.S. 239, and Science-and-Math Boarding School 45 affiliated with Leningrad State University—became the most popular in Leningrad and well-known even outside of the city.[18]

For the next 25–30 years these three schools played the defining role in educating new science and math elite coming out of Leningrad high schools. More than a half of all Leningradians accepted as undergrads to the Faculty of Mathematics and Mechanics of LSU[19] in

---

[18] There was also P.S. 38, which has eventually merged with P.S. 30.
[19] Often simply called Math-Mech.

the later years came from these schools. Nearly everyone who graduated from a Leningrad high school after 1963 to become a professional mathematician was an alum of the *Big Three*.

PH.M.S. 45, CIRCA 1980S

However, these schools had significant differences in their general instructional approach; namely, in the style of how the advanced and extracurricular math and science education was shaped in these schools.

P.S. 239, CIRCA 1990S

While the kids of P.S. 239 and, to the lesser degree, of P.S. 30 received most of their advanced math knowledge in olympiad-oriented math circles, at Ph.M.S. 45 there was virtually no mathematical seminars dedicated to olympiad problem solving. On the contrary, the main accent was made on teaching better and more advanced math during regular curriculum lessons. Also there was a system of so-called "facultatives", that is, after-school seminars on various areas of advanced mathematics such as number theory, linear and higher algebra, analytic geometry, combinatorics, graphs, topology, etc.

The fourth and the final factor was that at exactly the same time, namely around 1960, (it is unlikely to have been a mere coincidence) in Leningrad was created yet another network of math circles affiliated with Leningrad State University and Leningrad State Pedagogic Institute—it was called the Youth School of Mathematics. Mostly these math circles were less advanced than the ones at the Big Three or at the Young Pioneers Palace, and the intensity of kids' involvement there was on average somewhat lower. However, the number of these seminars was relatively high—in the 1970s there were 20–25 of those—and they were where many future mathematicians and scientists, especially those who did not have big olympiad achievements in high school, received their first glimpse of non-standard, advanced mathematics and science. Igor Daugavet was appointed as the first director of YMS, soon to be replaced by Alexey Verner (at the time, both were Assistant Professors at the Faculty of Mathematics and Mechanics at Leningrad State University).

All these factors, as well as the presence of a large number of enthusiasts among university students and professors (mostly from LSU) resulted, by the beginning of 1970s, in the creation of a unique educational system. It virtually guaranteed that in Leningrad any middle or high school student who was genuinely interested in mathematics could easily find an opportunity to join a math circle or a school math seminar, etc. If he or she was interested in math contests, they were also open for everyone; those who were more inclined to engage in slower research-style work could do that as well by enrolling in specialized schools or working under the guidance of numerous enthusiasts from the Leningrad universities and colleges. The three components—math circles, teachers-enthusiasts and specialized schools—completed and complemented each other, making the overall experience richer and deeper.

The natural competition between the Big Three schools expressed itself mostly in the form of mathematical contests. In 1967–1988 they have held more than a dozen of so-called math battles between these schools' teams; among them—six triple-team contests with participants representing each of the three schools solving together the same set of problems.

Another arena was, of course, the Leningrad Math Olympiad. The quantity and quality of the diplomas gained by the students of each school was compared and appreciated not only by the contestants but by their teachers as well. Each year the organizing committee awarded the annual trophy cup to the school with the best results at the LMO. Sometimes the superiority of one of the schools was especially evident becoming the talk of the community for the next few months. For example, in 1978, Ph.M.S. 45 swept all the 1st diplomas in Grades 8, 9, and 10, while in 1982, 1985, and 1986, students from P.S. 239 took home all the 1st diplomas in Grades 9 and 10.

Alas, in those same years this wonderful system was slowly but surely working to undermine itself. The fact that the sports side of mathematical education was acquiring higher status, as well as attracting more attention, led to a rather unhealthy situation when Leningrad became the cradle of another remarkable homegrown Soviet phenomenon; mathematical contests (and, as an unfortunate consequence, often the underlying science itself) as professional sport.

It did not help that the Big Three has basically ceased to exist as such. P.S. 30 had moved to a different location which was not conveniently accessible to the students from other city districts, and then it slowly but surely stagnated there. A few years after that, Ph.M.S. 45 was literally thrown out of the city. The management of the Leningrad State University decided to move it to the suburb of Leningrad called Old Peterhof (ostensibly to have it conveniently located next to the main campus of the University), after which the students from the city basically stopped enrolling there. The only one staying put was P.S. 239 whose central location became a huge advantage but surprisingly enough, that did not serve it well. Almost all high school students who wanted to study mathematics flocked there but for various reasons some of the school's best teachers had left (and some were forced out), while ideological indoctrination and hiking club became more popular than the educational side of things. As a result Leningrad gradually shifted to the other end of the spectrum in terms of pre-college math education.

USSR TEAM TO IMO-1974; FOUR OUT OF EIGHT MEMBERS—MIKHAIL GUSAROV [# 4; P.S. 30], SERGEY FOMIN [# 6; PH.M.S. 45], IGOR ANANIEVSKY [# 8; PH.M.S. 45], AND IGOR SIVITSKY [# 10; PH.M.S. 45]—WERE FROM LENINGRAD ("INDEXING" FROM LEFT TO RIGHT)

Nearly all advanced math education for high school students has concentrated in just a few mathematical circles. Unfortunately, in such an elite seminar it is usually relatively difficult to address the individual interests and individual issues for every student who attends it. Usually the teachers end up going at the speed of the more advanced members of the circle, and the best they can do for the others is to try and encourage them to make an extra effort and play catch up. An educator in charge often does not have time to fill all the gaps left by the regular school curriculum.

Also the motivation in a math circle oriented toward "math as sport" achievements, is built on short-term results and is skewed in favor of the kids who can think really fast. Those whose brains are not organized like that or those who do not like problems of a certain type will soon feel neglected or pushed to the side. Disappointment and general feeling of not doing the things in the expected or popular manner will not be long behind. These students then often quit the seminar or even, unfortunately, math and science altogether.

There was still one obvious plus to this system, which we have already mentioned. If a student liked math or science, had some skills and interest in learning more than she already knew, then the opportunity to develop and learn was certainly readily available. She could easily find a math circle or a school seminar suitable for her specific interests. The existing system of after-school math and science education was still maintained at a rather high level.

NIKOLAI GÜNTHER, ALEKSANDR ALEKSANDROV, LEONID KANTOROVICH, SOLOMON MIKHLIN—THESE ARE JUST A FEW AMONG THE RENOWNED MATHEMATICIANS INVOLVED IN LENINGRAD MATHEMATICAL EDUCATION MOVEMENT OF 1930s–1950s

The city still had enough math circles and after-school seminars to guarantee that the kids would be able to make their first steps in advanced mathematics. One of the main places in this system was occupied by the math circles in the Young Pioneers Palace which was renamed the City Palace of Youth Creativity in 1993. It was founded in 1937 and during the sixty years of its existence has accumulated vast experience in extracurricular education.

Numerous young mathematicians read lectures and held sessions of the math circles at the "Palace" and around the city. Before WWII (in 1935–1940) among them were M. L. Verzhbinsky, G. E. Tsvetkov, M. K. Gavurin, D. I. Fuchs-Rabinovich, S. P. Olovyanishnikov, M. A. Yavets. Immediately after the war (1945–1950) that list included Z. I. Borevich, V. A. Zalgaller, G. V. Epifanov, A. S. Sokolin. Then the system has grown, and in 1950–1960, it added multiple math seminars led by I. Ya. Bakel'man, E. N. Sokiryanskaya, O. G. Fayans, M. Z. Solomyak, A. A. Zinger, A. L. Verner, E. M. Gol'berg, A. V. Rukolaine, N. M. Mitrofanova, Y. D. Burago, A. V. Yakovlev, B. B. Lurie, O. N. Bondareva, O. M. Kalinin, B. Z. Moroz, L. A. Petrosyan, V. V. Zhuk, N. A. Suslina. Many of these math circles were organized by the students and mathematicians of the Faculty of Mathematics and Mechanics of Leningrad State University; some enthusiasts even worked with more than one of these "competing" outfits.

In addition, quite a few professors from Leningrad universities were invited to read specially prepared lectures to olympiad participants. Among these lecturers during the first thirty years we should mention such prominent names as N. M. Günther, Y. S. Bezikovich, V. I. Smirnov, G. M. Fichtenholz, O. K. Zhitomirsky, R. O. Kuzmin, V. A. Krechmar, V. A. Tartakovsky, A. G. Pinsker, I. P. Natanson, D. K. Faddeev, S. G. Mikhlin,

A. D. Aleksandrov, L. V. Kantorovich, K. F. Ogorodnikov, Y. V. Linnik, V. A. Zalgaller, S. V. Vallander, V. A. Rokhlin, N. A. Shanin, V. P. Havin, B. A. Venkov, G. S. Tseitin.

MATHEMATICAL CIRCLE AT THE PIONEERS PALACE, 1951. TEACHER ILYA YA. BAKELMAN (LOWER LEFT) AND STUDENTS VLADIMIR SUDAKOV, VERONIKA GUMAN, NINA URALTSEVA (LEFT-TO-RIGHT).[20]

Then, in 1960, the so-called Youth School of Mathematics was born—the intent was to gather into a slightly more formal system all the FMM-affiliated math circles. The list of its teachers in 1960s includes M. I. Bashmakov, I. Ya. Verebeichik, M. L. Gromov, Y. I. Ionin, V. S. Itenberg, R. V. Peisakhov, A. I. Plotkin, V. P. Odinets, M. Ya. Rozinsky, M. L. Gol'din, Y. A. Davydov, B. A. Plamenevsky, N. K. Nikolsky, V. P. Orevkov, A. P. Oskolkov, S. M. Belinsky, S. A. Vinogradov, A. V. Skitovich, A. O. Slisenko, I. M. Deniskina, S. E. Kozlov, S. S. Vallander, Y. V. Matiyasevich, I. B. Frenkel, B. A. Lifshits.

There were so many math circles in Leningrad during the next decade (1970s) that it is next to impossible to list here all the educators who led those seminars. Here are just a few of them (in addition to the previous list): A. L. Likhtarnikov, V. P. Fedotov, V. V. Nekrutkin, V. Ya. Gershkovich, I. E. Molochnikov, I. S. Rubanov, O. A. Ivanov, O. Ya. Viro, T. A. Bratus, N. N. Vasiliev, A. D. Yatsenko, S. V. Fomin, S. E. Rukshin, M. N. Gusarov.

Then, the enthusiasm started to flag and the number of seminars stabilized at a somewhat lower level, so that except for the Young Pioneers Palace, in 1978–1991 there were only about 10–15 city-level math circles working in Leningrad. Circles at the Palace were led by S. E. Rukshin, A. S. Golovanov, G. Ya. Perelman, A. V. Bogomolnaya, F. L. Nazarov, E. V. Abakumov, E. S. Dubtsov, L. D. Parnes, S. K. Smirnov, M. Ya. Pratusevich, M. M. Roginskaya. Other math circles of note, affiliated with Youth School of Mathe-

---

[20] All four will soon become very well known in their respective fields of mathematics and science.

matics at LSU, were run by S. A. Genkin, V. E. Kozyrev, I. A. Panin, A. L. Smirnov, A. A. Borichev, I. A. Chyornaya, D. Yu. Burago, D. V. Fomin, I. V. Itenberg, A. L. Kirichenko, K. P. Kokhas, I. B. Zhukov, A. Yu. Burago, A. G. Frolova, E. E. Domanitskaya, E. B. Boguta, I. A. Binder.

MIKHAIL GROMOV, YURI MATIYASEVICH, AND STANISLAV SMIRNOV—WINNERS OF LMO, LENINGRAD MATH CIRCLES' TEACHERS, AND WORLD-FAMOUS MATHEMATICIANS

Alas, we cannot list here all the enthusiasts who, during all these years, have contributed to the noble cause of extracurricular mathematical education in Leningrad. We can only hope that we mentioned here the most important events of that era while naming the people whom we remember as well as those, whose names we found in the archives. The possible (quite possible!) omissions are by no means the result of author's personal bias but are the unfortunate result of the errors of human memory and destruction or inaccessibility of too many archival materials of the past.

Olympiad: The Inside Look

This is a translation of the corresponding chapter from the Russian edition, updated for this publication.

Everyone who ever took part in it would tell you that the Leningrad Mathematics Olympiad was held every year in February–March. However, the contestants knew of and saw only the tip of the iceberg. For the organizing committee and the problem committee (which is often called the *jury of the olympiad*) this contest would begin several months earlier in October–November, when the jury members started their meetings, and the organizing committee gathered the instructors from the Department of Schools whose job includes keeping the district's math teachers up to date on all the educational standards, changes in curriculum, and many other small things.[21] Leningrad's school system consisted of twenty-two such districts.

The first round of the olympiad is the school round, which many schools simply replaced with recommendations that the school teachers handed out to the students. Quite a few of them also offered the kids some training problems and taught them how to prepare for the second (or district-level) round. It was then up to the students whether they showed up for the district round—often the participation was encouraged by a promise of a good grade. On average, when the round was held (by tradition it was one of the weekends in the first half of February) almost ten thousand middle and high school students would take part.

The jury usually needed about one or two months to invite the enthusiasts, gather the problems, consider their statements and solutions, judge their "fitness" for a specific grade and round, and so on. By the end of the fall, they created the problem sets for the olympiad's second (District) round, as well as made some healthy advance on the road to the third (All-City) round.

Both the organizing committee and the jury would resume their work in January. This is when the jury began the most difficult part of their mission: preparation of the two final rounds. Usually the third (All-City) round was held at the end of February for the middle school, and the first half of March—for the high school students.[22] After the fourth

---

[21] Soviet Union had a comprehensive universal curriculum which was more or less standardized throughout the country, with only small local deviations.

[22] At the time middle school comprised Grades 5–7, and high school—Grades 8–10 with the exception of 1960–1966 and 1990–1991, when it was Grades 5–7 and 8–10, respectively.

(Elimination) round was added, its date was usually set two weeks after the end of the All-City stage.

By tradition, the third round for the middle schoolers was always held on the grounds of Leningrad Pedagogic Institute named after A. I. Herzen. Approximately 300–350 winners of the district round from Grades 5–7 were invited.[23]

LENINGRAD PEDAGOGICAL INSTITUTE, 48 MOYKA RIVER EMBANKMENT

There was, however, a "special" category of kids—they did not need to attend the district level contest because they would already have a so-called "personal invitation" to the All-City round. They were the winners of the last year olympiad; namely, the ones who received a diploma at the previous year's All-City round. Some "elite" contestants would not ever go to the district round after their first year—if they kept on getting diplomas, they received personal invitations until the very end of their "professional" olympiad career. However, at the high school level it was harder to obtain a personal invitation. For that, one had to hold the first or the second diploma from the previous year's competition.

When the students entered the institute's building, they were "distributed" to several classrooms, based on the first letter of their last name. In the room they saw the blackboard which already carried on it the statements of the first four *introductory* problems. The entire problem set consisted of six or seven problems, but that list was given in its entirety only to the kids who made it to the "advanced stage" classrooms by solving two or three of the introductory problems. Naturally, the additional (advanced) problems were on average somewhat more complicated than the first four. Also, in the "introductory" rooms the olympiad ended one hour earlier than it did for the advanced contestants.

This division of the full problem set into two parts also seems to be a uniquely Leningradian tradition, and its main motivation came from the specifics of the oral contest. There are some clear advantages to the two-stage system.

---

[23] Occasionally, a fourth grader would merit an invitation; indeed, it was simply required to make a good showing at the district competition stage.

First, at the beginning of the olympiad the contestants concentrate on the fewer and simpler problems.

Second, the work of the jury during the olympiad itself is significantly simplified and, as a result, becomes more effective.

Now, let us explain the *oral* olympiad and how it works.

Every classroom, depending on its size, accommodates ten to fifty students, and there are several solution checkers assigned to it—in addition to permanent members of the problem committee, there are dozens of math majors and University alums who are invited every year to come and help during the olympiad season.

Deep in thought: seniors Pavel Suvorov and Aivars Bērziņš (center; left-to-right) at LMO-1969

When a student, who had been solving the problem in his or her own notebook, decides that the solution to one of the questions is finally found, he raises his hand and one of the checkers takes him to one of the unoccupied rooms set aside just for that purpose. There they sit down and the student explains the solution together with (if necessary) all the minute details. The student does not have to write the full solution down, although he might do that for his own convenience. During that conversation, the checker will ask various questions trying to find holes in the offered solution. If they discover such a hole, and the student cannot close it or fix his solution right there within some reasonable amount of time (usually, one minute), then the contestant goes back to his seat and a minus sign is written into the corresponding square of the results sheet (it has the table, where rows bear the contestants' names, columns—the problems' numbers). That sign means an attempt to solve question Q had been made by contestant C and it was judged unsuccessful. Each participant is given three attempts at each question. If the third attempt fails then the third minus sign goes into the table, and the student is barred from offering any more solutions to that particular problem.

However, if an attempt succeeds and the checkers decided that the solution they were given is correct, then the corresponding square gets a plus sign. Thus, the result table's squares can carry symbols like these: $+$, $\pm$ (the first attempt failed but the next one succeeded), $\underline{\pm}$, $-$, $=$, and $\equiv$.

Sometimes jury members make mistakes as well. That is rare but still quite possible. If a jury member who accepted a solution later discovered some serious hole in it, or some other checker got interested in how exactly the problem had been solved and found an error, the student would be told about this and given a chance to clarify the solution or quickly fix it and so on. The mark then was adjusted accordingly. However, such a change was possible only before the last minute of the time allotted for the contest. If a jury member's mistake was found afterward, the mark was not changed. Here the principle similar to *"presumed innocent unless proven guilty"* was at work. Who knows—if shown the error during the olympiad, the contestant could have been able to fix the mistake or even come up with a completely different solution.

SIXTH GRADER ILYA MIRONOV IS TRYING TO EXPLAIN HIS SOLUTION TO THE JURY MEMBER EVGENY ABAKUMOV DURING LMO-1989

In order to lower the likelihood of such errors (a very serious issue in some situations), the *grade leaders* (six senior jury members in charge of each grade) need to maintain constant but careful control of everything that happens within their little fiefdoms. To simplify that work, every plus sign must be always marked by the initials of the checker(s) who accepted the solution. Thus, if a grade leader decides to make sure that a plus sign for some non-trivial question was not awarded in error, he can always talk to the checker(s) and verify the validity of their decision. For instance, in 1981 all the solutions for Question 35 were at some point re-checked. And that turned out to be quite important—some of the case-by-case analysis solutions were found out to be incomplete or even incorrect. And when the olympiad was over, the jury realized that the subset of the contestants who had a correct solution for that problem coincided with the subset of the first prize winners.

So we can say that the oral olympiad has the following advantages:

a) direct communication between the students and the jury members;

b) learning the proper mathematical language, which is especially important for the younger participants;

c) a possibility to fix an incorrect decision, or even completely change your approach and come up with a different solution;

d) contestants do not have to spend a lot of time writing down the full solution, covering all minute details and all the lemmas, big and small;

e) the jury is able to quickly tally the results and determine the winners just a few hours after the contest is over.

LMO protocol (results) sheets from 1986–1989

The drawbacks are there as well. The main one of them we have already discussed—the errors committed by the solution checkers sometimes are not discovered until it is too late. It is likely to have happened almost every year. However, in most cases it does not really amount to any serious damage to the results or the award distribution. The fact that it happens is not surprising: some of the problems are quite complicated, the work of solution checkers is not trivial and requires considerable skill. Each offered proof must be carefully looked into in order to find out if it has any hidden hole aka *"lipa"*[24] (that is how the Russian

---

[24] Pronounced "lee-pah", similar to *kippah*.

olympiad jargon calls a camouflaged, usually a not-very-obvious, error in the solution).[25]

If such a hole is discovered, it has to be demonstrated to the student but without revealing a way to fix it—he has to figure it out himself. Sometimes that is really difficult to do. Thus, the solution checkers must be familiarized in advance with standard errors that might arise when the problems of this olympiad are being solved, as well as with the best methods of demonstrating the incorrectness of those approaches. One of the most common ways is to present, if possible, a simple counterexample to an erroneous statement.

Another (and often insurmountable) problem with any oral competition is the necessity of having a lot of very competent, responsible, and enthusiastic helpers—members of the extended jury, so to speak. Therefore, it can only be done in a large city with a tradition of extracurricular math education, where you have a lot of math professors, math teachers and college majors, willing to donate their time and efforts. Each oral round of LMO required anywhere from thirty to fifty such helpers.

The olympiad for the high school students was held on the campus of the Faculty of Mathematics and Mechanics (FMM) of Leningrad State University in Petrodvorets (suburb of Leningrad which is about 45 minutes by commuter rail away from one of the city's three main railroad stations). Before FMM was moved to the city suburbs in 1979, the olympiad used the old building of the FMM on the Tenth Line of Vassilievsky Island[26], which is much closer to the city center and accessible by the subway.

There were no substantial qualitative differences between the organizational styles of the high school olympiad and the middle school olympiad. One important general difference is that the solution checkers knew they must be especially vigilant—here they dealt with the seasoned "professionals", and some of them could come up with ingeniously invalid solutions which on the surface looked like the real thing. Also the olympiad lasted a little longer—4 hours instead of 3.5 hours for the middle schoolers.

As for the structure of the olympiad, until 1984 it was identical to the one they had for the younger students. But in 1984–1990, the organizing committee decided to run a sort of an experiment. Namely, the third round (which at the same time stopped being the final round of the olympiad for the high schoolers) was split into two separate olympiads—one for the Science-and-Math specialized schools and the other for the regular schools (but only in top two Grades, 9 and 10). For instance, in 1984–1988, the students from specialized schools did not take part in the usual third (All-City) round; instead they had a special second (District) round held as a written contest. The students with the best results at that stage (approximately 25–30 kids per each grade) qualified for the final round, in addition to the personal invitees (the winners of the last year's olympiad).

The regular schools' students took part in the oral third round, whose problems were selected accordingly in order to determine the reasonable number of the best contestants

---

[25] Word "lipa" in Russian literally means the *linden tree*. A rather plausible explanation of the slang usage says that for many centuries the counterfeiters would use cheap and easily malleable wood like linden instead of more expensive oak, redwood, yew or some such when making icons, ornaments and high-quality furniture. From there the term trickled into the regular everyday speech, where people used the word to denote something fake, counterfeit or deceitful.

[26] Largest of the numerous islands in Neva river delta which comprise a significant portion of the historical center of St. Petersburg (Leningrad); the oldest streets there were named *Lines*.

from that "stratum" to invite to the final round. Usually between five and ten winners of that round qualified for the finals.

OLD BUILDING OF FMM, 33 TENTH LINE, VASSILIEVSKY ISLAND

In 1988–1989, the district round for the specialized schools was not held—it was replaced by the third round for the same kids after they passed the regular district round of the contest. It was a written contest which was held simultaneously with the oral third round for the other students.

The main purpose of all these numerous experiments was to try and come up with the contest structure allowing the larger number of the students from regular schools to come in touch with a different mathematical culture, to go through the oral olympiad, to see and learn things they did not commonly encounter in their standard curriculum.

In 1991, the committee decided to go back to the old, "pre-reform" system, abandoning the specialized schools' round, and once again moving the Elimination Round outside the formal structure of the LMO. This was caused by two factors. First, it was technically difficult to maintain the two-pronged organization of the contest. Second, the jury did not see any tangible results from the experimentation undertaken in the preceding six years. Also with the advent of *perestroika*, many new schools sprung up in Leningrad, some of them specialized, some being affiliated with colleges and universities, others calling themselves gymnasiums and claiming their educational approach superior to the regular schools. It was a mess, and the actual content of the term "specialized school" became vague and even somewhat suspect. Eventually, the committee decided that it was simply unreasonable, not to mention non-efficient, to continue to treat the students differently, based on the type of school they went to.

Now, let us talk about Elimination (or Selection) Round. It was a relatively small contest, with no more than 50 students (in 1984–1990, when this round was simultaneously the final fourth round of the LMO, the overall number of participants was close to 100). Since the errors by solution checkers were both more likely and extremely undesirable at this level, the jury members always worked in pairs. The grade leaders' responsibility was also noticeably higher, and they had to make an extra effort to ensure that the competition went smoothly and all possible issues were resolved before the end of the olympiad.

NEW BUILDING OF FMM, 28 UNIVERSITY AVE, PETRODVORETS

There was no division of this round's problem set into the first (introductory) stage and the second (advanced) stage portions. All the questions (with rare exceptions, there were eight of them) were given to the contestants right away at the start—each student received their own copy of the questions sheet. The time was extended as well—the competition lasted 5 hours.

During the first twenty years of its existence, the Elimination round was organized as a contest with the common problem set for all three high school grades; only in 1975 the problems given to the 8th graders were different from the ones offered to the Grades 9–10. However, the reform of 1984 changed that as well—starting with that year the problem sets for the three grades became different, although some of the problems were used in more than one grade. This was caused by the changes in the structure of Leningrad team to the All-Union Olympiad, forced by a decision of the Ministry of Education regarding the teams' setup. In particular, that decision required Leningrad to choose exactly one new participant per each of the three high school grades. Since it was hard to create just one problem set fitting every grade from 8 to 10 in order to properly "differentiate" just one winner per grade, the jury decided to switch to the separate problem sets.

That still did not always solve the selection problem. For example, in 1986 the LMO committee decided not to send a 10th grader to the All-Union Olympiad in Ulyanovsk because all the seniors, who did not already have personal invitations, performed poorly at the selection round. So the committee had to obtain a waiver from the All-Union Olympiad bureaucracy in order to send not one but two winners of the 8th Grade selection round—the best result belonged to a 7th grader from P.S. 536 Sergey Ivanov, while Sergey Berlov, the 8th grader from P.S. 533, took the second place.

Both of them later made the Soviet team to the IMO with the younger Sergey becoming the first Soviet high school student taking part in three International Math Olympiads—Cuba (1987), Australia (1988) and Germany (1989)—and winning three gold medals. Two years later that outstanding result was repeated by Zhenya (Eugenia) Malinnikova from P.S. 239, who also took home four first diplomas at the All-Union competitions (Sergey "only" got three golds and one silver there). Now, many years later, both of them are renowned mathematicians working in Riemannian and metric geometry, and in multidimensional calculus and functional analysis, respectively.

USSR TEAM TO IMO-1988; FOUR OUT OF SIX MEMBERS—SERGEY BERLOV [# 1], SERGEY IVANOV [# 4], YURI KHOKHLOV [# 5], AND NIKOLAY FILONOV [# 6]—REPRESENTED LENINGRAD (LEFT-TO-RIGHT "INDEXING")

Now, let us talk a little about the awards and team selection. After the third round ends, the grade leaders would gather each grade's jury to make a decision on the award distribution. At the elimination (or the fourth) round the decision was almost always made at the meeting of the entire jury including the invited non-permanent members.

The plus-minus system used for the oral contests was obviously different from the point system used at the written contests. For instance, at an oral olympiad a problem cannot be deemed half-solved so that half of the points can be awarded. Thus the results lack this subtle gradation and are simply integers showing how many problems a contestant has fully solved; the erroneous attempts do not count against the tally. This often presented a difficulty in determining the specified number of the winners, which was always necessary for the team selection.

This approach excluded cases like the one described in book [4], where Boltyansky and Yaglom, in the introduction chapter, wrote about young middle schooler Eric Balash, who only solved one (although relatively difficult) problem but was nevertheless awarded the first prize because of his in-depth investigation which went far beyond what was asked in the problem's statement. At the Leningrad olympiad Balash would not even reach the advanced-stage auditorium—but, obviously, he would have taken that into account and solved some simpler problems first, not to mention that the actual problem (investigating divisibility properties of Fibonacci numbers) probably would not have been offered as one of the introductory problems.

However, that does not mean that non-standard showings never got a proper encouragement. In 1989, at the middle school olympiad the 7th grader Dima (Dmitry) Karpov was the only one who solved Question **1989.18**, and despite the fact that his result was just below the second diploma's threshold he was elevated to that level and awarded a silver medal.

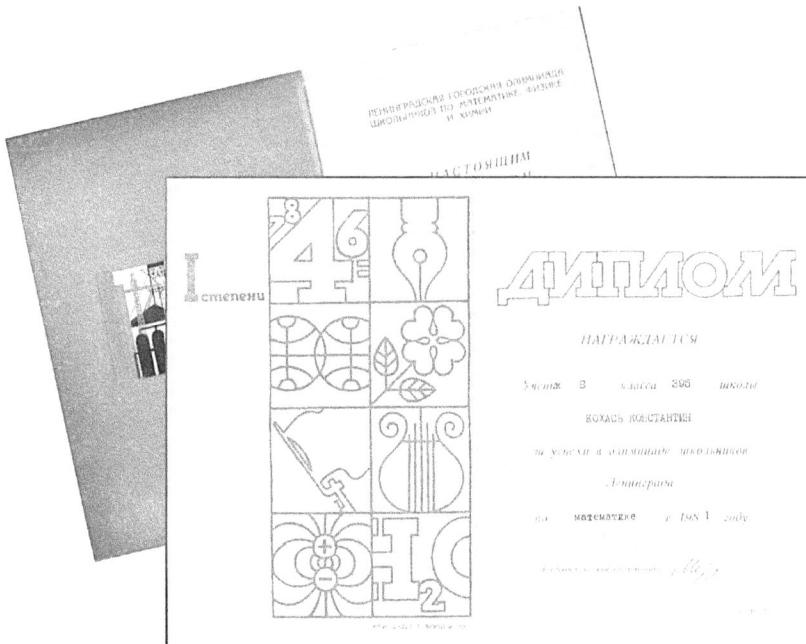

DIPLOMAS AWARDED AT LMO IN 1970–1980s

In the history of the LMO there were cases when the jury even had to resort to voting in order to determine the Leningrad team members. In 1987, two obvious candidates for that one much coveted place on the team—Lev Novik and Bella Sheftel—had identical results at both All-City and Elimination rounds. The committee was divided, and multiple rounds of voting created a rather tense atmosphere at the jury's late night session. It was even suggested to decide the outcome by drawing lots. The jury then made a "Solomonic" decision and chose... Masha (Maria) Roginskaya, who seemed to have had a somewhat worse result at the final round. The controversy stemmed from the fact that one of the problems at the Elimination round was Question **1987.44**—and with that question consisting of three items, the disagreement was about counting each one of these items as a separate question (partly because of how difficult the problem itself turned out to be). A very long and ultimately quite fruitless discussion lasted more than three hours and the vote then split 4 : 4. Finally, the decision had to be made by the jury chairman. It should be noted here that Masha did not let her advocates down—she became the winner at several All-Union Olympiads receiving one gold and two silvers there, and then was chosen for the Soviet team to IMO-1989, where she was awarded a silver medal. Her less successful "alternates" from that fateful day did not fare badly either: Bella received bronze medal at All-Union Math Olympiad in 1989, and Lev made the city team for the All-Union Informatics (Computer Science) Olympiad, where his first prize led to the silver medal at the very first International Olympiad in Informatics held in Bulgaria in 1989.

Another interesting case happened at LMO-1979, when all the first and second diplomas in Grade 9 were swept by the students from Ph.M.S. 45. By the then current All-Union Olympiad rules adapted in 1975 that school had its own team to the All-Russia (penultimate stage of the All-Union) Olympiad, and thus their students could not be included on the Leningrad city's team. So the Leningrad jury had to petition the Central Organizing Committee of the All-Union Olympiad, and as a result the winner of the All-City round Sasha (Alexander) Sivatsky from Ph.M.S. 45 went to Tbilisi[27] as a member of the Leningrad team.

LENINGRAD CITY TEAM TO ALL-UNION OLYMPIAD OF 1980 (LEFT TO RIGHT): ALEXANDER SIVATSKY, MIKHAIL OREL, SERGEY E. RUKSHIN (TEAM LEADER), GRIGORY PERELMAN, ANDREY MINARSKY, ALEXEY SOLOVYEV

History of the LMO knows many cases not clearly defined by instructions or rule books, when the jury or the organizing committee had to resort to voting instead of finding some easy consensus. There were, however, some rather standard situations when the jury hewed to well-established traditions. One of such traditions was the so-called *non-official participant* status. Sometimes... well, actually rather frequently, one of the jury members, a math circle leader or a school teacher petitioned the organizing committee to let a student to participate in the third round despite him or her not qualifying via the district round. The

---

[27] Each year the All-Union Math Olympiad was held in a different city; Tbilisi is the capital of Georgia, at the time one of the fifteen republics constituting the Soviet Union.

reasons could be quite diverse but usually it had to be something that objectively prevented the student from taking part in the second round (an illness or being out of the city at the day of the contest). Usually the committee granted such requests by giving the student the non-official participant status which allowed them to participate in the All-City round but barred them from receiving an award for their performance.

Sometimes on the day of the third round, the surprised jury members would discover a "stowaway" in their classroom—a student who somehow sneaked in to get a chance at solving the contest problems. This was a rare happenstance—and the traditional approach was again to grant the same non-official status. The only possible formal recognition of such a contestant's good showing could be a personal invitation to the next year's olympiad, provided their result was at the high enough level.

The LMO award ceremony circa 1952–54; in the center—Professor Isidor Natanson, the olympiad jury chairman; on the left, at the lectern—Elena Sokiryanskaya, director of the Young Pioneers Palace math circles

Another standard rule related to the possibility of the younger students participating in the upper grade's olympiad. Since the high school contest was held a week or two later than the middle school one, by tradition the first diploma winners of the Grade 7 olympiad received personal invitations to the Grade 8 All-City round. Despite the age handicap, some of the younger kids' results were often as good as their older colleagues' or even better. For instance, in 1981, seventh grader Anna Bogomolnaya won the 1st diploma at the Grade 8 olympiad while somehow "managing" to receive only the 3rd diploma at her own grade's contest. Nine years later the story repeated itself with Olga Plamenevskaya, who was at the time attending a math circle at Young Pioneers Palace led by none other but Anna Bogomolnaya, no longer a student but an assistant professor at the Leningrad Electro-Technical Institute of Communications. Then in 1988, two of the three first diplomas awarded at the Grade 8 olympiad were taken home by two seventh graders, Sasha Perlin and Zhenya

Malinnikova. But the most memorable and seemingly impossible event happened in 1963, when two sixth graders Andrey Suslin and Sasha Livshits (both later became well-known professional mathematicians) won first diplomas for the Grades 6, 8, and 10!

Now, about that part of the jury's work, which is mostly hidden behind the scenes. It consists of two components: one, creation and accumulation of the contest questions, and two, designing the olympiad's problem sets.

Realization of the former depends mostly on how active the jury members are in coming up with new problems and in involving the Leningrad mathematicians into the competition's orbit. Very often the new problems come to the jury quite unexpectedly and via such a long and winding path that the contents and sometimes the very authorship of the problem change on the way. Authors of the majority of the problems are the permanent members of the jury who have extensive experience in this creative art—the thing is that it is actually not that easy to come up with, or to invent a good math contest question.

To begin with, the new LMO question has to be actually new, virtually unknown to the students, although it might look very similar to the previously used problems. The solution should be neither too boring nor follow the same "*done to death*" standard tricks and ideas. The statement cannot be too long or confusing, and it should provoke interest. Ideal math contest problem has unusual and easily understood statement, with the fact itself being fresh and interesting, while the solution idea should be relatively non-standard and somewhat unexpected. It also helps if the problem were not too complicated but at the same time would not yield too much advantage to the professional problem-solving elite who know many things outside of standard school curriculum.

It is tough to satisfy all of these demands. Despite that, every year a few problems offered at the LMO come very close to that shining ideal.

What happens at a jury meeting? Usually it is attended by five to fifteen people; they discuss new problems, check and reexamine solutions for the problems suggested before, edit the problems sets, and make decisions on the qualification criteria (resulting in the compilation of the contestants' lists). Quite often during these brainstorming sessions the sets and the problems themselves change so much that some are hardly recognizable anymore, even by those who suggested these questions in the first place. At one of the jury sessions in 1987, one of the problems (Question **1987.12**) was suggested by a jury member who was convinced that the answer to it was negative. Right there, when the other participants tried to solve the problem, it turned out otherwise, and the duly updated problem with a completely different setup and answer was subsequently included into the sixth grade problem set.

The second part of the process has less to with mathematics but is rather akin to the art of interior design. There are a few natural requirements which the full problem set should comply with.

a) Questions must vary by topic. It would be quite inappropriate (not to mention, boring!) to cram three classic geometry questions into a six problem set. The jury should ensure that the final problem set represents arithmetic and geometry, combinatorics and computations, inequalities or estimations as well as "more precise" facts.

b) The diversity of the solution methods must be ensured as well. It would not be a good idea if three of the questions can be solved by induction, or if both geometry problems are about quick angle computation.

c) The set must also be balanced in its complexity. The jury has to take utmost care to make it, on one hand, easy enough so that almost every participant solves at least one problem, and, on the other hand, difficult enough so that the strongest contestants do not have an easy ride. At the same time, the set must satisfy one of the most important demands—it has to "statistically separate" the grade. That is, the results should turn out to be approximately one to five first prizes, five to ten second prizes, and ten to twenty third prizes.

Of course, that does not mean that the jury cannot offer a very difficult problem. That is often quite acceptable, but the average complexity level cannot be raised too much. Usually the vast majority of problems is solved by at least several students; only Elimination round of 1981 and the final round of Grade 9 in 1989 turned out to be exceedingly difficult for the contestants.

Obviously, the jury must possess a sort of clairvoyance to design the ideal problem set. Generally such a prediction is based not only on the rich experience of the members of the jury but also on the knowledge of what exactly is the approximate level of the students in this or that grade. Deviate a bit too much from that level and you get such extremely unwelcome results as 34 (!) first diplomas for Grade 11 contestants in 1966 or zero first prizes for Grade 6 students in 1991.

d) The requirement that the problems were unknown and never used before is especially important for the All-City and Elimination rounds. However, it is nearly impossible to prevent an occasional failure. For example, Question **1988.14** was given to the seventh graders literally on the next day after the subscribers found in their mailboxes an issue of "KVANT" magazine which featured that very problem in one of its articles.

One more instance (however not as obviously unlucky as the previous one) is Question **1983.27**, which was "created" by one of the jury members. Seven years later that problem was, quite accidentally, discovered among the questions of All-China Math Olympiad of 1962.

e) The questions should be oriented toward the "real" mathematics. The participants must understand that there is no such thing as a separate, purely contest-oriented, mathematics. Boris Delone, one of the founders of LMO, used to say that the real mathematical problem differs from the contest problem only in that *the real problem might take a thousand times longer to solve*. We should add one more important distinction, though—nobody can guarantee that even after all that time you will find a solution which, for all we know, might not even exist. At a competition, you know in advance that *all the contest questions are solvable* and, more than that, they have a solution that can be explained using terminology and methodology of the standard school curriculum.

At the same time, the higher, more advanced, mathematics lately became an important and rich source of new, often very interesting and beautiful, olympiad problems. Let me give you a few examples. The statements and the solutions of

Questions **1980.48** and **1988.60** can be interpreted in terms of the so-called Minkowski metric. Question **1981.46** can be solved and generalized using the standard methods of linear algebra. Question **1982.36** arose from the proof of the Fundamental Theorem of Algebra conceived by Karl-Friedrich Gauss. Traditionally, quite a few interesting questions come from the graph theory, such as **1983.12**, **1985.17**, **1986.25**, **1986.42**, **1988.40**, **1988.48** and many others. Question **1987.24** is explicitly connected to the multidimensional calculus, and Questions **1986.13** and **1987.60** represent variations of the famous combinatorial Sperner's Theorem; another well-known theorem (Sperner's Lemma), proved by the same mathematician, inspired Question **1988.18**.

The problem committee (aka the jury) was headed by the chairman and its executive secretary. For the entire sixty years of LMO the jury had only fifteen chairmen. Here they are (the list is arranged in chronological order; the exact dates are provided when known): Boris N. Delone (1934), Vladimir A. Tartakovsky, Grigory M. Fichtenholz, Isidor P. Natanson, Dmitry K. Faddeev, Solomon G. Mikhlin, Viktor A. Zalgaller, Gleb P. Akilov, Garald I. Natanson, Igor K. Daugavet (around 1965), Anatoly V. Yakovlev (1970–1979), Yuri A. Volkov (1980), Aleksey B. Aleksandrov (1983), Alexander S. Merkurjev (1981–82, 1984–2001).

LMO's JURY CHAIRMEN: VIKTOR ZALGALLER, GLEB AKILOV, GARALD NATANSON, IGOR DAUGAVET, ANATOLY YAKOVLEV, YURI VOLKOV, ALEKSEI ALEKSANDROV, AND ALEXANDER MERKURJEV

For a few years starting in 1969, the Grade 5 olympiad was organized by the faculty of Leningrad State Pedagogic Institute, and the problem sets were prepared by a separate jury, chaired by L. S. Livshits.

Most of the day-to-day mundane work of the jury—organizing the jury meetings, keeping the minutes and records, communication with the organizing committee etc—was the job of the executive secretary. Since early 1960s the LMO had only five executive secretaries: Yuri Ionin (≈1963–1967), Valery Fedotov (1968–1975), Sergey Rukshin (1976–1984), Nikita Netsvetaev (1984–1986), and Dmitri Fomin (1986–1994).

YURI IONIN, VALERY FEDOTOV, SERGEY RUKSHIN, NIKITA NETSVETAEV, DMITRI FOMIN—LMO'S JURY EXECUTIVE SECRETARIES, 1963–1991

The grade leaders, who are often assigned at the beginning of the annual cycle in November, were responsible for the preparation and design of the problem sets for their grades. During the olympiad itself the leaders selected the problem checkers for their respective grades, analyzed the problems with them, explaining the solutions and discussing various issues that are liable to arise during the solution checking. All of that was usually done on the day of the contest, immediately before its start.

During the olympiad itself, the grade leaders were responsible for maintaining the competition protocols, ensuring that the olympiad runs smoothly, dealing with all logistical hiccups which inevitably occur in the course of this event. After all, when you had to deal with more than three hundred schoolchildren and several dozens mathematicians, spending together three or four hours on the university campus, something completely unpredictable was surely bound to happen.

Obviously, the chairman, the secretary, and the grade leaders could not do much without dozens of the enthusiasts who comprised the problem committee, took part in its sessions, created or suggested dozens or even hundreds of various problems, read and checked the contestants' papers from the second round, helped in organizing and running the oral rounds of the olympiad.

Below we list just a few names of those who during the last thirty years played the most active role in all these endeavors: M. I. Bashmakov, A. G. Gol'dberg, I. V. Romanovsky, B. B. Lurie, Y. I. Ionin, A. I. Plotkin, I. Ya. Verebeichik, L. D. Kurliandchik, S. S. Vallander, A. N. Livshits, N. A. Shirokov, V. E. Lapitsky, V. P. Fedotov, S. V. Fomin, M. N. Gusarov, S. E. Rukshin, S. A. Genkin, N. Yu. Netsvetaev, O. T. Izhboldin, D. Yu. Burago, G. Ya. Perel'man, F. L. Nazarov, I. V. Itenberg, A. V. Bogomolnaya, E. V. Abakumov, K. P. Kokhas, S. K. Smirnov, I. A. Binder ... and many many others.

The First Olympiad

This short paragraph is dedicated to the participants (more precisely, to the winners) of that fateful first olympiad.

Just a few years ago all the information we had on the first Leningrad Math Olympiad was fully contained in only three small articles.

The first one was written in 1984 by Marianna Georg-Aleksandrova ([7]) who took part in that competition. In addition to her personal testimony, this article had the list of the winners with their schools (likely quoted from a newspaper clipping in her own archive).

The second article [8] by S. E. Rukshin and N. M. Matveyev was also mainly based on personal recollections; it contained some interesting specifics as well as one problem said to have been offered at LMO #1.

Finally, the third article by E. A. Belskaya ([29], pp. 8–10) presented some tiny biographical snippets for most of the first and the second prize winners.

However, in 2020, the author was lucky enough to unearth several publications, both online and printed, which contained new data some of which corrected or enhanced the information from the three sources above. Among them were articles [1] and [2], published in 1934 and 1935, respectively, within months of the event itself.

Marianna Aleksandrova recalled that about six hundred students took part in the second round, and about one hundred of them advanced to the final contest. However, article [2], published in 1935, states that those were the presumed numbers—meaning that the organizers had calculated there should be that many contestants. Due to the novelty of the entire event, many schools and work faculties did not qualify anyone or sent less students than requested. Some of them claimed their students were not good enough, while a few others even opined that it was not their job to promote a University-based math competition as those institutions were formally affiliated with other colleges or institutes. As a result, only 307 students came to the second round, and there were 49 winners (who solved all three problems presented to them). Thus, the final competition (the third round) had 48 participants (with just one qualified contestant missing due to either an illness or a scheduling conflict).

A few days after that, a local newspaper published the list of the winners of the very first official mathematical olympiad in the Soviet Union.[28] They were (listed here in Cyrillic alphabetical order):

- *Georgiy (Yuri) Ananov* (P.S. 23, Central district)
- *Aleksandr Bogomolov* (P.S. 2, Narva district)
- *Sergey Vallander* (P.S. 2, Narva district)
- *Marianna Georg* (work faculty, Leningrad State University)
- *Vladimir Kasatkin* (work faculty, Leningrad Electro-Technical Institute)
- *Boris Kizewalter* (work faculty, Institute of Hydrotechnology)
- *Yuri Kondrashov* (work faculty, Leningrad State University)
- *Beniamin Minzberg* (P.S. 15, Smolny district)
- *Sergey Olovyanishnikov* ("Red Chemist" factory)
- *Ivan Sanov* (P.S. 7, Volodarsky district)
- *Kirill Tagantsev* (work faculty, Institute of Hydrotechnology)

As the first place prize each of the winners received a briefcase with an engraved plaque which said "To the winner of the first mathematical olympiad"[29] as well as several books on mathematics.

It must be said that the fates of these people are so genuinely fascinating and, in some cases, tragic, that a separate book can be written about literally each one of them. Alas, this introductory chapter cannot cover all that information, and we will have to restrict ourselves to short (way too short) bios.

*   *   *

We were lucky to find pictures of almost all the first prize winners of LMO-1934. Some of them show our heroes circa 1934–1940 while others depict them at a more mature age. There is still hope we will be eventually able to learn more details about their lives.

---

[28] Other "competing" city olympiads were held in Tbilisi (just a few months later), and in Moscow (in 1935).

[29] To be precise, it said « *To the winner of the first scientific olympiad in mathematics as the award for the exhibited perseverance and industriousness. Remember that "There is no royal road to science, and only those who do not dread the fatiguing climb of its steep paths have a chance of gaining its luminous summits. (Karl Marx)"* », see [8].

Georgiy Ananov (1916–1976), graduated LSU in 1939, then completed Ph.D. program specializing in elasticity theory. Fought in WWII, surviving the infamous Rzhev encirclement; was medically discharged in 1944 after almost fatal case of epidemic typhus. After the war, worked at Leningrad Institute of Precise Mechanics and Optics, defended his Ph.D. (1945) and D.Sc. (1961) theses, promoted to full professor (1963), chaired three different departments at LIPMO, authored more than fifty papers in theoretical mechanics and descriptive geometry.

Aleksandr Bogomolov (1917–1999), completed LSU undergraduate and graduate programs. Fought in WWII as a navigator on naval bombers (same as his high school classmate Sergey Vallander), ending the war as a captain in the Soviet Navy air force. Awarded numerous Soviet military orders and medals (he was also nominated for the highest military award—The Hero of the Soviet Union—but the documents seem to have been lost and only recently found in the archives of the Defense Ministry). After the war, he defended Ph.D. thesis, then taught at Leningrad Navy Engineering College and other tech institutes, published several textbooks in mathematics.

Sergey Vallander (1917–1975), graduated LSU in 1939, then was accepted into Ph.D program in fluid and gas dynamics. During the war, served in Soviet Navy air force as a navigator. Awarded three orders of Red Banner, finished the war in the rank of captain. Defended his Ph.D. in 1946 while still serving in the navy. Became full professor in 1950, D.Sc. in 1959; was recognized as one of the top Soviet experts in fluid and gas dynamics. For many years, led the Institute of Mathematics and Mechanics at LSU, then served as dean of LSU's Faculty of Mathematics and Mechanics (FMM) in 1965–1973. Corresponding member of USSR Academy of Sciences (1966); awarded USSR State Prize in science (1973).

GEORGIY ANANOV, ALEKSANDR BOGOMOLOV, SERGEY VALLANDER

Marianna Georg (1916–2003), graduated LSU, majoring in physics. In 1937, married Aleksandr Aleksandrov,[30] taking his last name. Took up mountain climbing following her husband's example, but in 1939 suffered severe head trauma at Kichkinekol pass in Caucasus mountains which forced her to skip several months of her university studies and quit mountain climbing for good. Despite that accident, she successfully graduated and was

---

[30] Aleksandr D. Aleksandrov (1912–1999), a renowned Soviet geometer and physicist, full member of USSR Academy of Sciences, USSR Master of Sports in mountain climbing.

accepted into the Ph.D. program. After the war, worked at the Radium Institute of USSR Academy of Sciences, defended her Ph.D. thesis in physics of elementary particles in 1952.

Vladimir Kasatkin (1915–2001), the only first prize winner who did not enroll at LSU; instead he joined Leningrad Electro-Technical Institute, majoring in RF engineering. Worked in several military scientific research institutes, was the lead of several projects in the field of the submarine radiolocation. Subsequently he gained Ph.D. in electrical engineering (radiophysics). Awarded order of Labor Red Banner, other state medals, USSR State Prize in 1968.

Boris Kizewalter (1916–1982), while studying at LSU (majoring in mechanics) fell in love with geophysics. After the war, worked at LSU, then at Leningrad Institute of Mineral Processing, where he defended his Ph.D. thesis in geological sciences. Despite a serious hand injury received during the war, took up mountain climbing, and in 1959 became USSR Master of Sports. Received his Doctor of Sciences degree in 1978 for the introduction of methods of mathematical physics into theoretical foundations of mineral processing.

Yuri Kondrashov (1916–2007), enrolled at LSU, majored in chemistry. Survived the blockade of Leningrad while working at the State Institute of Applied Chemistry. In 1949, defended his Ph.D. thesis on chemistry of compounds of manganese. Published more than twenty articles and notes dedicated to research on borides, catalysts, passivity and corrosion of metals, and other areas of inorganic chemistry.[31]

Beniamin Minzberg (1917–2004) was still a junior in 1934, which gave him an opportunity to take part in LMO #2 in 1935 as well. Subsequently, he enrolled at LSU majoring in elasticity theory. Fought in WWII, finishing the war in the rank of senior lieutenant in the air force reconnaissance. After the war, he was accepted into Ph.D. program at Leningrad Polytechnic Institute where he gained his Ph.D. in 1948. For many years taught at Leningrad Navy Academy of Shipbuilding and Weaponry and at Leningrad Navy Engineering College.

MARIANNA GEORG, VLADIMIR KASATKIN, BORIS KIZEWALTER, BENIAMIN MINZBERG

Sergey Olovyanishnikov (1910–1941) was the oldest among the winners of the first olympiad, and that was not accidental. Since his father was an officer in czarist army and his mother a scion of wealthy merchant family in Yaroslavl, he was barred from college

---

[31] Yuri Kondrashov is the only winner whose photo we could not find.

education. His family moved to St. Petersburg where after finishing high school Sergey started working at a chemical factory. He also joined the Young Communists Union, and went to school for working youth. This "acquired identity" allowed him to get accepted to LSU where he majored in mathematics. However, within a year, his deception was discovered. He was expelled from university and exiled to Ufa with the rest of his family except for his father who was so depressed by this calamity that he killed himself. In 1936, Sergey managed to re-enroll at LSU, where he studied differential geometry under A. D. Aleksandrov. Some of his results were so significant that he was accepted into Ph.D. program in 1941 even before he formally finished his undergraduate studies. However, after the Great Patriotic War began on 22 June 1941, he joined the army and in December 1941, was killed in action during German troops' advance toward Leningrad.

Ivan Sanov (1919–1968) was the youngest of the first prize winners—he turned fifteen just a few days before the final round of LMO-1934. Same as Beniamin Minzberg, he repeated that feat at LMO #2. Majored in higher algebra; graduated from Faculty of Mathematics and Mechanics of LSU in 1940; while still an undergraduate student, he solved a partial case of famous Burnside problem in group theory. He was then accepted into Ph.D. program but the war put the studies on hold. Fought at the front, finishing the war as commander of air-defense squad in the rank of lieutenant; awarded Order of Red Star and several medals. Gained his Ph.D. in higher algebra (1946). In 1953, Sanov was "scouted" by KGB for their cryptography division; as a result he had to move to Moscow where he worked in probability theory (stochastic analysis) and algebra. For his work (most likely, it concerned application of stochastic analysis to cryptography) he was awarded the Order of Lenin; his D.Sc. thesis was classified. Died after a serious illness when he was only 49 years old.

SERGEY OLOVYANISHNIKOV, IVAN SANOV, KIRILL TAGANTSEV

Kirill Tagantsev (1916–2001) was a son of a prominent geographer Vladimir Tagantsev, a professor at Petrograd University who was sentenced to death by Bolshevik authorities and executed in 1921.[32] As a consequence, he was barred from enrolling in college for two years after he graduated high school. Finally, in 1936, due to changes in the Soviet educational policy and intercession by physicist Aleksandr Terenin (at the time Kirill worked as a lab

---

[32] Case of the so-called "Petrograd Military Organization"; his wife, as well as one of his alleged co-conspirators, the great Russian poet Nikolay Gumilyov, met the same fate.

assistant at Terenin's department) he was allowed to become a student at the Faculty of Physics of LSU. During Great Patriotic War, he served as a lieutenant in air-defense artillery. After the war, he returned to LSU and worked there for fifty more years. Never defended his Ph.D. thesis, and it is likely he would not be allowed to—in the Soviet times, he was considered tainted by his father's legacy as "the enemy of the people". Worked as a lecturer and as a lab assistant, was active in organization of science olympiads for high school students.

In addition to these eleven victors, ten more participants were awarded with second place diplomas. Unfortunately, significantly less is known about them, and we only have the names of five out of ten—*Anna Gokhberg, Georgiy Konnikov, Iosif Liberman, Aleksandr Smirnov,* and *Yakov Uflyand.*

Anna Gokhberg (1911–1991) tried to enroll at LSU several years earlier but was barred due to her "improper" social background (her father was a lawyer and her mother— a physician). It took her several years of working various menial jobs and the second prize at LMO to achieve her goal. She graduated from LSU in 1939 and then worked in Severodvinsk as a lecturer at a local technical college; during the war she was evacuated to Chelyabinsk where she married her college sweetheart Efim Goldin. After the war, they returned to Leningrad, and for more than 20 years, she taught mathematics at Leningrad Shipbuilding Institute.

Georgiy Konnikov (1917–1941) was accepted to LSU and majored in mechanics, specializing in elasticity theory. During the Great Patriotic War, he joined Kirov division of the People's Militia; killed in action near Leningrad in August of 1941.

Iosif Liberman (1918–1941) got interested in convex and differential geometry during his undergraduate studies at the LSU under guidance of A. D. Aleksandrov, and by the end of his Ph.D. program, he published several outstanding articles on the properties of geodesic curves. Iosif defended his thesis just before he was drafted at the very beginning of the war; surprisingly enough his thesis was not on geometry but on function theory where he solved a very non-trivial problem. He joined the Baltic Fleet's air defense artillery and within two months, in August of 1941, perished on the front lines in the vicinity of Tallinn.

Aleksandr Smirnov (1915–1944) graduated FMM with specialization in elasticity theory and subsequently worked as a math teacher in high school in Kyzyl, capital city of Tuva. During the war he served in air defense artillery, reaching the rank of first lieutenant. In August of 1944, he was killed in a battle near city of Narva.

Yakov Uflyand (1916–1991) graduated from LSU in 1939, then worked as a math teacher in high and middle school. After the war, he was accepted to the Ph.D. program at the Leningrad Polytechnic Institute specializing in mathematical physics. After finishing the program, he worked at the famous A. M. Ioffe Phys-Tech Institute of USSR Academy of Sciences. Gained his Ph.D. in 1948, D.Sc. in 1958, became a full professor in 1962. He also worked in numerical methods of applied mathematics and physics, publishing numerous articles and monographs on those subjects. For many years Yakov Uflyand taught at various Leningrad colleges; since 1971 he led the Phys-Tech's Computer Center.

ANNA GOKHBERG, IOSIF LIBERMAN, YAKOV UFLYAND

Other participants of the final round of the first olympiad included two more high school juniors **Nikolay Shanin** (1919–2011; professor, D.Sc., worked at the Mathematical Institute of USSR Academy of Sciences specializing in topology and then in mathematical logic and algorithm theory) and **Mikhail Perelman** (1919–1942; son of Yakov Perelman, the famous writer of immensely popular books in math and science; graduated FMM, then was accepted to the Ph.D. program; perished on the Leningrad front of the Great Patriotic War). One year after the first olympiad both of them became winners of LMO-1935.

PROBLEMS OF THE FIRST OLYMPIAD

For many many years, anyone who was ever interested in the problems of the pre-war Leningrad Math Olympiads (LMOs) pretty much knew that it was hopeless. Those problem sets either perished during the war and the blockade, or they were never preserved in the first place. The same fate befell even the first after-war olympiads; we were only able to find relatively complete records starting circa 1950.[33]

In 1984 an article [8] was published which quoted the recollections of one of the participants of the first LMO. That was followed by the claim that *"the history has saved for us the very first question of the first olympiad"*. And there it was.

Cube Coloring Problem. *How many ways are there to color the faces of the cube with six different colors? (Two colorings are considered the same if there exists a rotation of the cube which maps one to another.)*

The attribution of this statement was missing; we were given to understand that it came from a personal testimony of that contest's participant.

I was very excited—not only one of the questions from those pre-war olympiads was discovered, but also it was such a lovely question! It was in combinatorics and it was so very different from the known math contest problems of that epoch which all looked as if they were directly copied from the same dusty and boring school textbook.

However, after a while I started to get a nagging feeling that I had seen that problem before. And true enough, I opened book [4] and that question was there, in the problem set

---

[33] But not every year and not every grade.

of the first Moscow Math Olympiad of 1935.

Could the Moscow problem committee reuse a question from the previous year's contest in Leningrad? It goes without a saying that this would be absolutely impossible nowadays. But back then, when the era of the mathematical competitions just began, at the time when the exchange of the popular scientific information was still quite slow, such an issue was not taken as seriously. It is difficult to claim absolute certainty here, but it still seems to me that such a recycling is unlikely—and even if we assume that the Moscow jury loved the problem so much they just could not live without it, then it surely would have sufficed to make a small and very easy change in the problem's statement. Possibly, what happened was some sort of a mental overlap between personal memories about LMO-1934 and the problem set of MMO-1935.

There did not seem to have been any way to resolve this small, and generally speaking, not a very significant historical discrepancy. We could only hope that one day, when our grandchildren will have invented the time machine, one of them goes to Leningrad in the spring of 1934. Then the grateful humankind would finally become the proud owner of the full problem set of the very first city math olympiad.

Just a few years later, as I was writing the very first original Russian edition of this book, I talked to numerous veterans of the Leningrad math competitions. The consensus was overwhelming—finding the problems of the pre-war LMOs was an impossible, zero chance, forget-about-it kind of thing. And that is precisely what I wrote at the time: those problems sets were almost certainly lost to us, never to be discovered.

Now, jump thirty years into the future. As I was preparing a new, updated edition of the LMO book, I undertook new search on the internet, mostly looking for the photographs from 1930s through 1970s related to the math circles, olympiads and so forth.

Imagine my pleasant shock when I have discovered not just one but two articles [1] and [2] published by professor Ioasaf Chistyakov in two Moscow quarterly magazines dedicated to math and physics education in middle and high school back in 1935. They differed only slightly and contained some very interesting data on how the olympiad was organized. They also had the exact date of the final round (18 April 1934), the number of participants for the second and the third round (307 and 48, respectively). But most importantly, each of them was followed by an addendum containing the list of olympiad questions!

Both articles also skipped the issue of attribution. In this case, however, we are talking about the texts written just a few months after the competition. It is only natural to assume that the problem lists were sent to the author (or to the editors) directly by the organizers of the first LMO—we should add here that Boris Delone, the "founding father" of the Leningrad Math Olympiad, had moved to Moscow in January of 1935.

The first article had eight versions of two-question problem set. Apparently, ideally each contestant would get his or her unique set of problems in order to prevent cheating (yes, they had already thought about it back then; although that is not very surprising, kids were kids even ninety years ago). So the organizers have created multiple versions of the problem set (of course, they were supposed to be more or less equal in problem types and their difficulty). Then during the olympiad, they distributed the participants into several classrooms and made sure that the kids in the same auditorium had different versions.

The second article gave a list of eleven questions without specifying which of them were used at which round of the olympiad. However, eight of these questions were also present in the addendum to the first article; the remaining three must have come from the first two rounds.

The cube coloring problem was missing in either article, but it is still remotely possible that it was used in the first or the second round (by the way, many of the problems from the first round were also included in the Exercises sections of those magazines). I still doubt it very much because in all these lists not even one non-standard question is used; also, you cannot have just one such question in a multi-version contest. However, there is a distinct and very real possibility that this problem was used as the first one among the additional questions, to be given to the contestants who solved all problems of the main set.

I submit that the search for the main problem set of the first ever city math olympiad is practically over. These questions are back, once again, and this time, for good. And who knows, perhaps among those who read these lines right now, there are descendants of students, teachers, undergrads, or professors from Leningrad of 1930s. There might still exist some personal archives which preserved the texts or photos from the mathematical competitions of that long gone era. Let us hope that we will not have to wait ninety more years until the next such publication. Or, if that does not work out, we always have Plan B—the time machine.

* * *

Now, here is the problem list of the very first LMO—the eight questions from article [1]. In some places I have fixed the typos and replaced the outdated terminology. A few solutions were cleaned up and expanded.

For some totally unknown reason Question 5(a) is missing. Was it accidentally skipped? Omitted to fit the page space requirements? Lost? We can only guess. It is quite possible that one of the questions **1934.X**, **1934.Y**, **1934.Z** is, in fact, the missing problem.

LMO-1934. FINAL ROUND

1934.01. (a) Show that if a, b, and c are the sides of a triangle, then the roots of equation

$$b^2 x^2 + (b^2 + c^2 - a^2)x + c^2 = 0$$

are imaginary.[34]

(b) A fixed triangle is moved along the plane so that two of its sides always pass through the two fixed points. Show that there exists a point on the plane such that the distance from it to the third side of the triangle is always the same.

1934.02. (a) Show that if α and β are the roots of equation

$$x^2 + px + 1 = 0,$$

---

[34] Imaginary here means *not real*.

while γ and δ are the roots of equation

$$x^2 + qx + 1 = 0,$$

then

$$(\alpha - \gamma)(\beta - \gamma)(\alpha + \delta)(\beta + \delta) = q^2 - p^2.$$

(b) Intersect the given tetrahedron by a plane so that the cross-section is a rhombus.

1934.03. Exclude θ and φ from equations

$$a \sin^2 \theta + b \cos^2 \theta = m,$$
$$b \sin^2 \varphi + a \cos^2 \varphi = n,$$
$$a \tan \theta = b \tan \varphi.$$

(b) Two circumferences intersect at points A and B; through point A a secant is drawn which also meets the circumferences at points P and Q. What is the trajectory traced by the midpoint M of segment PQ, while the secant rotates around point A?

1934.04. (a) Find

$$\lim_{x \to 0} \frac{\tan(a + x) - \tan(a - x)}{\arctan(a + x) - \arctan(a - x)}.$$

(b) Two tangent lines to the circle are fixed, while the third one is moving. Prove that the segment of the third tangent line cut by the first two, is seen at the constant angle from the center of the circle.

1934.05. (b) Three faces of the trihedral angle with pairwise perpendicular edges intersect the given ball to produce three circles. Prove that the sum of the areas of these circles will not change if we rotate the angle around its vertex so that the faces still intersect the ball.

1934.06. (a) Solve the system of equations

$$x^2 = a + (y - z)^2,$$
$$y^2 = b + (z - x)^2,$$
$$z^2 = c + (x - y)^2.$$

(b) Two intersecting circles are given. Prove that the segments tangent to them which are drawn from an arbitrary point on their common secant have equal length.

1934.07. (a) Prove that

$$x + 2x^2 + 3x^3 + \cdots + nx^n = \frac{nx^{n+2} - (n+1)x^{n+1} + x}{(1 - x)^2}.$$

(b) Prove that the distance from an arbitrary point on a circle to some chord of that circle equals the mean proportional[35] of the distances from that point to the tangents drawn at the end of the chord.

---

[35] Same as *geometric mean*.

1934.08. (a) Given that $\sec\alpha\sec\beta + \tan\alpha\tan\beta = \tan\gamma$, show that $\cos 2\gamma \leqslant 0$.

(b) Prove that the four straight line segments connecting the vertices of a tetrahedron with the centers of gravity of the opposite faces have a common point which divides each of these segments in the ratio of $3:1$.

*   *   *

Below are the three problems from [2] not included on the previous list. They must have been offered at the previous rounds, or one of them could be the missing problem 5(a).

1934.X. Prove that equation

$$\frac{1}{x-a} + \frac{1}{x-b} + \frac{1}{x-c} = 0$$

cannot have imaginary[36] roots for real a, b and c.

1934.Y. Find the limit of

$$\left(\cos\frac{a}{x}\right)^x$$

when x tends to infinity, taking integer values $x = 1, 2, 3, \ldots$.

1934.Z. Prove that

$$\frac{n}{(n+1)!} + \frac{n+1}{(n+2)!} + \cdots + \frac{n+p}{(n+p+1)!} = \frac{1}{n!} - \frac{1}{(n+p+1)!}.$$

*   *   *

A few concluding comments.

First, from the vantage point of more than eighty years of math competition experience, we can immediately see certain glaring disadvantages of the multi-version olympiad, especially with a significant number of versions. It seems quite obvious that the difficulty levels of the versions of LMO-1934 were not even approximately equal. For instance, in version 1, the geometry problem **1934.01**(b) looks to be noticeably more complicated than the geometry problem from version 6. Similarly, problems in solid geometry always cause more difficulties than problems in planar geometry. Finding the limit in **1934.04**(a), more likely than not would present a higher obstacle than excluding unknowns in **1934.03**(a) or solving simultaneous equations in **1934.06**(a). Of course, we do not have any statistics but problem **1934.Y** certainly looks much more difficult than algebraic problems from other versions.

Second, this set demonstrates a curious contrast between school curricula of the twentieth century. For instance, we can see that the high school students from regular Soviet schools of the 1930s were offered questions on computation of limits and solving equations in complex numbers. Feel free to compare that to the math curriculum of the contemporary high school.

On the other hand, if we return our attention to the problem collection [4], then we will see that the problem sets of the Moscow Math Olympiads do not mention any limits

---

[36] Here it means *not real.*

or complex numbers, etc. Soon, these topics would disappear from the questions given at Leningrad contests as well. Was it just a very short and geographically localized experiment? It seems that within the next few years the school programs in the country were standardized and somewhat simplified—limits, complex numbers, or complicated solid geometry had been gradually phased out.

ANSWERS AND HINTS

Cube Coloring Problem. Answer: there are 30 ways to do that.

Hint: First, rotate the cube so that the lower face now has color one. Then there are five different ways to select the color for the opposite face. Prove that for each such choice there are exactly six different colorings of the remaining four faces of the cube.

1934.01. (a) **Hint**: These roots are imaginary if and only if the expression

$$4b^2c^2 - (b^2 + c^2 - a^2)^2 = (a + b + c)(b + c - a)(c + a - b)(a + b - c)$$

is positive.

Now, use the triangle inequality applied to the triangles with sides a, b, and c.

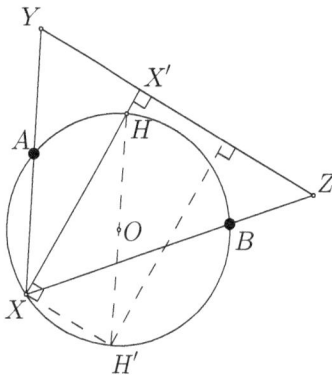

(b) Denote fixed points by A and B and assume that side XY of our moving triangle XYZ passes through point A, and side XZ—through point B. Since angle $\alpha = \angle YXZ$ is always the same, we have that vertex X lies on the fixed circumference σ passing through A and B (indeed, such is the locus of points from which segment AB is seen at angle α).

Drop altitude $h = XX'$ onto side YZ and define point H as intersection of this altitude and σ. Since angle $\angle BXH = \angle ZXX' = 90° - \angle YZX$, subtended by chord BH, must be always the same, point H does not change when the triangle moves. Find point H' which is diametrically opposite to H on σ. Then the distance between line YZ and point H' equals $|XX'|$, which is always the same—it equals to the length of altitude h in triangle XYZ.

1934.02. (a) **Hint**: Start with the equalities

$$\alpha + \beta = -p, \quad \alpha\beta = 1, \quad \gamma + \delta = -q, \quad \gamma\delta = 1.$$

(b) **Hint**: Find point M on edge BC of pyramid $ABCD$ such that ratio BM/MC is the same as ratio AB/CD. Then draw the plane through M parallel to both edges AB and CD.

1934.03. (a) **Answer**: the result of the exclusion is equation

$$(a - m)(n - b)b^2 = (b - n)(m - b)a^2.$$

Hint: Use identity $\sin^2 x + \cos^2 x = 1$ and the first equation to express squares of sine and cosine of θ, as well as the square of tangent of θ, via m, a, and b. Then do the same with the second equation, and finally, substitute the results into the third equation.

(b) **Answer**: denote the centers of the two given circumferences by O_1 and O_2. Then the trajectory of point M is the circumference passing through points A and B, with center at the midpoint of O_1O_2.

Hint: The angles of triangle PQB are always the same, since the angles at vertices P and Q are subtended by the same arc AB in the two given circumferences. Therefore, all triangles PQB are similar. It follows that angle $\angle AMB$ (or its supplementary angle) is always the same, and therefore all points M lie on the circumference passing through A and B.

1934.04. (a) **Answer**: the limit equals $(1 + a^2)/\cos^2 a$.

Hint: Use the formula for the difference of tangents and the formula for the difference of arctangents.

(b) Denote the center of the circumference by O, points of tangency of the two fixed tangent lines by A and B, and the variable point where the third tangent touches the circumference by X. Then it is easy to see that angle POQ equals one half of angle AOB, and therefore, does not depend on the location of point X.

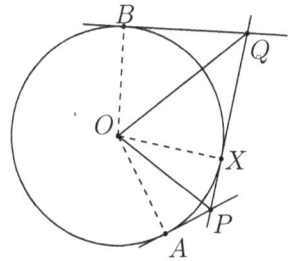

1934.05. (b) Suppose that the ball has radius R and center O located at distance D from the angle's vertex V. The area of the circle-intersection of a plane and the ball equals $\pi(R^2 - d^2)$, where d is the distance from O to this plane. Therefore, the sum of the areas from the statement is $S = \pi(R^2 - d_1^2 + R^2 - d_2^2 + R^2 - d_3^2)$, where d_1, d_2, d_3 are the distances from point O to the three faces of the angle. Since the faces are perpendicular to each other, the sum of the squares of these distances equals the square of the distance $|OV| = D$. Thus we obtain $S = \pi(3R^2 - D^2)$—this expression does not depend on the rotation of the trihedral angle.

1934.06. (a) **Answer**: there are two solutions

$$x = \pm\frac{a(b + c)}{2\sqrt{abc}}; \quad y = \pm\frac{b(a + c)}{2\sqrt{abc}}; \quad z = \pm\frac{c(a + b)}{2\sqrt{abc}}.$$

Signs must be either all top ones, or all bottom ones.

(b) **Hint**: Use the very well known (and easy) fact that for any point O outside of the given circumference σ and any secant line passing through O and intersecting σ at points P and Q, the product of lengths $|OP| \cdot |OQ|$ is equal to the square of the length of tangent from O to σ. (This is a so-called *tangent-secant theorem* or the property of the *degree of point* relative to a circle).

1934.07. (a) Transform the left-hand side as follows

$$x + 2x^2 + 3x^3 + \cdots + nx^n$$
$$= x(1 + x + \cdots + x^{n-2} + x^{n-1})$$
$$+ x^2(1 + x + \cdots + x^{n-2})$$

$$\ldots$$
$$+ x^{n-1}(1+x)$$
$$+ x^{n}(1)$$

and then use (multiple times!) the formula

$$1 + x + x^2 + \cdots + x^k = \frac{x^{k+1} - 1}{x - 1}.$$

(b) **Hint:** Consider triangle ABC, formed by chord AB and tangents AC and BC. Let M be an arbitrary point on arc AB, and C_1, A_1, B_1—the feet of the perpendiculars dropped from M onto AB, BC, and CA, respectively (C_1 here will represent that arbitrary point on the chord mentioned in the problem's statement). Prove and then use the fact that we have two pairs of similar triangles: $\triangle B_1 AM \sim \triangle C_1 BM$ and $\triangle A_1 BM \sim \triangle C_1 AM$.

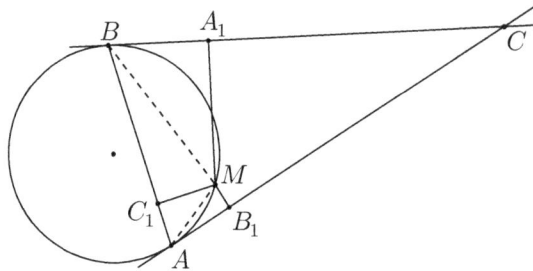

1934.08. (a) **Hint:** First, inequality $\cos 2\gamma \leqslant 0$ is equivalent to the inequality $\tan^2 \gamma \geqslant 1$. Use identity

$$\sec \alpha \sec \beta + \tan \alpha \tan \beta = \frac{1 + \sin \alpha \sin \beta}{\cos \alpha \cos \beta},$$

and now we need to prove that when the product $\cos \alpha \cos \beta$ is positive we have $1 + \sin \alpha \sin \beta \geqslant \cos \alpha \cos \beta$, and when it is negative then $1 - \sin \alpha \sin \beta \geqslant -\cos \alpha \cos \beta$. In either case it follows from the identity

$$\cos \alpha \cos \beta \pm \sin \alpha \sin \beta = \cos(\alpha \mp \beta),$$

which implies that this expression lies inside interval $[-1; 1]$.

(b) **Hint:** That common point is the center of gravity of the tetrahedron.

Using vector notation, we can use the fact that the center of gravity of the pyramid $ABCD$ is the end of the vector OM (O is the origin) defined by $\overrightarrow{OM} = \frac{1}{4}(\overrightarrow{OA} + \overrightarrow{OB} + \overrightarrow{OC} + \overrightarrow{OD})$. Write out the vectors to the centers of the faces, and then use the formula

$$\overrightarrow{OY} = \frac{q}{p+q}\overrightarrow{OX} + \frac{p}{p+q}\overrightarrow{OZ},$$

which is true for any point Y which lies on segment XZ and divides it in the ratio of $p : q$.

1934.X. Hint: Get rid of the denominators and combine the like terms. Consider the discriminant of the resulting quadratic equation

$$3x^2 - 2(a + b + c)x + (ab + bc + ca) = 0.$$

Another approach uses symmetry of the given expression; it allows us to assume without loss of generality that $a < b < c$. Now, consider the signs of $(x-b)(x-c)+(x-a)(x-c)+(x-a)(x-b)$ at points a, b and c; since they alternate, we can conclude that the equation has at least two real roots.

1934.Y. Answer: the limit is 1.
Hint: Prove the inequality

$$0 < 1 - \cos^x \frac{a}{x} < a \sin \frac{a}{2x} \, .$$

1934.Z. Hint: Use the easily verifiable (and well known) identity

$$\frac{n+k}{(n+k+1)!} = \frac{1}{(n+k)!} - \frac{1}{(n+k+1)!} \, ,$$

and add these equalities for $k = 0, 1, \ldots, p$.

Notation and Terminology

The questions' indexing is organized like this: first comes the olympiad's year and then the number of the question in the combined problem set of that year. For instance, **1973.14** points to the fourteenth question of the LMO-1973. If we refer to the question from the same year's contest, then we usually skip the year and simply write something like "*See Question 22*". In the same way we refer to the problems' numbers in the Solutions section of this book.

The most difficult problems are marked with an asterisk (e.g., **1961.25.***); a few very difficult problems carry the double-asterisk mark.

1) The symbol □ is used to mark the end of the proof for a lemma or an auxiliary proposition within a problem's solution. Traditionally, in mathematics, this symbol is equivalent to the Latin expression "*Quod Erat Demonstrandum*", which literally means "*(Which Is) What Was To Be Shown*", signifying the end of the proof.

2) $n!$ is called "*n factorial*" and denotes the product of all natural numbers 1 through n. By definition $0! = 1$.

3) $\overline{xy \ldots z}$ denotes the number written by the given decimal digits.

4) $[x]$ denotes the *integer part* of real number x, that is, the largest integer not exceeding x. For example, $[\pi] = 3$, $[-7.11] = -8$, and $[12] = 12$.

5) Abbreviations $\gcd(x, y)$ and $\text{lcm}(x, y)$ are used to denote the greatest common divisor (G.C.D) and lowest common multiple (L.C.M.), respectively, of two integers x and y.

6) $\mathbb{N}$ denotes the set of all natural numbers, i.e., positive integers. Also, $\mathbb{Z}$ denotes the ring of all integers, $\mathbb{Q}$—the field of rational numbers, and $\mathbb{R}$—the field of real numbers.

7) For the binomial coefficients $C_n^k = \dfrac{n!}{k!\,(n-k)!}$ we use the more internationally accepted notation $\binom{n}{k}$.

8) In many geometry problems, the distance between two points A and B is denoted by $|AB|$ (sometimes simply AB), the area of triangle ABC is denoted by $S(ABC)$ or S_{ABC}. The measure of angle XYZ is denoted by $\angle XYZ$, while the measure of arc XY or XYZ is denoted by $\overset{\frown}{XY}$ or $\overset{\frown}{XYZ}$.

9) The distance between two figures P and Q on the plane (or between two bodies in space) is defined as the minimum distance between points from P and Q. More formally,

$$\mathrm{dist}(P, Q) = \min\{|XY| \; : \; X \in P, Y \in Q\}.$$

10) Geometrical constructions, unless explicitly specified otherwise, are always to be performed using only straight edge and compasses.

11) All the tournaments mentioned in the problem statements are considered *round-robin* tournaments unless explicitly stated otherwise. A round-robin (or *all-play-all*) tournament is the one where each participant plays every other one the same number of times (usually once but sometimes twice or even more).

Also, contestants were presumed to know that in **chess** the win brings 1 point, the draw—$\frac{1}{2}$ points, and the loss—zero points. There are no draws in **volleyball** (or in **table tennis**), the winner gains one point, the loser—zero. In **soccer**, winner of the match gains two points, the loser—zero, while in the case of the draw each team gets one point (these rules were changed in 1994–1996, when the *three points for the win* rule was universally adopted).

Also assumed was the general knowledge of the most basic rules of chess, checkers, and dominoes.

12) AM-GM inequality is a well-known (and probably the most used) inequality for arithmetic and geometric means, which states that given any finite sequence of n non-negative numbers (x_k) their *arithmetic mean (AM)* is always greater than or equal to their *geometric mean (GM)*, with equality attained if and only if all numbers x_k are the same. In other words,

$$\frac{1}{n}(x_1 + \cdots + x_n) \geqslant \sqrt[n]{x_1 \ldots x_n}.$$

We recommend books [3] and [20] as excellent introductions into classic inequalities and more.

13) In a triangle, the *orthocenter* is the point of intersection of all three altitudes of the triangle. Almost all basic definitions and terms related to planar geometry can be found in Wikipedia article on triangle or by following the corresponding web links in that text.

The readers are also encouraged to peruse some good introductory book on classic planar and spacial geometry—we recommend [6], [5], and [9].

14) A *polyomino* is a connected planar figure formed by several unit squares with sides parallel to the coordinate axes so that (a) no two squares have a common interior point; and (b) if two squares have a common point A, then either A is their only common point and it serves as a vertex for both of them, or they share a common edge. A few examples are shown below.

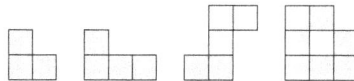

For more on polyominoes, see book [13].

15) Numerous problems in this collection use graph theory terminology. To familiarize yourself with the basic concepts and terms, we can recommend a very good introductory book [10].

16) Same can be said about the number theory and modular (or residue) arithmetic. Our recommendation is another book by the same author [30] from the same wonderful series of introductory books on various areas of mathematics called NEW MATHEMATICAL LIBRARY.

PART 2
Problems

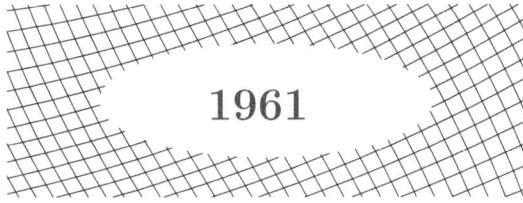

GRADE 6

1961.01. A specific job can be done by three workers. The second and the third together can do it twice as fast as the first worker alone, and the first and the third together can do it three times faster than the second worker alone. How much faster can the first and the second workers do the job than the third one alone?

1961.02. Prove that the greatest common divisor of the sum of two numbers and their least common multiple equals the greatest common divisor of the two original numbers.

1961.03. Solutions to twenty questions are to be presented after the contest with twenty participants. It turned out that each student solved exactly two problems and that each problem was solved by exactly two students. Prove that the presentation can be organized in such a way that each student presents exactly one solution and all the problems are discussed.

1961.04. Abby and Beth have to get from point M to point N which lies 15 km apart from M. Their foot speed is 6 km/h. They also have a bicycle which each of them can ride at 15 km/h. Both Abby and Beth start their trip from M simultaneously with Abby on foot and Beth riding the bike until she meets Claire who walks from N to M. After that Beth walks and Claire rides the bike until she meets Abby whereupon she gives up the bike which Abby then uses to ride to N. If Claire walks with the same speed as other two girls, when should she start her trip from N to ensure that Abby and Beth arrive to N simultaneously?

1961.05. Prove that among any six people, one can always find either three persons any two of whom know each other, or three persons any two of whom do not know each other.

GRADE 7

1961.06. See Question 1.

1961.07. Circle O, square K, and line L are given on the plane. Construct a segment of given length, parallel to L and such that its ends lie on contours of O and K correspondingly.

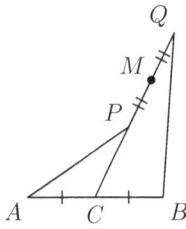

1961.08. Three-digit number $\overline{abc}$ is divisible by 37. Prove that the sum of $\overline{bca}$ and $\overline{cab}$ is also divisible by 37.

1961.09. Point C is a midpoint of segment AB. On a ray originating from C and not parallel to AB, three consecutive points P, M, and Q are selected in such a way that $|PM| = |MQ|$. Prove that $|AP| + |BQ| > 2\,|CM|$.

1961.10. There are $2n + 1$ different objects. Prove that the number of ways to choose an odd number of them is equal to the number of ways to choose an even number of these objects.

GRADE 8

1961.11. Construct a quadrilateral given the lengths of its sides and the distance between midpoints of its diagonals.

1961.12. Numbers a, b, and $\sqrt{a} + \sqrt{b}$ are rational. Prove that $\sqrt{a}$ and $\sqrt{b}$ are both rational as well.

1961.13. Solve the equation $x^3 - [x] = 3$.

1961.14. Prove that if in some triangle an angle bisector also bisects the angle between the corresponding median and altitude, then the triangle is either isosceles or right.

1961.15. Each of n numbers $x_1, x_2, \ldots, x_n$ is equal to either $+1$ or -1. Given that they satisfy the equation

$$x_1 x_2 + x_2 x_3 + \cdots + x_{n-1} x_n + x_n x_1 = 0\,,$$

prove that n is divisible by 4.

1961.16. There are n points selected on a circumference in such a way that for any two points, one of the arcs defined by them is shorter than $120°$. Prove that all points can be covered by an arc measuring $120°$.

GRADE 9

1961.17. Construct a triangle given one of its vertices, as well as the lines of all three of its bisectors.

1961.18. See Question 13.

1961.19. For any $2n$ points on a plane, prove that one can always connect them with n disjoint segments.

1961.20. Given angle of measure α and triangle ABC such that $\angle A + \angle B = \alpha$, find trajectory of vertex C when vertices A and B slide along the sides of the angle.[37]

1961.21. Prove that a 10×10 chessboard cannot be covered by T-shaped polyominoes

.[38]

1961.22. Three non-zero integers k, m, and n are such that k and m are co-prime. Prove that there exists an integer x such that $mx + n$ is divisible by k.

1961.23. (bonus)[39] Prove that all numbers 1156, 111556, 11115556, ... are perfect squares.

GRADES 10–11

1961.24. Trapezoid with bases m and n serves as a base for a pyramid with volume V. A plane cuts off of it a smaller pyramid with volume U, producing a section which is again a trapezoid with bases m_1 and n_1. Prove that

$$\frac{U}{V} = \frac{(m_1 + n_1)m_1 n_1}{(m + n)mn}.$$

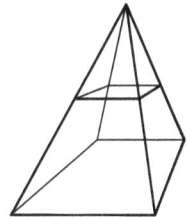

1961.25.* A school's curriculum covers $2n$ subjects. All the students' final grades are **A**s and **B**s only, and no two of them have the same set of grades. It is also known that for any two students, each one of them is better than the other in at least one subject. Prove that the number of the students does not exceed $\dfrac{(2n)!}{n!n!}$.

1961.26. Find the minimum value of expression $(a^2 + x^2)/x$, where $a > 0$ is a given constant and $x > 0$ is a variable.

1961.27. Let $f(x) = x^n + a_1 x^{n-1} + \cdots + a_n$ be a polynomial with integer coefficients and p—its rational root. Prove that p is an integer and that $f(m)$ is divisible by $p - m$ for any integer m.

1961.28. Prove that $\tan 20° - \tan 40° + \tan 80° = 3\sqrt{3}$.

1961.29. Each face of a cube can be split into two right triangles one of which is then painted black and the other—white. Prove that there are only two distinct colorings (that is, one cannot be obtained from the other by rotations of the cube) such that for any vertex the sums of white and black angles adjacent to it are equal.

---

[37] The triangle is always positioned in such a way that vertex C and the vertex of the given angle lie on the opposite sides of straight line AB.

[38] The definition of polyominoes can be found in Polyomino articles from Wolfram Mathworld or Wikipedia.

[39] It is possible that this Grade 9 bonus question was actually used at the contest instead of one of the previous problems. Some handwritten copies indicate that might have been the case.

1961.30. A snail crawls on the table's surface with constant velocity. Every 15 minutes, it makes a turn at 90°, and between these turns, it crawls along a straight line. Prove that it can return back to its original position only after a whole number of hours.

1962

Grade 6

1962.01. Three travelers in possession of one two-seat motorcycle simultaneously set out from town A to town B. How should they travel to guarantee that the time in which the last one of them reaches B is minimal? Find that time. Their speed on foot is 5 km/h, the motorcycle's speed is 45 km/h, and distance between A and B is 60 km.

1962.02. Numbers a and b are co-prime. What are the possible common divisors of numbers $a + b$ and $a - b$?

1962.03. The age of someone in 1962 exceeds by one the sum of all digits of their year of birth. How old is this person?

1962.04. Fifteen magazines lie on a table, fully covering the table. Prove that it is possible to remove eight of them so that the remaining magazines cover at least $\frac{7}{15}$ of the table's surface.

1962.05. Prove that the chess knight can travel through all squares of 201×201 board visiting each square exactly once.[40]

1962.06. Can an integer, two last digits of which are odd, be a perfect square?

Grade 7

1962.07. Prove that the sides of an arbitrary quadrilateral can form a trapezoid.[41]

1962.08. See Question 2.

1962.09. See Question 4.

1962.10. In a six-digit number divisible by 7, the last digit was moved to the beginning. Prove that the resulting number is also divisible by 7.

---

[40] To learn how the chess pieces move, visit the Chess.com website or consult any introductory book on chess.

[41] This is not formally correct. Obviously, the jury meant that "trapezoid" here denotes a quadrilateral some two sides of which are parallel.

1962.11. A circle is split into 49 regions such that no three of them have a common point. The resulting map is painted using three colors so that no two neighboring regions are of the same color. The common border of any two regions is considered to have both colors. Prove that on the circle's boundary, one can always find two diametrically opposite points of the same color.[42]

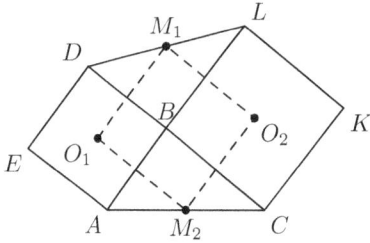

1962.12. Squares $ABDE$ and $BCKL$ with centers O_1 and O_2, respectively, are built on sides AB and BC of triangle ABC. Points M_1 and M_2 are the midpoints of DL and AC. Prove that $O_1M_1O_2M_2$ is a square.

GRADE 8

1962.13. Four circumferences are placed on a plane in such a way that each one is externally tangent to some two of the others (see the figure). Prove that their tangency points lie on a circumference.

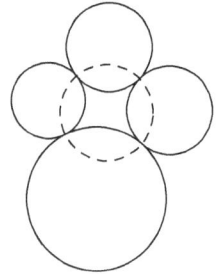

1962.14. Each of two numbers a and b can be represented as $x^2 - 5y^2$ where x and y are integers. Prove that ab can be represented in that form as well.

1962.15. Solve the equation $x(x+d)(x+2d)(x+3d) = a$.

1962.16. If $a^2 + b^2 + c^2 = 1$, $m^2 + n^2 + p^2 = 1$, then prove that $-1 \leqslant am + bn + cp \leqslant 1$.

1962.17. Inscribe a triangle of maximum possible area into the given semicircle.

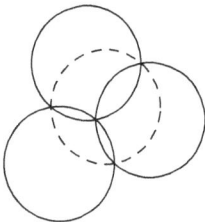

1962.18. Three identical circumferences have a common point. Prove that three other points where some two of them meet lie on a circumference of the same radius.

1962.19. Find the smallest circle that covers the given triangle.

1962.20. Given polynomial $x^{2n} + a_1x^{2n-2} + a_2x^{2n-4} + \cdots + a_{n-1}x^2 + a_n$ which is divisible by $x - 1$, prove that it is also divisible by $x^2 - 1$.

1962.21. Prove that for any prime number p different from 2 and 5 there exists natural number k such that the decimal representation of number pk contains digits 1 only.

---

[42] Question 11 is presented differently in different versions of jury archives. Some of the versions are extremely simple and some literally make no sense. In the above version, the question is valid although it seems to be too difficult for the 7-graders of that time.

GRADE 9

1962.22. See Question 16.

1962.23. See Question 20.

1962.24. One of the unit squares is selected on the infinite square grid. Prove that one of the distances from an arbitrary grid node to the vertices of the square is irrational.

1962.25. The sum of ten numbers is zero as well as the sum of all possible pairwise products of these numbers. Prove that the sum of their cubes is also zero.

1962.26. Inside an acute-angled triangle a point A is chosen such that all segments connecting it to the triangle's vertices are shorter than the shortest side of the triangle. Prove that the sum of the distances from A to the vertices does not exceed three quarters of the triangle's perimeter.

1962.27. See Question 21.

1962.28. There are n points on the plane which do not all lie on one straight line. Prove that one can find a closed broken line without self-intersections whose vertices are the given points.

1962.29. Cargo with total weight of 13.5 tons is distributed into several boxes each one of them not heavier than 350 kg. Prove that the entire cargo can be shipped off using eleven trucks with 1.5 ton capacity each.

GRADE 10

1962.30. A quadrilateral is inscribed into the unit square so that each side of the square has a vertex of the quadrilateral on it. Prove that one of the quadrilateral's sides has length of at least $1/\sqrt{2}$.

1962.31. The sum of ten numbers is zero as well as the sum of all pairwise products of these numbers. Find the sum of their fourth powers.

1962.32. Get rid of irrationality in the denominator: $\dfrac{1}{\sqrt[3]{a} + \sqrt[3]{b} + \sqrt[3]{c}}$.[43]

1962.33. Solve the equation

$$(3x + 2)^4 + (2x - 4)^4 = (2x + 3)^4 + (4x - 2)^4 .$$

1962.34. Point O lies within an angle, each of whose sides contains a segment. The endpoints of these two segments are connected to O and the sum of the areas of two resulting triangles is S. Find a locus of points M such that the sum of the corresponding areas equals S.

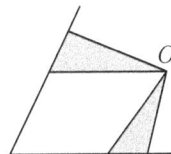

---

[43] In some archival versions Question 32 is presented with specific values, namely, $a = 2$, $b = 3$, $c = 4$.

1962.35. Find the sum $\sum\limits_{k=0}^{n}(-1)^k\binom{n}{k}k$.

1962.36. Solve the equation in natural numbers:

$$1-\cfrac{1}{2+\cfrac{1}{3+\cfrac{1}{4+\cfrac{1}{\cdots+\cfrac{1}{n}}}}}=\cfrac{1}{x_1+\cfrac{1}{x_2+\cfrac{1}{x_3+\cfrac{1}{\cdots+\cfrac{1}{x_n}}}}}.$$

1962.37. Polyhedron P and a natural number $n>1$ are such that for all faces of the polyhedron save one the number of their sides is divisible by n. Prove that the faces of P cannot be painted with two colors in such a way that any two neighboring faces are colored differently.

1962.38. All the sides and diagonals of a convex polygon with n sides are extended to form straight lines. No two of these lines are parallel and no three of them meet at the same point. How many points of their intersection are inside the polygon and how many are outside?

GRADE 11

1962.39. See Question 30.

1962.40. Given polynomial $x^{kn}+a_1x^{k(n-1)}+a_2x^{k(n-2)}+\cdots+a_{n-1}x^k+a_n$ which is divisible by $x-1$ prove that it is also divisible by x^k-1.

1962.41. See Question 33.

1962.42. Find the sum $\sum\limits_{k=0}^{n}(-1)^k\binom{n}{k}k^2$.

1962.43. See Question 32.

1962.44. See Question 34.

1962.45. See Question 37.

1962.46. See Question 36.

1962.47. A convex polygon with area greater than $\frac{1}{2}$ lies inside a unit square. Prove that one can find a chord of the polygon which is longer than $\frac{1}{2}$ and is parallel to the given side of the square.

ELIMINATION ROUND

1962.48. Prove that seven circles with radii equal to 1 can be positioned to cover a circle of radius 2, but they are not enough to cover a circle of any greater radius.

1962.49. Nine polygons, each of area 1, are placed inside a square of area 5. Prove that some two of them have intersection whose area is at least $\frac{1}{9}$.

1962.50. A sphere is split into three regions which are then colored, each into its own color (borders of the regions are considered to be of both adjacent colors). Prove that there exists a diameter whose ends are of the same color.

1962.51.*** Prove that a number written only with 0s and 1s which uses least two 1s cannot be a perfect square.

1962.52. Prove that out of any k integers one can always choose several whose sum is divisible by k.

1962.53. Solve the equation $(x^{n(n-1)/2} - x^{[n/2]})/(x-1) = 0$.

1962.54.* For any two infinite sequences consisting only of numbers $1, 2, \ldots, n$ we will call two terms in these sequences "identical" if they are equal and have the same index. How many such sequences can be constructed so that each two of them have exactly k identical terms?

Commentary.

(1) *Despite the fact that the only existing version of Grade 9 problem set has eight questions, it is likely that there were nine of them. Quite possibly Question 1962.37 was the missing problem.*

(2) *The version of the Elimination Round presented here is likely a preliminary draft of the "final edition". We claim this because two problems here look rather suspicious.*

    (a) *First, Question 51 is so difficult that we were not able to produce or find an elementary solution. Is it possible that the question was given without a known solution? or perhaps the proposed "solution" had a hole in it? We will never know.*

    (b) *Second, Question 53, in the form given above, is extremely simple, especially for the Elimination round. This is, likely, a typo.*

1963

Grade 6

1963.01. Two travelers started out from point A to point B. The first one used a highway and walked at 5 km/h while the second one walked along a forest path at 4 km/h. The first traveler traveled 6 km more than the second one and reached B one hour later. Find the highway distance between A and B.

1963.02. A pedestrian walks along the highway at 5 km/h. Buses travel along the same highway with fixed constant velocity meeting each other every 5 minutes. At 12 pm a pedestrian noticed that two buses met exactly next to him, and while he kept on walking he started to count the buses going past him in both directions. At 2 pm, some two buses again met exactly next to him and by that moment the number of buses that went toward him exceeded by four the number of buses that passed him. Find the velocity of the buses.

1963.03. Prove that the difference $43^{43} - 17^{17}$ is divisible by 10.

1963.04. Two unit squares at the edge of the chessboard are cut out from the board. In what cases the rest of the chessboard can be covered by dominoes ⊏⊐ without overlapping and in what cases it cannot?

1963.05. Distance from city A to city B (measured along the straight line) is 30 km, distance from B to C is 80 km, from C to D—236 km, from D to E—86 km, from E to A—40 km. Find the distance from E to C.

1963.06. Is it possible to write numbers 1 through 1963 in a row so that any two neighboring numbers, as well as any two numbers with a common neighbor, were co-prime?

Grade 7

1963.07. The area of a quadrilateral is 3 cm^2 while its diagonals are 6 cm and 2 cm long. Find the angle between the diagonals.

1963.08. Prove that number $1 + 2^{3456789}$ is composite.

1963.09. Twenty players participated in a round-robin[44] chess tournament where the

---

[44] That is, a tournament in which each participant played every other participant exactly once.

competitor who won undisputed 19th place had 9.5 points. How could the points be distributed between all other participants?

1963.10. The sum of the distances between the midpoints of opposite sides of a quadrilateral equals half of its perimeter. Prove that this quadrilateral is a parallelogram.

1963.11. Forty passengers who only have coins of value 10, 15 and 20 kopecks[45] ride in a bus with trip fare of 5 kopecks.[46] Altogether they have 49 coins. Prove that they cannot pay the required amount of money to the driver so that each one of them ends up paying exact fare.

1963.12. A natural number a was divided (with remainder) by all natural numbers fewer than a. The sum of all different remainders obtained in this process turned out to be equal to a. Find a.

1963.13. Two of the chessboard's squares were cut out. In what cases the rest of the chessboard can be covered by dominoes ⊏⊐ without overlapping and when it cannot be so covered?

GRADE 8

1963.14. Point A is selected on a triangle's median between two of its sides with lengths x and y. Given that the sum of distances from A to these sides equals s, find these distances.

1963.15. Decimal fraction $0.abc\ldots$ is constructed by the following rule: a and b are two arbitrary digits; every next digit is the remainder obtained when the sum of the two preceding digits is divided by 10. Prove that this fraction is periodic.

1963.16. Two convex polygons with m and n sides, respectively, are drawn on the plane $(m > n)$. What is the maximum possible number of parts they can divide the plane into?

1963.17. The sum of three perfect squares is divisible by 9. Prove that the difference between some two of them is also divisible by 9.

1963.18. Given $k + 2$ integers prove that there exist two among them such that either their sum or their difference is divisible by $2k$.

1963.19. A right angle is rotating about its vertex. Find the locus of midpoints of the segments connecting the points of intersection of the angle's sides and a given circle.[47]

---

[45] Russian currency is *ruble* which is equal to 100 kopecks.
[46] This was the actual city bus fare in Leningrad circa 1960s–1980s.
[47] Obviously, it was implied that the vertex of the right angle lies inside of the given circle.

GRADE 9

1963.20. Five-digit number $\overline{abcde}$ is divisible by 41. Prove that if we shift its digits cyclically, then the result will also be divisible by 41.[48]

1963.21. Given two monotonically decreasing sequences of numbers

$$a_1 \geqslant a_2 \geqslant \cdots \geqslant a_n \geqslant 0, \quad b_1 \geqslant b_2 \geqslant \cdots \geqslant b_n \geqslant 0,$$

prove that

$$a_1 b_1 + a_2 b_2 + \cdots + a_n b_n \geqslant a_1 b_n + a_2 b_{n-1} + \cdots + a_n b_1,$$

1963.22. Construct a trapezoid given its diagonals and sides.

1963.23. Two circumferences of radius $R = \frac{1}{2\pi}$ are given. On one of them 20 points are selected, and on the other—several arcs with total length less than $\frac{1}{20}$. Prove that one circumference can be superimposed upon the other so that no selected point lies on a selected arc.

1963.24. See Question 18.

1963.25. Solve the equation $(z + 1)^x - z^y = -1$ in natural numbers.

GRADES 10–11

1963.26. Prove that if two quadratic equations with integer coefficients

$$x^2 + p_1 x + q_1 = 0, \quad x^2 + p_2 x + q_2 = 0$$

have a common non-integer root, then $p_1 = p_2$, $q_1 = q_2$.

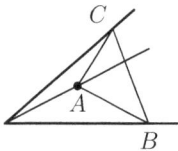

1963.27. Point A is selected on an edge of a trihedral angle. Points B and C move freely along the two other edges. Find the locus of the centers of gravity of triangles ABC.

1963.28. Prove that equation

$$a_1 x_1 + a_2 x_2 + \cdots + a_n x_n = b$$

cannot be satisfied by more than half of all possible n-tuples $(x_1, x_2, \ldots, x_n)$, where each of x_i equals either 0 or 1, if not all of the numbers $a_1, a_2, \ldots, a_n, b$ are zeroes.

1963.29. In a face of a polyhedron we will call angle α adjacent to an edge if this edge is one of the sides of angle α. Prove that in any tetrahedron one can find an edge such that all angles adjacent to it are acute.

---

[48] Question 1963.20 seems to be an undisputed leader with respect to the number of times it was used at the LMOs. The very first time, as far as we know, was 1958. It was again reused at the LMO-1978.

1963.30. Prove that for any k infinite sequences of natural numbers

$$x_{1,1}, \; x_{1,2}, \; x_{1,3}, \; \ldots$$

$$x_{2,1}, \; x_{2,2}, \; x_{2,3}, \; \ldots$$

$$\ldots \ldots \ldots \ldots \ldots \ldots$$

$$x_{k,1}, \; x_{k,2}, \; x_{k,3}, \; \ldots$$

there exist indexes p and q such that for any $1 \leqslant i \leqslant k$ we have $x_{i,p} \geqslant x_{i,q}$.

1963.31. Prove that it is impossible to place ten equal squares on the plane in such a way that they do not have common interior points, but one of them touches all the others.

ELIMINATION ROUND

1963.32. In decimal representation of $(\sqrt{26}+5)^{1963}$ find the first 1963 digits after period.

1963.33. How do you turn a cereal box in space so that the area of its projection onto the given plane is maximal?

1963.34. Five circles on the plane are such that each two of them intersect. Prove that some three of them have a common point.

1963.35. Natural number a is divisible by 1, 2, 3, $\ldots$, and 9. Prove that if $2a$ is represented as a sum of several numbers each one of which is 1, 2, 3, $\ldots$, or 9, then among them one can always find a few with their sum equal to a.

1963.36.* Four positive numbers are such that the sum of any three of them is greater than the fourth. Prove that there exists a tetrahedron such that these numbers represent the areas of its faces.

1963.37. Solve the equation $x^4 - 2y^4 - 4z^4 - 8t^4 = 0$ in integers.

1963.38. Prove that any planar closed broken line of length 1 can be covered by a circle of radius $\frac{1}{4}$.

1963.39. Does there exist a triangle with all sides, altitudes, and angle bisectors of integer length?

1963.40. For which n number $2^n + 1$ is a non-trivial power of an integer?

1963.41. On a sheet of graph paper two parallel lines are drawn through some two of the grid nodes.[49] Prove that there are infinitely many other nodes in the strip between the lines (the strip by definition includes both of its boundaries).

---

[49] Obviously, it was implied that each of these two lines passes through one of these two nodes, respectively.

Commentary.

(1) As we have mentioned above, Question 20 was already used only five years before that, in 1958; then it was used again, in 1978.

    This reveals that (a) the jury records were not meticulously kept and passed down between generations (a shame, really); and (b) the age of professionalization of mathematical contests was yet to come. It would be absolutely impossible for something like that to happen after mid-1970s.

(2) It is worth mentioning that Question 10 was also given in 1980 as a problem for Grade 9. Such coincidences can provide interesting insight into possible comparison of school syllabi several years apart or of the level of students' preparedness.

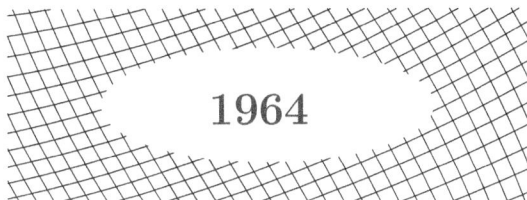

1964

Grade 6

1964.01. Adam, Bob, and Charles made six shots at one target and got the same amount of points. Adam scored 43 points in his first three shots, and Bob scored only 3 points with his first shot. How many points did each of them score with every shot if we know there was one hit of 50 point area, two of 25, three of 20, three of 10, two of 5, two of 3, two of 2, and three of 1?

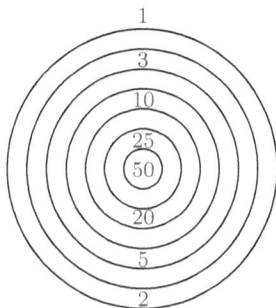

1964.02. Prove that a 10×10 chessboard cannot be covered by 25 tetrominoes of type ⌐⌐⌐.

1964.03. Chessboard squares are filled with natural numbers so that every number equals the arithmetic mean of its neighbors.[50] Sum of the numbers in the corners of the chessboard is 16. Find the number in square $e2$.

1964.04. A table with dimensions 100×100 is given. What is the minimum number of different letters that can be written in its squares so that no two identical letters are written next to each other?

1964.05. A squad of cadets is arranged on the parade ground in rectangular formation. In every row, the tallest cadet is selected, and John Smith is the shortest one of them. Then in every column, the shortest cadet is selected, and Jack Brown is the tallest one of them. Who is taller—John or Jack?

1964.06. Find the product of three numbers whose sum is the same as the sum of their squares and the same as the sum of their cubes which is equal to 1.

Grade 7

1964.07. In the given n-gon all angles are obtuse. Prove that the sum of all diagonals in it is greater than the sum of all sides.

---

[50] Here we call the squares *neighboring* if they share a common side.

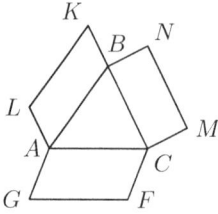

1964.08. Find all integers x and y such that $x^4 + 4y^4$ is prime.

1964.09. On the sides of triangle ABC, parallelograms $ABKL$, $BCMN$, and $ACFG$ are built. Prove that segments KN, MF, and GL can form a triangle.

1964.10. See Question 2.

1964.11. Find the maximum number of different positive integers less than 50 such that each two of them are co-prime.

1964.12. In triangle ABC, points D and E are the midpoints of sides AB and BC. Point M lies on AC and $|ME| > |EC|$. Prove that $|MD| < |AD|$.

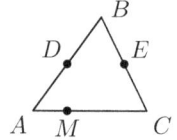

GRADE 8

1964.13. Find all prime numbers p, q, and r such that $pqr = 5(p + q + r)$.

1964.14. Prove that if $\overline{ab/bc} = a/c$, then
$$\overline{abb\ldots bb} \, / \, \overline{bb\ldots bbc} = a/c$$
(each number has n digits).

1964.15. Construct triangle knowing its perimeter, one angle at its side, and an altitude to the same side.

1964.16. Prove that the square of the sum of n different non-zero perfect squares is itself a sum of n non-zero perfect squares.

1964.17. In quadrilateral $ABCD$, the circles inscribed within triangles ABC and ADC touch each other. Prove that the circles inscribed within triangles BAD and BCD touch as well.

1964.18. Natural numbers a and n are co-prime. Prove that there exist integers x and y such that $|x| < \sqrt{n}$, $|y| < \sqrt{n}$, and $ax - y$ is divisible by n.

GRADE 9

1964.19. See Question 15.

1964.20. Fifty-one points lie within a unit square. Prove that some three of them can be covered by a circle of radius $\frac{1}{7}$.

1964.21. Prove that if for some natural n
$$\left[\frac{n}{1}\right] + \left[\frac{n}{2}\right] + \cdots + \left[\frac{n}{n}\right] = 2 + \left[\frac{n-1}{1}\right] + \left[\frac{n-1}{2}\right] + \left[\frac{n-1}{3}\right] + \cdots + \left[\frac{n-1}{n-1}\right],$$
then n is a prime number.

1964.22. See Question 16.

1964.23. In triangle ABC vertex A is connected with point D which lies on side BC.

a) Prove that the centers of circles circumscribed around triangles ABD, ADC, ABC, and point A all lie on the same circumference.

b) Find point D for which radius of that circumference is the maximum possible one.

1964.24. See Question 18.

GRADES 10–11

1964.25. Convex polygon P with n sides lies inside of a unit square. Prove that one can always find three vertices A, B, and C of P such that $S(ABC) \leqslant 100/n^2$.

1964.26. Denote by N the number of points on the plane whose coordinates (x, y) are integers satisfying inequality $x^2 + y^2 < n$, where n is some integer. Prove that $N \geqslant 3(n-1)^2$.

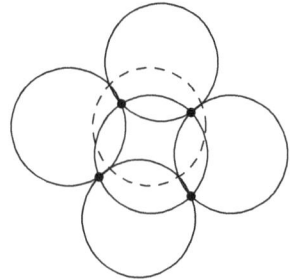

1964.27. Four points are selected on a circumference of unit radius. Through every two neighboring points we draw another circumference of unit radius. Prove that four other points of intersection between every pair of two consecutive circumferences lie on the same circumference.

1964.28. Let α_n be the exponent of 2 in prime factorization of n. Also let $S_n = \alpha_1 + \alpha_2 + \cdots + \alpha_n$. Prove that in sequence S_1, S_2, S_3, ..., S_{2^n-1}, amounts of even and odd numbers differ by exactly one.

1964.29. Given positive number d, points A and B, and line L on the plane, find point E on L such that $|AE| + |BE| = d$.

1964.30. Let S be a set of natural numbers fewer than the given prime number p and such that if numbers a and b belong to S, then the same is true for the remainders of ab, a, and b modulo p. Prove that if S has at least two numbers, then the sum of all numbers in S is divisible by p.

ELIMINATION ROUND

1964.31. Vertices of two acute-angled triangles belong to two different pairs of opposite sides of a unit square. Prove that the area of the intersection of these triangles is not greater than the quadrupled product of their areas.

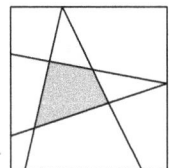

1964.32. Sequence $a_1, a_2, \ldots, a_n$ consists of natural numbers, some of which can be equal. Let f_k denote the number of terms in this sequence which are greater than or equal to k. Prove that $f_1 + f_2 + \cdots = a_1 + a_2 + \cdots + a_n$.

1964.33. A unit square is covered by n figures in such a way that each of its points belongs to at least q figures. Prove that one of these figures has area greater than or equal to q/n.

1964.34. All real numbers are split into two groups. Prove that for any $r > 0$ there exist three numbers $a < b < c$ from one group such that $(c - b)/(b - a) = r$.

1964.35. Inside a regular unit triangle place a triangle with the given angles so that it has the largest possible area.

1964.36.* There are n coin makers; some of them manufacture only counterfeit coins, others—only genuine ones. The weight of a counterfeit coin is always the same, and different from the weight of a genuine coin. You are allowed to take as many coins as necessary from any coin maker. Using a balance with a full set of weights and one coin which is known to be genuine, find all the counterfeiters after conducting only three weighings.

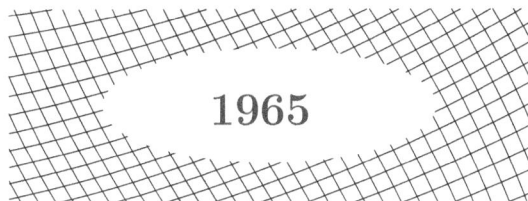

1965

GRADE 6

1965.01. The printer shop has 92 sheets of white paper and 135 sheets of red paper. Each book's cover uses one sheet of white paper and one sheet of red paper. After several book covers were produced, the number of white sheets became twice as small as the number of red sheets. How many books have been bound?

1965.02. Prove that if you multiply all integers 1 through 1965, then the result will be a number whose last non-zero digit is even.

1965.03. Front tires of an automobile wear out after 25,000 km, and the rear tires—after 15,000 km. When do you need to rotate the tires so that they will wear out simultaneously?

1965.04. Rectangle with dimensions 19 cm × 65 cm is dissected by straight lines parallel to its sides into squares with sides of 1 cm. How many parts will the dissection consist of if you also drew one of the rectangle's diagonals?

1965.05. Find dividend, divisor, and quotient in the following example of long division.

$$
\begin{array}{r|l}
\text{XXXXXX} & \text{XXX} \\
\underline{\text{XXXX}} & \text{XXX} \\
\text{XXX} \\
\underline{\text{XXX}} \\
\text{XXXX} \\
\underline{\text{XXXX}} \\
0
\end{array}
$$

1965.06. Odd numbers from 1 through 49 are written out in the form of the table.

$$
\begin{array}{ccccc}
1 & 3 & 5 & 7 & 9 \\
11 & 13 & 15 & 17 & 19 \\
21 & 23 & 25 & 27 & 29 \\
31 & 33 & 35 & 37 & 39 \\
41 & 43 & 45 & 47 & 49
\end{array}
$$

Five numbers are then chosen with no two of them being in the same row or in the same column. What is their sum?

GRADE 7

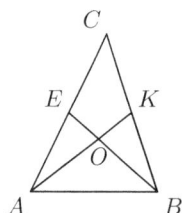

1965.07. Prove that a natural number with an odd number of divisors is a perfect square.

1965.08. In triangle ABC with area S, medians AK and BE meet at point O. Find the area of quadrilateral $CKOE$.

1965.09. Front tires of an automobile wear out after $25,000$ km, and the rear tires—after $15,000$ km. When do you need to rotate the tires so that the car could drive the longest possible distance?

1965.10. Rectangle with dimensions 24×60 cm is dissected by straight lines parallel to its sides into unit squares. How many parts altogether will the dissection have if you also drew one of the rectangle's diagonals?

1965.11. Solve the equation[51]

$$\left[\frac{5 + 6x}{8} \right] = \frac{15x - 7}{5} .$$

1965.12. Black ink was spilled onto the white plane. Prove that one can find two points of the same color exactly 1965 meters apart.

GRADE 8

1965.13. Rectangle with dimensions 24×60 is dissected by straight lines parallel to its sides into unit squares. Draw one more line so that the rectangle will be dissected into the maximum possible number of parts.

1965.14. Scientists always tell the truth and lawyers always lie. Fred and Gina are scientists. Alan declares that Beth claims that Chris affirms that Dana says that Ethan insists that Fred denies that Gina is a scientist. Also Chris announces that Dana is a lawyer. If Alan is a lawyer, then how many lawyers are there in this group of people?

1965.15. Straight road cuts through the field. A hitchhiker stands on the road at point O. He can walk down the road at 6 km/h or through the field at 3 km/h. Find the locus of all points that the hitchhiker can reach by walking for one hour.

1965.16. See Question 11.

---

[51] A reminder: $[a]$ denotes the integral part function—maximum integer number not exceeding a.

1965.17. In some country, every two towns are connected by a one-way road. Prove that one can find a town such that you can start your trip from it and then proceed to visit each town in the country exactly once.

1965.18. Find all octuplets of prime numbers such that the sum of the squares of these numbers is 992 less then four times the product of all eight primes.

GRADE 9

1965.19. Given an angle on the plane find two segments of unit length cutting a quadrilateral of maximum possible area off the angle.

1965.20. See Question 14.

1965.21. A parallelogram is dissected by straight lines parallel to its sides into several parts, with one of its sides split into m parts and the other one—into n. What is the maximum possible number of parts you can dissect the parallelogram into if you add one more line?

1965.22. The lengths of the sides of triangle ABC satisfy the inequality $|AB| \cdot |BC| \cdot |AC| \leqslant 60$. Points C', A', and B' are selected on sides AB, BC, and AC, respectively. Prove that

$$|AC'| \cdot |C'B| \cdot |BA'| \cdot |A'C| \cdot |CB'| \cdot |B'A| < |AB| \cdot |BC| \cdot |AC| \,.$$

1965.23. Let a_1, a_2, $\ldots$, a_n and b_1, b_2, $\ldots$, b_n be two permutations of sequence 1, 2, 3, $\ldots$, n. Prove that if n is even, then some two numbers from sequence $a_1 + b_1$, $a_2 + b_2$, $\ldots$, $a_n + b_n$ have equal remainders modulo n.

1965.24.* Several circles on the plane cover unit area. Prove that one can always choose several disjoint circles with sum of their areas at least $\frac{1}{9}$.

GRADES 10–11

1965.25. Find all integer solutions of the equation

$$6xy - 4x + 9y - 366 = 0 \,.$$

1965.26. Find the sum

$$\frac{1}{\cos \alpha \cos 2\alpha} + \frac{1}{\cos 2\alpha \cos 3\alpha} + \cdots + \frac{1}{\cos(n-1)\alpha \cos n\alpha} \,.$$

1965.27. See Question 23.

1965.28.

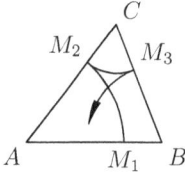

Grade 10: Point M_1 is selected on side AB of triangle ABC. Arc $M_1 M_2$ with center A and radius AM_1, lies inside ABC and connects M_1 with side AC ($M_2 \in AC$). Next, arc $M_2 M_3$, with center C and radius CM_2, lies inside ABC and connects M_2 with side BC ($M_3 \in BC$). Then arc $M_3 M_4$ with center B and radius BM_3 connects M_3 with side AB, and so on until these arcs form a closed curve. Find the length of this curve if you know the sides and angles of triangle ABC.

Grade 11: Prove that

$$(a + b)^n \leqslant 2^{n-1}(a^n + b^n),$$

if a and b are non-negative numbers.

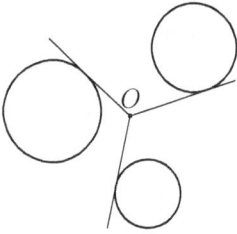

1965.29. A cube with dimensions $12 \times 12 \times 12$ is dissected by planes parallel to its sides into unit cubes. How many parts will there be if we add a plane that produces a cross-section of the cube congruent to a regular hexagon?

1965.30. Given three circles on the plane, find point O such that the three "right" tangents, drawn from O to the given circles, form equal angles with each other.

ELIMINATION ROUND

1965.31.[*] Non-negative integers are written into the squares of $n \times n$ table in such a way that if some square S contains zero, then the sum of all numbers which share either the row or the column with S is at least n. Prove that the sum of all numbers in the table is at least $n^2/2$.

1965.32. Positive numbers a, b, c, x, y, z are such that $a < b < c$; $a \leqslant x < y < z \leqslant c$; $abc = xyz$; $a + b + c = x + y + z$. Prove that $a = x$, $b = y$, $c = z$.

1965.33. Among all polynomials of form $x^2 + ax + b$ find the one for which its maximum absolute value on segment $[-1; 1]$ is minimal.

1965.34. Prove that the area of a square inside a triangle does not exceed half of the area of the triangle.

1965.35. A figure of area less than 1 is placed on a sheet of graph paper. Prove that there exists a translation such that the translated image of the figure contains no grid nodes.

1965.36.[*] Find the locus of the centers of regular triangles circumscribed around the given triangle.

Commentary.

1) *In 1965, the All-Russia Olympiad was organized using a very interesting and rather unique approach. Both Math and Physics Olympiads were held simultaneously and at the same location in Moscow. On the first day, all contestants were given five problems in mathematics, and on the next day—five problems in physics. This competition was called All-Russia Sci-Math Olympiad.*

2) *Because of this unusual setup, the city of Leningrad had to field a combined team consisting of its best high school mathematicians and physicists (with possible overlap between those two categories). The team size was accordingly increased from four to six members. It turned out that after the regular elimination round, the organizing committee still needed one more stage in order to select additional members of the team. Problems from that extra round are included in section "Additional problems".*

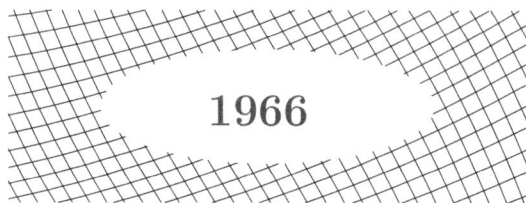

1966

GRADE 6

1966.01. Which number is greater

$$\underbrace{1000\ldots001}_{1965 \text{ zeros}} / \underbrace{1000\ldots001}_{1966 \text{ zeros}} \quad \text{or} \quad \underbrace{1000\ldots001}_{1966 \text{ zeros}} / \underbrace{1000\ldots001}_{1967 \text{ zeros}} ?$$

1966.02. Thirty teams participate in a soccer championship. Prove that at any given moment some two teams have played equal number of games.

1966.03. All integers 1 through 1966 are written on the blackboard. We are allowed to erase any two numbers and write their difference instead. Prove that using such operations we cannot end up with several zeros.

1966.04. Black ink was spilled onto the white plane. Prove that one can find three points of the same color that belong to the same straight line and such that one of them lies exactly in the middle between the other two.

1966.05. More than three players took part in a chess tournament where each participant plays every other the same number of times. Altogether there were 26 rounds, and after the 13th round, one of the participants realized that he had an odd number of points while every other player had an even number of points. How many people took part in the tournament?

GRADE 7

1966.06. See Question 3.

1966.07. Prove that the radius of a circle equals the difference of two chords, one of which spans $\frac{3}{10}$ of the circle, and the other one spans $\frac{1}{10}$ of the circle.

1966.08. Prove that for each natural n the product $n(2n+1)(3n+1)\cdots(1966n+1)$ is divisible by every prime number fewer than 1966.

1966.09. Which number must stand in place of $*$ so that the following textbook problem had exactly one solution: *"There are n straight lines on the plane intersecting in $*$ points. Find n."*?

1966.10. See Question 4.

1966.11. There are n points on the plane such that the area of any triangle with vertices on these points is less than 1. Prove that all these points can be covered by one triangle of area 4.[52]

GRADE 8

1966.12. See Question 9.

1966.13. See Question 8.

1966.14. See Question 11.

1966.15. Prove that the sum of all divisors of number n^2 is odd.

1966.16. Three angles of a quadrilateral are obtuse. Prove that its longest diagonal issues from its only acute angle.

1966.17. Sequence of numbers x_1, x_2, ... is constructed according to the following rule: $x_1 = 2$, $x_2 = (x_1^4 + 1)/5x_1$, $x_3 = (x_2^4 + 1)/5x_2$, and so on. Prove that all the numbers in this sequence will always be greater than or equal to $\frac{1}{5}$, and less than or equal to 2.

GRADE 9

1966.18. Given a rectangle, find inside of it a rhombus of maximum possible area.

1966.19. How many integer solutions does equation $\sqrt{x} + \sqrt{y} = \sqrt{1960}$ have?

1966.20. Prove that it is possible to color the plane with nine colors so that the distance between any two points of the same color is different from 1966.

1966.21. Two prime numbers p and q are such that $q^3 - 1$ is divisible by p, and $p - 1$ is divisible by q. Prove that $p = 1 + q + q^2$.

1966.22. Using sides of triangle ABC as hypotenuses, isosceles right triangles ABD, BCE, and ACF are constructed on the outside of triangle ABC. Prove that segments DE and BF are equal and perpendicular to each other.

1966.23.* Given k different colors, how many ways are there to paint the sides of the given regular n-gon so that any two neighboring sides were of different color (the polygon is fixed in place and cannot be rotated)?

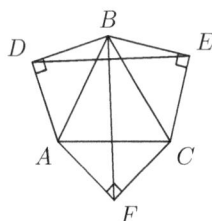

---

[52] One of the archival versions for Grade 7 lists Question 5 instead of Question 11.

GRADES 10, 11

1966.24. See Question 18.

1966.25. See Question 19.

1966.26. See Question 20 for 11 colors.

1966.27. See Question 23.

1966.28. For which values of ε one can always dissect segment of length $2a$ into n smaller segments each of them having length not greater than a, so that it were impossible to put some of them together to create a segment whose length differs from a by less than ε?[53]

1966.29. Find all complex solutions of the system of equations

$$\begin{cases} x_1 + x_2 + \cdots + x_n = n \\ x_1^2 + x_2^2 + \cdots + x_n^2 = n \\ \quad \cdots \\ x_1^n + x_2^n + \cdots + x_n^n = n. \end{cases}$$

ELIMINATION ROUND

1966.30. Numbers m and n are positive integers, m is odd. Prove that numbers $2^n + 1$ and $2^m - 1$ are co-prime.

1966.31.* Let T_n denote the area of the largest (by area) n-gon which lies inside the given convex polygon with k sides $(3 < n < k)$. Prove that for any $n < k$ we have $T_{n-1} + T_{n+1} \leqslant 2 T_n$.

1966.32. From sequence of numbers 1, 2, 3, 4, ..., 2^n, some $[(2^n - 2)/3]$ terms are removed. Prove that it is possible to find two among the remaining numbers one of which is twice the other.

1966.33. For the given square we will call any dissection of it into finite number of rectangles with sides parallel to the sides of the square a *tiling*. We will call a tiling *primitive* if it cannot be obtained from another tiling by dissecting some of its rectangles into more rectangles. For which values of n there exists a primitive tiling with exactly n rectangles?

1966.34.* On the plane, n points in general position[54] are given. Some of them are connected by line segments so that for any two of the points there is exactly one way to pass from one to another.[55] Prove that there are exactly n^{n-2} such connections.

---

[53] Formulation of Question 28 is somewhat ambiguous. Most likely, a proper clarification was communicated to the contestants during the competition.
[54] Which means that no three of them lie on the same straight line.
[55] The path must consist of whole segments.

1966.35.* A polygon with n sides a_1, a_2, ..., a_n is inscribed into a circle whose center lies inside the polygon. Prove that this circle can be completely covered by n circles with radii $na_1/6$, $na_2/6$, ..., $na_n/6$.

> **Commentary.** *This olympiad set another record: in Grade 11 contest no less than 34 (thirty-four!) first prizes were awarded.*

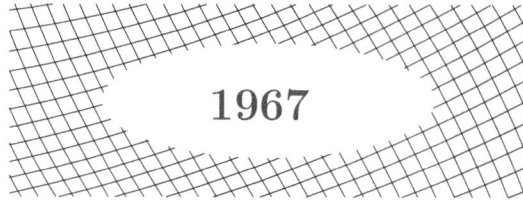

1967

GRADE 6

1967.01. Volumes of three cubic jars relate as $1 : 8 : 27$ while volumes of water in them relate, respectively, as $1 : 2 : 3$. Some amount of water was poured out from the first jar into the second, and then from the second one into the third in such a way that the water levels became the same in all three jars. Then $128\frac{4}{7}$ liters was poured from the first jar into the second, and after that some water was poured from the second jar back into the first so that the water level in the first jar became twice the level in the second. It turns out that the first jar then contained 100 liters less water than at the very beginning. How much water was there originally in each jar?

1967.02. How many times a day do all three hands on the clock (that is, hour hand, minute hand, and second hand) point in the same direction?

1967.03. Prove that there are two persons in Leningrad who have the same number of Leningrad acquaintances.

1967.04. Each of eight different natural numbers is less than 16. Prove that among their pairwise differences one can always find three equal numbers.

1967.05. Distance $|AB|$ equals 100 km. Two bicycle riders start simultaneously from A and B riding toward each other with velocities of 20 km/h and 30 km/h, respectively. A bee whose speed is 50 km/h flies out of A together with the first rider; when it meets the second rider, it turns around and flies in opposite direction; when it meets the first rider, it turns around again, etc. How many kilometers will the bee fly altogether in the direction from A to B by the time the two riders meet?

GRADE 7

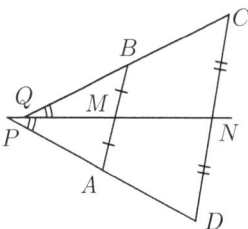

1967.06. Build a trapezoid given the lengths of its four sides.

1967.07. Prove that $(1 + x + x^2 + \cdots + x^{100})(1 + x^{102}) - 102x^{101} \geqslant 0$.

1967.08. In quadrilateral $ABCD$, point M is the midpoint of AB, and point N is the midpoint of CD. Lines AD and BC intersect MN at P and Q, respectively. Prove that if $\angle BQM = \angle APM$, then $|BC| = |AD|$.

1967.09. See Question 4.

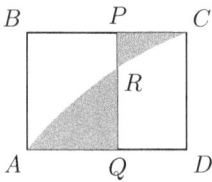

1967.10. A circular arc lying inside rectangle $ABCD$ passes through its vertices A and C. Construct line parallel to AB which intersects line BC at point P, line AD at point Q, and arc AC at point R so that the sum of areas of figures AQR and CPR is minimal.

1967.11. See Question 5.

GRADE 8

1967.12. Numbers x and y are the roots of equation $t^2 - ct - c = 0$. Prove inequality $x^3 + y^3 + (xy)^3 \geqslant 0$.

1967.13. Two circumferences touch internally at point A. Point B, different from A, lies on the smaller circumference. The tangent line at B intersects the larger circumference at points C and D. Prove that AB bisects angle CAD.

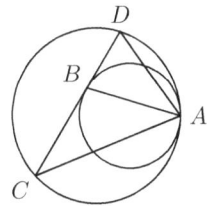

1967.14. Prove that $2^{3^{100}} + 1$ is divisible by 3^{101}.

1967.15. See Question 10.

1967.16. In certain group of people, everyone has exactly one enemy and exactly one friend. Prove that it is possible to split that group into two subgroups so that everyone has neither friends nor enemies within their subgroup.

1967.17. Numbers a_1, a_2, ..., a_{100} satisfy inequalities

$$a_1 - 2a_2 + a_3 \leqslant 0$$
$$a_2 - 2a_3 + a_4 \leqslant 0$$
$$\ldots$$
$$a_{98} - 2a_{99} + a_{100} \leqslant 0$$

and $a_1 = a_{100} \geqslant 0$. Prove that all these numbers are non-negative.

GRADE 9

1967.18. Given any two consecutive odd integers p and q, prove that $p^p + q^q$ is divisible by $p + q$.

1967.19. Circumference with its center at point B touches side AC of triangle ABC. Segments AM and CP are tangents to this circumference drawn from vertices A and C. Line MP intersects lines AB and BC at points E and H, respectively. Prove that AH and CE are altitudes in triangle ABC.

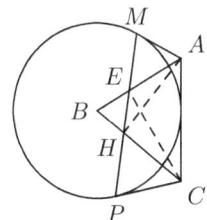

1967.20. The sequence of k numbers is written on the blackboard. It is permitted to replace any number with the sum of all numbers positioned to the right of it. Prove that if this operation is performed indefinitely, then the sequence on the board will eventually repeat.

1967.21. Among 106 points, no three of which lie on the same straight line, four are the vertices of a square, and all the others lie inside that square. Prove that there exist at least 107 triangles with vertices on these points whose area is not greater than 0.01.

1967.22. Let $a_1 \geqslant a_2 \geqslant \cdots \geqslant a_n \geqslant 0$ be a sequence of n numbers, $b_1, \ldots, b_n$—some other n numbers, and B—the maximum of $|b_1|, |b_1 + b_2|, \ldots, |b_1 + b_2 + \cdots + b_n|$. Prove that
$$|a_1 b_1 + a_2 b_2 + \cdots + a_n b_n| \leqslant Ba_1.$$

1967.23. A tourist got lost in the forest shaped like a convex polygon of area 1 km². Prove that he can always find his way out after walking no more than 2507 meters.

GRADE 10

1967.24. Given any two consecutive odd integers p and q prove that $p^q + q^p$ is divisible by $p + q$.

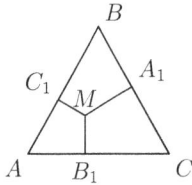

1967.25. Point M lies inside equilateral triangle ABC, and points A_1, B_1, C_1 are its projections onto sides BC, AC, and AB, respectively. Prove that $|AB_1| \cdot |BC_1| + |BC_1| \cdot |CA_1| + |CA_1| \cdot |AB_1| = |AC_1| \cdot |BA_1| + |BA_1| \cdot |CB_1| + |CB_1| \cdot |AC_1|$.

1967.26. See Question 20.

1967.27. See Question 22.

1967.28. See Question 23.

1967.29. Prove that there exists a finite set of points on the plane such that for every point A in that set there are at least one hundred points in the set equidistant from A.

ELIMINATION ROUND

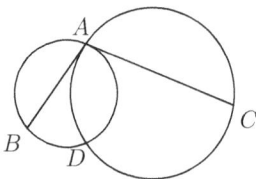

1967.30. Find a figure on the graph paper, consisting of the smallest possible number of squares, such that if two people are playing tic-tat-toe on its squares then the first player always wins.

1967.31. Two circumferences intersect at point A; AB and AC are their respective chords which are also tangent to the other circumference at point A. Let D be the other point of the intersection of these two circumferences. Prove that $|AB|^2/|AC|^2 = |BD|/|CD|$.

1967.32. Two polynomials with real coefficients take integer values in the same points. Prove that either their sum or their difference is constant.

1967.33.* Three points are selected on the three sides of the triangle and then connected by segments, which resulted in splitting the triangle into four smaller ones. Given that the perimeters of these four triangles are equal, prove that the selected points are the midpoints of the corresponding sides.

1967.34. Eighteen people attended a party. Prove that either there are four people who are all pairwise acquainted with each other or four people who do not know each other.

1967.35.* A table with dimensions $n \times n$ is filled with non-negative numbers in such a way that the sum of the numbers in every row, as well as in every column, is equal to 1. Prove that it is possible to choose n positive numbers in this table so that no two of them lie in the same row or in the same column.

1967.36. Sequence of numbers $0 < a_0 < a_1 < a_2 < \cdots < a_{25} < a_{26}$ satisfies equation

$$a_n = 2\sqrt{a_{n-1}^2 - a_{n-1}^4}$$

for all $0 < n \leqslant 26$. Prove that $a_0 < 7 \cdot 10^{-8}$.

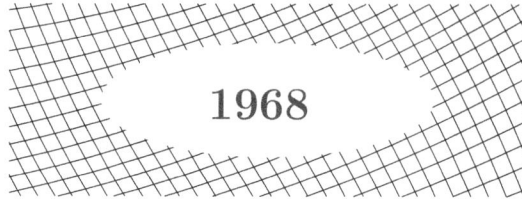

1968

GRADE 6

1968.01. A student bought a backpack, a pen, and a book. If the backpack were to cost him five times less than the actual price, the pen—two times less, and the book—two and a half times less, then the price of his entire purchase would have been 2 rubles. If the backpack were to cost him two times less, the pen—four times less, and the book—three times less, then the entire purchase would have cost him 3 rubles. What was the actual purchase price?

1968.02. Which number is greater:

$$\underbrace{888\ldots88}_{\text{19 digits}} \cdot \underbrace{333\ldots33}_{\text{68 digits}} \quad \text{or} \quad \underbrace{444\ldots44}_{\text{19 digits}} \cdot \underbrace{666\ldots67}_{\text{68 digits}}$$

and by how much?

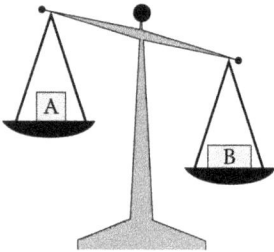

1968.03. The distance between Luga and Volkhov is 194 km, between Volkhov and Lodeinoe Pole—116 km, between Lodeinoe Pole and Pskov—451 km, and between Pskov and Luga—141 km. What is the distance between Pskov and Volkhov?[56]

1968.04. There are four objects with pairwise different weights. Using five weighings on two-pan scales without weights, how do you order these objects by weight?

1968.05. Several teams took part in a volleyball tournament.[57] Team A is considered *stronger* than team B if either A defeated B or there exists team C such that A defeated C and C defeated B. Prove that if team T is the winner of the tournament, then it is stronger than all other teams.

1968.06. In Question 1, determine which is more expensive, the backpack or the pen.

GRADE 7

1968.07. A non-square rectangle is inscribed into a square. Prove that the rectangle's semi-perimeter equals the length of the square's diagonal.

---

[56] Luga, Volkhov, Pskov and Lodeinoe Pole are four cities in Russia's Northwest, not far from Leningrad.
[57] Remember, there are no draws in volleyball.

1968.08. Find five numbers whose pairwise sums are 0, 2, 4, 5, 7, 9, 10, 12, 14, and 17.

1968.09. In a 1000-digit number, all digits except one are fives. Prove that this number is not a perfect square.

1968.10. See Question 5.

1968.11. In pentagon $ABCDE$, point K is the midpoint of AB, L is the midpoint of BC, M is the midpoint of CD, N is the midpoint of DE, P is the midpoint of KM, and Q is the midpoint of LN. Prove that segment PQ is four times shorter than and parallel to side AE.

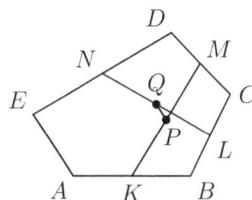

1968.12. Several circles with the sum of radii equal to 25 lie inside a circle of radius 3. Prove that there exists a straight line that intersects at least nine of the smaller circles.

GRADE 8

1968.13. In parallelogram $ABCD$, diagonal AC is longer than diagonal BD. Point M on diagonal AC is such that quadrilateral $BCDM$ is inscribed. Prove that BD is tangent to the circumcircles of both triangles ABM and ADM.

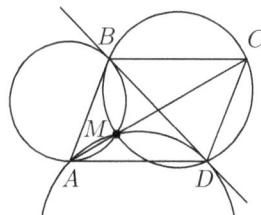

1968.14. Let a be an odd integer a. Numbers x and y are the roots of equation $t^2 + at - 1 = 0$. Prove that $x^4 + y^4$ and $x^5 + y^5$ are two co-prime integers.

1968.15. Equilateral triangle is reflected with respect to one of its sides. The resulting triangle is again reflected in the same manner, and that operation is repeated several times. It turns out that the final triangle coincides with the original one. Prove that the number of performed operations is even.

1968.16. See Question 12.

1968.17. All two-digit numbers not ending in zero are written one after another so that each number begins with the same digit the previous number ends with. Prove that it can be done, and find the sum of the largest and the smallest of such numbers.

1968.18.* All ten-digit numbers written by digits 1, 2, and 3 are written one under another. Then to the right of each number, one extra digit 1, 2, or 3, which we will call a *signature* of that number, is written. It turned out that the signature for $111\ldots11$ is 1, the signature for $222\ldots22$ is 2, and the signature for $333\ldots33$ is 3. It is also known that if any two numbers differ from each other in every decimal place, then their signatures are also different. Prove that the column formed by the signatures coincides with one of the other ten columns.

GRADE 9

1968.19. Solve the equation $x = 1 - 1968(1 - 1968x^2)^2$.

1968.20. Construct a quadrilateral given the midpoints of its three sides if all sides have equal length. (It is also known which one of the three given points belongs to the "middle" side).

1968.21. An infinite sheet of graph paper is colored with 9 colors so that each square is painted one color, and all the colors are used. We will call two colors *neighbors* if there exist two neighboring squares painted with these colors. What is the smallest possible number of pairs of neighbors?

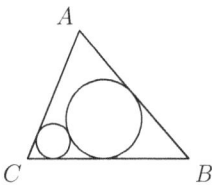

1968.22. Two circles placed inside acute-angled triangle ABC touch sides AC and BC, and sides AB and BC, respectively, as well as each other. Prove that the sum of their radii is greater than the inradius of triangle ABC.

1968.23. Prove that

$$1!(n-1)! + 2!(n-2)! + \cdots + (n-1)!1! \leqslant \frac{2}{3}n!$$

for any natural number n.

1968.24. *Prove that a regular triangle cannot be dissected into several pairwise different regular triangles.

GRADE 10

1968.25. Triangles ABC, $A_1B_1C_1$, and $A_2B_2C_2$ with areas S, S_1, and S_2, respectively, are such that $|AB| = |A_1B_1| + |A_2B_2|$, $|BC| = |B_1C_1| + |B_2C_2|$, and $|AC| = |A_1C_1| + |A_2C_2|$. Prove that $S \geqslant 4\sqrt{S_1 S_2}$.

1968.26. How many solutions does the system of equations

$$\begin{cases} \cos x_1 = x_2 \\ \cos x_2 = x_3 \\ \cdots \\ \cos x_n = x_1 \end{cases}$$

have?

1968.27. Let ABC and $A_1B_1C_1$ be two equilateral and identically oriented triangles. Angle between AA_1 and BB_1 is α, $|AA_1| = a$, $|BB_1| = b$. Find $|CC_1|$.

1968.28. See Question 21 for ten colors.

1968.29. *Prove that it is impossible to cut a regular tetrahedron into several pairwise different regular tetrahedrons.

1968.30.[*] Given positive numbers $a_1, a_2, \ldots, a_n$ and natural number k prove that

$$(\sqrt{2} - 1)(a_1 + a_2 + \cdots + a_n) \leqslant \sqrt[k]{a_1^k \cdot 2 + a_2^k \cdot 2^2 + \cdots + a_n^k \cdot 2^n}.$$

ELIMINATION ROUND

1968.31. Let AB and AC be two chords in some given circle. A perpendicular is dropped from the midpoint M of arc BAC onto the longer of these two chords. Prove that its foot divides broken line BAC into two parts of equal length.

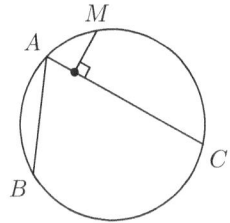

1968.32. Given several points on the circle, two players play the following game. A move consists in drawing a segment connecting the last point with another one.[58] Players take turns, and it is prohibited to draw the same segment twice. The player who does not have a valid move loses. Prove that in the errorless game, the first player always wins.

1968.33. Suppose that m points are given in the plane, not all of them lie on the same straight line. Prove that one can find at least $\frac{1}{2}(m-1)(m-2)$ triangles with vertices on these points.

1968.34. Given n real numbers $a_1, a_2, \ldots, a_n$, denote by M the largest and by m the smallest number among them. Prove that

$$(n-1)(M-m) \leqslant \sum_{i<j} |a_i - a_j| \leqslant n^2 \frac{(M-m)}{4}.$$

1968.35. Prove that the sum of the distances from a point inside tetrahedron to its vertices is less than its perimeter.[59]

1968.36.[*] Find natural number n such that one can find at least 1968 consecutive zeros among digits of 5^n.

1968.37.[*] In convex polyhedron P, each vertex is common to exactly three faces. Each face of P is painted in one of four colors so that any two neighboring faces are of different color. Prove that the number of the faces of the first color that have odd number of sides has the same parity as the number of the faces of the second color that have odd number of sides.

1968.38. Let us call a *word* any sequence of letters A and B. Consider two word operations:

1) Insert A at any position and add B at the end of the word.

---

[58] In the beginning, the first player can choose the "last" point arbitrarily.
[59] A polyhedron's perimeter is the sum of the lengths of its edges.

2) Insert AB at any position.

Prove that the words which can be obtained from AB using only the first operation are the same words that can be obtained from AB using only the second operation.

> **Commentary.** *Only one copy of this year's elimination round has been dis-covered in the archives—therefore, we cannot be absolutely certain in its au-thenticity.*

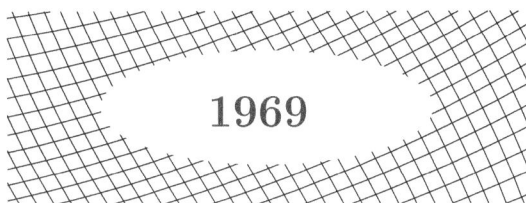

1969

GRADE 6

1969.01. Eight rooks are placed on the chessboard in such a way that neither of them attacks another. Prove that the number of the rooks standing on the black squares is even.

1969.02. The 3×3 table is filled with natural numbers—see the figure. Nick and Pete both have erased four numbers. It turned out that the sum of the numbers erased by Pete is three times the sum of the numbers erased by Nick. What is the last remaining number?

4	12	8
13	24	14
7	5	23

1969.03. Mike and Alex started riding their bicycles from town A to town B at noon. At the same moment Johnny started his ride from B to A. All three boys ride with constant but different velocities. At 1 pm Alex was exactly at the midpoint between Mike and Johnny, and at 1.30 pm Johnny was exactly at the midpoint between Mike and Alex. When will Mike be exactly at the midpoint between Alex and Johnny?

1969.04. There are thirty-five piles of nuts on the table. You can add one nut simultaneously to any twenty-three piles. Prove that, using such operations, you can make all the piles equal.

1969.05. Surface of a circular cylinder consists of 64 vertical strips. We need to write all sixty-four different six-digit numbers that only use digits 1 and 2 into these strips so that each two neighboring numbers differ in exactly one decimal place. How do we do that?

1969.06. Two genius mathematicians were each given their own secret natural number, and they were told that the difference of these two numbers equals 1. After that they took turns asking each other the same question "Do you know my number?". Prove that sooner or later one of them answers "yes".[60]

---

[60] Obviously, we also assume that these mathematicians always tell the truth.

GRADE 7

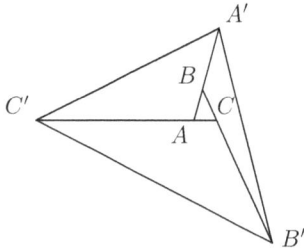

1969.07. See Question 1.

1969.08. Sides of triangle ABC were extended as shown in the figure. Given that $|AA'| = 3\,|AB|$, $|BB'| = 5\,|BC|$, and $|CC'| = 8\,|CA|$, find the ratio between the area of triangle $A'B'C'$ and the area of triangle ABC.

1969.09. Prove the identity

$$\frac{2}{x^2 - 1} + \frac{4}{x^2 - 4} + \frac{6}{x^2 - 9} + \cdots + \frac{20}{x^2 - 100}$$
$$= 11 \left[\frac{1}{(x-1)(x+10)} + \frac{1}{(x-2)(x+9)} + \cdots + \frac{1}{(x-10)(x+1)} \right].$$

1969.10.* In the center of the square-shaped field sits a wolf, while in each vertex of the square—a wolfhound. The wolf is able to run anywhere while the dogs can only run along the sides of the square. In a fight, the wolf will kill a dog, but two dogs will kill the wolf. Maximum speed of each dog is 1.5 times the maximum speed of the wolf. Prove that the dogs can keep the wolf inside the field.

1969.11. A collective farm consists of four villages located in the vertices of the square with the side equal to 10 km. The farm can afford to build 28 km of roads. Can they build a system of roads that will allow the farmers to drive from each village to any other?

1969.12. See Question 6.

GRADE 8

1969.13. Point P on the base AD of trapezoid $ABCD$ is such that the perimeters of triangles ABE, BCE, and CDE are equal. Prove that $|BC| = |AD|/2$.

1969.14. Given a convex pentagon with sides of equal length, find a point on its longest diagonal such that all the sides are seen from that point at angles not greater than $90°$.

1969.15. In some country the air transportation system is built in such a way that each city is directly connected with no more than three other cities, but one can travel from every city to any other with no more than one stopover. What is the maximum possible number of cities in this country?

1969.16. See Question 10.

1969.17. Sum of four different three-digit numbers beginning with the same digit is divisible by three of these numbers. Find these numbers.

1969.18. Finite sequence of zeros and ones possesses the following properties.

1) If you select five consecutive digits in two arbitrary places of the sequence (they can overlap), then the selected quintuples will be different.
2) If you extend this sequence by one more digit then the first property will cease to be true.

Prove that the first four digits in the sequence coincide with the last four digits.

GRADE 9

1969.19. Let A and B be two points on circumference S. Point C is the midpoint of arc AB, and point P lies inside S. Given that $AP < BP$, prove that $\angle APC > \angle BPC$.

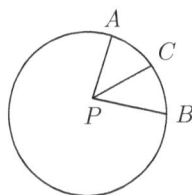

1969.20. A tram ticket is called "Leningrad-style lucky" if the sum of its first three digits equals the sum of its last three digits. The ticket is called "Moscow-style lucky" if the sum of its digits on the odd-numbered decimal places is the same as the sum of the digits on the even-numbered places. How many tickets are deemed "lucky" in both cities (ticket 000000 also counts)?[61]

1969.21. Natural numbers m and n are such that $\frac{m}{n} < \sqrt{2}$. Prove that

$$\frac{m}{n} < \sqrt{2}\left(1 - \frac{1}{4n^2}\right).$$

1969.22. Find the locus of the centers of all rectangles circumscribed around the given convex quadrilateral.

1969.23. Given natural number $k > 1$, sequence (x_n) is constructed as follows: $x_1 = 1$, $x_2 = k$, $x_n = kx_{n-1} - x_{n-2}$ for every $n > 2$. Prove that for every natural n there exists $m > n$ such that x_m is divisible by x_n.

1969.24.* A tourist group of 60 people arrived at the hotel to stay there for the next 15 days. The hotel serves four meals every day, and there are exactly 61 seats at the dining table. One of these places is always occupied by the hotel manager. He wishes to get acquainted with every tourist, and he wants to make sure that the tourists will get acquainted among themselves as well. To achieve that, he intends to sit the tourists differently for every meal so that none of them will ever sit at the same place and also that each time everyone (including himself) will have a new neighbor on the right. How can he do that?

---

[61] In those times each tram ticket carried a six-digit index from 000000 through 999999.

GRADE 10

1969.25. In tetrahedron $ABCD$, edge AB is perpendicular to edge CD. Given arbitrary point O in space prove that the sum of the squares of the distances from O to the midpoints of edges AC and BD equals sums of the squares of the distances from O to the midpoints of AD and BC.

1969.26. See Question 20.

1969.27. See Question 24.

1969.28. Given natural number $k > 1$, sequence (x_n) is constructed as follows: $x_1 = 1$, $x_2 = k$, $x_n = kx_{n-1} - x_{n-2}$ for every $n > 2$. Prove that for every natural n there exists $m > n$ such that $x_m - 1$ and x_n are co-prime.

1969.29. Construct a sequence of natural numbers such that among its pairwise differences each natural number occurs exactly once.

1969.30.* Segments AA', BB', and CC' are the altitudes of acute triangle ABC with side lengths a, b, and c, respectively. A_1 and A_2 are projections of point A' onto sides AB and AC, respectively, B_1 and B_2—projections of point B' into BC and BA, respectively, C_1 and C_2—projections of point C' onto CA and CB, respectively. Prove that

$$a^2 S(A'A_1A_2) + b^2 S(B'B_1B_2) + c^2 S(C'C_1C_2) = \frac{S^3}{R^2},$$

where S is the area of triangle ABC, and R is its circumradius.

ELIMINATION ROUND

1969.31. What is the minimum number of unit circles that can be positioned to cover the circle of radius $\frac{3}{2}$?

1969.32.* Fifty gangsters met on the market square. Simultaneously they all fire off one shot—each one shoots at the gangster nearest to him (or one of the nearest). They are all perfect shots and they shoot to kill. What is the smallest possible number of casualties?

1969.33. The sum of positive numbers a_1, a_2, ..., a_n equals 1. Prove that

$$\sum_{i<j} \frac{a_i a_j}{a_i + a_j} \leqslant \frac{n-1}{4}.$$

1969.34. The plane is colored using three colors. Prove that there exists a triangle of area 1 with all vertices of the same color.

1969.35.* A corner square with the side of 0.001 is cut from the larger square with side of $1,000,000$. The remaining part is dissected into ten rectangles. Prove that for at least one of them the ratio of the sides is greater than 9.

1969.36. Centers of four equal circumferences are located in the vertices of a square. Construct a quadrilateral of the maximum possible perimeter with its vertices lying on respective circumferences.

1969.37. Let F_1, F_2, $\ldots$, F_n be polynomials with integer coefficients. Prove that for some integer a all numbers $F_1(a)$, $F_2(a)$, $\ldots$, $F_n(a)$ are composite.

1969.38.* Given natural numbers $x_0 < x_1 < \cdots < x_n$, prove the inequality

$$\frac{\sqrt{x_1 - x_0}}{x_1} + \cdots + \frac{\sqrt{x_n - x_{n-1}}}{x_n} < 1 + \frac{1}{2} + \frac{1}{3} + \cdots + \frac{1}{n^2}.$$

Commentary.

(1) *Grade 8 problem set is identical to the one offered at the All-Union Math Olympiad of that year. The reason for this "amazing" coincidence is that due to some misunderstanding the city olympiad was scheduled to run virtually simultaneously with the All-Union competition—thus, the Leningrad team had to be selected beforehand. Hence, the LMO committee were able to "borrow" the All-Union problem set.*

(2) *Problem sets for Grades 9 and 10 were found in just one personal archive; thus it is possible that the reconstructed versions might be somewhat different from the actual ones.*

(3) *The year of 1969 was the first one when Grade 5 city olympiad was held. However, as we have mentioned in the preface, for several years its organization was entrusted to a separate committee. As a result we could not find most of the problem sets from that time.*

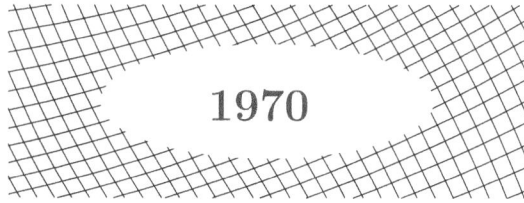

1970

GRADE 6

1970.01. A nine-digit number is written using all nine different non-zero digits. Some rearrangement of its digits resulted in the number decreasing eight-fold. Find all such numbers.

1970.02. Lengths of a triangle's sides are consecutive integers. Find these numbers if one of the triangle's medians is perpendicular to one of its angle bisectors.

1970.03. Small village has population of 1970. From time to time some two villagers exchange a dime for two nickels. Is it possible that during one week each person in the village handed over exactly ten coins?

1970.04. A three-digit number was decreased by its own sum of digits. Then the new number was subjected to the same procedure and so on, 100 times altogether. Prove that the final number is zero.

1970.05. In some country every two cities are connected by an air route or by a water route. Prove that either each city can be reached from any other one by air or each city can be reached from any other one by water.

1970.06. Twelve teams took part in a volleyball tournament. None of them ended up gaining exactly 7 points. Prove that there are three teams A, B and C such that A defeated B, B defeated C, and C defeated A.

GRADE 7

1970.07. Find angle B in triangle ABC if altitude CH equals half of the side AB and angle A equals $75°$.

1970.08. See Question 4.

1970.09. Given the isosceles triangle with $20°$ angle at its vertex.

  (a) Prove that the length of the sides is greater than twice the length of the base.
  (b) Prove that the length of the sides is less than three times the length of the base.

1970.10. See Question 5.

1970.11. Solve the system of equations

$$\begin{cases} x^2 = 2y - 1 \\ x^4 + y^4 = 2 \,. \end{cases}$$

1970.12. Thirty-six teams participate in a round-robin soccer tournament where every team plays each other exactly once. At some moment, each team has played at least thirty-four matches. Prove that one can divide the teams into three groups with twelve teams in each one so that inside every group all the matches have already been played.

GRADE 8

1970.13. Two circumferences are inscribed into angle ABC so that one of them touches side AB at point A and the other one touches side BC at point C. Prove that these circles cut equal segments from straight line AC.

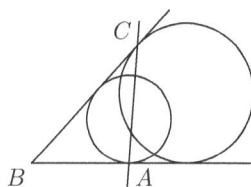

1970.14. Solve the system of equations

$$\begin{cases} x^2 + xy + y^2 = 7 \\ x^2 + xz + z^2 = 21 \\ y^2 + yz + z^2 = 28 \,. \end{cases}$$

1970.15. A pentagon whose sides have integer length with both its first and third sides equal to 1, has the inscribed circle. Find the segments into which this circle divides the second side.

1970.16. In a 5×5 table some 16 squares are painted over. Prove that it is possible to find a 2×2 subtable which has at least three painted squares.

1970.17. A circle and a square are inscribed in the triangle (all vertices of the square lie on the sides of the triangle). Prove that the ratio between the side of the square and the radius of the circle lies between $\sqrt{2}$ and 2.

1970.18. Thirty-five points on the plane are such that no three of them lie on the same straight line. Some of them are connected by line segments, 100 segments altogether. Prove that some two of these segments intersect.

GRADE 9

1970.19. Two equal circumferences O_1 and O_2 touch circumference O from the inside at points A and B, respectively. Let M be an arbitrary point on O, C, and D—the points of intersection of AM and BM with circumferences O_1 and O_2, respectively. Prove that lines AB and CD are parallel.

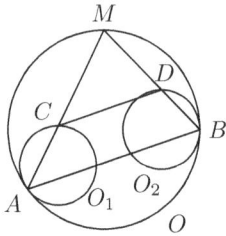

1970.20. See Question 16.

1970.21. See Question 15 for a polygon with nine sides.

1970.22. Positive numbers a, b, c, and d satisfy equalities

$$a^2 + b^2 = c^2 + d^2 \,, \ a^3 + b^3 = c^3 + d^3 \,.$$

Prove that $ab = cd$.

1970.23. Find all natural numbers x, y, z, and t such that

$$31(xyzt + xy + xt + zt + 1) = 40(yzt + y + t) \,.$$

1970.24. A 100×100 table is split into three vertical strips. The first (left) strip is 19 squares wide, and the second strip is 70 squares wide. The first row of the table is filled with numbers from 1 through 100 written in ascending order. Now, we fill the second row from left to right. First we write the numbers from the third strip portion of the first row without changing anything in their order, then we write the numbers from the second strip, and finally, the numbers from the first strip. Then the same procedure is applied to the numbers in the second row to produce the third row and so on until the entire table is filled. Prove that for every column the numbers in it do not repeat.

GRADE 10

1970.25. A triangular pyramid contains a sphere and a cube, both inscribed in it (all vertices of the cube lie on the faces of the pyramid). Prove that the ratio of the edge of the cube to the radius of the sphere lies between 2 and $1 + \sqrt{2}$.

1970.26. Prove the inequality

$$\sin 2 + \cos 2 + 2(\sin 1 - \cos 1) \geqslant 1 \,.$$

1970.27. In an equilateral pentagon, all angles are less than 120 degrees. Prove that they are all obtuse.

1970.28. See Question 22.

1970.29. Find function $f(x)$ such that for any real x, except 0 and 1, equality

$$f\left(\frac{1}{x}\right) + f(1 - x) = x$$

holds true.

1970.30.* A $n \times n$ table is split into three vertical strips. The first (left) strip is k squares wide, and the third strip—m squares wide, where numbers $n - k$ and $n - m$ are co-prime. The first row of the table is filled with numbers from 1 through n written in ascending order. Now, we fill the second row from left to right. First, we write the numbers from the third strip portion of the first row without changing anything in their order, then we write the numbers from the second strip, and finally, the numbers from the first strip. Then the same procedure is applied to the numbers in the second row to produce the third row and so on until the entire table is filled. Prove that every column contains all the numbers from 1 through n.

ELIMINATION ROUND

1970.31. Natural number is such that crossing out any of its digits will result in a number divisible by 7. Prove that either all or none of its digits are 4.

1970.32. Numbers t_1, t_2, ..., t_n are positive and their product is 1. Prove that there exists index $k < n$ such that $t_k(t_{k+1} + 1) \geqslant 2$ (t_{n+1} is the same as t_1).

1970.33. A flat convex polygon is made of wire. Prove that one cannot bend that same piece of wire into another (not congruent to the first one) flat polygon so that the distance between any two fixed points on the wire would not increase.

1970.34. Given sequence of natural numbers a_1, a_2, ..., such that $0 < a_1 < a_2 < \cdots$ and $a_n < 2n$ for every n, prove that any natural number can be represented either as a difference of some two terms of this sequence or as a term of the sequence.

1970.35.* Let D and C be some arbitrary points on a circumference that touches segment AB at its midpoint. Lines AD and BC intersect the circumference at points X and Y, respectively, and lines CX and DY intersect AB at points M and N, respectively. Prove that $|AM| = |BN|$.

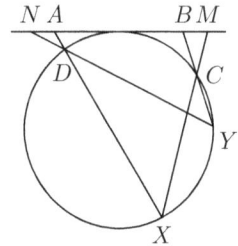

1970.36.* We will call numbers 2, 23, 234, and 2345 *heads* for the number 2345, and numbers 5, 45, 345, 2345—its *tails*. In the same way, we define these two terms for all natural numbers. Let a_1, a_2, ..., a_k be a sequence of natural numbers which have n_1, n_2, ..., n_k digits respectively. Each of these numbers is neither a head nor a tail for any other number in the sequence. Also none of these numbers' decimal representations include any other number's decimal representation as a contiguous part. If $n_i < 2n_j$ for any i and j, prove that

$$\frac{n_1}{10^{n_1}} + \frac{n_2}{10^{n_2}} + \cdots + \frac{n_k}{10^{n_k}} < 1.$$

1970.37.* A real number is written into every square of the infinite sheet of graph paper in such a way that the absolute value of the sum of the numbers inside any square with sides parallel to the grid lines does not exceed 1. Prove that the absolute value of the sum of the numbers inside any rectangle with sides parallel to the grid lines does not exceed 10,000.

1970.38.* Same as Question 37 but with the following addition. It is also known that in one of the rectangles the absolute value of the sum of the numbers reaches its maximum. Prove that this maximum value does not exceed 4.

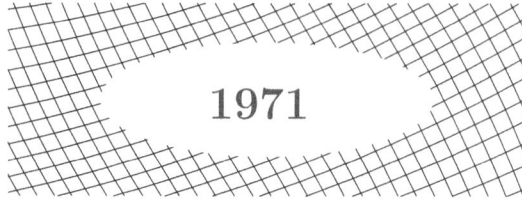

1971

GRADE 6

1971.01. Two bikers, John and Peter, started travelling from town A to town B, and at the same moment two other bikers, Johnson and Peterson, rode out from B travelling to A. John rides two times faster than Peter, and Johnson—three times faster than Peterson. John met Peterson at the same moment as Peter met Johnson. Which of the two events happened closer to A: meeting between John and Johnson or meeting between Peter and Peterson?

1971.02. Among all the triangles that have the given sum of the medians, find the one which has the largest sum of the altitudes.

1971.03. Nick, Eugene, and Nadia faced several entrance exams. At each exam, regardless of the subject, a student receives a certain number of points depending on the ranking (1st, 2nd, or 3rd place) they achieve; the better the ranking the higher the result. After all the exams, Nick's total was 22 points, while Eugene and Nadia had 9 points. If Eugene was first in algebra, who was second in physics?

1971.04. Find a positive integer given that the decimal representation of its sixth power is written with digits 0, 1, 2, 2, 2, 3, 4, 4 in some order.

1971.05. Two players take turns rolling a cubic die that has 1 or 2 stamped on its every face. There is also a blackboard with number 200 written on it. After rolling the die, a player can either increase or decrease the number on the board by the value of the roll, or can leave the number unchanged (take a pass). The player who is the first to obtain a four-digit number, wins. The game is declared a draw if the number on the board has only two digits, or if there were three passes in a row. Prove that the first player can guarantee himself at least a draw if he plays the errorless game.

1971.06. Is it possible to dissect the plane into squares among which only two are equal?

GRADE 7

1971.07. Solve the system of equations

$$\begin{cases} x^2y^2 + xy^2 + x^2y + xy + x + y + 3 = 0 \\ x^2y + xy + 1 = 0 . \end{cases}$$

1971.08. Among all the triangles that have the given sum of the angle bisectors, find the one which has the largest sum of the altitudes.

1971.09. Point K lies inside square $ABCD$. Perpendiculars are dropped from vertices A, B, C, and D onto lines BK, CK, DK, and AK, respectively. Prove that these four perpendiculars have a common point.

1971.10. See Question 5.

1971.11. Prove that for every convex polygon except parallelogram, one can find three of its sides such that the triangle formed by the straight lines defined by these sides contains the entire polygon.

1971.12. The bottom of the rectangular box was covered without overlapping by rectangular tiles with dimensions 1×4 and 2×2. The tiles were taken out of the box whereupon one of the 2×2 tiles was lost and therefore replaced by a 1×4 tile. Prove that now the bottom of the box cannot be fully covered by these tiles without overlapping.

GRADE 8

1971.13. Points A and B move along two intersecting straight lines with constant equal speed. Prove that there exists a point on the plane which at any moment is equidistant from A and B.

1971.14. Numbers 5^{1971} and 2^{1971} are written one after the other. What is the total number of digits?

1971.15. One hundred integers whose sum equals 1 are written around the circle. We will call several numbers a *chain* if they are adjacent to each other, thus forming a contiguous sequence. Find the number of all chains with positive sum of their numbers.

1971.16. Two piles of matches have exactly 100 matches each one in them. Two people play the following game making their moves in turn. The move consists of discarding one of the piles and then dividing the other one into two non-empty, possibly unequal, piles of matches. The player who cannot make a move loses. Can the first player win, and if the answer is positive, then what is his strategy?

1971.17. Natural number n is such that $n+1$ is divisible by 24. Prove that the sum of all divisors of n, both 1 and n included, is also divisible by 24.

1971.18. Let BD be one of the angle bisectors in triangle ABC. Length of AB is 15; length of BC is 10. Prove that the length of BD does not exceed 12.

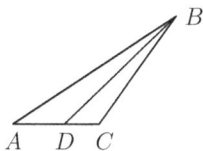

GRADE 9

1971.19. Solve the system of equations

$$
\begin{cases}
x_0 = x_0^2 + x_1^2 + \cdots + x_{100}^2 \\
x_1 = 2(x_0 x_1 + x_1 x_2 + \cdots + x_{99} x_{100}) \\
x_2 = 2(x_0 x_2 + x_1 x_3 + \cdots + x_{98} x_{100}) \\
x_3 = 2(x_0 x_3 + x_1 x_4 + \cdots + x_{97} x_{100}) \\
\cdots \\
x_{100} = 2 x_1 x_{100}.
\end{cases}
$$

1971.20. See Question 13.

1971.21. Given three-digit prime number $\overline{abc}$, prove that $b^2 - 4ac$ cannot be a perfect square.

1971.22. See Question 16. **Additional question**: what must be the original size of the piles, to guarantee that the first player cannot win in the errorless game?

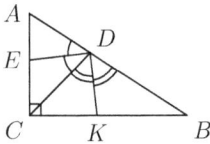

1971.23. Let CD be the bisector of the right angle in triangle ABC; DE and DK—the angle bisectors in triangles ADC and BDC. Prove that $|AD|^2 + |BD|^2 = (|AE| + |BK|)^2$.

1971.24. Is it possible to dissect the plane into pairwise different triangles whose sides all have rational length?

GRADE 10

1971.25. See Question 17.

1971.26. In a tetrahedron, one of the altitudes intersects two other altitudes. Prove that all four altitudes have a common point.

1971.27. Two flies crawl with constant equal speed along two non-parallel non-intersecting lines in space. Prove that there is a point in space which at any moment is equidistant from these flies.

1971.28. Prove that

$$
\left(\frac{1}{2}\right)^{200} + \left(\frac{1}{2}\right)^{199} \cdot \frac{1}{2} + \left(\frac{1}{2}\right)^{198} \cdot \frac{1}{3} + \cdots + \frac{1}{2} \cdot \frac{1}{200} + \frac{1}{201} < \frac{1}{90}.
$$

1971.29. Prove that if you write numbers $2^1, 2^2, 2^4, 2^8, \ldots$, one after another, following the decimal point, then the resulting decimal fraction is irrational.

1971.30. * Several straight lines no two of which are parallel lie in the plane. Point P is an arbitrary point on the plane which is projected onto all these lines. Then the same is done with all the projections, and so on. Prove that all the points we ever obtain in this manner can be covered by one circle.

ELIMINATION ROUND

1971.31. Given positive real numbers $a_1, a_2, \ldots, a_n$, prove that the equation

$$x^n + a_1 x^{n-1} - a_2 x^{n-2} - \cdots - a_n = 0$$

has no more than one positive solution.

1971.32. A collection of n real numbers $a_1, a_2, \ldots, a_n$ is given. For every $1 \leqslant k \leqslant n$ we denote by A_k the largest among numbers

$$a_k, \quad \frac{a_{k-1} + a_k}{2}, \quad \frac{a_{k-2} + a_{k-1} + a_k}{3}, \quad \ldots, \quad \frac{a_1 + a_2 + \cdots + a_k}{k}.$$

Prove that the smallest of numbers $A_1, A_2, \ldots, A_n$ does not exceed the arithmetic mean of $a_1, a_2, \ldots, a_n$.

1971.33. An arbitrary straight line L is drawn through vertex A of triangle ABC; B_1 and C_1 are projections of points B and C onto L; B_2 is projection of B_1 onto line AC; C_2 is projection of C_1 onto line AB. Prove that the intersection of lines $B_1 B_2$ and $C_1 C_2$ lies on one of the altitudes of triangle ABC or on its extension.

1971.34.* Sum of several integers $b_1, b_2, \ldots, b_n$ equals 1. For any $1 \leqslant k \leqslant n$ let us denote by N_k the number of positive sums in the sequence

$$b_k, \quad b_k + b_{k+1}, \quad b_k + b_{k+1} + b_{k+2}, \quad \ldots, \quad b_k + b_{k+1} + \cdots + b_n + b_1 + \cdots + b_{k-1}.$$

Prove that all numbers N_k are different.

1971.35. Square S with side $n-1$ and rectangle R with sides $a-1$ and $b-1$ (such that $ab = n^2$) are both dissected into unit squares. Some correspondence is established between n^2 nodes (vertices of unit squares) of S and ab nodes of R; if two nodes of S are different, then the corresponding nodes of R must be different as well. Also for every 2×2 square K inside S, the vertices in R that correspond to the vertices and the center node of K are, respectively, the four vertices and the center node of some parallelogram (possibly, a trivial one, that has all five points lying on the same straight line). Prove that $a = b$, if $a, b > 2$.[62]

1971.36.* Edges of complete graph with $2n + 1$ vertices are colored with three colors. Prove that it is possible to choose one of the colors and some $n + 1$ vertices of the graph so that for any two of these vertices, there is a path connecting them that consists only of edges of the chosen color.[63]

---

[62] This additional restriction $(a, b > 2)$ was not present in the archival printout of the problem set, but it is necessary; without it the statement is false.

[63] It was not stated explicitly that the path must pass only through the $n+1$ selected vertices; however, it is quite likely that this was the presumed meaning—the statement is true even with this stricter requirement.

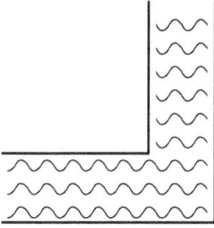

1971.37. Prove that the equation $x^3+y^3+z^3 = 2$ has infinitely many solutions in integer numbers.

1971.38.* A toy boat floated through both vertical and then horizontal parts of the river shown in the figure. Prove that it is possible to affix to this toy boat a wooden raft of such shape that if the boat repeats the exact same trip, then every point on the boundary of the raft will at some moment touch the river shore.

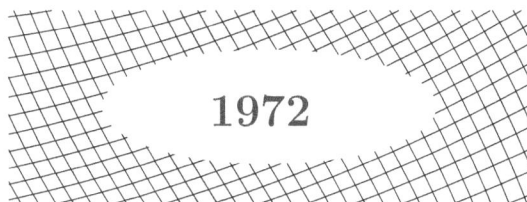

1972

GRADE 6

1972.01. Two identical planes simultaneously left city A. The first plane's route is $A \to B \to D \to C \to A \to D \to B \to C \to A$, and the second plane flies $A \to B \to C \to D \to A \to B \to C \to D \to A \to B \to C \to D \to A$. Which of the planes will finish the trip first?

1972.02. Only the members who participated in the last two outings were invited to the hiking club's meeting (some of the members took part in both trips). In the first outing, sixty percent of the participants were men, in the second—seventy-five. Prove that there were at least as many men at the meeting as women.

1972.03. Draw six points on the plane so that every three of them form an isosceles triangle.

1972.04. All zeros were erased from the decimal representation of the product of all integers from 1 through 100. Is the last digit of the result even or odd?

1972.05. Students had final exams on seven subjects, and all passed them receiving only grades of A and B, with each student getting no more than two B's. It turned out that no student can say that she did better than some other student on all the subjects. Prove that there are at least 21 students.

1972.06. Is it possible to place numbers 1 through 12 on the edges of the cube in such a way that for every vertex the sum of the numbers on the edges coming out of that vertex is the same?

GRADE 7

1972.07. Is it possible, using two straight cuts passing through vertices of a triangle, to dissect it into four parts such that three of them are the triangles of the same area?

1972.08. See Question 2.

1972.09. In trapezoid $ABCD$, top base BC is parallel to AD. Point M is the intersection of angle bisectors from A and B, and N is the intersection of angle bisectors from C and D. Knowing the sides of the trapezoid, find the length of segment MN.

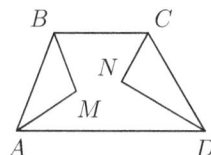

1972.10. White and black tokens are placed in the vertices of regular dodecagon. Prove that there are three tokens of the same color positioned in the vertices of an isosceles triangle.

1972.11. See Question 5.

1972.12. Prove that if some even number can be represented as a sum of two perfect squares, then its half can also be represented in the same way.

GRADE 8

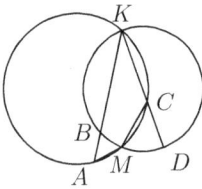

1972.13. Two circumferences intersect at points K and M. Two rays issue from point K. One of them intersects the first circumference at point A and the second one at point B; the second ray intersects the first circumference at point C and the second one at point D. Prove that angles MAB and MCD are equal.

1972.14. Does there exist a natural number with the sum of digits of its square equal to 1972?

1972.15. Prove that a quadrilateral whose diagonals divide it into four triangles with equal perimeters is a rhombus.

1972.16. Several small non-overlapping circles are drawn on the plane, then some of them are connected by segments. Prove that one can label these circles with integer numbers so that two circles are connected if and only if their numbers are co-prime.

1972.17.* The regular triangle with side of 32 is given. A unit triangle is cut off from one of its corners, then the rest is dissected into regular triangles with integer sides. Prove that there are at least 15 of them.

1972.18. Natural numbers m, n, a, b, k, l are such that

$$\frac{m}{n} < \frac{a}{b} < \frac{k}{l}, \quad |ml - kn| = 1.$$

Prove that $b \geqslant n + l$.

GRADE 9

1972.19. See Question 14.

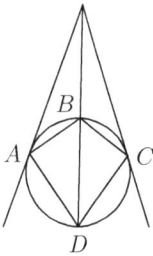

1972.20. Quadrilateral $ABCD$ is inscribed into a circumference. Prove that if the point of intersection of the lines tangent to the circumference at vertices A and C lies on line BD, then $|AB| \cdot |CD| = |BC| \cdot |AD|$.

1972.21. See Question 18.

1972.22. Let AC be the largest side of triangle ABC. Points A_1 and C_1 on that side are such that $|AC_1| = |AB|$ and $|CA_1| = |CB|$.

Points A_2 and C_2, respectively, are chosen on sides AB and CB in such a way that $|AA_2| = |AA_1|$ and $|CC_2| = |CC_1|$. Prove that points A_1, A_2, C_1, and C_2 lie on one circumference.

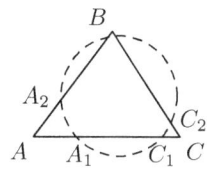

1972.23. Let m and n be any natural numbers. Prove that

$$\frac{1}{\sqrt[m]{n+1}} + \frac{1}{\sqrt[n]{m+1}} \geq 1.$$

1972.24. Several integers are written in a row. Then under each number we write how many times it has occurred in this sequence. From this new sequence, the third one is built using the same rule, and so on. Prove that eventually there will be two identical sequences written one under the other.

GRADE 10

1972.25. Numbers a, b, and c lie between 0 and 1. Prove that

$$a + b + c - 2\sqrt{abc} \geq ab + bc + ac - 2abc.$$

1972.26. A three-dimensional closed broken line is called *regular* if all its segments have equal length and all the angles between adjacent segments are equal. Prove that for any $N > 5$ there exists a regular broken line with N edges which is not flat, i.e., does not entirely lie inside some plane.

1972.27. Let p be a prime number other than 3. Prove that $4p^2 + 1$ can be represented as a sum of three squares of natural numbers.

1972.28. See Question 22.

1972.29.* A regular triangle with area 1 lies inside convex heptagon whose area is equal to 1.0000001. Prove that at least one of the heptagon's angles is greater than $139°$.

1972.30. Several matches have been played in a volleyball tournament, resulting in each team having ten wins and ten losses. Prove that it is possible to choose several matches so that in these selected matches every team has exactly one win and exactly one loss.

ELIMINATION ROUND

1972.31. Find maximum of the ratio $k^2/(1.01)^k$ for natural k, and find all k for which it is attained.

1972.32. Prove that any straight line that divides area of the triangle in half divides its perimeter with ratio not exceeding $3:1$.

1972.33. Ninety-nine digits 9 are written in a row. Prove that it is possible to write 100 more digits to the right of them so that the resulting number will be a perfect square.

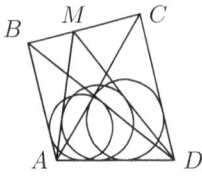

1972.34. Point M lies on side BC of convex quadrilateral $ABCD$. Prove that some two of the circumferences inscribed in triangles ABD, ACD, and AMD have different radii.

1972.35. In a strictly monotonically increasing sequence of natural numbers, each term except the first two equals the sum of some two of the preceding terms of the sequence. Prove that this sequence contains infinitely many composite numbers.

1972.36. Each side of the regular triangle is divided into 30 equal parts. Straight lines drawn through the division points parallel to the sides of the triangle dissect the triangle into 900 small triangles, creating triangular lattice grid. What is the maximum number of the grid points, no two of which lie on the same grid line?

1972.37.* The subway system in Metropolis has 1972 stations, and each of its subway lines connects exactly two stations. If any nine stations were closed, the subway system would still be connected.[64] It is also known that to get from station A to station B at least 99 line transfers are necessary. Prove that one can split all subway stations into 1000 groups in such a way that no two stations in the same group are directly connected.

1972.38.* The city map is a 100×100 square with the side of each block being a 500 meters long one-way street. Given that it is impossible to drive more than 1 km around the city without a driving violation, prove that there are at least 1300 "dead-end" intersections. A "dead-end" intersection is the one which cannot be exited without breaking the traffic rules; the corners of the city are counted as intersections as well.

---

[64] Meaning that it would be possible to ride from any open station to any other open station.

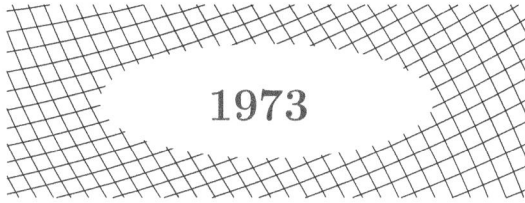

1973

GRADE 6

1973.01. Three book stores had 1973 textbooks on their shelves. In the first three days of the clearance sale, the first store sold $\frac{1}{47}$, $\frac{1}{7}$, and $\frac{1}{2}$ of its textbooks, respectively. The second store—$\frac{1}{41}$, $\frac{1}{5}$, and $\frac{1}{3}$, and the third—$\frac{1}{25}$, $\frac{1}{20}$, and $\frac{1}{10}$. How many textbooks was there originally in each store?

1973.02. A flat chocolate bar has two long and three short grooves dividing it into twelve smaller pieces. What is the smallest possible number of breaks (along the grooves) necessary to break the bar into the pieces without grooves if you are allowed to break several pieces together, putting them atop each other (that operation would be considered as one break)?

1973.03. Prove that a number written with six hundred sixes and several zeros cannot be a perfect square.

1973.04. Prove that a square can be dissected into 1973 smaller squares.

1973.05. Three columns of numbers are written on the blackboard, with none of the numbers repeating inside the same column. In the 4th column, we then write all the numbers which occur exactly once in the first two columns, in the 5th column—the numbers that occur exactly once in the 3rd and the 4th columns, in the 6th column—the numbers that occur exactly once in the 2nd and the 3rd columns, and in the 7th column—the numbers that occur exactly once in the 1st and the 6th columns. Prove that the number of entries in the 5th column is the same as in the 7th column.

1973.06. In an isosceles triangle, one of the angles equals 108 degrees. Prove that this triangle can be dissected into several acute-angled triangles.

GRADE 7

1973.07. See Question 2.

1973.08. See Question 3.

1973.09. Points E and K are the midpoints of rectangle $ABCD$'s sides AD and BC, point H is an arbitrary point on segment EK, point M is the intersection of lines AH and BC,

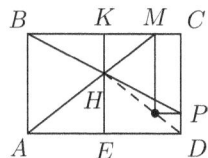

point P is the intersection of lines BH and CD. Line L_1, parallel to CD, passes through M; line L_2, parallel to AD, passes through P. Prove that the intersection of L_1 and L_2 lies on DH.

1973.10. Prove that $2^{10} + 5^{12}$ is a composite number.

1973.11. On each side of a parallelogram, a point is selected, and the area of the quadrilateral formed by these four points equals half of the area of the parallelogram. Prove that one of the diagonals of this quadrilateral is parallel to one of the parallelogram's sides.

1973.12. Sides a, b, and c of a triangle satisfy the equality $2(a^8 + b^8 + c^8) = (a^4 + b^4 + c^4)^2$. Prove that this triangle is right.

GRADE 8

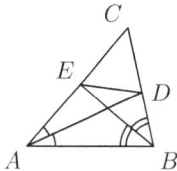

1973.13. In triangle ABC, segments AD and BE are the angle bisectors. Prove that if $|AC| > |BC|$, then $|AE| > |DE| > |BD|$.

1973.14. The sum of the digits of a ten-digit number equals 4. What could be the sum of the digits of its square?

1973.15. For any two points A and B on the plane denote by $A * B$ the point symmetric to A with respect to B. Three vertices of a square are given. Using the $*$ operation only, can one obtain the missing vertex of the square?

1973.16. A triangle is dissected into several convex polygons. Prove that unless one of them is a triangle, two of these polygons have the same number of sides.

1973.17. Several natural numbers are written around a circle. Between each two of the neighboring numbers their G.C.D. is written, and after that the old numbers are erased. Then this operation is repeated and so on. Prove that after several steps all the numbers will become the same.

1973.18. A finite number of points is given on the plane, and some pairs of them are connected with segments in such a way that one can reach any point from any other via these connections. Is it always possible to delete one of these points together with adjacent segments so that the remaining points-and-segments structure is still fully connected?

GRADE 9

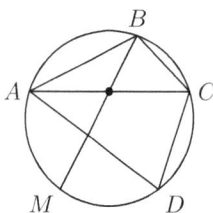

1973.19. Quadrilateral $ABCD$ inscribed in circumference S satisfies equality $|AB|/|BC| = |AD|/|DC|$. Line that passes through vertex B and the midpoint of diagonal AC, intersects S also at point M. Prove that $|AM| = |CD|$.

1973.20. The sum of the digits of a nine-digit number is 3. What could be the sum of the digits of this number's cube?

1973.21. On sides AB and CD of parallelogram $ABCD$, find points K and M such that the area of the quadrilateral obtained by intersecting triangles AMB and CKD is the largest possible.

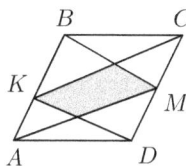

1973.22. See Question 17.

1973.23. See Question 16.

1973.24. Ten white and twenty black tokens are placed around the circle. It is permitted to swap any two tokens which are separated by three others. Two arrangements of these thirty tokens (in the same thirty locations) will be called equivalent if one can be obtained from the other by a series of such swaps. How many non-equivalent arrangements are there?

GRADE 10

1973.25. See Question 13.

1973.26. Infinite arithmetic progression with the first term A and difference $d \neq 0$ contains an infinite geometric progression. Prove that A/d is rational.

1973.27. Prove the inequality

$$\cos\frac{\pi}{4}\cdot\left(1-\cos\frac{\pi}{4}\right) + 2\cos\frac{\pi}{8}\cdot\left(1-\cos\frac{\pi}{8}\right) + \cdots + 256\cos\frac{\pi}{1024}\cdot\left(1-\cos\frac{\pi}{1024}\right) < \frac{1}{2}.$$

1973.28. Prove that in a convex polyhedron some two faces have the same number of sides.

1973.29. Let P be a convex polygon with 1973 vertices. Point O lies inside the polygon. For some fixed acute angle α, the common part of P and any angle equal to α with its vertex at O always has the same area. Prove that P is a regular polygon.

1973.30. See Question 24.

ELIMINATION ROUND

1973.31. The sum of absolute values of the pairwise differences of some five non-negative numbers equals 1. Find the minimum possible value of the sum of these five numbers.

1973.32. On the plane, $2k + 3$ points in general position[65] are selected so that no four of them lie on the same circumference. Prove that there is a circumference passing through some three of these points which contains exactly k selected points inside of it.

1973.33. All the coefficients of polynomial $P(x) = a_0x^n + \cdots + a_{n-1}x + a_n$ are integers. If all equations $P(x) = 1$, $P(x) = 2$, $P(x) = 3$ have integer roots, prove that equation $P(x) = 5$ cannot have more than one integer root.

---

[65] A set of distinct points on the plane is called *points in general position* if no three of these points are collinear.

1973.34. Several volleyball teams played a round-robin tournament in which every team played every other team exactly once. For any two teams A and B such that A defeated B, there exists team C that lost to A but defeated B. What is the minimum possible number of teams in such a tournament?

1973.35. A quadrilateral is inscribed into a unit square. Another square with the sides parallel to the sides of this unit square is then inscribed into the quadrilateral. Prove that if the sides of the smaller square equal $\frac{1}{2}$, then the vertices of the smaller square are the midpoints of the sides of the inscribed quadrilateral.

1973.36. Area of a convex polygon equals 9. The polygon is intersected by nine parallel lines which are uniformly spaced at distance 1 from each other. Prove that the sum of the lengths of segments-intersections is not greater than 10.

1973.37.* Two people are playing the following game. The first player writes down (in secret) a ten-digit number. Then the second player will ask him questions of the following form "*tell me the digits in the* 10^{a_1}, 10^{a_2}, ..., 10^{a_k} *places*" and the first player will reveal the digits but without specifying which digits occupy which places. What is the minimum number of questions which will always allow the second player to deduce the secret number?

1973.38.* A cubic die with edge length of a is tossed onto a sheet of graph paper. Prove that this die cannot cover more than $(a+1)^2$ nodes of the grid.

> **Commentary.** *Some archival problem sets had Questions 39 and 40 (see below) being used instead of Questions 16 and 23, respectively. It is also possible that these two questions were given—as an additional challenge—to the contestants who solved most of the regular problems.*
>
> **1973.39.*** *Two beetles and two flies crawl along the contour of the convex polygon—they all move with the same speed and in the same direction. What should be the initial position of the beetles so that for any initial position of the flies the minimum distance between the flies would not exceed the minimum distance between the beetles?*
>
> **1973.40.** *A unit square contains 1973 figures with the sum of their areas greater than 1972. Prove that all these figures have a common point.*

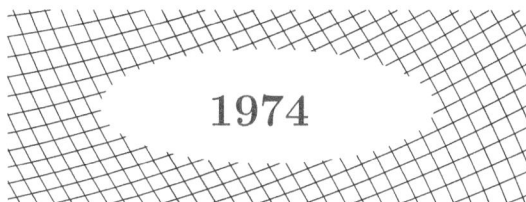

1974

GRADE 6

1974.01. Find all three-digit numbers $\overline{abc}$ such that
$$\overline{abc} = 2(\overline{ab} + \overline{bc} + \overline{ac}).$$

1974.02. Does there exist a convex polygon with exactly 1974 diagonals?

1974.03. Several congruent regular triangles were cut from a sheet of construction paper. Vertices of each triangle are marked with numbers 1, 2, and 3, and then the triangles are placed one on top of the other forming a triangular prism stack. Is it possible that the sum of the numbers along each edge of the stack equals 55?

1974.04. Can all the sums in Question 3 be equal to 50?

1974.05. Three short distance runners A, B, and C competed in the race. Runner C was the last one out of the starting blocks and then during the race he either passed another runner or was passed by another runner exactly six times. We also know that runner B was second to A at the beginning, that runner A passed or was passed by another runner exactly 5 times, and that B finished before A. What were the results of the race?

1974.06. The country of Aire has 1974 cities. Its capital is connected by air flights with 101 other cities, while the city of Fartown—only with one. Every other city in the country is connected to exactly 20 others. Prove that one can fly (possibly with some stopovers) from the capital to Fartown.

GRADE 7

1974.07. Three numbers a, b, and c satisfy equalities
$$a + b + c = 7, \quad \frac{1}{a+b} + \frac{1}{b+c} + \frac{1}{a+c} = \frac{7}{10}.$$
Find $\frac{a}{b+c} + \frac{b}{a+c} + \frac{c}{a+b}$.

1974.08. Point O is the center of equilateral triangle ABC. Find the locus of points X such that any straight line that passes through X intersects at least one of the segments AB and OC.

1974.09. Find all four-digit numbers $\overline{ABCD}$ such that
$$\overline{ABCD} = \overline{AD} \cdot \overline{ADA}.$$

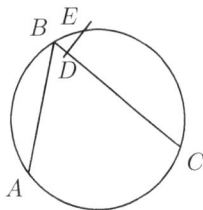

1974.10. Points A, B, and C lie on circumference S. Point D on segment BC is the midpoint of the broken line ABC, and E is the midpoint of arc ABC. Prove that lines ED and BC are perpendicular.[66]

1974.11. See Question 6.

1974.12. In an isosceles triangle T, angle at the vertex is obtuse, and its bisector's length is half of the other angle's bisector. Find the angles of T.

GRADE 8

1974.13. Two circles on the plane do not overlap. Does there exist a point on the plane which is not covered by these circles, yet any line that passes through this point intersects at least one of them?

1974.14. Solve the equation in integers

$$x^{x^{x^x}} = (19 - y^x)y^{x^y} - 74.$$

1974.15. A corner square was cut from an 8×8 board. Is it possible to dissect the remaining part of the board into 17 triangles of equal area?

1974.16. A white pawn stands on $a1$ square of the chessboard, and a black pawn—on $h8$. The white pawn is only allowed to move up or right, while the black pawn—only down or left. It is prohibited to move to the square already occupied by the other pawn, but any pawn can skip a move as many times as needed. After several moves, the pawns have switched their places. Prove that at some moment the line connecting the centers of the squares occupied by the pawns is perpendicular to the line connecting centers of $a1$ and $h8$.

1974.17. Prove that it is impossible to find a set of $n > 4$ points on the plane such that for any three points from that set, there exists another point from the set which, together with these three, forms a parallelogram.

1974.18. A few minutes ago, a lonely bacterium inside the test tube divided in two. Then one of its halves also divided in two and so on. Right now there are 1000 bacteria in the tube. Prove that at some moment there existed a bacterium whose progeny consists of at least 334 but no more than 667 bacteria.

GRADE 9

1974.19. See Question 15.

1974.20. See Question 14.

---

[66] This is virtually the same problem as Question **1968.31**.

1974.21. See Question 16 for 9×9 board.

1974.22. Triangle ABC has circumradius R, while circumference S has radius equal to $\frac{R}{2}$. Prove that there exists point T such that S intersects all three segments TA, TB, and TC exactly at their midpoints.

1974.23. Point X lies on circumference centered at point O. Point Y is selected on diameter drawn from point X so that O lies between X and Y. Find chord AB passing through Y so that angle AXB has the smallest possible value.

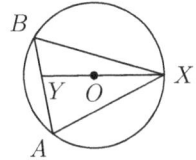

1974.24.[*] Martian language has three words A, B, and C such that word $AABB$ is identical to CC. Prove that there exists word D from which any of the words A, B, and C can be obtained by writing several copies of D next to each other.

GRADE 10

1974.25. Does there exist a twenty-digit number which is a perfect square beginning with eleven digits 1?

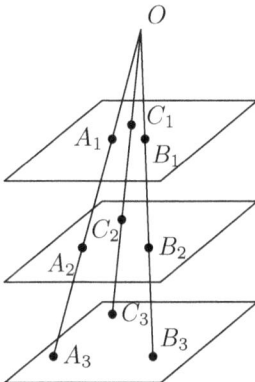

1974.26. Three rays issuing from the same point O intersect three parallel planes at points A_1, B_1, and C_1; A_2, B_2, and C_2; A_3, B_3, and C_3, respectively. Denote the volume of pyramid $OA_1B_2C_3$ by V, and volumes of pyramids $OA_1B_1C_1$, $OA_2B_2C_2$, and $OA_3B_3C_3$ by V_1, V_2, and V_3, respectively. Prove that

$$V \leqslant (V_1 + V_2 + V_3)/3.$$

1974.27. Find the maximum value of expression

$$x_1 + x_2 + x_3 + x_4 - x_1x_2 - x_1x_3 - x_1x_4 - x_2x_3 - x_2x_4 - x_3x_4$$
$$+ x_1x_2x_3 + x_1x_2x_4 + x_1x_3x_4 + x_2x_3x_4 - x_1x_2x_3x_4$$

for $|x_1| \leqslant 1$, $|x_2| \leqslant 1$, $|x_3| \leqslant 1$, $|x_4| \leqslant 1$.

1974.28. Prove that in a three-dimensional space there is no finite set of n points $(n > 4)$ such that for any three of its points the set contains the fourth point which, together with these three, forms a parallelogram.

1974.29. Let O be point on the plane. Twelve rays R_1, R_2, $\ldots$, R_{12} issue from that point (they are indexed in the clockwise order) with any two neighboring rays forming angle of measure less than $\frac{\pi}{4}$. Point A_1 is selected on R_1 so that distance $|OA_1|$ equals 729. We draw a line passing through A_1 and parallel to R_{12} until it intersects R_2 at point A_2. Then we draw a line passing through A_2 and parallel to R_1 until it intersects R_3 to produce point A_3 and so on. Finally point A_{13} on R_1 is obtained. Prove that $|OA_{13}| \leqslant 1$.

1974.30. Number 2^n is written on the blackboard. Two natural numbers whose sum equals this number are written in a row under it. Then the same is done for every number in that row which is different from 1, so that the third row of numbers is formed. This goes on until all the "available" numbers are ones. Prove that the sum of all the numbers on the board is greater than or equal to $n2^n$.

ELIMINATION ROUND

1974.31. Thirty-seven satellites orbit some spheroidal planet. Prove that at any moment there is a point on the planet's surface from which one cannot see more than 17 satellites.

1974.32. In some group, any two persons who know the same number of people within that group do not have any common acquaintances. Prove that either no two people are acquainted or someone has exactly one acquaintance within that group.

1974.33. Prove that any convex polygon with even number of vertices possesses a diagonal which is not parallel to any of the sides.

1974.34. Natural numbers are written into the squares of a rectangular table. It is permitted to either double all numbers in one column, or to subtract one from all numbers in one row. Prove that, using these operations, it is possible to produce the table with all zeros.

1974.35. The sides of the square are indexed in the clockwise direction by numbers 1, 2, 3, and 4. For arbitrary point A and side k, we will denote by A_k the point obtained from projection of A onto side k (or its extension) by symmetry with respect to A. Find all points A such that every point in the infinite sequence A_1, A_{12}, A_{123}, A_{1234}, A_{12341}, $\ldots$ lies inside the square.

1974.36. Each one of one hundred natural numbers is less than 100, and their sum equals 200. Prove that the sum of some of these numbers equals 100.

1974.37. Find all natural numbers k satisfying the following condition: there is no polygon with k vertices such that the line passing through any side of that polygon contains some other side of the same polygon (we consider only polygons whose adjacent sides are not parallel).

1974.38.* Prove that for any prime number p and any $p+1$ pairwise different natural numbers, one can always select two of these numbers so that the ratio of the larger one to their greatest common divisor is at least $p+1$.

Commentary. *In various archives, we have found slightly different versions of the Grade 7 problem set. In one of them, for instance, Question 8 was substituted with Question 5. Here we present the version taken from the archive of A. G. Goldberg.*

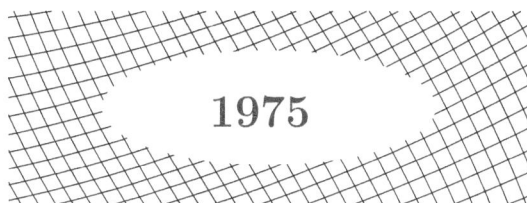

1975

GRADE 6

1975.01. Nick has a secret two-digit number, and Vera is trying to discover it. To do that, she writes two-digit "trial" numbers on the blackboard, and for each number, Nick writes next to it either a plus sign if that number coincides with the secret one, a minus sign if it has one correct digit in the same position as the secret number, or nothing. Prove that Vera needs no more than ten trials to determine the secret number.

1975.02. Twenty-six dominoes are laid out in a row in accordance with the domino rules; then each of the two remaining tiles is sawed in half. Prove that some two of the four halves show the same value.

1975.03. Prove the inequality

$$\frac{1}{2} - \frac{3}{4} + \frac{5}{6} - \frac{7}{8} + \cdots + \frac{97}{98} - \frac{99}{100} = -\frac{1}{2}\left(\frac{1}{26} + \frac{1}{27} + \cdots + \frac{1}{49} + \frac{1}{50}\right).$$

1975.04. Five lines on the plane all intersect the same unit circle. Point X located at a distance of 11 from the center of the circle, is transformed five times by reflecting it across these lines. Prove that the final position of X does not lie inside the circle.

1975.05. Nick and Vera are writing out a 20-digit number using only digits 1, 2, 3, 4, and 5. Nick writes its first digit, Vera—its second digit, Nick—the third digit, and so on. Vera's objective is to ensure that the final number is divisible by 9. Can Nick prevent that?

1975.06. For every possible four-digit number written with digits 1, 2, and 3, its "signature", which is one of the numbers 1, 2, and 3, is computed using some secret rule. If any two numbers have no identical digits in the same decimal place, then their signatures are different. Also for each one of the numbers 1111, 2222, 3333, and 1222, its signature coincides with its first digit. Prove that for any other number its signature is also equal to that number's first digit.

GRADE 7

1975.07. Nick and Vera are writing out 30-digit number using only digits 1, 2, 3, 4, and 5. Nick writes its first digit, Vera—its second digit, Nick—its third, and so on. Vera's objective is to ensure that the final number is divisible by 9. Can Nick prevent that?

1975.08. Both p and $p^{p+1} + 2$ are primes. Find p.

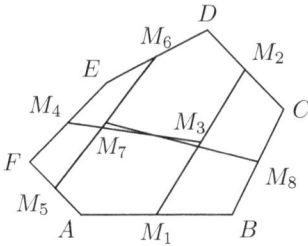

1975.09. A three-digit number has no zeros in its decimal representation. Find the maximum possible value of the product of this number and the sum of its digits' reciprocals.

1975.10. In convex hexagon $ABCDEF$, point M_1 is the midpoint of AB, M_2—the midpoint of CD, M_3—the midpoint of M_1M_2, M_4—the midpoint of EF, M_5—the midpoint of AF, M_6—the midpoint of DE, M_7—the midpoint of M_5M_6, and M_8—the midpoint of BC. Prove that the segment M_3M_4 intersects segment M_7M_8.

1975.11. For every possible seven-digit number written with digits 1, 2, and 3 its "signature", which is one of the numbers 1, 2, and 3, is computed using some secret rule. If two numbers have different digits in every decimal place, then their signatures must be different. Also for each one of the numbers 1111111, 2222222, 3333333 and 1222222 its signature coincides with its first digit. Prove that for every number its signature is equal to that number's first digit.

1975.12. Find a chord of the given unit circle such that if we build the square with this chord as its side, then the distance from the center of the circle to the other vertices of the square is the largest possible.

GRADE 8

1975.13. Point of intersection of the altitudes in an isosceles triangle lies on its inscribed circumference. Find the ratios between the sides of the triangle.

1975.14. Given five geometric progressions with integer terms, prove that there exists a natural number that does not belong to any of these progressions.

1975.15. In triangle ABC, point F is the common point of angle bisectors AD and CE. Points B, D, E, and F lie on one circumference. Prove that the radius of that circumference is greater than or equal to the triangle's inradius.

1975.16. Prove that the points of intersection of two parabolas $y = x^2 + x - 41$ and $x = y^2 + y - 40$ lie on one circumference.

1975.17. Find the number of positive integers $N \leqslant 1\,000\,000$ such that N is divisible by $[\sqrt{N}]$.

1975.18. Seven tokens of seven different colors are placed in seven consecutive vertices of regular polygon with 100 vertices. It is permitted to move any token in the clockwise direction jumping over the next ten vertices if the destination vertex is unoccupied. The objective is to move these tokens into the next seven vertices of the polygon using the described operations. How many different arrangements of the tokens in those final seven vertices are possible?

GRADE 9

1975.19. Let AB and CD be two perpendicular diameters of a circle. Point X is selected on arc BD; lines AX and CX intersect CD and AB, respectively, at points E and F. Prove that if ratio $|CE|/|ED|$ is rational then so is ratio $|AF|/|FB|$.

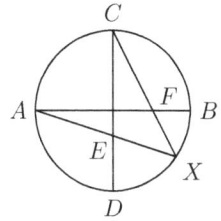

1975.20. See Question 16.

1975.21. Among all triangles that fit inside the given triangle, find the one with the maximum ratio of area vs perimeter.

1975.22. See Question 18.

1975.23. What is the maximum number of obtuse angles formed on the plane by fifteen rays with the same origin point?

1975.24. In a 100×100 table, some squares are marked so that each marked square is the only such square either in its column or in its row. What is the maximum possible number of the marked squares?

GRADE 10

1975.25. Given 1975 geometric progressions with integer terms, prove that they do not cover the entire set of natural numbers.

1975.26. See Question 21.

1975.27. Solve the equation $x^2 + 2 = 4\sqrt{x^3 + 1}$.

1975.28. Four different balls of the same radius are positioned in space. Prove that no three of them fully cover the remaining fourth ball.

1975.29. See Question 24.

1975.30. A convex n-gon lies inside 1×1 square. Prove that there exist three consecutive vertices of that polygon such that the area of the triangle formed by these vertices is at most $8/n^2$.

ELIMINATION ROUND

GRADE 8

1975.31. Numbers a, b, c, and d satisfy equalities $a^2 + b^2 = c^2 + d^2 = 1$ and $ac + bd = 0$. Find $ab + cd$.

1975.32. Which of the two numbers is greater:

$$2^{2^{2^{.^{.^{.^{2}}}}}} \quad \text{or} \quad 3^{3^{3^{.^{.^{.^{3}}}}}} \quad ?$$

(100 twos)   (99 threes)

1975.33. Find natural numbers a, b, and c such that all three numbers $a^b + c$, $b^c + a$, and $c^a + b$ are prime.

1975.34. Given a set of several points on the plane, it is permitted to add a point which is symmetric to one of the points with respect to the perpendicular bisector of a segment connecting any two other points of that set. If we start with a set of three points all distances between which are smaller than 1, can we obtain a set which has two points with distance between them greater than 1?

1975.35. There are thirty people in the room. Each one of them likes exactly k other people in the room. What is the smallest possible value of k which would allow us to claim with absolute certainty that there must always be two people in the room who like each other?

1975.36. Eight vertices of a regular polygon with 35 sides are selected. Prove that some four of them form either a trapezoid or a rectangle.

1975.37.* Some cities in the country are connected with highways, each one not longer than 500 km, in such a way that for any two cities it is possible to get from one to another by highway having traveled no more than 500 km. One of the roads was closed for repair, but it is still true that one can travel from any city to any other via the highways. Prove that now it can be done by driving no more than 1500 km.

1975.38.* Centers of all 64 centers of the chessboard squares are marked. Is it possible to dissect the chessboard with thirteen straight cuts so that every piece of the board will have no more than one marked point in it?

GRADES 9–10

1975.39. See Question 31.

1975.40. Do there exist four pairwise disjoint balls in space not containing the given point O, and such that any ray starting from O intersects at least one of these balls?

1975.41. See Question 37.

1975.42.* Does there exist a one-to-one mapping $f : \mathbb{R} \to \mathbb{R}$ such that

$$f(x) + f^{-1}(x) = -x$$

for any x?

1975.43. See Question 38.

1975.44.* Sequence of integers x_0, x_1, x_2, ..., is such that $x_0 = 0$, and $|x_n| = |x_{n-1} + 1|$ for every natural n. What is the minimum possible value of expression $|x_1 + x_2 + \cdots + x_{1975}|$?

1975.45. Points A_1, B_1, and C_1 on sides BC, AC, and AB of triangle ABC are chosen in such a way that segments AA_1, BB_1, and CC_1 have common point D. Point E is the intersection of A_1C_1 and BB_1. Prove that if $|BD| = 2\,|B_1D|$, then $|BE| = |B_1E|$.

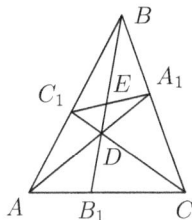

1975.46.* Each one of numbers x_0, x_1, ..., x_n is equal to either 0 or 1. Prove that

$$x_0 + \frac{x_1}{(\sqrt{2})} + \frac{x_2}{(\sqrt{2})^2} + \cdots + \frac{x_n}{(\sqrt{2})^n} \leqslant (1 + \sqrt{2})\sqrt{x_0 + \frac{x_1}{2} + \cdots + \frac{x_n}{2^n}}.$$

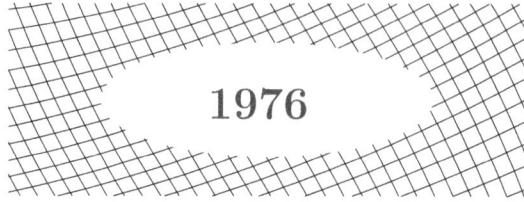

1976

GRADE 6

1976.01. A flea sits in one of the three hundred points arranged around the circle. It starts jumping from point to point in the counterclockwise direction. With the first jump, it moves to the next point, then it jumps over one point, then over two, and so on. Prove that one of those 300 points will never be visited by the flea.

1976.02. Numbers from 1 through 9 are split into three groups, three numbers in each. For every group we compute the product of its three numbers, with a being the largest of the three products. What is the smallest possible value of a?

1976.03. Villages A, B, and C are situated in the vertices of an equilateral triangle. There are 100 schoolchildren living in A, 200 schoolchildren in B, and 300 schoolchildren in C. Where do we build the school so that the total distance traveled by the schoolchildren every day has the smallest possible value?

1976.04. Two tokens are placed onto the end squares of 1×30 strip. Two people play the game making their moves in turn. Each move consists of moving one's own token[67] one or two squares in either direction; it is not allowed to jump over the other player's token or to place the tokens on the same square. The player who cannot make a move loses. What is the winning strategy for the first player?

1976.05. Table with dimensions 11×11 is drawn on the sheet of graph paper. The centers of some of the unit squares are selected in such a way that the center of every other square belongs to a segment connecting some two of the selected points which both lie in either the same row or in the same column. What is the minimum number of selected centers required to satisfy that condition?

1976.06. A square plot of land has a fence erected along its boundary. There are also some other fences inside the plot which split it into several smaller square-shaped plots with integer dimensions. Prove that the total length of all the fences is divisible by 4.

---

[67] Clearly, it is assumed that the players "own" different tokens.

GRADE 7

1976.07. A flea sits in one of the 101 points arranged around the circle. It starts jumping from point to point in the counterclockwise direction. With the first jump it moves to the next point, then it jumps over one point, then over two and so on. Prove that one of those 101 points will never be visited by the flea.

1976.08. See Question 2.

1976.09. A grasshopper made several jumps around the plane using the following rules. The first jump was 1 cm long, the second one was 2 cm long, the third was 3 cm long, and so on. After every jump, the grasshopper changed the direction of the jump by 90 degrees. At some moment, the grasshopper decided to come back to the point of origin. Is it always possible?

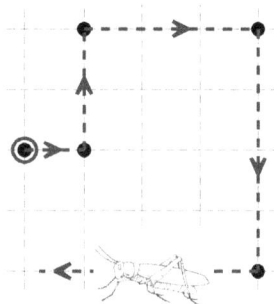

1976.10. The perimeter of a five-pointed star with the vertices coinciding with those of the given convex pentagon F, the perimeter of F itself, and the perimeter of the smaller pentagon formed inside the star are all prime numbers. Prove that their sum is greater than or equal to 20.

1976.11. See Question 5.

1976.12. See Question 6.

GRADE 8

1976.13. Given sequence of integers $x_1, x_2, \ldots, x_{25}$, another sequence $y_1, y_2, \ldots, y_{25}$ is obtained from it by some permutation of the terms. Prove that the product $(x_1 - y_1)(x_2 - y_2) \ldots (x_{25} - y_{25})$ is even.

1976.14. Two three-digit numbers a and b are written one after another to form two six-digit numbers in two different ways: first—a then b, second—b then a. Prove that the difference between these six-digit numbers is not divisible by 1976 unless $a = b$.

1976.15. A race runner, two bicycle riders, and a biker run, ride, and drive, respectively, on the closed circular highway; each one of them moves with their own constant velocity; the runner and one of the cyclists move in the clockwise direction, while the other cyclist and the biker—in the counterclockwise direction. The runner meets the second cyclist every 12 minutes, the first cyclist passes the runner every 20 minutes, and the biker passes the second cyclist every 5 minutes. How often does the biker meet the first cyclist?

1976.16. Point A lies inside regular polygon P with 1976 sides, and point B lies outside of it. $\vec{X}_A$ is the sum of all vectors from A to the vertices of P, and $\vec{X}_B$ is the similar sum for point B. Could the length of vector $\vec{X}_A$ be greater than the length of vector $\vec{X}_B$?

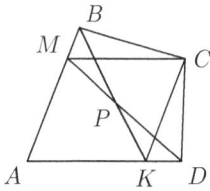

1976.17. Angle C is the largest in quadrilateral $ABCD$, point K is the intersection of line AD and the line parallel to AB passing through C; similarly, M is the intersection point of line AB and the line parallel to AD passing through C; finally, P is point of intersection of lines BK and MD. Prove that quadrilaterals $AMPK$ and $BCDP$ have equal areas.

1976.18. Three disjoint non-intersecting subsets of $\mathbb{N}$ are given. Prove that it is possible to find numbers x and y belonging to two different subsets such that their sum $x + y$ does not belong to the third subset.

GRADE 9

1976.19. In triangle ABC, side AC equals half-sum of the other two sides. Prove that the inradius of this triangle equals one third of one of its altitudes.

1976.20. Prove that for any two different natural numbers m and n, the smaller of the numbers $\sqrt[n]{m}$ and $\sqrt[m]{n}$ is less than $\sqrt[3]{3}$.

1976.21. See Question 14.

1976.22. Prove that for any $n \geqslant 5$ there exists convex polygon P with n sides of different length such that the sum of distances from point X to its sides (or their extensions) is the same for any point X inside P.

1976.23. King Bureaucrat the Mighty has twelve viceroys, and out of them he forms numerous committees following the two simple rules. First, any two committees must have at least one member in common. Second, there must be no identical committees. All in all, the king has created 1000 committees. Prove that he can form one more committee still satisfying the two rules above.

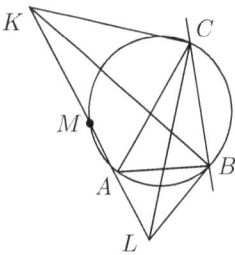

1976.24.* Point K in triangle ABC is the common point of the internal angle bisector for vertex B and the external angle bisector for vertex C. Similarly, L is the common point of the internal angle bisector for vertex C and the external angle bisector for vertex B. Prove that the midpoint of segment KL is the midpoint of the arc CAB on the circumcircle of triangle ABC.

GRADE 10

1976.25. Four segments of equal length form a closed broken line L in three-dimensional space. Prove that for any point X the distance from X to any of the vertices of L is less than the sum of the distances from X to the three other vertices of L.

1976.26. Find function f, defined on the set of all real numbers and satisfying the equality $f^2(x+y) = f^2(x) + f^2(y)$ for any numbers x and y.[68]

1976.27. Find all real solutions of the equation

$$\sqrt{a+bx} + \sqrt{b+cx} + \sqrt{c+ax} = \sqrt{b-ax} + \sqrt{c-bx} + \sqrt{a-cx},$$

if the solution set is known to be non-empty.

1976.28. Sides of a triangle equal a, b, and c. Prove the inequality

$$2 < \frac{a+b}{c} + \frac{b+c}{a} + \frac{c+a}{b} - \frac{a^3+b^3+c^3}{abc} \leqslant 3.$$

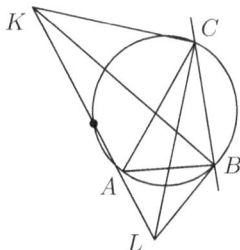

1976.29. A 5×5 table is filled with zeros and ones. The upper left and the lower right corner squares contain ones, two other corner squares—zeros. Prove that it is possible to find two different (but possibly intersecting) 2×2 subtables which carry identical number arrangements.

1976.30.* Point K in triangle ABC is the common point of the internal angle bisector for vertex B and the external angle bisector for vertex C. Similarly, L is the common point of the internal angle bisector for vertex C and the external angle bisector for vertex B. Prove that the midpoint of KL lies on the circumcircle of triangle ABC.

ELIMINATION ROUND

1976.31. All sides in convex pentagon $ABCDE$ have equal length; angle ACE equals half of angle BCD. Find angle ACE.

1976.32. A three-dimensional space is dissected into five non-empty subsets. Prove that there exists a straight line that intersects at least three of them.

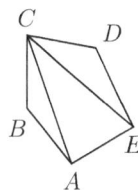

1976.33. Solve the following system of equations in real numbers.

$$\begin{cases} x_1 + x_2 = x_3^2 \\ x_2 + x_3 = x_4^2 \\ x_3 + x_4 = x_5^2 \\ x_4 + x_5 = x_1^2 \\ x_5 + x_1 = x_2^2. \end{cases}$$

1976.34. Two people play a game with a finite number of possible positions, where players in turn move the pawn from one position to another, and the player who cannot

---

[68] It is implicitly implied here that we are not looking for just one such function; we want them all!

make a move loses. For any position X, the sum of the number of all positions where the pawn can move from X and the number of all positions from which the pawn can move to X, is always the same. Prove that the number of losing positions is less than or equal to half of the total number of positions (position is called "losing" if the player who makes the first move from that position loses in the errorless game).

1976.35. Squares of a 100×100 table are painted in three colors. It is permitted to repaint any 2×2 square inside the table entirely into the color which prevails in it; if there is no such color, then the square can be repainted entirely into the color which is absent in it. Prove that it is possible to obtain the table with all squares of one color using these operations.

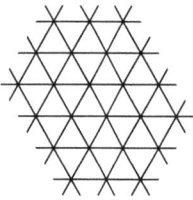

1976.36. The plane is dissected into equal equilateral triangles forming triangular grid. One hundred of them are colored black with these black triangles forming a connected figure. Prove that it is possible to find in this figure at least 33 non-overlapping rhombuses, each consisting of two adjacent triangles.

1976.37. Prove that among 1976-digit numbers, whose decimal representation consists of 1975 ones and one seven, at least 658 are composite.

1976.38.* Let F be a one-to-one mapping of the plane onto itself possessing the following property: if $ABCD$ is any convex non-degenerate quadrilateral, then quadrilateral $F(A)F(B)F(C)F(D)$ is also convex and non-degenerate. Prove that if points A, B, and C are collinear, then points $F(A)$, $F(B)$, and $F(C)$ are collinear as well.

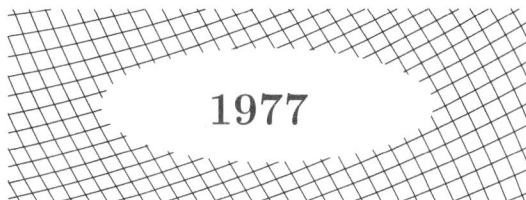

1977

GRADE 6

1977.01. Is it possible to dissect a square into 1977 triangles so that none of the triangles' vertices was interior to any of their sides, and every side of the square had the same number of these vertices?

1977.02. Twenty rooks are positioned on the chessboard so that each unoccupied square is under attack. Prove that twelve of them can be removed in such a way that the remaining eight rooks still attack the entire chessboard.

1977.03. A regular octagon is dissected into four parts of equal area by two straight cuts. Prove that the cuts are perpendicular.

1977.04. Number $111\ldots11$ (100 ones) is represented as

$$a_0 + a_1 \cdot 10^1 + a_2 \cdot 10^2 + \cdots + a_{99} \cdot 10^{99},$$

where $a_0, a_1, \ldots, a_{99}$ are non-negative integers whose sum does not exceed 100. Prove that they are all equal to 1.

1977.05. Sequence of integers 2, 3, 4, 6, 9, 13, 19, 28, 42, 63, 94, ..., begins with 2, and every subsequent number is obtained by multiplying the previous one by $\frac{3}{2}$ and rounding down. Prove that this sequence contains a six-digit integer.

1977.06. Two players in turn write digits into the 1×12 table until they end up with a twelve-digit number; use of digits 0 and 9 is prohibited. Prove that the second player can always ensure that the final result is divisible by 77.

GRADE 7

1977.07. See Question 1.

1977.08. Number 197719771977 is represented as

$$a_0 + a_1 \cdot 10^1 + a_2 \cdot 10^2 + \cdots + a_{11} \cdot 10^{11},$$

where $a_0, a_1, \ldots, a_{11}$ are non-negative integers whose sum does not exceed 72. Find these numbers.

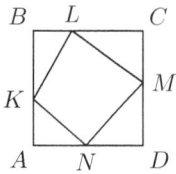

1977.09. Points K, L, M, N lie on sides AB, BC, CD, and DA of square $ABCD$, respectively. Prove that

$$|KL| + |LM| + |MN| + |NK| \geqslant 2\,|AC|.$$

1977.10. See Question 6.

1977.11. See Question 5.

1977.12. Point O inside convex polygon P is such that any line that passes through it divides P into two parts of equal area. Prove that O is the center of symmetry for P.

GRADE 8

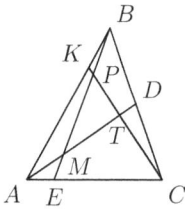

1977.13. Point D lies on side BC of triangle ABC, E is an arbitrary point on side AC, and K—on side AB. Lines AD and BE intersect at point M, and lines BE and CK—at P, lines CK and AD—at T. Prove that if $|BM| = |PE|$ and $|AT| = |MD|$, then $|CP| > |TK|$.

1977.14. Let A_1, A_2, ..., A_n be some sets of natural numbers. Prove that there exist natural numbers x and y such that each one of these sets either contains both x and y, or contains none of them.

1977.15. A grid point on the plane is a point whose coordinates are integers. Consider a convex polygon with all vertices being grid points which has no other grid points inside of it or on its boundary. How many vertices can such a polygon have?

1977.16. Consider all broken lines going along the grid lines and connecting the opposite corners of 100×100 sheet of graph paper by one of the shortest possible paths. What is the minimum number of such broken lines we need to select in order to cover all the vertices on the sheet?

1977.17. Segments A_1, A_2, ..., A_{1977} and B_1, B_2, ..., B_{1977} lie on the same straight line. Each segment A_k has a point in common with each one of segments B_{k-1}, B_{k+1}; also segment A_{1977} has a point in common with B_1, and A_1—with B_{1977}. Prove that for some index k, segments A_k and B_k intersect.

1977.18. Set of pairwise non-collinear vectors $\vec{a}_1$, $\vec{a}_2$, ..., $\vec{a}_n$ on the plane is such that for any two different indexes i and j, some vector in the set can be represented as $x\vec{a}_i + y\vec{a}_j$ with negative coefficients x and y. Prove that n is odd.

GRADE 9

1977.19. See Question 13.

1977.20. See Question 16.

1977.21. See Question 17.

1977.22. See Question 18.

1977.23. Function f is defined on $[0;1)$ by the following formula.

$$f(x) = \begin{cases} x + \frac{2-\sqrt{2}}{2}, & \text{if } x \in \left[0; \frac{\sqrt{2}}{2}\right) \\[2mm] x - \frac{\sqrt{2}}{2}, & \text{if } x \in \left[\frac{\sqrt{2}}{2}; 1\right). \end{cases}$$

Prove that for any segment $(a;b) \subset [0;1)$ there is point x on this segment and such natural number n that point $f(f(...f(x)...))$ (f is applied n times) lies inside $(a;b)$.

1977.24.* Several points are connected with arcs so that it is possible to get to any of these points from any other via these arcs. Prove that one can mark every arc with two arrows—one red and the other blue, pointing into opposite directions—so that it will still be possible to get from any point to any other by moving along the arcs in such a way that the color of the move direction arrow changes no more than once.

GRADE 10

1977.25. Prove that if $0 < x < \frac{\pi}{2}$, then $\sin x \cdot \tan x > x^2$.

1977.26. A grid point in three-dimensional space is a point all of whose coordinates are integers. Consider a convex polyhedron with all vertices being grid points which has no other grid points inside of it or on its boundary. How many vertices can such a polyhedron have?

1977.27. We are given three natural numbers n, k, p such that $p > 1$. Prove that at least one of the numbers $\binom{n}{k}$, $\binom{n+1}{k}$, ..., $\binom{n+k}{k}$ is not divisible by p.

1977.28. Prove that the sum of all dihedral angles of a triangular pyramid is greater than $360°$.

1977.29. Real numbers a_1, a_2, ..., a_{1977} satisfy inequalities $a_1 \geqslant a_2 \geqslant \cdots \geqslant a_{1977}$. Prove that

$$a_1^2 - a_2^2 + \cdots + (-1)^{1976} a_{1977}^2 \geqslant \left(a_1 - a_2 + a_3 - \cdots + (-1)^{1976} a_{1977}\right)^2.$$

1977.30. See Question 23.

Commentary.

a) *In 1977, the elimination round was not held; the Leningrad city team for the All-Union Olympiad was selected based on the results of the all-city round.*

b) *Question 26 in different archives was formulated slightly differently. It was either the one we have here (see above) or exactly the same as Question 15.*

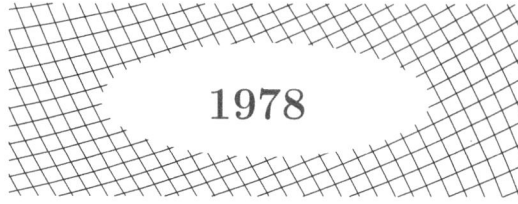

1978

GRADE 6

1978.01. Vertices of regular triangle are marked with numbers 1, 2, and 3. Can we assemble several such triangles on top of each other so that the sum of the numbers over each vertex equals 55?[69]

1978.02. A polygon is dissected by several diagonals into triangles which are colored black and white so that any two neighboring triangles have different colors. Prove that the number of black triangles does not exceed triple the number of white triangles.

1978.03. Is $57,599$ a prime number?

1978.04. What is the minimum number of chess kings such that for any placement of all of them on the 8×8 chessboard one can always find a square under attack by at least two of these kings?

1978.05. Can we arrange natural numbers 1 through 1978 in a row so that any two numbers standing next to each other as well as any two numbers with the same neighbor are co-prime?

1978.06. Angle B at the vertex of isosceles triangle ABC equals $20°$. Prove that (a) $3\,|AC| > |AB|$; (b) $2\,|AC| < |AB|$.

GRADE 7

1978.07. Five integers produce ten pairwise sums. Can these sums be ten consecutive integers?

1978.08. Natural number a was divided with remainder by all natural numbers less than itself. All these remainders were added up, and their sum is equal to a. Find a.

1978.09. See Question 4.

1978.10. An arbitrary point inside a square is connected by segments with all vertices of the square. Then from each vertex we drop a perpendicular onto the segment issuing from the vertex following this one in the clockwise direction. Prove that these four perpendiculars intersect at the same point.[70]

---

[69] This is the same problem as Question **1974.03**.
[70] This is the same problem as Question **1971.09**.

1978.11. One hundred three-digit integers are placed on the vertices of an inscribed regular polygon with 100 sides. Prove that there exists a diameter separating these numbers into two groups with the absolute value of the difference of the sums of the numbers in these groups not exceeding 900.

1978.12. Given convex polygon with $n \geqslant 5$ sides, prove that it possesses three sides whose extensions form the triangle containing the entire polygon.

GRADE 8

1978.13. Prove that a quadrilateral whose diagonals divide it into four triangles with equal perimeters is a rhombus.[71]

1978.14. A five-digit number is divisible by 41. Prove that any five-digit number that can be obtained from it by a circular transposition of its digits is also divisible by 41.[72]

1978.15. A bicycle track is a hexagon whose angles are all equal to $120°$, and the sides' lengths in kilometers are integers. A bicycle relay race was organized with the six sides serving as the race's stages. The women cyclists rode the first, the third, and the fifth stages, and men cyclists rode the second, the fourth, and the sixth stages. After the race, the women claimed that the total length of their stages was 3 km longer than the total length of the men's stages. The men then counterclaimed that in fact the total length of their stages was longer by 5 km. Which team is definitely wrong?

1978.16. Can integers be written into every square on the infinite sheet of graph paper in such a way that the sum of the numbers within every 1918×1978 rectangle equals 60?

1978.17. Six circumferences are drawn on the plane so that the first one touches the sixth and the second, the second one touches the first and the third, the third touches the second and the fourth, and so on. Prove that there exists a circumference which intersects all the above-mentioned six.

GRADE 9

1978.18. See Question 14.

1978.19. A polynomial with positive leading coefficient takes prime values only for the prime natural arguments. Prove that for any prime argument the value of the polynomial is prime.

1978.20. Positive numbers a_1, a_2, a_3, b_1, b_2, b_3 satisfy inequalities

$$\sum_{i \leqslant j} a_i a_j \leqslant 1, \quad \sum_{i \leqslant j} b_i b_j \leqslant 1.$$

---

[71] Same problem as Question **1972.15**.
[72] Same problem as Question **1963.20**.

Prove that

$$\sum_{i \leqslant j}(a_i - b_i)(a_j - b_j) \leqslant 1.$$

1978.21. See Question 16.

1978.22. (a) See Question 17. (b) Is the same statement true for eight circumferences?

GRADE 10

1978.23. See Question 19.

1978.24. What is the maximum possible number of equilateral triangles that can be formed by six straight lines on the plane?

1978.25. See Question 14.

1978.26. See Question 20.

1978.27. Every long diagonal of a convex hexagon divides the hexagon in two parts of equal area. Prove that these diagonals have a common point.

ELIMINATION ROUND

1978.28. Squares of a 100×100 table are colored into four colors in such a way that any two squares which have a common vertex are colored differently. Prove that all the corner squares have different colors.

1978.29. Let A and B be two finite sets on the plane. Denote $d_H(A, B) = \max(d_1; d_2)$, where d_1 is the largest distance from a point of A to set B, and d_2 is the largest of the distances from a point in B to set A. Prove that d_H complies with the triangle inequality; that is, prove that for any three finite planar sets X, Y, and Z we have

$$d_H(X, Y) + d_H(Y, Z) \geqslant d_H(X, Z).$$

1978.30. An integer is placed into every vertex of a regular polygon with 100 sides. Every minute, each number is replaced with the difference between this number and the one which follows it in the clockwise direction. Prove that in five minutes the sum of all the numbers will be divisible by 5.

1978.31. No two of 1978 segments given on the straight line have a common endpoint. Prove that these segments cannot be indexed in such a way that for every k from 1 through 1978, segment number k contains exactly k endpoints of other segments.

1978.32.* Sequence (a_n) consists of zeros and ones, and for any k, n such that $k < 2^n$, number a_k is different from a_{k+2^n}. Prove that this sequence is not periodic.

1978.33. Let M be a convex polygon. If H is a homothety with coefficient $(-\frac{1}{2})$, then prove that there exists a translation T such that polygon $T(H(M))$ lies inside M.

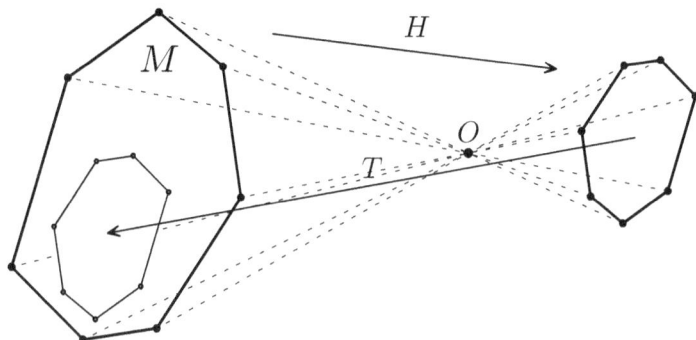

1978.34.* All vertices of a finite graph are colored using two colors. Each second every vertex changes its color to the one prevalent among its neighbors.[73] Prove that for every vertex there is a moment of time after which the vertex either stops changing the color or changes it every second.

1978.35.* All the sides and diagonals of convex polygon M have integer lengths. Let K be a square. Prove that there exists a finite collection of polygons congruent to M such that their union contains K, and every point of the square which does not lie on a side of these polygons is covered by the same number of the polygons from the collection.

Commentary.

a) *In 1978, contrary to the tradition, the all-city round was organized as a written contest. The prizes, however, were awarded based on the results of the elimination round (held, as always, in the form of oral competition); the same round also determined the team of the city for the All-Union Olympiad.*

b) *It is rather obvious that the problem committee encountered a certain dearth of introductory level problems for that year's contest. This is the only logical explanation of why so many such problems were borrowed from the LMOs of the previous years.*

---

[73] The likeliest version of the correct statement used at the olympiad should insert the following clause: *... but only if such a prevalent color exists. Otherwise the statement is false.*

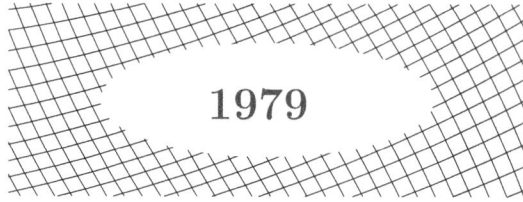

1979

Grade 5

1979.01. A 13×7 rectangle is drawn on graph paper. Show how to cut from it fifteen rectangles with dimensions 2×3.

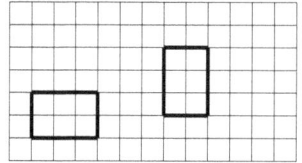

1979.02. What year was a person born if we know that her age this year will become equal to the sum of the digits of her birth year?

1979.03. Twenty-five squares in a white 6×7 rectangular table are painted black. Prove that some 2×2 square inside that rectangle has at least three black squares.

1979.04. Ten cards with digits 0, 1, 2, ..., 9 are put into the bag.

a) Three random cards are taken out of the bag. Prove that some of these cards can be arranged so that the number they form (with 1, 2, or 3 digits) will be divisible by 3.

b) What is the number of cards, to be randomly taken out of the bag, that will guarantee us that some of them can be arranged to form a number divisible by 9?

1979.05. In the class, every boy is friends with exactly three girls, and every girl is friends with exactly two boys. There are 19 desks and 31 young pioneers in the class.[74] How many students are there?

Grade 6

1979.06. Integers a, b, and c are such that $a + b$ and ab are divisible by c. Prove that $a^3 - b^3$ is also divisible by c.

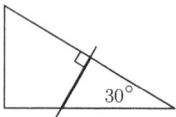

1979.07. One of the angles in the right triangle is $30°$. A perpendicular is erected from the midpoint of its hypotenuse. Prove that the length of the segment cut from this perpendicular by the triangle equals one third of the length of the larger leg of the triangle.

---

[74] At the time almost every student between the ages of 9 and 14 in the Soviet Union was a member of Young Pioneers organization; each school desk accommodated one or two students.

1979.08. How do you count off 9 minutes using a 5-minute hourglass and a 7-minute hourglass?

1979.09. A natural number is written into every square of a rectangular table. It is permitted to double all the numbers in any row, or to subtract one from all the numbers in any column. Prove that using these operations one can obtain a table with all zeros.

1979.10. What is the maximum number of positive integers less than 50 that can be selected in such a way that any two of them are co-prime?

1979.11. Find the sum of the digits of the cube of a number which is written by three ones and several zeros.

GRADE 7

1979.12. Solve the system of equations

$$\begin{cases} 1 + a + b = ab \\ 2 + a + c = ac \\ 5 + b + c = bc. \end{cases}$$

1979.13. In triangle ABC, segments AA_1, BB_1, and CC_1 are its altitudes, and AA_0, BB_0, CC_0—its medians. Prove that the length of broken line $A_0B_1C_0A_1B_0C_1A_0$ equals the perimeter of ABC.

1979.14. A 1000×1979 rectangle is dissected into unit squares by the grid lines. If we also make a cut along the rectangle's main diagonal, then how many parts will be there?

1979.15. Find all integers a such that a^6 is an eight-digit number written with digits 0, 1, 2, 2, 2, 3, 4, 4 in some order.

1979.16. Triangle ABC is inscribed in circumference S. A_1 is a midpoint of arc BC, B_1—of arc AC, and C_1—of arc AB. From segments A_1B_1, B_1C_1, A_1C_1 the sides of ABC cut smaller segments with midpoints M_1, M_2, M_3, respectively. Prove that points B_1, C_1, M_1, and M_3 lie on the same circumference.

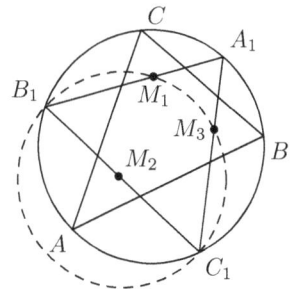

1979.17. Prove that there are infinitely many natural numbers which cannot be represented as a sum of a perfect square and a prime number.

GRADE 8

1979.18. Prove that for any natural $k > 1$, number $1010\ldots0101$ ($k + 1$ ones and k zeros) is composite.

1979.19. Sides of a triangle have lengths a, b, c. Prove that

$$\left| \frac{a}{b} + \frac{b}{c} + \frac{c}{a} - \frac{a}{c} - \frac{c}{b} - \frac{b}{a} \right| < 1 .$$

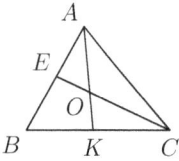

1979.20. In triangle ABC, angle B equals $60°$, segments AK and CE are angle bisectors, and point O is the intersection of AK and CE. Prove that $|OK| = |OE|$.

1979.21. $2n$ vectors connect the center of regular polygon with $2n$ sides to its vertices. How many of them do we need to add up to obtain the vector with the maximum possible length?

1979.22. Sequence a_1, a_2, ..., a_n consists of n pairwise co-prime natural numbers which satisfy inequalities $1 < a_i < (2n - 1)^2$. Prove that one of these numbers is prime.

1979.23. Number 1 is written in each square of $n \times m$ table. It is permitted to take any 2×2 subtable and change the sign of all the numbers in it. Is it possible, using these operations, to obtain a table with "chessboard" sign arrangement? The answer must depend on m and n.

GRADE 9

1979.24. Prove that if a k-digit number ($k \geqslant 2$) is prime, then either all numbers obtained from it by circular shifts of its digits are different or this number's digits are all ones.

1979.25. Consider all pairwise non-congruent triangles with the vertices on the points that divide the given circumference into n equal arcs ($n > 2$). For which values of n isosceles triangles constitute exactly half of these triangles?

1979.26. Pairwise co-prime natural numbers a_1, a_2, ..., a_n, where $n \geqslant 12$, are all greater than 1 and less than $9n^2$. Prove that one of these numbers is prime.

1979.27. A square hole with dimensions 1×1 is made in the center of the bottom of the box which has the shape of 5×5 square.[75] What is the minimum area of the convex figure (a "flap") which will entirely cover the hole regardless of its position on the bottom of the box?

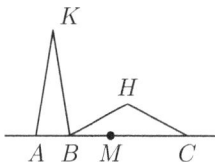

1979.28. Point B belongs to segment AC. Points K and H which lie to the same side of line AC are such that $|AK| = |KB|$, $|BH| = |HC|$, $\angle AKB = \alpha$, $\angle BHC = \pi - \alpha$. If point M is the midpoint of ACA, find the angles of triangle KHM.

---

[75] It was implied that the hole's sides are parallel to the sides of the box.

1979.29. Solve the system of equations

$$\begin{cases} x + \dfrac{x+2y}{x^2+y^2} = 2 \\[3mm] y + \dfrac{2x-y}{x^2+y^2} = 0. \end{cases}$$

GRADE 10

1979.30. See Question 24.

1979.31. See Question 27.

1979.32. For which natural values of y number $y^2 + 3^y$ is a perfect square?

1979.33. In a circle, some n sectors with angular value of each one of them less than $\frac{\pi}{n^2-n+1}$ are painted black. Prove that it is possible to rotate the circle so that all the painted sectors will be inside the unpainted portion of the circle.

1979.34. Given a convex polyhedron, prove that if all of its faces except one are centrally symmetric polygons, then this one remaining face must also be centrally symmetric.

1979.35.* Let AB and CD be the diameters of circumference S. Prove that for any points E and F on S, the point of intersection of lines AE and DF, the center of S, and the point of intersection of lines CE and BF all lie on the same straight line.

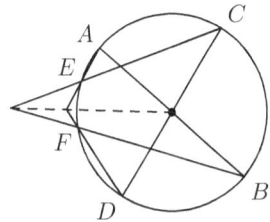

Commentary.

a) *At the 6th and the 7th grade contests the students who solved all the problems were offered two additional questions. In the 6th grade—Questions 17 and 36 (see below) and in the 7th grade—Questions 36 and 9. Incidentally, Question 36 is identical to Question **1970.02**.*

b) *In some archives, the problem set for the 9-10th grade contests had Question 18 instead of Question 24; Question 24 instead of Question 30; Question 26 instead of Question 31.*

c) *In 1979, there was no elimination round. The Leningrad city team for the All-Union Olympiad was determined by the results of the all-city round.*

1979.36. *In triangle ABC, all sides have integer lengths, and one of its angle bisectors is perpendicular to one of its medians. Find the sides of the triangle.*

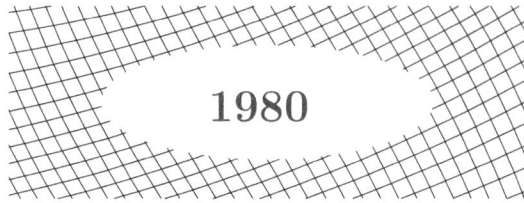

GRADE 5

1980.01. Is it possible to write numbers from 1 through 30 into the 5×6 table so that all the sums in the columns are equal?

1980.02. Ten-, eleven-, twelve- and thirteen-year-old boys attend a boy scout camp. There are 23 of them altogether and their total age is 253 years. How many twelve-year-old boys are there if there are 50% more of them than the thirteen-year-olds?

1980.03. Two hundred points are positioned on segment AB symmetrically with respect to the segment's midpoint. One hundred of them are colored red, the others—blue. Prove that the sum of the distances from red points to A equals the sum of the distances from blue points to B.

1980.04. We have nine coins, two of which are counterfeit. Genuine coins all weigh the same. Counterfeit coins also weigh the same—they are heavier than the genuine ones. Find both of them using four weighings on the two-pan scales.

1980.05. Dissect a square into convex pentagons.

1980.06. Bus stops in the city are located on the vertices and on the intersections of the diagonals of a convex polygon (no three streets-diagonals intersect at the same interior point). Some of the streets are pedestrian-only and others have one-way bus traffic established so that each stop has at least one bus line passing through it. Prove that one can get by bus from any stop to any other with no more than two transfers.

GRADE 6

1980.07. See Question 1.

1980.08. See Question 3.

1980.09. Is it possible to write integers 1 through 1980 in some order so that the sum of any two numbers which are separated by exactly one other number were divisible by 3?

1980.10. See Question 4.

1980.11. A bag of $2n$ candies is somehow distributed among n boxes. A girl and a boy start taking candies doing that in turns, each time taking one candy from one of the boxes;

the girl starts first. Prove that the boy can do it in such a manner that the last two candies will be taken from the same box.

1980.12. Find the way to color a sheet of graph paper into five colors so that every figure of shape A will contain all five colors but every figure of shape B will not.

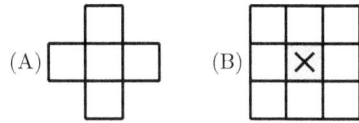

GRADE 7

1980.13. Find all trios of integers a, b, and c such that

$$a^2 - b^2 - c^2 = 1 \quad \text{and} \quad b + c - a = 3 \ .$$

1980.14. See Question 9.

1980.15. A polygon with n vertices is circumscribed around a circle. An arbitrary point inside the circle is connected by line segments to all vertices and all points of tangency. The resulting triangles are indexed in clockwise order by numbers 1 through $2n$. Prove that the product of the areas of all even-numbered triangles equals the product of the areas of all odd-numbered triangles.

1980.16. Square $ABCD$ with dimensions 9×9 is dissected into unit squares. All their vertices are colored in four colors, with 25 points of every color. Consider all vectors $\overrightarrow{M_1 A}$, where M_1 is a point of the first color, $\overrightarrow{M_2 B}$, where M_2 is a point of the second color, $\overrightarrow{M_3 C}$, where M_3 is a point of the third color, and $\overrightarrow{M_4 D}$, where M_4 is a point of the fourth color. Prove that the sum of all these vectors is zero vector.

1980.17. Prove that $53 \cdot 83 \cdot 109 + 40 \cdot 66 \cdot 96$ is composite.

1980.18. See Question 11.

1980.19. See Question 12.

GRADE 8

1980.20. Sum of four positive numbers a, b, c, and d equals 1. Prove the inequality

$$\sqrt{4a + 1} + \sqrt{4b + 1} + \sqrt{4c + 1} + \sqrt{4d + 1} < 6 \ .$$

1980.21. Point O is the incenter of triangle ABC. Points M and K are selected on sides AC and BC, respectively, in such a way that $|BK| \cdot |AB| = |BO|^2$ and $|AM| \cdot |AB| = |AO|^2$. Prove that points M, O, and K lie on the same straight line.

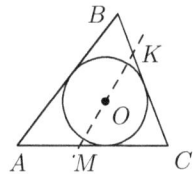

1980.22. Three athletes play table tennis using one table, and when a player loses, she yields her place at the table to the player who was sitting that game out, and so on.[76] At the end of the day it turned out that the first athlete played 10 games and the second one—21 games. How many games did the third athlete play?

1980.23. Is it possible to fill the squares of 8×8 board with natural numbers 1 through 64 so that the sum of the numbers in any T-shape tetromino ⬜ is divisible by 5?

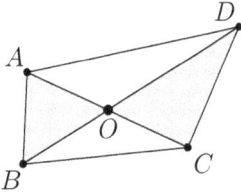

1980.24. A convex quadrilateral $ABCD$ is split into four triangles by its diagonals which intersect at point O. Given that
$$S(AOB)^2 + S(COD)^2 = S(BOC)^2 + S(DOA)^2,$$
where S denotes the area of a figure, prove that point O is a midpoint of at least one of the diagonals.

1980.25. What is the minimum value of n such that the decimal representation of some proper fraction m/n can contain sequence $\ldots 501 \ldots$?

1980.26. Two out of the nine coins are counterfeit. Genuine coin weighs 10 g, and counterfeit—11 g. Determine the counterfeit coins with five weighings on a one-pan scale.

GRADE 9

1980.27. Angle A in triangle ABC equals two times angle B. Prove that $|BC|^2 = (|AC| + |AB|) \cdot |AC|$.

1980.28. Three athletes play table tennis using one table, and when a player loses a game, she yields her place at the table to the player who was sitting that game out, and so on. At the end of the day it turned out that the first athlete played 10 games, the second one—15 games, and the third one—17 games. Who was the second game's loser?

1980.29. Find two unequal natural numbers whose arithmetic mean and geometric mean are two-digit numbers written with the same digits but in different order.

1980.30. Prove that if any value of x from segment $[0; 1]$ satisfies inequality $|ax^2 + bx + c| \leqslant 1$, then $|a| + |b| + |c| \leqslant 17$.

1980.31. A *midline* in a quadrilateral is one of the two segments connecting the midpoints of two opposite sides. Prove that if the sum of the midlines' lengths equals the quadrilateral's half-perimeter, then this quadrilateral is a parallelogram.

1980.32. See Question 25.

1980.33. Squares of a table with dimensions 8×8 are filled with real numbers. It is permitted to replace any two numbers with two copies of their arithmetic mean. Prove that, using these operations, it is possible to obtain the table where all numbers are the same.

---

[76] There are no draws in table tennis.

GRADE 10

1980.34. Three athletes play table tennis using one table, and when a player loses a game, she yields her place at the table to the player who was sitting that game out, and so on. At the end of the day it turned out that the first athlete won 10 games, the second one—12 games, and the third one—14 games. How many games did each one of them play?

1980.35. How many different numbers occur in the sequence

$$\left[\frac{1^2}{1980}\right], \left[\frac{2^2}{1980}\right], \ldots, \left[\frac{1980^2}{1980}\right] ?$$

1980.36. Do there exist numbers a and b such that for any $x \in [0; 2\pi]$ inequality

$$f^2(x) - f(x)\cos x < \frac{1}{4}\sin^2 x,$$

where $f(x) = ax + b$, holds true?

1980.37. See Question 31.

1980.38. Squares of a table with dimensions $n \times n$ are filled with real numbers. It is permitted to replace any two numbers with two copies of their arithmetic mean. Find all numbers n possessing the following property: starting from any initial collection of numbers it is possible, using these operations, to obtain the table where all numbers are the same.

1980.39.* Prove that any two points on the surface of a regular tetrahedron with edges of length 1 can be connected with a broken line of length not exceeding $\frac{2}{\sqrt{3}}$ that lies entirely on the surface of the tetrahedron.

1980.40.* Two cockroaches sit on the contour of a convex polygon. At the same moment they start moving around the contour in the same direction with the same speed. What is the initial position of the cockroaches for which the minimum distance between the cockroaches attained during this motion is the largest possible?[77]

ELIMINATION ROUND

1980.41. High school students from the grades 8, 9, and 10 take part in the All-Union Olympiad. By the regulations, the Leningrad team to the All-Union contest consists of k winners of this year's All-City Olympiad and the winners of the previous year All-Union Olympiad. What is the maximum possible size of the Leningrad team?

1980.42. Prove that any numbers x, y, z from segment $[0; 1]$ satisfy inequality

$$3(x^2y^2 + x^2z^2 + y^2z^2) - 2xyz(x + y + z) \leqslant 3.$$

1980.43. Inside the given triangle find the point with the maximum product of the lengths of perpendiculars dropped from that point onto the sides of the triangle.

---

[77] Compare this problem with Question **1973.39**.

1980.44. A survey was conducted in which people around the country voted for their favorite writer, artist, and composer. It turned out that every celebrity, who has been voted for at least once, received exactly k votes. Prove that all the survey participants can be split into $3k - 2$ groups so that in each group any two people have completely different preferences.[78]

1980.45. Prove that the length of the angle bisector which divides the largest side of a triangle does not exceed the length of the altitude dropped at the shortest side of that triangle.

1980.46. The vertices of a convex polygon with odd number of sides are colored in three colors so that every two neighboring vertices have different color. Let us call a triangle *diverse* if its vertices carry all three colors. Prove that the polygon can be dissected into diverse triangles by several diagonals without common interior points.

1980.47. What is the maximum number of non-overlapping rectangular parallelepipeds $1 \times 1 \times 4$, with their faces parallel to the cube's faces, that can be placed inside a $6 \times 6 \times 6$ cube?

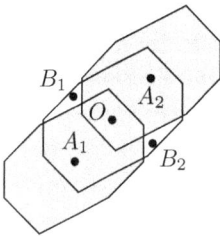

1980.48.* Point O is the center of symmetry for convex polygon F; A_1 and A_2, B_1 and B_2 are pairs of points in the polygon, symmetric with respect to O. The union of polygons obtained from F by translations by vectors $\overrightarrow{OA_1}$ and $\overrightarrow{OA_2}$ does not contain points B_1 and B_2. Prove that the union of polygons obtained from F by translations by vectors $\overrightarrow{OA_1}$, $\overrightarrow{OA_2}$, $\overrightarrow{OB_1}$, and $\overrightarrow{OB_2}$ fully covers entire polygon F.

---

[78] Meaning that these two people voted for different people in each of the three categories.

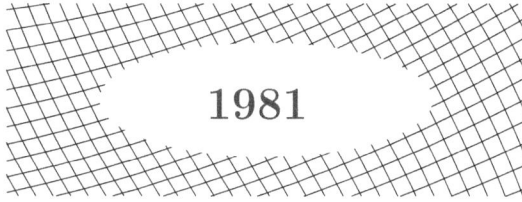

1981

GRADE 5

1981.01. Fifth graders Nick and Vera counted the grades they received during the last school semester. It turned out that Nick received as many 5s as Vera got 4s, as many 4s as Vera had 3s, as many 3s as Vera had 2s, and as many 2s as Vera had 5s. Also they found out that both of them received 54 grades, and their point averages were the same.[79] Prove that they have made a mistake somewhere in their calculations.

1981.02. Is it possible to arrange digits 1, 2, ..., 9 in a row so that there is an odd number of digits standing between 1 and 2, an odd number of digits between 2 and 3, ..., and an odd number of digits between 8 and 9?

1981.03. Points A, B, and C do not lie on the same straight line. Place nine more points inside triangle ABC and connect some of these twelve points by non-intersecting line segments in such a way that $\triangle ABC$ is dissected into triangles, and each of the twelve points is connected to exactly five others.

1981.04. A 12×12 square is drawn on a sheet of graph paper. What is the minimum number of its squares which can be painted black to make it impossible to fit an L-shape trimino into the unpainted portion of the square?

1981.05. Does there exist a natural number n such that the decimal representation of n^2 begins with digits 123456789?

1981.06. The sum of four rubles was paid out by coins of 1 kopecks, 2 kopecks, 5 kopecks, and 10 kopecks. Prove that it is possible to pay three rubles out of the same set of coins.

GRADE 6

1981.07. Prove that for any natural number n, $\dfrac{10^n - 1}{81} - \dfrac{n}{9}$ is an integer.

1981.08. Point C lies inside right angle XOY. Point A is selected on ray OX, and point B is selected on ray OY. Prove that the perimeter of triangle ABC is greater than $2\,|OC|$.

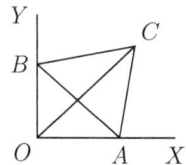

---

[79] In the Soviet school point system, 5 is equivalent to A, 4—to B, 3—to C, and 2—to D.

1981.09. See Question 4.

1981.10. See Question 6.

1981.11. Number n is a positive integer such that $n^2 + 1$ is a ten-digit number. Prove that there are two equal digits in its decimal representation.

1981.12. Every vertex of a cube is labeled by an integer. It is permitted to increase by 1 any two numbers written at the endpoints of any edge of the cube. Using these operations, is it possible to make all these numbers divisible by three if at the beginning one of the numbers is 1 and all the others are zeros?

Grade 7

1981.13. Does there exist a natural number which would be decreased by the factor of 1981 were the first digit in its decimal representation removed?

1981.14. See Question 8.

1981.15. See Question 4.

1981.16. Let m be an arbitrary natural number greater than 3, and S is the sum of all natural numbers x which do not exceed m and such that $x^2 - x + 1$ is divisible by m. Prove that S is divisible by $m + 1$ (if such numbers do not exist, we assume that $S = 0$).

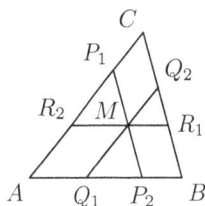

1981.17. Point M is chosen inside triangle ABC. Segments $P_1 P_2$, $Q_1 Q_2$, and $R_1 R_2$ all pass through M, have their endpoints on the sides of the triangle, and are parallel to sides BC, AC, and AB, respectively. Prove that

$$\frac{|P_1 P_2|}{|BC|} + \frac{|Q_1 Q_2|}{|AC|} + \frac{|R_1 R_2|}{|AB|} = 2.$$

1981.18. See Question 12.

Grade 8

1981.19. Does there exist a set of 1981 consecutive integer numbers whose sum is a perfect cube?

1981.20. A square is dissected into several rectangles with sides parallel to the sides of the square. For each rectangle, we compute the ratio of the smaller side to the larger side. Prove that the sum of these ratios is greater than or equal to 1.

1981.21. Prove that if $x^2 + xy + xz < 0$, then $y^2 > 4xz$.

1981.22. Points C_1 and C_2 are selected on side AB of triangle ABC, points A_1 and A_2—on side BC, points B_1 and B_2—on side AC in such a way that

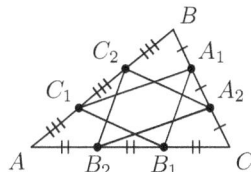

$$|AC_1| = |C_1C_2| = |C_2B| = |AB|/3,$$
$$|BA_1| = |A_1A_2| = |A_2C| = |BC|/3,$$
$$|CB_1| = |B_1B_2| = |B_2A| = |CA|/3.$$

Prove that intersecting triangles $A_1B_1C_1$ and $A_2B_2C_2$ produces six congruent triangles.

1981.23. Triangle ABC on an infinite sheet of graph paper with the square side 1 has all its vertices on the grid lines intersections. Prove that if $|AB| > |AC|$, then $|AB| - |AC| > 1/p$, where p is the perimeter of ABC.

1981.24.* Squares of the 8×8 board are painted into two colors using the standard chess coloring. A checker can stand only on the black squares, and it can move to any square which is a diagonal neighbor of its current position. What is the minimum number of moves necessary to visit all the black squares?

GRADE 9

1981.25. In some convex quadrilateral, the sum of the distances from any point inside of it to its sides (or their extensions) is always the same. Prove that this quadrilateral is a parallelogram.

1981.26. See Question 20.

1981.27. See Question 24 for 9×9 board (the lower left corner is black).

1981.28. Sequence of natural numbers $1 < a_1 < a_2 < \cdots < a_n < \cdots$ is such that $a_{n+a_n} = 2a_n$ for any natural n. Prove that there exists natural number c such that $a_n = n+c$ for any n.

1981.29. Integers a, b, c, d, and A satisfy equalities $a^2 + A = b^2$ and $c^2 + A = d^2$. Prove that number $2(a + b)(c + d)(ac + bd - A)$ is a perfect square.

1981.30. See Question 23.

GRADE 10

1981.31. In each of the two given congruent regular hexadecagons[80] some seven vertices are marked. Prove that it is possible to place one of them on top of the other so that at least four of the marked vertices of one of them will coincide with the marked vertices of the other.

---

[80] *Hexadecagon* is a polygon with sixteen vertices.

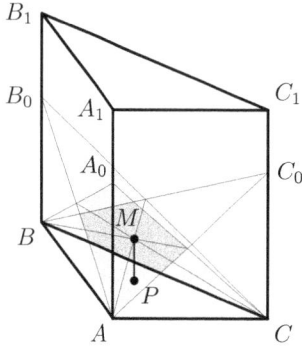

1981.32. See Question 23.

1981.33. See Question 29.

1981.34. Points A_0, B_0, and C_0 are selected on the side edges AA_1, BB_1, CC_1 of triangular prism $ABCA_1B_1C_1$, respectively. Let us denote $a = |AA_0|$, $b = |BB_0|$, $c = |CC_0|$. Point M is the intersection of planes A_0BC, B_0AC, and C_0AB. Point P on the prism's base ABC is such that segment MP is parallel to the side edges. Denoting $|MP|$ by d, prove the equality

$$\frac{1}{d} = \frac{1}{a} + \frac{1}{b} + \frac{1}{c}.$$

1981.35. See Question 24 for 10×10 board.

1981.36.* Does there exist a power of number 5 which has at least 30 zeros in the lowest one hundred places of its decimal representation?

ELIMINATION ROUND

1981.37. Find all natural numbers p for which number

$$2^{2^{\cdot^{\cdot^{\cdot^2}}}} + 9$$

(p digits 2) is prime.

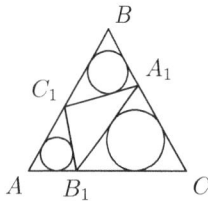

1981.38. Points C_1, B_1, and A_1 are selected on sides AB, AC, and BC, respectively, of regular triangle ABC with the side length 2. What is the maximum value of the sum of the inradii of triangles AB_1C_1, A_1BC_1, and A_1B_1C?

1981.39. There is a pile with m stones, and two people play the game making their moves in turn. On the kth move, a player is allowed to remove from 1 through k stones from the pile. The player who removes the last stone wins the game. Find all values of m for which the first player has a winning strategy.

1981.40. Prove that there are no rational numbers a, b, c and d such that

$$(a + b\sqrt{3})^4 + (c + d\sqrt{3})^4 = 1 + \sqrt{3}.$$

1981.41. Sixteen squares on the white 8×8 board are painted black, so that every column, as well as every row, contains exactly two black squares. Prove that it is impossible to reduce the total number of black squares using operations of repainting of columns and rows (by *repainting* we mean changing the color of all squares in a column or in a row).

1981.42.* Do there exist six six-digit numbers using digits 1 through 6 without repetitions such that any three-digit number which uses digits 1 through 6 without repetitions, can be obtained from one of these numbers by crossing out some of its digits?

1981.43. On the board, p numbers, each one equal to $+1$ or -1, are written in a row. It is permitted to change the sign of several consecutive numbers.[81] What is the minimum number of such operations which will allow to turn any such collection into the one consisting of $+1$'s only?

1981.44.* Is it possible to find five points in space such that all the pairwise distances between them are different, but the perimeter of any spatial pentagon with these points as vertices is always the same?

1981.45. Numbers a, b, and c lie inside interval $[0; 1]$. Prove the inequality

$$\sqrt{a(1-b)(1-c)} + \sqrt{b(1-a)(1-c)} + \sqrt{c(1-a)(1-b)} \leqslant 1+\sqrt{abc}.$$

1981.46.* Let $a_1, a_2, \ldots, a_7$; $b_1, b_2, \ldots, b_7$; $c_1, c_2, \ldots, c_7$ be three integer sequences. Prove that it is possible to remove several triples of numbers with identical indexes (but not all of them) so that in each sequence the sum of the remaining numbers is divisible by 3.

---

[81] For example, you can change the sign of any one specific number.

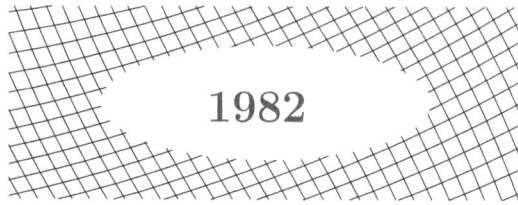

1982

GRADE 5

1982.01. A six-digit phone number is given. How many seven-digit phone numbers are there which will produce this given number if one of their digits is crossed out?

1982.02. A grasshopper, sitting on a straight line, jumps 1 cm, then 3 cm in the same or in the opposite direction, then 5 cm in the same or in the opposite direction, and so on. Could it end up at the point where it started after 25 such jumps?

1982.03. Some of the squares in a 5×5 table are colored red, others—blue. Prove that one can find four squares of the same color which lie on the intersection of some two rows and some two columns.

1982.04. Prove that if the sum of two natural numbers is 770, then their product is not divisible by 770.

1982.05. Place numbers 1, 2, 3, ..., 12 on the edges of a cube so that the sums of the numbers on every face of the cube are the same.

1982.06. Two people play the following game making their moves in turn. There are several piles of stones. With each move, a player must split one of the piles which has more than one stone in it into two smaller piles. When every group contains exactly one stone, the game stops, and the player who made the last move wins. Prove that if originally there is only one pile with 1000 stones, then the first player wins in the errorless game.

GRADE 6

1982.07. Two different two-digit numbers a and b end with the same digit. When dividing a by 9, the integer quotient is the same as the remainder from integer division of b by 9, and when dividing b by 9, the integer quotient is the same as the remainder from integer division of a by 9. Find all such pairs a and b.

1982.08. See Question 3.

1982.09. Prove that if the sum of two natural numbers is 30,030, then their product is not divisible by 30,030.

1982.10. A unit circumference and 1982 points are drawn on the plane. Prove that there is a point on this circumference such that the sum of the distances from this point to all the given points is greater than 1982.

1982.11. A grasshopper, sitting on the plane, makes a 1 cm long jump. Then it turns by 90 degrees and makes a 2 cm long jump, again turns at the right angle and makes a 3 cm long jump, and so on. Is it possible that after 1982 jumps it finds itself at the starting point?

1982.12. At a school discotheque, none of the boys danced with all the girls; however, every girl there danced with at least one boy. Prove that there are two pairs B_1, G_1 and B_2, G_2, such that boy B_1 danced with girl G_1, and boy B_2 danced with girl G_2, but B_1 did not dance with G_2, and B_2 did not dance with G_1.

GRADE 7

1982.13. Two positive numbers a and b satisfy inequality $\left(\dfrac{1+ab}{a+b}\right)^2 < 1$. Prove that one of the numbers, a or b, is greater than 1, and the other one is less than 1.

1982.14. Prove that for any four angles in a convex polygon the sum of any two of them is greater than the difference of the two others.

1982.15. Prove that number $222\ldots22$ (1982 twos) cannot be represented as $xy(x+y)$, where x and y are integers.

1982.16. Several sixth and seventh graders participated in a round-robin table tennis tournament. The number of the sixth graders was twice the number of the seventh graders. The number of games won by the seventh graders was 40% higher than the number of games won by the sixth graders. How many students participated in the tournament?

1982.17. Write nine different natural numbers not exceeding 40 into the squares of a 3×3 table so that all the products of the numbers in every row, in every column, and on two main diagonals are the same.

1982.18. See Question 12.

GRADE 8

1982.19. Let p_1, p_2, and p_3 be quadratic polynomials with positive leading coefficients. Prove that if every two of them have a common real root, then quadratic polynomial $p_1 + p_2 + p_3$ has a real root.

1982.20. In triangle ABC, the measure of angle C is twice that of angle A and $|AC| = 2|BC|$. Prove that ABC is a right triangle.

1982.21. Sequence of digits begins with 1, 9, 8, 2, and every next digit is the last digit of the sum of the previous four digits. In this sequence, will we ever encounter quadruple 3, 0, 4, 4 in exactly this order?

1982.22. The angle between any two diagonals of a convex polygon with 180 sides has a value equal to some integer number of degrees. Prove that this polygon is regular.

1982.23. Squares of 5×41 rectangle are colored in two colors. Prove that it is possible to choose three rows and three columns such that all the nine squares on their intersections have the same color.

1982.24. The plane is dissected into several parts by $2n$ ($n > 1$) straight lines in general position (no two of these lines are parallel, and no three of them intersect at the same point). Prove that there are at least $2n - 1$ angles among these parts.

GRADE 9

1982.25. See Question 20.

1982.26. See Question 21.

1982.27. See Question 22.

1982.28.

$$\frac{a^2 + b^2 - c^2}{2ab} + \frac{a^2 + c^2 - b^2}{2ac} + \frac{b^2 + c^2 - a^2}{2bc} = 1 \, .$$

Prove that two of these fractions are equal to 1 and the third—to (-1).

1982.29. Prove that for any natural number k there exists natural number n such that $\sqrt{n + 1981^k} + \sqrt{n} = (\sqrt{1982} + 1)^k$.

1982.30.* Several small chips are placed on the plane so that not all of them lie on the same straight line. It is allowed to move any chip to the point symmetric to its position with respect to any other chip. Prove that it is possible, using these operations, to get all the chips into the vertices of a convex polygon.

GRADE 10

1982.31. See Question 21.

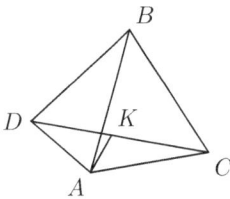

1982.32. See Question 23.

1982.33. See Question 28.

1982.34. Sum of the angles BAC and BAD in tetrahedron $ABCD$ equals $180°$. AK is the angle bisector of angle CAD. Find angle BAK.

1982.35. Prove that it is possible to label every vertex of regular polygon with n sides, using non-zero numbers, in such a way that for every subset of vertices, which forms a regular polygon with k vertices ($1 < k \leqslant n$), the sum of the numbers in these vertices is zero.

1982.36.[*] Some $4n$ points on the circumference are alternatingly colored red and blue. Then points of each color are somehow divided into pairs, and the points in every pair are connected with a line segment which is colored the same as its endpoints. No three of these segments intersect at the same point. Prove that there are at least n points where a red segment intersects a blue one.

ELIMINATION ROUND

1982.37. Does there exist natural number k such that out of any 180 vertices of a regular polygon with 360 sides centered at point O, one can always find two vertices A and B such that $\angle AOB$ equals k degrees?

1982.38. Square is dissected into $9801 = 99^2$ equal unit squares; all their centers, except for one corner square, are marked. These points are then divided into pairs with each pair defining a vector. Prove that the sum of these vectors is not zero.

1982.39. Ten points are selected on a circumference. What is the maximum number of line segments with ends in these points that can be drawn so that no three of them form a triangle with vertices in the selected points?

1982.40. Pete bought eight dumplings stuffed with rice or with cabbage and paid one ruble for them. Vicky bought nine dumplings and paid one ruble and one kopeck. How much is the rice dumpling if it is more expensive than the cabbage dumpling and both cost more than one kopeck?

1982.41. Let D and E be the points where incircle of triangle ABC touches sides BC and AC, respectively. Perpendicular BK is dropped from vertex B onto the bisector of angle A. Prove that points D, E, and K lie on the same straight line.

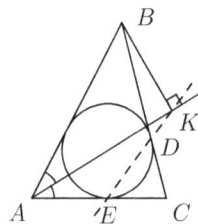

1982.42. Infinite sequence (a_n) consists of natural numbers such that for any n and k, difference $a_{n+k} - a_k$ is divisible by a_n. Let b_n denote the product $a_1 a_2 \cdots a_n$. Prove that b_{n+k} is divisible by $b_n b_k$ for any n and k.

1982.43.[*] Some n points on the plane are such that no three of which lie on the same straight line and no four of them—on the same circumference. Prove that there are at most $n - 2$ circumferences passing through some three of these points and containing all other points inside.

1982.44.[*] Prove that the set of all natural numbers greater than 1 cannot be split into two non-empty disjoint sets such that if any two numbers a and b belong to one of these sets, then number $ab - 1$ belongs to the same set.

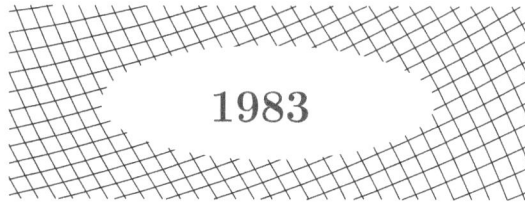

1983

GRADE 5

1983.01. Thirty people played the chess tournament. A qualification prize was awarded to anyone who obtained at least 60% of the best possible result. What is the maximum possible number of qualified participants?

1983.02. Each of the ten tiles with dimensions $10\,\text{cm} \times 20\,\text{cm}$ was cut into two triangular pieces. Arrange these twenty triangles to form a square.

1983.03. Each one of the four dwarfs—Blick, Flick, Glick, and Plick—either always tells the truth or always lies. One day the following conversation between them took place:
Blick (to Flick): You are a liar.
Glick (to Blick): You are a liar yourself.
Plick (to Glick): They are both liars. (Then, after a pause.) And so are you.
Which of the dwarfs are liars and which ones always tell the truth?

1983.04. Eight tokens were taken from a large bag with red and black tokens and then arranged around the circle. It is permitted to take any three consecutive tokens and replace each one of them with a token of the different color (taken from the bag). Prove that by using these operations, it is possible to have all the tokens in red.

1983.05. The time machine allows one to instantaneously transfer themselves from March 1 to November 1 of any other year, from April 1 to December 1, from May 1 to January 1, and so on. It is impossible to use the machine twice within one day. Baron Munchhausen[82] started his time travel on April 1 and immediately reappeared, returning back from his adventure. Then he announced that his trip took him 26 months.[83] Prove that he is wrong (as usual).

1983.06. Two numbers—the largest possible and the smallest possible—were formed from four different digits without repetitions. Sum of these numbers is $10,477$. Find the four original digits.

---

[82] A fictional German nobleman, famous for his wild exaggerations and fantastic lies, especially about his purported travels in Russia; see [28].
[83] That, of course, is understood to mean 26 months of his subjective time.

GRADE 6

1983.07. Even integer a possesses the following property: if it is divisible by a prime number p, then $a - 1$ is divisible by $p - 1$. Prove that a is a power of two.

1983.08. A two-meter stick is cut into five smaller sticks, each one of them at least 17 cm long. Prove that some three of them can form a triangle.

1983.09. In five-pointed star $ABCDE$, angle A is equal to angle D, angle B is equal to angle C, and segment AB is equal to segment CD. Prove that segments AE and DE are equal.

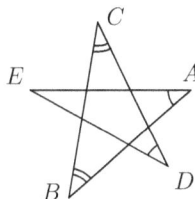

1983.10. Eight numbers, each one of them either $+1$ or -1, are placed around the circle. In one move you are allowed to change the signs for any three consecutive numbers. Prove that using these operations you can obtain any arrangement of ± 1 from any other.[84]

1983.11. See Question 5.

1983.12. Can we mark ten blue and twenty red points inside a triangle with blue vertices so that no three blue points lie on the same straight line and that every triangle with blue vertices contains at least one red point?

GRADE 7

1983.13. See Question 9.

1983.14. Each square on the sheet of graph paper with dimensions 100×100 is colored black or white so that the number of white squares is the same as the number of black ones. Prove that this sheet can be cut along the grid lines into two polygons so that each one of them contains equal numbers of black and white squares.

1983.15. A two-meter stick is cut into five smaller sticks, each one of them at least 19 cm long. Prove that some four of them can form a quadrilateral.

1983.16. See Question 12.

1983.17. Prove that number $2^{58} + 1$ can be represented as a product of three natural numbers greater than 1.

1983.18. A $+1$ or -1 is written into every square of a 24×24 table. It is permitted to change the signs of any one of these numbers together with all other numbers which share the same column or the same row with it. Prove that, using these operations, it is possible to obtain any sign arrangement from any other.

---

[84] This is, obviously, almost the same problem as Question 4.

GRADE 8

1983.19. Integer a is obtained from integer b by some permutation of its digits. Prove that the sum of the digits of $5a$ is equal to the sum of the digits of $5b$.

1983.20. Two grandmasters played 24 games of chess. None of the odd-numbered games ended in a draw, and no three consecutive games were won by the same grandmaster. What is the maximum possible number of points accumulated by the winner of the match?

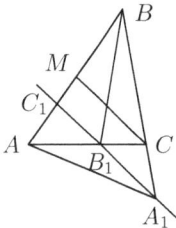

1983.21. A point that lies inside the given regular hexagon H is connected by line segments with all vertices of H. Prove that these six segments can be used to form another hexagon whose area is at least two thirds of the area of H.

1983.22. Line P is parallel to median CM of triangle ABC. Lines AB, BC, and AC intersect line P at points C_1, A_1, and B_1, respectively. Prove that the area of triangle AA_1C_1 is equal to the area of triangle BB_1C_1.

1983.23. Regular polygon with 400 sides is dissected into parallelograms. Prove that at least one hundred of them are rectangles.

1983.24.* Several points are positioned on a circumference. These points start moving along the circumference with equal velocities but possibly different directions, some—clockwise, others—counterclockwise. If two points collide, they bounce off each other, reversing their directions but keeping the same velocity and staying on the circumference. Prove that at some moment, all the points will be at exactly their original positions.

GRADE 9

1983.25. Let AH and CP be the altitudes in triangle ABC, $|AC| = 2|PH|$. Find angle B.

1983.26. We will call a point on the plane an *integer point* if both its coordinates are integers. We will call an integer point a *prime point* if both its coordinates are prime numbers. We will call a square whose vertices are all integer points a *prime square* if every integer point on its contour is a prime point. Find all prime squares.

1983.27. Several kids stand around the circle, each one of them holding a few candies. At a signal from the teacher, each one of them passes half of their candies to the kid standing to his or her right (every kid who has an odd number of candies gets one extra candy from the teacher before the signal is given). This is repeated again and again. Prove that sooner or later all the kids will have the same amount of candies.

1983.28. For which natural $n \geqslant 2$ inequality
$$x_1^2 + x_2^2 + \cdots + x_n^2 \geqslant x_n(x_1 + x_2 + \cdots + x_{n-1})$$
holds true for any sequence of real numbers (x_i)?

1983.29. Two circles K_1 and K_2 have no common interior points. Lines ℓ_1 and ℓ_2 touch these circles externally in four points which lie on the vertices of a circumscribed quadrilateral. Prove that circles K_1 and K_2 touch each other.

1983.30. See Question 24.

GRADE 10

1983.31. Two sequences of numbers (x_n) and (y_n) are defined by the following formulas: $x_1 = x_2 = 10$, $y_1 = y_2 = -10$, and

$$x_{n+2} = (x_n + 1)x_{n+1} + 1, \quad y_{n+2} = (y_{n+1} + 1)y_n + 1$$

for any natural n. Prove that any natural number occurs in no more than one of these two sequences.

1983.32. See Question 27.

1983.33. See Question 28.

1983.34. Trihedral angle T in space is such that some two angle bisectors of its face angles are perpendicular. Prove that the plane passing through these bisectors is perpendicular to the third face of T.

1983.35. Regular pyramid has a regular polygon with 20 sides as a base. The center of its inscribed sphere coincides with the center of the circumscribed sphere. Find the angles (in degrees) in any triangle that serves as a side face of the pyramid.

1983.36. See Question 24.

ELIMINATION ROUND

1983.37. Nine vertices are selected in a regular polygon with 20 sides. Prove that there exists an isosceles triangle with its vertices on some three of the selected points.

1983.38. Triangle ABC is given on the plane. Prove that for any point P on the plane the area of the polygon obtained by intersecting ABC with the triangle symmetric to ABC with respect to P does not exceed two thirds of the area of ABC.

1983.39. Two people play the game writing X's and O's, respectively, into the squares of an infinite sheet of graph paper, making their moves in turn. The first player's objective is to have four X's positioned in the vertices of a square with the sides parallel to the grid lines, and the second player's objective is to prevent that from happening. Does the first player have a winning strategy?

1983.40. Vertex angle A in isosceles triangle ABC equals $100°$. Point M is selected on ray AB so that segment AM equals base BC. Find angle BCM.

1983.41. For which n is it possible to write n numbers a_1, a_2, $\ldots$, a_n—not all of them equal to zero—around the circle so that for any $k \leqslant n$, the sum of k consecutive numbers starting from a_k equals zero?

1983.42. Infinite sheet of graph paper is covered by one layer of 1×2 rectangular tiles with sides on the grid lines. Prove that it can be covered by three more layers of such tiles with none of these tiles positioned exactly above some other tile.

1983.43.* Increasing sequence of natural numbers (a_n) is such that every natural number which is not a term of this sequence can be represented as $a_k + 2k$ for some natural k. Prove that $a_k < \sqrt{2k}$ for any k.

1983.44. In the country of Eldorado, there are n^2 towns positioned in the nodes of a rectangular grid in the form of a square. Distance between neighboring towns is 10 km. Towns are connected by the roads which are parallel to the sides of the square (but they do not have to go along the grid lines) so that one can drive from any town in Eldorado to any other town. What is the minimum total length of this road system?

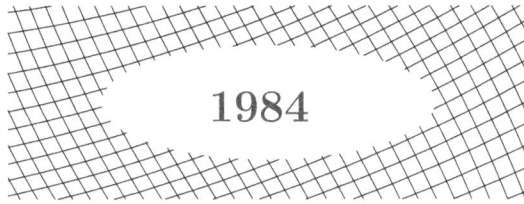

1984

GRADE 5

1984.01. Prove that it is possible to cross out a few digits at the beginning and a few digits at the end of the 400-digit number 84198419...8419 so that the sum of the remaining digits equals 1984.

1984.02. Place non-zero numbers into every square of 4×4 table so that the sum of the numbers in the corners of every 2×2, every 3×3, and every 4×4 square equals zero.

1984.03. Forty-five points are selected on line AB, all of them outside segment $[AB]$. Prove that the sum of the distances from these points to A cannot be equal to the sum of the distances from these points to B.

1984.04. Squares of a large sheet of graph paper are colored using eight different colors. Prove that there exists a figure ⌐⊔ which contains two squares of the same color.

1984.05. The circle is divided into six sectors with numbers written into them as it is shown in the figure. It is allowed to increase any two neighboring numbers by one. Prove that, using these operations, it is impossible to make all six numbers equal.

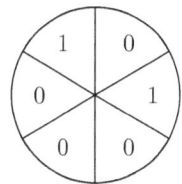

1984.06. Numbers 1 through 100 are written in a row in their natural order. Two players in turn insert signs "+", "−", and "×" in one of the spaces between these numbers (move consists of choosing both a space which has not been yet filled, and an arbitrary operator sign). Prove that the first player can ensure that after the last move is made the result of the computation will be odd.

GRADE 6

1984.07. See Question 3.

1984.08. Diagonals AC and BD of quadrilateral $ABCD$ intersect at point O. The perimeters of triangles ABC and ABD are equal, same as the perimeters of triangles ACD and BCD. Prove that $|AO| = |BO|$.

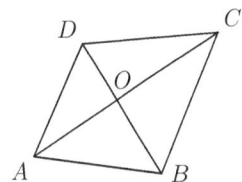

1984.09. (a) See Question 2. (b) Prove that for any such arrangement the sum of the numbers in each column must be equal to zero.

1984.10. In the gift shop, 175 Humpties cost more than 125, but less than 126 Dumpties. Prove that to buy three Humpties and one Dumpty, one ruble is not enough.

1984.11. See Question 6.

1984.12. Ninety-nine square tiles, each consisting of four unit squares, were cut out from the 29×29 sheet of graph paper. Prove that it is possible to cut out one more such tile.

GRADE 7

1984.13. Can segments with lengths 2, 3, 5, 7, 8, 9, 10, 11, 13, 15 serve as a set of sides and diagonals of a convex pentagon?

1984.14. Find the value of expression

$$\frac{1}{x^2 + 1} + \frac{1}{y^2 + 1} + \frac{2}{xy + 1},$$

where the sum of the first two summands equals the third, and x is not equal to y.

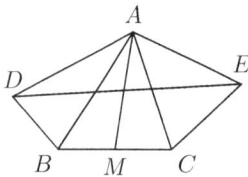

1984.15. Isosceles triangles ADB ($|AD| = |AB|$) and ACE ($|AC| = |AE|$) are built on sides AB and AC of triangle ABC with angle DAE equal to the sum of angles ABC and ACB. Prove that segment DE is twice as long as the median AM of triangle ABC.

1984.16. Forty-five points are selected on line AB, all of them outside segment AB. Several points are colored red, all the others—blue. Prove that the sum of the distances from red points to point A plus the sum of the distances from blue points to point B cannot be equal to the sum of the distances from red points to point B plus the sum of the distances from blue points to point A.

1984.17. See Question 12 for the sheet with dimensions 31×31.

1984.18. What is the minimum number of colors needed to color the squares of infinite sheet of graph paper so that in every figure ⊔⊓ all four squares were of different color?

GRADE 8

1984.19. Square table 100×100 is filled by 1×2 tiles without overlapping. Prove that some two of them form a 2×2 square.

1984.20. Difference of two six-digit numbers $\overline{abcdef}$ and $\overline{fdebca}$ is divisible by 271. Prove that $b = d$ and $c = e$.

1984.21. Real numbers x_1, x_2, $\ldots$, x_{100} are such that

$$x_1^3 + x_2 = x_2^3 + x_3 = \cdots = x_{100}^3 + x_{101} = x_{101}^3 + x_1 .$$

Prove that all these numbers are equal.

1984.22. Prove that the sum of the distances from an arbitrary point in the plane to some three vertices of an isosceles trapezoid is greater than the distance from that point to the fourth vertex.

1984.23. For some interior point K of median BM in triangle ABC, angles BAK and BCK are equal. Prove that triangle ABC is isosceles.

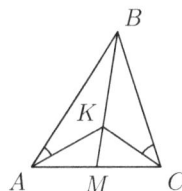

1984.24. Set A, consisting of natural numbers, is such that any one hundred consecutive natural numbers contain an element of A. Prove that there exist four different elements of A—a, b, c, and d—such that $a + b = c + d$.

GRADE 9

1984.25. See Question 21.

1984.26. See Question 22.

1984.27. Find the smallest natural number which can be represented as $13x + 73y$, where x and y are natural numbers, in at least three different ways.

1984.28. Segments AA_1, BB_1, and CC_1 are altitudes of triangle ABC. Find angles of this triangle if it is similar to triangle $A_1B_1C_1$.

1984.29. See Question 24.

1984.30. Sum of integers a, b, and c is zero. Prove that $2(a^4 + b^4 + c^4)$ is a perfect square.

GRADE 10

1984.31. See Question 22 for an arbitrary point in space.

1984.32. See Question 24.

1984.33. Prove that the difference of the squares of the lengths of adjacent sides of a parallelogram is less than the product of its diagonals.

1984.34. After several operations of differentiation and multiplication by $x + 1$ performed in some order, polynomial $x^8 + x^7$ was transformed into $ax + b$. Prove that the difference of integers a and b is divisible by 49.

1984.35. What is the maximum possible value of
$$|x_1 - 1| + |x_2 - 2| + \cdots + |x_{63} - 63|,$$
if sequence $\{x_1, x_2, \ldots, x_{63}\}$ is a permutation of $\{1, 2, 3, \ldots, 63\}$?

1984.36. Numbers 1, 2, 3, ..., 100 are placed on the vertices of 50-gonal prism. Prove that the prism has an edge such that the numbers at its endpoints differ by at most 48.

FINAL ROUND

GRADES 8,9

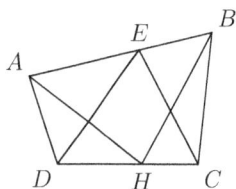

1984.37. Points E and F are selected on sides AB and CD of convex quadrilateral $ABCD$. Prove that if triangles ABH and CDE have the same area, and $|AE| : |BE| = |DH| : |CH|$, then line BC is parallel to line AD.

1984.38. Two people play the following game. The first player writes a digit on the board. Then the second player adds another digit on either side of the first digit. Then the first player adds another digit on either side of the resulting two-digit number and so on. Prove that the first player can play in such a way that after every move by the second player the number on the board is not a perfect square.

1984.39. Four points with integer coordinates are selected on the plane. It is permitted to replace any of these points by the point symmetric to it with respect to any other of the current four points. Is it possible, using these operations, to go from points with coordinates $(0; 0)$, $(0; 1)$, $(1; 0)$, $(1; 1)$ to the points with coordinates $(0; 0)$, $(1; 1)$, $(3; 0)$, $(2; -1)$?

1984.40. There are ten numbers: 1 and nine zeros. It is permitted to choose any two numbers and replace both of them with their arithmetic mean. What is the minimum possible value of the number which occupies the original position of 1 after several such operations?

1984.41. Positive numbers $a_1, a_2, \ldots, a_k$; $b_1, b_2, \ldots, b_n$ are such that
$$a_1 + a_2 + \cdots + a_k = b_1 + b_2 + \cdots + b_n.$$
Prove that it is possible to place non-negative numbers in the squares of the $k \times n$ table so that their row sums are $a_1, a_2, \ldots, a_k$, their column sums are $b_1, b_2, \ldots, b_n$, and at least $(k-1)(n-1)$ of these numbers are zeros.

1984.42. A poker chip is placed on one of the 169 points $(x; y)$, where x and y are integers, $0 \leqslant x \leqslant 12$, $0 \leqslant y \leqslant 12$. The chip then moves from point to point as follows. Move from (x_1, y_1) to (x_2, y_2) is permitted if and only if each of the numbers $|x_1 - x_2|$, $|x_1 - y_2|$, $|y_1 - x_2|$, and $|y_1 - y_2|$ is at least 2 and at most 9. Prove that the chip cannot visit each of the 169 points exactly once.

1984.43. Two diametrically opposite points A and B are chosen on a circumference which touches two sides of an angle with vertex O (these two points are both different from the points of tangency). Line L, tangent to the circumference at point B, intersects the sides of the angle at points C and D. Also L intersects line OA at point E. Prove that $|BC| = |DE|$.

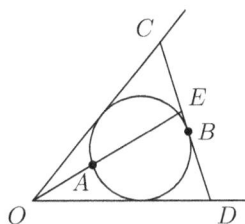

1984.44.* Prove that there exists a set of natural numbers A such that any natural number not belonging to A is equal to the arithmetic mean of some two different numbers from A, while any element of A does not possess that property.

GRADE 10

1984.45. See Question 40.

1984.46. Prove that if the sum of the face angles at the vertex of a pyramid is greater than $180°$, then any side edge of the pyramid is shorter than a half-perimeter of the pyramid's base.

1984.47. Integer a is such that $3a$ can be represented as $x^2 + 2y^2$, where x and y are integers. Prove that number a can be represented in that form as well.

1984.48. See Question 39.

1984.49. See Question 43.

1984.50. See Question 44.

1984.51.* Integers a, b, c, d, and e are such that sums $a + b + c + d + e$ and $a^2 + b^2 + c^2 + d^2 + e^2$ are divisible by some odd number n. Prove that $a^5 + b^5 + c^5 + d^5 + e^5 - 5abcde$ is also divisible by n.

1984.52. In infinite sequence 1, 0, 1, 0, 1, 0, ... each term, beginning with the seventh, equals the last digit of the sum of the six previous terms. Prove that numbers 0, 1, 0, 1, 0, 1 (in this specific order) do not occur in this sequence.

Commentary. *In 1984, the structure of the olympiad was slightly reformed. Namely, the elimination round was now formally included into the All-City olympiad and became its final fourth round, whose results now determined how the medals and diplomas were awarded. In 1984–1991, the number of the participants in the final round totaled somewhere between 50 and 100 high school students.*

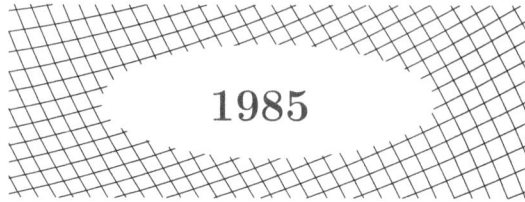

1985

GRADE 5

1985.01. There are 68 coins, all different in weight. In 100 weighings on two-pan scales, find the heaviest and the lightest coins.

1985.02. A forty-five-digit number is written with one digit 1, two digit 2s, three digit 3s, ..., and nine digit 9s. Prove that this number is not a perfect square.

1985.03. All the distances between towns in the country of Too-Far-Away are different. A traveler goes from his native town A to town B lying farther away from A than any other town in the country. Then he goes from B to town C located farther away from B than any other town and so on. Prove that if C is different from A, then the traveler will never return home.

1985.04. Find 1000 natural numbers whose sum is equal to their product.

1985.05. The warehouse has 300 boots of three sizes—size 9, size 10, and size 11. The number of right boots is the same as the number of left boots, 150 of each type. Prove that the warehouse will be able to assemble at least 50 usable pairs of boots so that every pair contains right and left boots of the same size.

1985.06. See Question 18 for $n = 10$.

GRADE 6

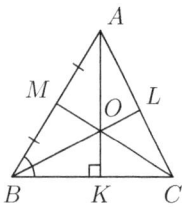

1985.07. Number 1 is written on the blackboard. Every second, the number on the board is increased by its sum of the digits. Will number 123456 ever appear on the board?

1985.08. Altitude AK, angle bisector BL, and median CM of triangle ABC intersect at point O. Prove that if $|AO| = |BO|$, then triangle ABC is equilateral.

1985.09. Prime numbers p, q, and natural number n satisfy equality

$$\frac{1}{p} + \frac{1}{q} + \frac{1}{pq} = \frac{1}{n}.$$

Find these numbers.

1985.10. See Question 5.

1985.11. Three grasshoppers sit on the straight line. They start playing leapfrog jumping one over another (but not over two at once). Is it possible that after 1985 jumps, they end up in their original positions?

1985.12. See Question 18 for $n = 10$.

GRADE 7

1985.13. Points P and Q are selected on sides BC and CD, respectively, of square $ABCD$ in such a way that triangle APQ is equilateral. A straight line passing through P and perpendicular to AQ intersects AD at point E. Point F outside of triangle APQ is such that triangles PQF and AQE are congruent. Prove that segment FE is twice as long as segment FC.

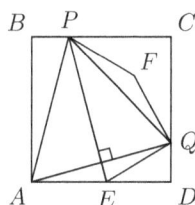

1985.14. A natural number is squared and then diminished by 600, then the same operation is performed with the result and so on. What could be the value of the original number if after several such operations we obtained it again?

1985.15. Two points, selected on the opposite sides of a rectangle, are connected by line segments to the vertices of this rectangle. Prove that the areas of the seven parts into which it is dissected by these segments cannot all be the same.

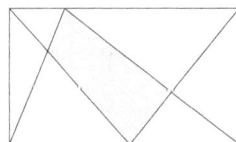

1985.16. What is the maximum number of integers between 1 and 1985 that can be selected so that the difference between any two of them is not a prime number?[85]

1985.17. Some 1985 points on the plane, no three of which lie on the same straight line, are colored red, blue, and green. Several segments connect these points, with each segment's ends colored differently. It turned out that every point has the same number of segments coming out of it. Prove that there exists a red point connected both with a blue point and with a green point.

1985.18. There are n containers, indexed by numbers 1 through n, stored in the warehouse where they are arranged in two stacks. A forklift can drive up to a stack, take a few containers off the top of the stack, and then deposit them on top of the other stack. Prove that it is possible to arrange all the containers in one stack in the ascending order (#1 at the very bottom, then #2 etc.) using no more than $2n - 1$ forklift operations described above.

---

[85] The end numbers 1 and 1985 are included.

GRADE 8

1985.19. Arithmetic progression of natural numbers contains at least one perfect square. Prove that it contains infinitely many perfect squares.

1985.20. Solve the system of equations

$$\begin{cases} (x+y)^3 = z \\ (y+z)^3 = x \\ (z+x)^3 = y. \end{cases}$$

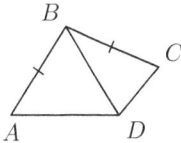

1985.21. Angle A in convex quadrilateral $ABCD$ equals $65°$, angle CBD equals $35°$, angle ADC equals $130°$, and $|AB| = |BC|$. Find the angles of quadrilateral $ABCD$.

1985.22. Positive numbers a_1, b_1, c_1, a_2, b_2, c_2 satisfy inequalities $b_1^2 \leqslant a_1 c_1$ and $b_2^2 \leqslant a_2 c_2$. Prove that

$$(a_1 + a_2 + 5)(c_1 + c_2 + 2) > (b_1 + b_2 + 3)^2.$$

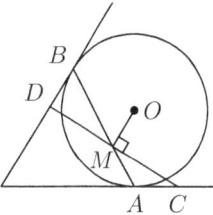

1985.23. A circumference with center O touches the sides of an angle at points A and B. Line L, drawn through some arbitrary point $M \in [AB]$ different from points A and B, is perpendicular to OM. Points C and D are intersections of L with the sides of the angle. Prove that $|MC| = |MD|$.

1985.24. Two people play a game on a rectangular board with dimensions 1×25, using 25 tokens. A move consists of either placing a new token onto one of the unoccupied squares or moving one of the tokens to the nearest unoccupied square on its right. The game ends when all the squares are occupied; the player who made the last move, wins. In the beginning, the board is empty. Who wins in the errorless game, the player who makes the first move or his opponent?

GRADE 9

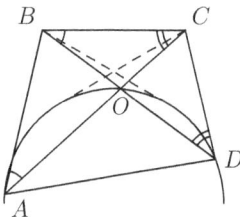

1985.25. In convex quadrilateral $ABCD$ whose diagonals intersect at point O, angles BAC and CBD are equal, as are angles BCA and CDB. Prove that the tangent segments from points B and C to the circumference AOD have equal length.

1985.26. Real numbers a, b, c, x, y, z satisfy equalities $x = by + cz$, $y = cz + ax$, $z = ax + by$. Given that at least one of the numbers x, y, and z is not zero, prove that $2abc + ab + ac + bc = 1$.

1985.27. Lengths of all sides of a convex quadrilateral do not exceed 7. Prove that four circles with radius 5, centered at the vertices of the quadrilateral, cover the entire quadrilateral.

1985.28. Real numbers a, b, and c satisfy inequalities

$$a + b + c > 0, \ ab + ac + bc > 0, \ abc > 0.$$

Prove that numbers a, b, c are positive.

1985.29.* Sequence x_1, x_2, $\ldots$ is defined by formulas $x_1 = 0.001$, $x_{n+1} = x_n - x_n^2$. Prove that $x_{1001} < 0.0005$.

1985.30. See Question 24.

GRADE 10

1985.31. Real numbers a, b, c, x, y, z satisfy equalities

$$a^x = bc, \ b^y = ca, \ c^z = ab.$$

Given that a, b, c are positive and at least one of them is different from 1, prove that $x + y + z - xyz$ is an integer.

1985.32. See Question 23.

1985.33. For every face F of a tetrahedron, we construct the straight line that passes through the incenter of triangle F and is perpendicular to the plane of F. Prove that if these four lines intersect at one point, then the sums of opposite edges of the tetrahedron are the same.

1985.34. Prove the inequality

$$9 < \int_0^3 \sqrt[4]{x^4 + 1} \, dx + \int_1^3 \sqrt[4]{x^4 - 1} \, dx < 9.0001 \,.$$

1985.35. See Question 29.

1985.36. See Question 24.

FINAL ROUND

GRADE 8

1985.37. Number 1584 possesses the following properties:

a) it is not a perfect square;
b) it is different from its "reverse" number 4851;
c) the product of the number and its reverse (i.e., 1584×4851) is a perfect square.

Find a 20-digit number which possesses the same three properties.

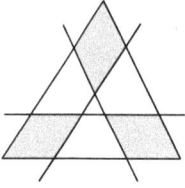

1985.38. The perimeter of a triangle is 100 cm, and its area is 100 cm². Three straight lines, drawn parallel to the sides of the triangle at the distance of 1, dissect the triangle into seven parts, three of which are parallelograms. Prove that the sum of the areas of these parallelograms is less than 25 cm².

1985.39. Fifteen volleyball teams played a round-robin tournament in which every team won exactly seven matches. How many trios of teams are there such that in the matches among them each team has exactly one win?

1985.40. Each square of a 100×100 table is colored into one of four colors, so that each 2×2 subtable contains all four colors. Prove that the corner squares of the table are all colored differently.

1985.41. The inscribed circumference of a trapezoid with bases a and b has radius R. Prove that $ab \geqslant 4R^2$.

1985.42. Sequence of natural numbers a_1, a_2, ... is such that $a_1 = 1$ and $a_{k+1} - a_k$ equals 0 or 1 for every k. Prove that if $a_m = m/1000$ for some m, then there exists n such that $a_n = n/500$.

1985.43. In sequence f_1, f_2, ... the first two terms are equal to 1, and every other term is the sum of the two previous ones.[86] Prove that

$$\frac{1}{f_1 f_3} + \frac{1}{f_2 f_4} + \frac{1}{f_3 f_5} + \cdots + \frac{1}{f_{98} f_{100}} < 1.$$

1985.44. Each natural number $k \leqslant 100$ is mapped to a natural number $f(k)$ which also does not exceed 100. We construct sequence $\{a_i\}$ using formulas $a_1 = 1$, $a_2 = f(a_1)$, $a_3 = f(a_2)$, and so on. Prove that there exists index $n \leqslant 100$ for which $a_n = a_{2n}$.

GRADES 9–10

1985.45. See Question 38.

1985.46. See Question 40.

1985.47. Twenty volleyball teams played a round-robin tournament. Let T be the number of trios of teams such that in matches among them each team has exactly one win. Prove that:

    a) if every team won at least 9 and at most 10 matches, then $T = 330$;
    b) $T \leqslant 330$.

1985.48. See Question 41.

1985.49. See Question 42.

---

[86] Obviously, this is none other than the famous Fibonacci series.

1985.50. See Question 43.

1985.51. The sequence of natural numbers $a_1, a_2, \ldots$ satisfies formula $a_{n+2} = a_{n+1}a_n + 1$ for any n. Prove that for any $n > 10$, number $a_n - 22$ is composite.

1985.52.* In a school attended by an equal number of girls and boys, each boy is friends with an even number of girls. Prove that it is possible to choose a group of several boys such that each girl is friends with an even number of boys from that group.

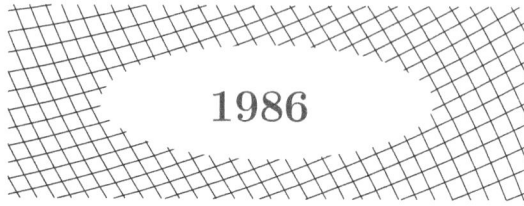

1986

GRADE 5

1986.01. Nine cards are laid out in a row, showing the numbers 7, 8, 9, 4, 5, 6, 1, 2, and 3. It is permitted to take a few consecutive cards and reverse their order. How can we obtain the arrangement 1, 2, 3, 4, 5, 6, 7, 8, 9 using three such operations?

1986.02. Forty-four queens are placed on the chessboard. Prove that each one of them attacks some other queen.

1986.03. Natural numbers a and b are such that $34a = 43b$. Prove that $a + b$ is a composite number.

1986.04. How do you position several nickels on the surface of the table so that each one of them touches exactly three other coins?

1986.05. Fifty-five numbers are written around the circle in such a way that each one of them is equal to the sum of its two neighbors. Prove that all these numbers are zeros.

1986.06. (a) Find a seven-digit number whose digits are all different, and each one of them divides this number. (b) Does there exist an eight-digit number possessing the same properties?

GRADE 6

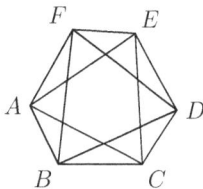

1986.07. See Question 1.

1986.08. See Question 2.

1986.09. In convex hexagon $ABCDEF$, triangles ABC, ABF, FEA, FED, CDB, and CDE are congruent. Prove that diagonals AD, BE, CF have equal length.

1986.10. See Question 5.

1986.11. The snail started out from point O on the plane and then crawled with the constant speed, turning by $60°$ every half-hour. Prove that the snail can return to point O only after an integer number of hours.

1986.12. See Question 6.

1986.13. Eleven students attend five school clubs. Prove that there are two of them, A and B, such that all the clubs attended by A are attended by B as well.

GRADE 7

1986.14. See Question 5.

1986.15. Natural numbers a, b, and c are such that $\frac{1}{a} + \frac{1}{b} + \frac{1}{c} < 1$. Prove the inequality $\frac{1}{a} + \frac{1}{b} + \frac{1}{c} \leqslant \frac{41}{42}$.

1986.16. The angle at vertex B in an isosceles triangle ABC is $108°$. The angle bisector of ACB intersects side AB at point D. The line passing through point D and perpendicular to this bisector intersects base AC at point E. Prove that $|AE| = |BD|$.

1986.17. See Question 6.

1986.18. Points K and H are taken on sides BC and CD of square $ABCD$ so that $|KC| = 2\,|KB|$ and $|HC| = |HD|$. Prove that angles AKB and AKH are equal.

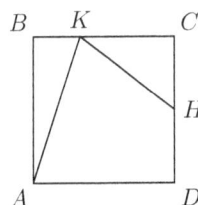

1986.19. A pile of 25 stones is arbitrarily divided into two smaller piles, then one of these piles is again somehow divided into two piles, and so on until every pile consists of one stone. During each operation, we write down the product of the numbers of stones in the two resulting smaller piles. Prove that the sum of all these numbers equals 300.

GRADE 8

1986.20. Find all three-digit numbers which are 11 times larger than the sum of their digits.

1986.21. In square $ABCD$, point E is selected on side BC, point K and M—on side CD, point H—on side AD, so that $|CE| = |CK|$ and $|DM| = |DH|$. Prove that the quadrilateral formed by intersection of angles HBM and EAK is cyclic.

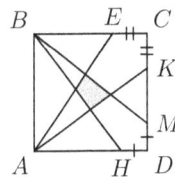

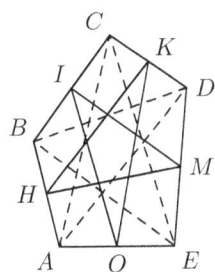

1986.22. Find some integers a and b such that
$$\frac{a}{999} + \frac{b}{1001} = \frac{1}{999999}.$$

1986.23. In convex pentagon $ABCDE$, points H, I, K, M, O are the midpoints of sides AB, BC, CD, DE, and EA, respectively. Prove that the length of broken line $HKOIMH$ is less than the length of broken line $ACEBDA$.

1986.24. Quadratic polynomials $x^2 + b_1 x + c_1$ and $x^2 + b_2 x + c_2$ have integer coefficients and a common non-integer root. Prove that they are identical.

1986.25.* Two hundred soccer teams are playing in a tournament. On the first day, each team played exactly one match, on the second day, each team again played exactly one match, and so on. Prove that at the end of the sixth day of the tournament, one can select 34 teams such that no two of them have yet played each other.

GRADE 9

1986.26. See Question 20.

1986.27. See Question 21.

1986.28. Solve the system of equations

$$\begin{cases} a = bcd \\ a + b = cd \\ a + b + c = d \\ a + b + c + d = 1 \,. \end{cases}$$

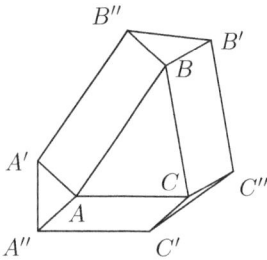

1986.29. Parallelograms $AA'B''B$, $BB'C''C$, $CC'A''A$ are built on sides of triangle ABC with equal lengths of the other sides $|AA'| = |BB'| = |CC'| = a$. Find a if $|A'A''| = 3$, $|B'B''| = 4$, $|C'C''| = 5$.

1986.30. See Question 24.

1986.31. Find some integers a, b, and c such that

$$\frac{a}{999} + \frac{b}{1000} + \frac{c}{1001} = \frac{1}{999 \cdot 1000 \cdot 1001} \,.$$

GRADE 10

1986.32. See Question 20.

1986.33. See Question 21.

1986.34. What is the minimum value of the product of two positive numbers a and b that satisfy equality $ab = a + b$?

1986.35. Find all positive solutions of equation

$$x^{1986} + 1986^{1985} = x^{1985} + 1986^{1986} \,.$$

1986.36. See Question 31.

1986.37. Find the angle between edge AB and face ACD in trihedral angle $ABCD$ with vertex A if $\angle BAC = 45°$, $\angle CAD = 90°$, and $\angle BAD = 60°$.

FINAL ROUND

GRADE 8

1986.38. Let $a_1 = 2$ and $a_{n+1} = a_1 a_2 \cdots a_n + 1$ for $n = 1, 2, \ldots$. Prove that

$$\frac{1}{a_1} + \frac{1}{a_2} + \cdots + \frac{1}{a_n} < 1.$$

1986.39. Prove that any polygon possesses side BC and vertex A, different from B and C, such that the foot of the perpendicular dropped from A onto line BC lies within segment BC.

1986.40. A Martian is always born at midnight and then lives for exactly 100 days. The Martian civilization is currently extinct, and the number of Martians who ever lived is odd. Prove that there were at least 100 days when the population of Mars was odd.

1986.41. Point K is selected on side AB of square $ABCD$, point H—on side CD, and point M—on segment KH. Prove that the second (different from M) point of intersection of circumcircles of triangles AKM and MHC lies on the square's diagonal AC.

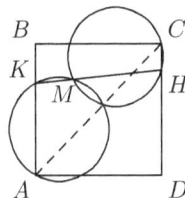

1986.42.* In the Land of Oz, airports are connected by several non-stop routes so that you can fly from any airport in Oz to any other landing no more than n times (landing at the destination counts as well). One route was canceled, but you can still travel by air from any airport to any other one. Prove that now this can be done with no more than $2n$ landings.

1986.43. The sequence of 36 zeros and ones begins with five zeros. It turns out that if you consider all quintuples of five consecutive digits in this sequence, they will cover all 32 possible variations. Find the last five digits in the sequence.

1986.44.* Prove that it is possible to draw several straight lines and select several points on the plane so that every line contains exactly four selected points, and every selected point belongs to exactly four lines.

1986.45.* Two people play on the 30×45 sheet of graph paper making their moves in turn. The player's move consists of making a cut along a unit segment connecting two neighboring grid points. The first player starts cutting from the edge of the sheet, and every next move must continue the line formed by the previous cuts. A player, after whose move the sheet falls apart, wins. Who wins in the errorless game?

GRADE 9

1986.46. See Question 38.

1986.47. See Question 39.

1986.48. See Question 40.

1986.49. See Question 41.

1986.50. A set of positive numbers A is such that the sum of any two of its elements again belongs to this set. Also any segment $[a; b]$, $0 < a < b$, contains a segment which consists entirely of elements of A. Prove that A contains all positive real numbers.

1986.51. For any natural number m, consider the following algorithm:

Step 1. Set $n = m$.
Step 2. If n is even, divide it by 2; if n is odd, increase it by 1.
Step 3. If $n > 1$, go to Step 1. If $n = 1$, exit the algorithm.

How many natural numbers m are there such that if this algorithm is run, then **Step 1** is executed exactly fifteen times?

1986.52. See Question 44.

1986.53.* A chess king moved around a 9×9 board visiting each square exactly once. Its route is not closed and can self-intersect. What is the maximum possible length of such a route if the length of the king's diagonal move is $\sqrt{2}$, and the length of a vertical or horizontal move is 1?

Grade 10

1986.54. For any set on the plane, we will call its *diameter* the maximum distance between two points of the set (if such maximum exists). The sum of the diameters of polygons M_1, M_2, ..., M_n is less than the diameter of their union. Prove that there exists a straight line which does not intersect any of these polygons but has at least one of them on either side of it.

1986.55. Function $F \colon \mathbb{R} \to \mathbb{R}$ is continuous, and for any real x there exists natural n such that

$$F(F(\ldots F(x)\ldots)) = 1$$

(F is applied n times). Prove that $F(1) = 1$.

1986.56. Prove that

$$\frac{1}{1+\sqrt{3}} + \frac{1}{\sqrt{5}+\sqrt{7}} + \cdots + \frac{1}{\sqrt{9997}+\sqrt{9999}} > 24\,.$$

1986.57. Parallelogram $ABCD$ is not a rhombus. Points X and Y are intersections of the bisector of angle BAD with lines BC and CD, respectively. Prove that the center of the circumference passing through points C, X, and Y lies on the circumference passing through points B, C, and D.

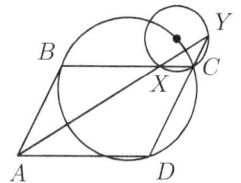

1986.58. Compute the integral

$$\int_{-1}^{1} \frac{dx}{1 + x^3 + \sqrt{1 + x^6}}.$$

1986.59.* Prove that it is possible to draw several circles on the plane which do not have common interior points, but every circle touches exactly five others.

1986.60. Given real numbers a, b, c, x, y, and z, numbers u_1, u_2, u_3, v_1, v_2, and v_3 are defined by formulas

$$u_1 = ax + by + cz, \ v_1 = ax + bz + cy,$$
$$u_2 = ay + bz + cx, \ v_2 = az + by + cx,$$
$$u_3 = az + bx + cy, \ v_3 = ay + bx + cz.$$

Prove that if $u_1 u_2 u_3 = v_1 v_2 v_3$, then triple (v_1, v_2, v_3) can be obtained from (u_1, u_2, u_3) by permutation.

1986.61.* See Question 45 for the sheet of paper with dimensions 30×30.

Commentary. *In 1986, it took the jury two more additional elimination rounds to finally determine one last member for the Leningrad team to the All-Union Math Olympiad.*

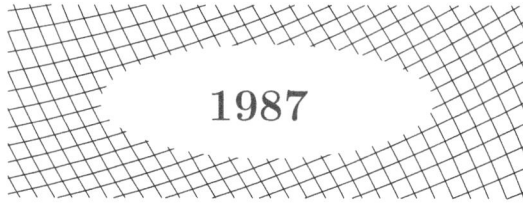

1987

GRADE 5

1	2	3	4
5	6	7	8
9	10	11	12
13	14	15	16

1	5	9	13
2	6	10	14
3	7	11	15
4	8	12	16

1987.01. Numbers 1 through 16 are written into the squares of 4×4 table (see the left table in the figure). It is permitted to add 1 simultaneously to all numbers in any row, or to subtract 1 from all numbers in any column. How can we obtain the right table in the figure, using these operations?

1987.02. In Republic of Anchuria,[87] banknotes have four denominations—1 dollar, 10 dollars, 100 dollars, and 1000 dollars. Is it possible to pay one million Anchurian dollars with exactly half a million bills?

1987.03. The king wants to build six fortresses and connect each two of them with a road. Draw the construction diagram on the plane so that it has only three road intersections, each one with exactly two roads meeting there.

1987.04. If every boy in the class buys a doughnut and every girl—a muffin, altogether they would spend one kopeck more than if every boy bought a muffin and every girl—a doughnut. There are more boys than girls in the class. By how many?

1987.05. A bus ticket is called *lucky* if the sum of its first three digits is the same as the sum of its last three digits. How many consecutive tickets do you need to buy on the bus so that at least one of them were a "lucky" one?

1987.06. Two players—Ann, who makes the first move, and Bob—write X's and O's into the squares of 9×9 board (Ann writes X's and Bob writes O's.) When the board is filled, they count the rows and columns where X's are prevalent—this number is Ann's result. Similarly, the number of rows and columns where O's are prevalent is Bob's total. How does Ann play to achieve the highest possible result?

GRADE 6

1987.07. See Question 1.

---

[87] If you are not familiar with this exotic location, read "Cabbages and Kings" by O.Henry.

1987.08. In an acute-angled triangle ABC, altitude CH and median BK are drawn. Given that $|BK| = |CH|$ and $\angle KBC = \angle HCB$, prove that triangle ABC is equilateral.

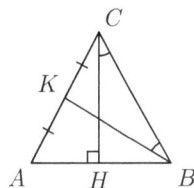

1987.09. See Question 4.

1987.10. In neighboring countries of Dillia and Dallia, units of currency are *diller* and *daller*, respectively. In Dillia, the exchange rate is one diller for ten dallers, but in Dallia, it is one daller for ten dillers. A sneaky financier has one diller, and he can freely move from one country to the other while exchanging his money without paying any fees. Prove that he will never have the same number of dillers and dallers.

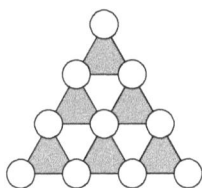

1987.11. See Question 5.

1987.12. Is it possible to place numbers 0 through 9 (using each one exactly once) in the white circles (see the figure) in such a way that all the sums of the numbers in the vertices of the shaded triangles are the same?

Grade 7

1987.13. Segments connecting the midpoints of opposite sides of convex quadrilateral $ABCD$ divide it into four quadrilaterals with equal perimeters. Prove that $ABCD$ is a parallelogram.

1987.14. See Question 4.

1987.15. See Question 10.

1987.16. Vertices of a closed non-self-intersecting broken line with eight segments coincide with the vertices of a cube. Prove that at least one of the segments coincides with one of the cube's edges.

1987.17. The Long Slog road construction company was contracted to build a 100 km long highway from Tinseltown to Toonville. Their plan is to build 1 km of the road in the first month, and then during each following month to build $1/a^{10}$ km more, where a is the length of the road already built by the beginning of that month. Will they ever complete the highway?

1987.18. A *set square* is a formal geometric construction tool that allows you to do exactly the two following operations:

1) for any two given points A and B, it constructs the line passing through A and B;
2) for any given line L and any point $P \in L$, it constructs line perpendicular to L passing through P.

Using this tool, drop a perpendicular to the given line L from some given point A which does not lie on L.

Grade 8

1987.19. Fifty checkers are placed on the 10×10 board; twenty-five of them are in the lower left quarter and twenty-five others—in the upper right quarter. In one move, any checker can jump over any other adjacent checker (vertically, horizontally, or diagonally) onto the next square if it is not occupied. Is it possible to move all the checkers into the left half of the board?

1987.20. The bank has unlimited supply of coins of value 1, 2, 5, 10, 20, 50 kopecks and 1 ruble. Given that the sum of A kopecks can be paid out with B coins, prove that the sum of B rubles can be paid out with A coins.

1987.21. Let a, b, c, and d be arbitrary real numbers. Prove that

$$(1 + ab)^2 + (1 + cd)^2 + (ac)^2 + (bd)^2 \geqslant 1.$$

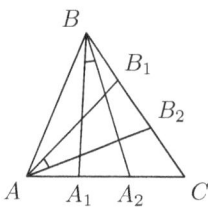

1987.22. In triangle ABC, points A_1 and A_2 divide side AC into three equal parts; points B_1 and B_2 do the same with side BC. Prove that if angles A_1BA_2 and B_1AB_2 are equal, then triangle ABC is equilateral.

1987.23. Several crows sit on a tall oak tree, with all of its branches at different heights. Every minute, a few crows change their positions in the following manner: One of the crows gets "evicted" by its neighbors, and so she flies over to the next (higher) branch of the tree. If there are no higher branches, the crow flies away.

Prove that the time after which this process will stop (that is, each branch will have at most one crow on it) does not depend on the order of the "evictions" but only on the initial position of the crows.

1987.24. Sixty-four unit cubes were laid down on the plane in the shape of a 8×8 square. Could the same cubes be arranged as a $4 \times 4 \times 4$ cube in such a way that any two-unit cubes that were neighbors before are still neighbors now? We call two-unit cubes neighbors if they have a common face.

Grade 9

1987.25. Squares of a 8×8 table are colored into black and white in the chessboard pattern. It is permitted to swap any two columns or any two rows. Is it possible, using these operations, to obtain a table with the entire left half black and entire right half white?

1987.26. Find the value of

$$\cfrac{1}{2 - \cfrac{1}{2 - \cfrac{\ddots}{2 - \cfrac{1}{2 - \cfrac{1}{2}}}}} \qquad (100 \text{ twos}).$$

1987.27. Two circumferences intersect at points A and B, and the tangents to these circumferences at each of these points are perpendicular. Let M be an arbitrary point on one of the circumferences such that M lies inside the other circumference. Extend segments AM and BM beyond point M until they intersect that other circumference at points X and Y. Prove that XY is the diameter of that circumference.

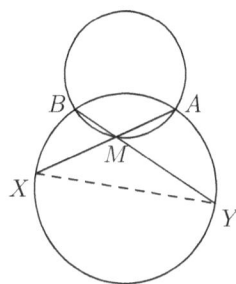

1987.28. See Question 21.

1987.29. Find the largest natural number such that each of its digits except for the first one and the last one is less than the arithmetic mean of its neighbors.

1987.30. As an astronomer was observing fifty stars in the night sky, he computed sum S of all of their pairwise distances. Then a cloud obscured twenty-five of those stars. Prove that the sum of all pairwise distances between the visible stars is less than $\frac{S}{2}$.

GRADE 10

1987.31. See Question 20.

1987.32. See Question 26.

1987.33. See Question 27.

1987.34. See Question 21.

1987.35. Does there exist a natural number n such that $n^n + (n+1)^n$ is divisible by 1987?

1987.36. See Question 30.

FINAL ROUND

GRADE 8

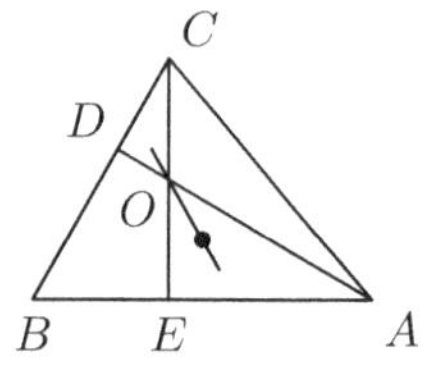

1987.37. In acute-angled triangle ABC, angle B equals $60°$, and altitudes CE and AD intersect at point O. Prove that the circumcenter of ABC lies on the common bisector of angles AOE and COD.

1987.38. Positive numbers a, b, c, d are such that $cd = 1$. Prove that the interval with endpoints ab and $(a + c)(b + d)$ contains at least one perfect square.

1987.39. Two people play the game using the "playing field" shown below.

$$((((((((\underline{\ \ } * \underline{\ \ }) * \underline{\ \ }) * \underline{\ \ }) * \underline{\ \ }) * \underline{\ \ }) * \underline{\ \ }) * \underline{\ \ }) * \underline{\ \ }).$$

They make their moves in turn; the first player starts by writing any digit into the first (the leftmost) space. Every move after that consists of writing some digit that has not been already used into the next (the leftmost) unoccupied space and replacing the asterisk * to the left of it with a sign of either addition or multiplication. At the end of the game the value X of the resulting expression is computed. If X is even, the first player wins, otherwise—if X is odd—the second player does. Who wins in the errorless game?

1987.40. In Paradise City, people are allowed to swap their apartments, but they cannot sell them or perform any other operations with them. Two families who have swapped their apartments cannot participate in any other swap on the same day. Prove that any apartment ownership rearrangement can be achieved within at most two days.

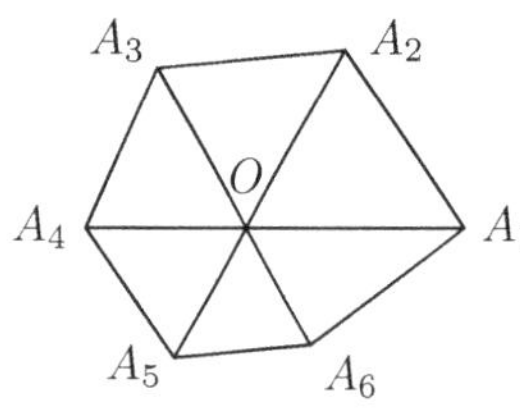

1987.41. Point O inside hexagon $A_1A_2A_3A_4A_5A_6$ is such that all sides of the hexagon are seen from O at the angle of $60°$. Prove that if $OA_1 > OA_3 > OA_5$ and $|OA_2| > |OA_4| > |OA_6|$, then $|A_1A_2| + |A_3A_4| + |A_5A_6| < |A_2A_3| + |A_4A_5| + |A_6A_1|$.

1987.42. Find the prime factorization of number $989 \cdot 1001 \cdot 1007 + 320$.

1987.43.* The plane is cut along several circumferences some of which intersect but not one of them encircles any other. Prove that the pieces that were cut off the plane cannot be rearranged to form several complete circles.

1987.44.* The Thief of Baghdad sneaked into the sultan's palace where n guards are trying to catch him. A guard can catch the thief if and only if they happen to be in the same room. The palace consists of 1000 rooms connected with doors, and the palace's plan is such that it is impossible to get into the adjoining room in any other way except for walking through the (always unique) door that connects them.[88]

---

[88] In other words, the graph of the palace's plan is a *tree*.

(a) Prove that there exists a strategy that guarantees the guards will catch the thief if $n = 10$.
(b) Prove that statement of item (a) is not true for $n = 5$.
(c) Prove item (a) if $n = 6$.

GRADE 9

1987.45. See Question 37.

1987.46. See Question 38.

1987.47. Eight non-negative real numbers whose sum equals 1 are placed into the vertices of a cube. For each edge, we compute the product of the numbers in the endpoints of that edge. Prove that the sum of these products does not exceed $\frac{1}{4}$.

1987.48. Sequence of natural numbers a_1, a_2, ..., is such that $a_1 < 1987$ and $a_k + a_{k+1} = a_{k+2}$ for each natural k. Prove that if for some n numbers $a_1 - a_n$ and $a_2 + a_{n-1}$ are divisible by 1987, then n is odd.

1987.49. See Question 41.

1987.50. See Question 42.

1987.51. Given the deck of $2n + 1$ indexed cards, you are permitted to perform the two following operations.

(a) Take several cards off the top of the deck and put them underneath the stack without changing their order.
(b) Take the upper n cards and insert them (without changing their order) into the n gaps between the lower $n + 1$ cards.

Prove that using these operations you cannot obtain more than $2n(2n + 1)$ different arrangements of the deck.

1987.52. See Question 44.

GRADE 10

1987.53. See Question 37.

1987.54. Continuous functions f, $g\colon [0; 1] \to [0; 1]$ satisfy the equality $f(g(x)) = g(f(x))$ for any $x \in [0; 1]$. Prove that if f is an increasing function, then there exists $a \in [0; 1]$ such that $f(a) = g(a) = a$.

1987.55. See Question 47.

1987.56. Let x_1, x_2, x_3, ... be a sequence of real numbers, and T —a natural number. It is known that there are no more than T pairwise different ordered T-tuples of form $(x_{k+1}$, x_{k+2}, ..., $x_{k+T})$. Prove that sequence x_1, x_2, x_3, ... is periodic.

1987.57. Diagonals of inscribed quadrilateral $ABCD$ intersect at point O. Prove that

$$\frac{|AB|}{|CD|} + \frac{|CD|}{|AB|} + \frac{|BC|}{|AD|} + \frac{|AD|}{|BC|} \leq \frac{|OA|}{|OC|} + \frac{|OC|}{|OA|} + \frac{|OB|}{|OD|} + \frac{|OD|}{|OB|}.$$

1987.58. See Question 42.

1987.59. See Question 51.

1987.60.* In a set with m elements, some s subsets contain a_1, a_2, ..., a_s elements, respectively. None of these subsets contains another one of them. Prove that

$$\frac{1}{\binom{m}{a_1}} + \frac{1}{\binom{m}{a_2}} + \cdots + \frac{1}{\binom{m}{a_s}} \leq 1,$$

where $\binom{m}{a_k}$ is the binomial coefficient $m!/(a_k!(m - a_k)!)$.

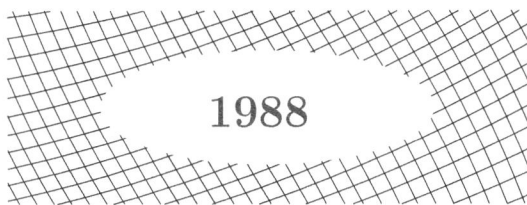

1988

GRADE 5

1988.01. Nine zeros are written into the squares of a 3 × 3 table. It is permitted to choose any 2 × 2 square in that table and increase all the numbers in it by 1. Prove that, by using these operations, it is impossible to obtain the table in the figure.

4	9	5
10	18	6
6	13	7

1988.02. Thirty players and the gamemaster all write down numbers 1 through 30 in some order. Then the gamemaster compares the players' sequences to his own—for every coincidence (same number in the same place) player gets one point. It turned out that all the players have accumulated different number of points. Prove that somebody's sequence is identical to the gamemaster's.

1988.03. Is it possible to write natural numbers 1 through 100 in a row in some order so that the difference between any two neighbors is at least 50?

1988.04. Do there exist non-zero integers a and b such that one of them is divisible by their sum and the other—by their difference?

1988.05. A pile of 1001 stones lies on the table. You are allowed to choose any pile with more than one stone, throw away one of its stones, then divide any pile on the table into two non-empty piles. Is it possible that after several such operations, each pile on the table contains exactly three stones?

1988.06. The king's castle consists of 64 identical square rooms arranged in the form of a large 8 × 8 square. Every internal wall separating two rooms has a door in it. All the rooms' floors are painted white. Every day the painter walks through several rooms and repaints the floor in every room he visits—white to black and vice versa. Is it possible that on some day the floors will be colored in the chessboard pattern?

GRADE 6

1988.07. See Question 3.

1988.08. See Question 1.

1988.09. Each one of the natural numbers a, b, c, and d is divisible by $ab - cd$. Prove that $ab - cd = 1$.

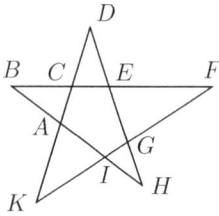

1988.10. Prove that a five-pointed star cannot be drawn in such a way that five inequalities $|AB| < |BC|$, $|CD| < |DE|$, $|EF| < |FG|$, $|GH| < |HI|$, $|IK| < |KA|$ hold true.

1988.11. Fifty cards have numbers 1 through 25 written on them, with each number occurring exactly twice. Each one of the twenty-five students, sitting around a circular table, is given two of these cards, and every minute on the minute, each student passes his or her card with the smaller number to the neighbor on the right. As soon as someone has two cards with identical numbers, this process ends. Prove that eventually that will happen (that is, this process cannot go on forever).

1988.12. Five hundred matches lie on the table. Two people play a game making their moves in turn. The move consists of taking 1, 2, 4, 8, ... (any power of two) matches from the table. Whoever cannot make a move loses. Who wins in the errorless game: the player who makes the first move or his opponent?

GRADE 7

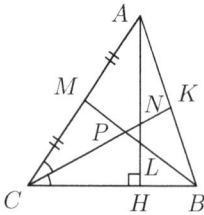

1988.13. Real numbers x and y are such that $0 \leqslant x, y \leqslant 1$. Prove that

$$\frac{x}{1+y} + \frac{y}{1+x} \leqslant 1.$$

1988.14. In an acute-angled triangle ABC, altitude AH intersects median BM at point L, and angle bisector CK—at point N. Median BM and angle bisector CK intersect at point P. Given that the points L, N, and P are all different, prove that triangle LNP is not equilateral.

1988.15. See Question 9.

1988.16. See Question 11.

1988.17. See Question 10.

1988.18.* On the sheet of graph paper with dimensions 21×21, all the grid points are colored red and blue. All the vertices on the upper edge of the sheet are colored red; same is true for all the vertices on the right edge, except the lower right corner. All other vertices on the sheet's boundary are colored blue. Prove that there is a square which has two red and two blue vertices with vertices of the same color being the endpoints of one of the square's sides.

GRADE 8

1988.19. Let a, b, and c be real numbers such that $abc = 1$, $a + b + c = \frac{1}{a} + \frac{1}{b} + \frac{1}{c}$. Prove that one of numbers a, b, and c is equal to 1.

1988.20. In an acute-angled triangle ABC, angle A is equal to $30°$. BB_1 and CC_1 are the triangle's altitudes; B_2 and C_2 are the midpoints of sides AC and AB, respectively. Prove that segments B_1C_2 and B_2C_1 are perpendicular.

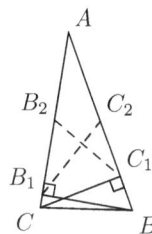

1988.21. Find a 100-digit number which has no zero digits and is divisible by the sum of its digits.

1988.22. A stack of n differently colored bricks is placed on the table. It is permitted to take several bricks from the bottom and put them on top of the stack without changing their order, and then to turn over the entire stack. Prove that the number of different color arrangements that one can get, using these operations, does not exceed $2n$.

1988.23. A building with 120 apartments has 119 tenants. We will call an apartment *crowded* if it is inhabited by 15 or more people. Every day, the inhabitants of one of the crowded apartments have a fight after which they all move to the different apartments. Is it true that eventually the moves will stop?

1988.24. (a) x, y, $z \geqslant 0$; $x + y + z = \frac{1}{2}$. Prove that

$$\frac{1-x}{1+x} \cdot \frac{1-y}{1+y} \cdot \frac{1-z}{1+z} \geqslant \frac{1}{3}.$$

(b) x_1, x_2, $\ldots$, $x_n \geqslant 0$; $x_1 + \cdots + x_n = \frac{1}{2}$. Prove that

$$\frac{1-x_1}{1+x_1} \cdot \frac{1-x_2}{1+x_2} \cdot \ldots \cdot \frac{1-x_n}{1+x_n} \geqslant \frac{1}{3}.$$

GRADE 9

1988.25. Integers a, b, c, and d satisfy the equalities

$$ab + cd = -1, \quad ac + bd = -1, \quad ad + bc = -1.$$

Find a, b, c, and d.

1988.26. Squares $ABDE$ and $BCFG$ are built outside of triangle ABC on its sides AB and BC. Prove that if line DG is parallel to line AC, then ABC is isosceles.

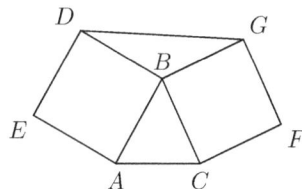

1988.27. Given numbers $a < b < c$, prove that the equation

$$\frac{1}{x-a} + \frac{1}{x-b} + \frac{1}{x-c} = 0$$

has two roots x_1 and x_2 such that $a < x_1 < b < x_2 < c$.

1988.28. a) See Question 24 (b) for $n = 2$. (b) See Question 24 (b) for $n = 4$.

1988.29. Natural numbers a, b, and c are such that a^3 is divisible by b, b^3 is divisible by c, and c^3 is divisible by a. Prove that $(a + b + c)^{13}$ is divisible by abc.

1988.30. All diagonals of a parallelepiped P are equal. Prove that P is rectangular.

GRADE 10

1988.31. See Question 28 (a).

1988.32. See Question 20.

1988.33. See Question 29.

1988.34. Real-valued functions $f(x)$ and $g(x)$ are defined for all real x. For any x and y, equality $f(x + g(y)) = 2x + y + 5$ holds true. Find function $g(x + f(y))$—that is, express it explicitly through x and y.

1988.35. Given one hundred of consecutive natural numbers, is it possible that they can be arranged around the circle so that the product of any two neighboring numbers were a perfect square?

1988.36. In a regular hexagonal pyramid, the circumcenter lies on the surface of the inscribed sphere. Find the ratio of the circumradius of this pyramid to its inradius.

FINAL ROUND

GRADE 8

1988.37. A line that contains side AC of acute-angled triangle ABC is reflected with respect to lines AB and BC. Two resulting lines intersect at point K. Prove that line BK passes through point O—circumcenter of triangle ABC.

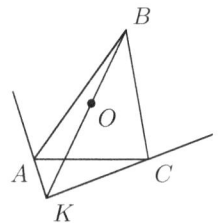

1988.38. Real numbers x_1, x_2, x_3, x_4, x_5, x_6 lie on segment $[0; 1]$. Prove that

$$(x_1 - x_2)(x_2 - x_3)(x_3 - x_4)(x_4 - x_5)(x_5 - x_6)(x_6 - x_1) \leqslant \frac{1}{16}.$$

1988.39. Find two co-prime four-digit natural numbers a and b such that for any natural m and n, numbers a^m and b^n differ by at least 4000.

1988.40. There are N towns connected with each other by $2N - 1$ one-way roads. This road system is such that one can drive from any town to any other complying with the traffic rules. Prove that there is a road which can be closed and the system will still possess that property.

1988.41. In trapezoid $ABCD$ (with bases BC and AD), points K and L are selected on sides AB and CD, respectively. Prove that if angles BAL and CDK are equal, then so are the angles BLA and CKD.

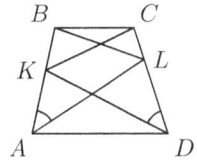

1988.42. There are two piles of matches: one has 100 matches, and the other—252. Two people play a game, making their moves in turn. A move consists of taking several matches from one of the piles in such a way that the number of the taken matches divides the number of matches in the other pile. The player who takes the last match, wins. Who will win in the errorless game, the player who makes the first move or his opponent?

1988.43. We will call any finite sequence of zeros and ones a *word*. The *triplicate* of word A is the word obtained from A by writing it next to itself three times, that is, AAA. For instance, if $A = 101$, then its triplicate is word 101101101. The two following word operations are permitted:

a) insert a triplicate of an arbitrary word at any place (that also covers adding that triplicate at the very beginning or at the end of the word);

b) cross out a triplicate of an arbitrary word.

(For example, from 0001 we can obtain words 0111001 and 1.) Is it possible to obtain 01 from 10?

1988.44.* Baron Munchhausen's forest has firs and birches which grow in such a way that there are exactly 10 birch trees at the distance of 1 km from every fir tree. Baron claims that his forest has more firs than birches. Could it be true?

GRADE 9

1988.45. See Question 37.

1988.46. Several checkers are placed on the chessboard. A move consists of shifting one of the checkers to the neighboring square (horizontally or vertically). After several such moves, it turned out that every checker has visited all the squares exactly once and returned to its original position. Prove that there was a moment when none of the checkers were standing on their original squares.

1988.47. For positive real numbers a, b, c, and d prove that

$$\frac{1}{a} + \frac{1}{b} + \frac{4}{c} + \frac{16}{d} \geqslant \frac{64}{a+b+c+d}.$$

1988.48. Every street in the city of Faux-Minsk is a one-way street which connects two street intersections. Mayor has announced a tender for the construction of a network of gas stations such that it is always possible to reach a gas station from any street intersection, but at the same time no gas station can be reached from any other gas station. Prove that all viable submissions for this tender must plan the construction of the same number of gas stations.

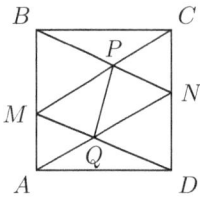

1988.49. Points M and N are selected on sides AB and CD of square $ABCD$. Segments CM and BN intersect at point P, and segments AN and MD intersect at point Q. Prove that $|PQ| \geqslant |AB|/2$.

1988.50.* Sequence a_1, a_2, ... consists of natural numbers not exceeding 1988. Given that for any n and m, number $a_m + a_n$ is divisible by a_{m+n}, prove that this sequence is aperiodic.

1988.51. See Question 43.

1988.52. See Question 44.

GRADE 10

1988.53. A snail crawls along the plane turning $90°$ after every meter of travel. At what maximum distance from its initial position can it find itself after travelling for 300 meters with 99 left and 200 right turns?

1988.54. Function $f \colon \mathbb{R} \to \mathbb{R}$ is continuous, and equality $f(x) \cdot f(f(x)) = 1$ holds true for any real x. Given t hat $f(1000) = 999$, find $f(500)$.

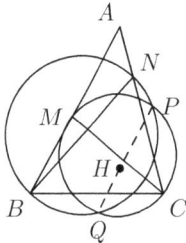

1988.55. See Question 39.

1988.56. See Question 40.

1988.57. Points M and N lie on sides AB and AC, respectively, of acute-angled triangle ABC. Circumferences built on segments BN and CM as diameters intersect at points P and Q. Prove that P, Q, and triangle's orthocenter[89] H lie on the same straight line.

1988.58.* Polynomial $P(x)$ has real coefficients. Prove that $P(x)$ is always non-negative if $P(x) - P'(x) - P''(x) + P'''(x) \geqslant 0$ for any real x.

1988.59. See Question 43.

1988.60.* In convex polygon M with n sides, the length of its kth side is a_k, and the length of the projection of M onto the line passing through that side is d_k ($k = 1, 2, \ldots, n$). Prove that

$$2 < \frac{a_1}{d_1} + \frac{a_2}{d_2} + \cdots + \frac{a_n}{d_n} \leqslant 4.$$

---

[89] Intersection of the triangle's altitudes.

Commentary. *The following three additional questions were offered to those contestants in Grades 7 and 8 who solved all the problems from the official set. However, these additional questions were not counted for the final tally and did not affect the awards.*

1988.61. *(a) See Question 6. (b) Prove that the painter can change any coloring of the castle with dimensions $n \times n$ to any other in no more than $2n^2$ steps (one step consists in moving from one room to an adjacent room).*

1988.62. *Is it possible to split numbers 1 through 1000 into several groups so that each group contains a number equal exactly to one third of the sum of all other numbers in the group?*

1988.63. *Suppose that $n - 1$ beetles sit on the squares of $n \times n$ board so that no two of them occupy neighboring squares (i.e., the squares with a common side). Prove that one of these beetles can crawl to a neighboring square so that this property will keep.*

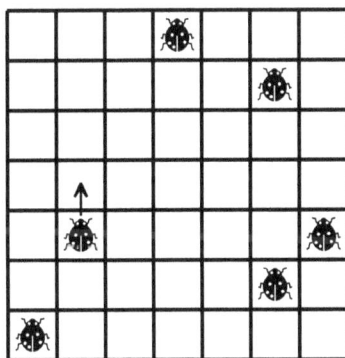

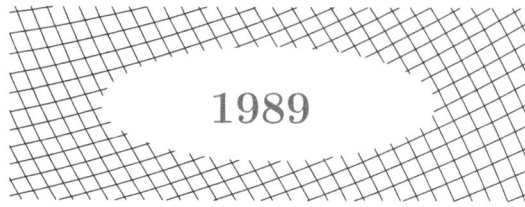

1989

GRADE 5

1989.01. A committee is tasked with creating problem sets for the all-city math olympiads of Grades 5, 6, 7, 8, 9, and 10. Members of the committee decided that each grade's set will consist of seven questions with exactly four of them not used in any other grade. What is the maximum possible number of questions used for the entire olympiad?

1989.02. Bus tickets have numbers $000\,000$ through $999\,999$. A number is called *lucky* if the sum of its first three digits is the same as the sum of its last three digits. Prove that the number of lucky tickets is the same as the number of bus tickets with the sum of the digits equal to 27.

1989.03. A toy railroad set consists of several rails of Type 1 and Type 2 (see the figure below), with each one of them marked by the arrow. The railroad must be built so that the directions of all arrows coincide with the direction of the train's motion. A closed track can be assembled from all the rails in the provided set. Prove that if one of Type 1 rails is replaced by a Type 2 rail, then it will become impossible to assemble such a closed track.

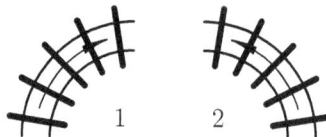

1989.04. There are 32 stones of different weights. Prove that it is always possible to find the two heaviest stones using no more than 35 weighings on the two-pan scales.

1989.05. Find at least one pair of different six-digit numbers such that if you write them next to each other, then the resulting twelve-digit number is divisible by the product of the two original numbers.

1989.06. Two people play a game on 10×10 board, writing in turn X's and O's into the board's squares; each player can write either one of the two symbols. The player after whose move three identical symbols form a contiguous sequence (horizontal, vertical, or diagonal) wins the game. Can one of the players guarantee himself a win, and if yes, then which one: the player who has the first move or his opponent?

GRADE 6

1989.07. See Question 1 for five grades (6 through 10).

1989.08. In a convex pentagon $ABCDE$, diagonals BE and BD intersect diagonal AC at points K and M, respectively. Prove that if $|AE| = |EK| = |KB|$ and $|AK| = |MC|$, then $|EM| = |BC|$.

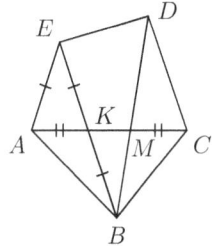

1989.09. See Question 4 for 64 stones and 68 weighings.

1989.10. Find all trios of integers a, b, and c such that

$$\begin{cases} a^2 + 2b^2 - 2bc = 100 \\ 2ab - c^2 = 100 \,. \end{cases}$$

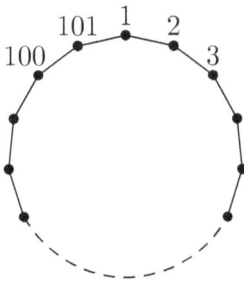

1989.11. Vertices of a regular polygon with 101 sides are indexed in clockwise order by numbers 1 through 101. Is it possible to stack up 99 copies of this polygon (both rotations and line reflections of the polygon are permitted) so that all the sums along each edge of the stack are the same?

1989.12. Find the smallest natural number greater than 1 which exceeds every one of its prime divisors by the factor of at least 600.

1989.13. Several (at least two) non-zero numbers are written on the blackboard. It is permitted to replace any two numbers a and b with pair $a + \frac{b}{2}$ and $b - \frac{a}{2}$. Prove that after several such operations the numbers on the board cannot be the same as the original collection.

GRADE 7

1989.14. At a conference, $2n$ scientists—n physicists and n mathematicians—sit around the round table. Some of them always tell the truth, and the others always lie; also, the number of liars among mathematicians is the same as the number of liars among physicists. When asked "Who is your neighbor on the right?", all scientists answered "A mathematician". Prove that n is even.

1989.15. In quadrilateral $ABCD$, diagonals AC and BD intersect at point O. Prove that if $|AB| = |OD|$, $|AD| = |CO|$, and $\angle BAC = \angle BDA$, then $ABCD$ is a trapezoid.

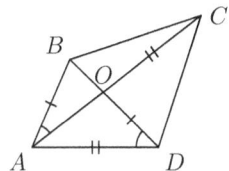

1989.16. Prove that if $x+y+z \geqslant xyz$, then $x^2+y^2+z^2 \geqslant xyz$.

1989.17. See Question 5.

1989.18. The library has a bookshelf with the eight-volume collected works of Leonhard Euler. Every minute a librarian comes over and swaps some two neighboring volumes. Can she perform these operations in such a way that at some moment all possible volume arrangements were realized, each one—exactly once?

1989.19. Two people play a game making their moves in turn. A 10×10 square is drawn on the sheet of graph paper, and a pawn is placed on the central grid node of this square. A player's move consists in moving the pawn to some other node of the square so that the length of the move (that is, the distance by which the pawn is shifted) is greater than that of the previous move, made by the opponent. The player who cannot make a move loses. Who wins in the errorless game?

1989.20. Does there exist a set of one hundred different natural numbers such that for any five of them their product is divisible by their sum?

GRADE 8

1989.21. Prove that the system of equations

$$\begin{cases} x + y + z = 0 \\ \dfrac{1}{x} + \dfrac{1}{y} + \dfrac{1}{z} = 0 \end{cases}$$

does not have any solutions in real numbers.

1989.22. Natural number a is greater than 1, and b is a natural divisor of number a^2+1. Prove that if $b - a > 0$, then $b - a > \sqrt{a}$.

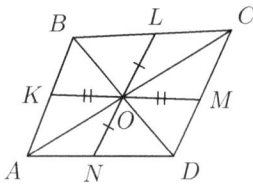

1989.23. In convex quadrilateral $ABCD$, diagonals AC and BD intersect at point O. Points K, L, M, and N lie on sides AB, BC, CD, and DA. Given that point O is the midpoint of both segments KM and LN, prove that $ABCD$ is a parallelogram.

1989.24. Suppose that m pawns are placed into the squares of infinite chessboard, and for each pawn we compute the product of the number of pawns in its column by the number of pawns in its column. Prove that the number of pawns for which this number is at least $10m$ cannot exceed $\frac{m}{10}$.

1989.25. Upon the completion of a k-round chess tournament, the players' results formed a geometric progression with integer quotient greater than 1. How many participants could there be if (a) $k = 1989$? (b) $k = 1988$?

1989.26. Set of n straight lines on the plane is such that no two of them are parallel and no three intersect at the same point. It is required to label every intersection point with a natural number smaller than n so that every line carried all $n - 1$ possible labels. What are the values of n for which such a labeling is possible?

GRADE 9

1989.27. See Question 21.

1989.28. Point X is selected on side AC of triangle ABC. Prove that if the inscribed circumferences of triangles ABX and BCX touch each other, then X lies on the incircle of triangle ABC.

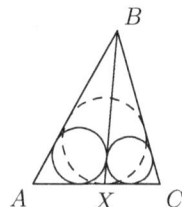

1989.29. See Question 22.

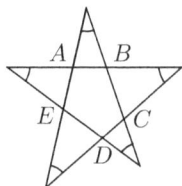

1989.30. A closed five-segment broken line forms the five-pointed star whose five vertex angles are all equal. What is the perimeter of the internal pentagon $ABCDE$, if the length of the broken line is 1?

1989.31. Is it possible to write numbers $+1$ and -1 into the squares of a 10×10 table so that all twenty sums of these numbers in rows and columns are different?

1989.32. See Question 25.

GRADE 9 (SCI-MATH)[90]

1989.33. See Question 21.

1989.34. Chords XK and XM of circumference divide its diameter AB into three equal parts. Prove that $5\,|KM| \leqslant 3\,|AB|$.

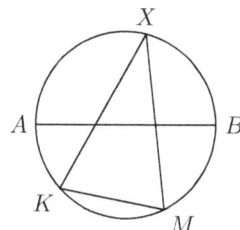

1989.35. See Question 25.

1989.36. See Question 30.

1989.37. Find out if operation $*$ which maps every pair of two natural numbers x and y to some natural number $z = x * y$ can possess all three following properties:

   a) $a * b = |a - b| * (a + b)$, $\forall a, b \in \mathbb{N}$ if $a \neq b$;
   b) $(ac) * (bc) = (a * b)(c * c)$, $\forall a, b, c \in \mathbb{N}$;
   c) $a * a = a$, if $a \in \mathbb{N}$ is odd.

GRADE 10

1989.38. Prove that if a, b, and c are real, then numbers $(b-c)(bc-a^2)$, $(c-a)(ca-b^2)$, $(a-b)(ab-c^2)$ cannot be all positive at the same time.

1989.39. See Question 28.

---

[90] For specialized Science-and-Math schools.

1989.40. Find the locus of all points $(x; y)$ on the coordinate plane that possess the following property: there exist two non-negative numbers a and b such that the largest of numbers a^2 and b equals x, and the smallest of numbers b^2 and a equals y.

1989.41. Pyramid has a regular polygon as its base. Prove that if all the face angles at the vertex are equal, then some two of the pyramid's side faces are congruent.

1989.42. See Question 25.

GRADE 10 (SCI-MATH)[91]

1989.43. See Question 21.

1989.44. Operation $*$ maps every pair of integers x and y to some integer number $z = x * y$. Every integer is equal to $x * y$ for some integers x and y. Prove that this operation cannot possess two following properties at the same time.

1) $a * b = -(b * a)$;
2) $(a * b) * c = a * (b * c)$.

1989.45. See Question 41.

1989.46. See Question 25.

1989.47. Prove that if equation $Ax^2 + (C - B)x + (E - D) = 0$ has a real root greater than 1, then equation $Ax^4 + Bx^3 + Cx^2 + Dx + E = 0$ has at least one real root.

FINAL ROUND

GRADE 8

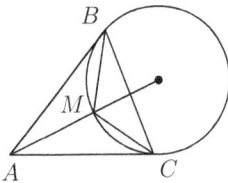

1989.48. Point M inside triangle ABC is such that $\angle BMC = 90° + \frac{1}{2}\angle BAC$, and line AM passes through the circumcenter of triangle BMC. Prove that M is the incenter of triangle ABC.

1989.49. All prime factors of all elements of a finite collection of different natural numbers do not exceed the given number n. Prove that the sum of reciprocals of these natural numbers does not exceed n.

1989.50. Natural number k is greater than 1. Prove that it is impossible to place numbers 1 through k^2 into the squares of $k \times k$ table so that the sum of the numbers in every row as well as in every column is a power of two.

---

[91] For specialized Science-and-Math schools.

1989.51. Ninety-one white pawns are somehow placed onto the squares of 10×10 board. A painter takes one of them, repaints it black, then places it back on any free square of the board. Then he again takes one of the white pawns, repaints it, puts it back on the board and so on, until all the pawns are black. Prove that at some moment two pawns of the different color stood next to each other (i.e., their squares had a common side).

1989.52. What is the maximum possible area of a quadrilateral with the sides 1, 4, 7, 8?

1989.53. Number 2 is written on the board. Two people play the following game, making their moves in turn. The player's move consists of replacing number n on the board with number $n + d$ where $d < n$ is some arbitrary divisor of n. The player who writes a number greater than 19891989 loses. Who wins in the errorless game, the player who makes the first move or her opponent?

1989.54.* In the language of Twiddle-Dee-Dum tribe, any sequence of ten digits 0 and 1 is a word. Two words are synonyms if and only if one of them can be obtained from the other by a series of operations of the following kind: several consecutive digits whose sum is even are removed from the word and then inserted back but in the reverse order. How many words with different meanings are there in the Twiddle-Dee-Dum language?

1989.55.* Professor Smith stands in the square room with mirror walls. Professor Jones would like to place a few of her students in the room so that Smith would not be able to see any of his own reflections (consider all these people as points; students can be placed at the walls or in the corners of the room). Is that possible?

GRADE 9

1989.56. See Question 48.

1989.57. All possible seven-digit sequences from 0000000 to 9999999 are written next to each in some order. Prove that the resulting $70,000,000$-digit number is divisible by 239.

1989.58. Real numbers x, y, z belong to $[0; 1]$. Prove that

$$2(x^3 + y^3 + z^3) - (x^2 y + y^2 z + z^2 x) \leqslant 3.$$

1989.59. See Question 51.

1989.60. A number triangle whose first row consists of n ones, and the second row—of $n - 1$ arbitrary integers (see below an example for $n = 6$), possesses the following property. For any four numbers that form rhombus $\begin{smallmatrix} & a & \\ b & & c \\ & d & \end{smallmatrix}$ (b and c are neighbors in their row), we have $bc = ad + 1$. If all the numbers in this triangle are different from zero, prove that each of them is an integer.

$$
\begin{array}{cccccccc}
1 & & 1 & & 1 & & 1 & & 1 & & 1 \\
& 2 & & 1 & & 3 & & 5 & & 2 \\
& & 1 & & 2 & & 14 & & 9 \\
& & & 1 & & 9 & & 25 \\
& & & & 4 & & 16 \\
& & & & & 7
\end{array}
$$

$$\boxed{2 \cdot 14 = 3 \cdot 9 + 1}$$

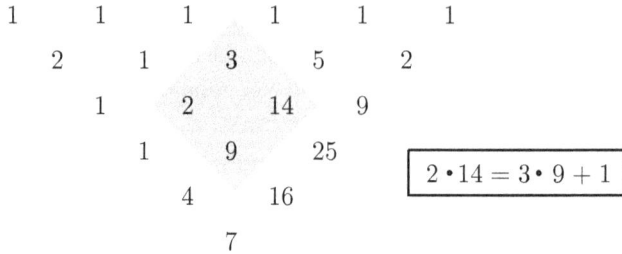

1989.61. Sequence of real numbers a_1, a_2, a_3, ... is such that for every natural k we have

$$a_{k+1} = \frac{ka_k + 1}{k - a_k}.$$

Prove that this sequence contains infinitely many positive and infinitely many negative numbers.

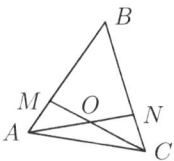

1989.62.* Point M lies on side AB of triangle ABC, point N lies on side BC, and O is intersection of segments CM and AN. Prove that if $|AM| + |AN| = |CM| + |CN|$, then $|AO| + |AB| = |CO| + |CB|$.

1989.63.* For which integers k is it possible to place 100 arcs on a circumference so that each arc intersects exactly k others?

1989.64.* Prove that if all numbers in the triangle from Question 60 are positive integers, then it contains at least $\frac{n}{4}$ different numbers.

GRADE 10

1989.65. See Question 48.

1989.66. See Question 49.

1989.67. See Question 60.

1989.68. See Question 52.

1989.69. How many solutions in real numbers does the equation

$$\sin(\sin(\sin(\sin(\sin(x))))) = \frac{x}{3}$$

have?

1989.70. Microcalculator "**FN-89**" performs only two operations: $X \mapsto 2X - 1$ and $X \mapsto 2X$. Some arbitrary natural number is typed into the calculator. Prove that by pushing the operation buttons you can obtain a number which is a perfect fifth power of an integer.

1989.71. Sequence of real numbers a_1, a_2, a_3, ... satisfies the inequality

$$|a_m + a_n - a_{m+n}| \leqslant \frac{1}{m + n}$$

for any indexes m and n. Prove that this sequence is an arithmetic progression.

1989.72.* There is a blackboard with number 1000 written on it, and there is a pile of 1000 matches. Two people are playing the game, making their moves in turn. The player's move consists of either taking from or putting back into the pile one, two, three, four, or five matches (at the beginning of the game players do not have any matches), after which the updated number of matches in the pile is written on the blackboard in addition to all the previously written numbers. The player who writes the number already present on the blackboard loses. Who will win in the errorless game, the player who makes the first move or his opponent?

1989.73. See Question 64.

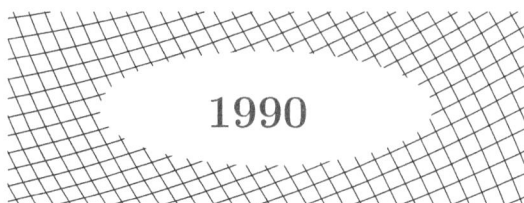

1990

GRADE 6

1990.01. Pete bought a notebook with 96 sheets and indexed all of its pages in order by numbers 1 through 192. Nick tore out some 25 sheets and added up all their page numbers. Prove that the result cannot be equal to 1990.

1990.02. Among one hundred and one coins, one is counterfeit with its weight different from that of a genuine coin; the rest are genuine identical coins. Using two weighings on two-pan scales, determine whether the counterfeit coin is heavier or lighter than the genuine ones.

1990.03. Is it possible to dissect a 55×39 rectangle into rectangles with dimensions 5×11?

1990.04. Tom and Jerry play the following game making their moves in turn. An integer is written on the board and the move consists of subtracting one of its non-zero digits from it, then replacing the number on the board with the result. The player who obtains zero wins. In the beginning the number on the board is 1234, and Tom makes the first move. Who will win in the errorless game?

1990.05. Pete, Nick, and Vera together solved one hundred math problems, each one of them solving sixty. We will call a problem *difficult* if it was solved by only one of the students, and we will call a problem *easy* if it was solved by all three of them. Prove that the number of difficult problems exceeds the number of easy ones by exactly twenty.

1990.06. In the village of Marmosette[92] all the girl acquaintances of any boy know each other. Also for any girl, the number of boys she knows is greater than the number of girls she knows. Prove that the number of boys in Marmosette is at least as high as the number of girls there.

GRADE 7

1990.07. John and Mary live in a skyscraper which has 10 apartments on every floor. John's floor number equals Mary's apartment number, and the sum of the numbers of their apartments is 239. Find John's apartment number.[93]

---

[92] This rather unusual name is a literal translation of the name of a small village next to the Leningrad University suburban campus. Surprisingly, it has nothing to do with cute little monkeys.
[93] Standard apartment numbering, using the ascending order of the floors, is assumed.

1990.08. Thirty chairs are arranged in a row. From time to time, a person comes in and sits on one of the unoccupied chairs. If one of the neighboring chairs is already occupied, then one of the neighbors stands up and leaves. What is the maximum possible number of simultaneously occupied chairs if they were all free at the beginning?

1990.09. The computer screen shows number 123. Each minute the computer increases the number on the screen by 102. Hacker Theo can at any moment change the number on the screen by reordering its digits in some arbitrary way. Can Theo do that indefinitely in such a way that the screen always shows a three-digit number?

1990.10. In quadrilateral $ABCD$, we have $|BC| = |AD|$; point M is the midpoint of AD, and point N is the midpoint of BC. Perpendicular bisectors of segments AB and CD intersect at point P. Prove that P also lies on the perpendicular bisector of segment MN.

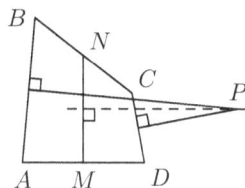

1990.11. A 2×2 square is dissected into several (more than one) rectangles. Prove that you can shade some of them so that the length of the projection of the shaded figure onto one side of the square does not exceed 1, while the length of its projection onto the other side is at least 1.

1990.12. See Question 6.

1990.13. Numbers $+1$ and -1 are written into some squares of 50×50 table in such a way that the absolute value of the sum of all numbers in the table does not exceed 100. Prove that for some square subtable with side 25, the absolute value of the sum of the numbers in it does not exceed 25.

Grade 8

1990.14. Diane bought a notebook with 96 sheets and indexed all of its pages in order by numbers 1 through 192. Samantha tore out some 24 sheets and added up all the page numbers on them. Could the result be equal to 1990?

1990.15. See Question 10.

1990.16. Find all trios of natural numbers (a, b, c) such that

$$\begin{cases} a^2 + b - c = 100 \\ a + b^2 - c = 124. \end{cases}$$

1990.17. Republic of Faraway has 101 towns, and they are connected with one-way roads so that every two towns are directly connected by at most one road. It is also known that there are exactly 40 roads coming out of every town, and that there are exactly 40 roads going into every town in Faraway. Prove that it is possible to travel from any town to any other having used no more than three roads.

1990.18. Among 103 identically looking coins exactly two are counterfeit with equal weight different from that of a genuine coin. Using three weighings on two-pan scales, determine whether a counterfeit coin is heavier or lighter than a genuine one.

1990.19. On the island of Logika, each person is either a "knave" who always lies or a "knight" who always speaks the truth. Each islander uttered the following two sentences.

   a) All of my acquaintances know each other.
   b) I know at least as many knaves as I know knights.

Prove that the number of knaves on the island does not exceed the number of knights.

1990.20. How many pairs of natural numbers (m, n) with $m, n \leqslant 1000$ satisfy inequalities

$$\frac{m}{n+1} < \sqrt{2} < \frac{m+1}{n}\ ?$$

GRADE 9

1990.21. Let x and y be some arbitrary natural numbers. Can the sum $x! + y!$ end with digits $\ldots 1990$?

1990.22. Does there exist a triangle whose sides' lengths are integers, and the length of one of the medians is 1?

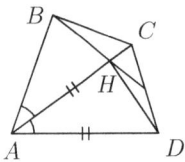

1990.23. Prove that any arithmetic progression consisting of natural numbers contains two terms with equal sums of the digits.

1990.24. In a convex quadrilateral $ABCD$, angle B is right, and the length of diagonal AC, which also happens to be the bisector of angle A, is equal to the length of side AD. If DH is an altitude in triangle ADC, then prove that line BH divides segment CD in half.

1990.25. Real numbers a, b, and c lie inside interval $[0; 1]$. Prove that

$$\frac{a}{1+bc} + \frac{b}{1+ac} + \frac{c}{1+ab} \leqslant 2\,.$$

1990.26. See Question 37 (a) and (b).

GRADES 10–11

1990.27. Find all solutions of the system of equations

$$\begin{cases} x^2 + y^2 = 6z \\ y^2 + z^2 = 6x \\ z^2 + x^2 = 6y\,. \end{cases}$$

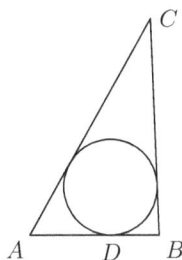

1990.28. The circumference inscribed into triangle ABC divides its side AB into segments AD and DB with lengths 5 and 3, respectively. Measure of angle A is $60°$. Find the length of side BC.

1990.29. See Question 23.

1990.30. See Question 22.

1990.31. For any four different natural numbers, prove that their product is greater than half of the sum of all pairwise products of these numbers.

1990.32. Is it possible to cover the plane with squares whose side lengths are 1, 2, 4, 8, 16, ... without overlap if it is permitted to use every square no more than (a) 10 times? (b) once?

GRADE 10 (SCI-MATH)[94]

1990.33. Is it possible to form a six-digit number divisible by 11 from digits 1 through 6 without repetitions?

1990.34. See Question 24.

1990.35. Polynomial F with integer coefficients is such that $F(2)$ is divisible by 5, and $F(5)$ is divisible by 2. Prove that $F(7)$ is divisible by 10.

1990.36. See Question 25.

1990.37. A board with dimensions 10×10 is covered by n squares with dimensions 2×2 whose sides lie on the grid lines of the board. Prove that it is possible to remove one of the squares so that the remaining ones still cover the entire board, if

a) $n = 55$;
b) $n = 45$;
c) Try to find the smallest value of n for which the statement is true (of course, n must be large enough so that it is possible to cover the board with n two-by-two squares).

GRADE 11 (SCI-MATH)

1990.38. See Question 33.

1990.39. See Question 24.

1990.40. See Question 35.

---

[94] For specialized Science-and-Math schools.

1990.41. Positive real number x is a root of the equation

$$x^n = x^{n-1} + x^{n-2} + \cdots + x + 1.$$

Prove that $2 > x > 2 - \dfrac{1}{n}$.

1990.42. Prove that the three-dimensional space can be dissected into regular octahedrons and tetrahedrons with integer lengths of their edges so that this dissection does not contain ten identical polyhedrons.

Final Round

Grade 9

1990.43. Natural numbers a and b are such that $a^2 + ab + 1$ is divisible by $b^2 + ba + 1$. Prove that $a = b$.

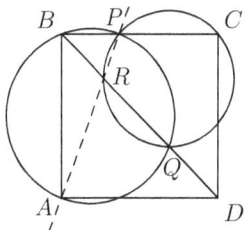

1990.44. Segment S contains several smaller segments which completely cover it. Prove that their left halves cover at least a half of segment S.

1990.45. Point P is selected on side BC of square $ABCD$. A circumference passing through points A, B, and P intersects diagonal BD also at point Q. A circumference passing through C, P, and Q which intersects BD also at point R. Prove that points A, R, and P lie on the same straight line.

1990.46. Consider all possible subsets of set $\{1, 2, \ldots, n\}$ which do not contain any two neighboring numbers, and for each subset compute the product of all numbers in the subset. Prove that the sum of the squares of these products is equal to $(n+1)! - 1$.

1990.47. Vertices of an inscribed quadrilateral $ABCD$, that is not a trapezoid, lie on the grid nodes of a graph paper sheet whose squares have unit side length. Prove that $\big| |AC| \cdot |AD| - |BC| \cdot |BD| \big| \geqslant 1$.

1990.48. In the country of Humdrum, consisting of two states Hum and Drum where each road connects two cities from different states. It is also known that there are no more than ten roads coming out of every city. Prove that the road map of Humdrum can be colored in ten colors so that any two roads coming from the same city have different colors.

1990.49.* A hostess has baked a pie for the guests at her birthday party. The number of guests is expected to be either p or q, where p and q are co-prime. The hostess wants to cut the pie in advance into several pieces so that regardless of how many guests will attend she will always be able to give everyone their equal share of the pie. What is the smallest possible number of pieces (not necessarily equal in size) that will guarantee that she can achieve her objective?

1990.50.* Twenty numbers are written around the circle. It is allowed to replace any trio of consecutive numbers x, y, z with trio $x + y$, $-y$, $z + y$ (in this exact order). Can sequence

$$\{10,\ 9, 8,\ \ldots,\ 2,\ 1,\ -10,\ -9,\ -8,\ \ldots,\ -2,\ -1\}$$

be obtained from the original sequence

$$\{1,\ 2,\ 3,\ \ldots,\ 9,\ 10,\ -1,\ -2,\ -3,\ \ldots,\ -9,\ -10\}$$

(in both cases numbers are listed in the clockwise direction)?

GRADE 10

1990.51. See Question 43.

1990.52. See Question 44.

1990.53. See Question 45.

1990.54. Alex and Serge play the following game on a 25×25 board. They make their moves in turn, with Alex having the first move. Each move consists of a player coloring one of the yet unpainted squares—Alex uses the white paint and Serge—the black paint. Can Alex play in such a way that regardless of Serge's moves, at the end of the game a chess king will be able to travel through all the white squares (possibly visiting some squares more than once)?

1990.55. Vertices of quadrilateral $ABCD$ lie on the grid nodes of a graph paper sheet whose squares have unit side length. Angles A and C are equal while angles B and D are different. Prove that $\big||AB| \cdot |BC| - |CD| \cdot |DA|\big| \geqslant 1$.

1990.56.* The collected works of Leo Tolstoy in one hundred volumes was placed on the shelf in some random order. A *swap* is the operation of taking any two volumes whose numbers have the opposite parity and switch their positions. What is the minimum number n such that you can always restore the correct order by using no more than n swaps?

1990.57.* Polynomial $F(x)$ with integer coefficients and finite sequence of integers $A = \{a_1, a_2, \ldots, a_m\}$ are such that for any integer n value $F(n)$ is divisible by at least one of the numbers from sequence A. Prove that one of the numbers in A divides $F(n)$ for all integer values of n.

1990.58.* Twenty-two points are selected in segment $[0; 1]$. It is permitted to replace any two of these points by the midpoint of the segment connecting them. Prove that it is possible to perform twenty such operations so that the distance between the two final points will not exceed 0.001.

Grade 11

1990.59. Natural numbers a and n are greater than 1. Prove that the number of positive integers which are less than $b = a^n - 1$ and co-prime with b is divisible by n.

1990.60. Segment S contains several smaller segments which completely cover it. From each of these smaller segments, a half is removed—either right one or left one. Prove that the remaining halves cover at least the third of segment S.

1990.61. Does there exist a planar hexagon (possibly, non-convex) all of whose diagonals except one have the same length?

1990.62. A board with dimensions 100×100 was made of graph paper. Its upper edge was then glued to the lower edge, and its right edge—to the left one, after which the board became doughnut-shaped. Is it possible to place 50 chess rooks of three colors—red, blue, and green—on its squares so that each red rook attacks at least two blue ones, each blue rook attacks at least two green ones, and each green rook attacks at least two red ones?

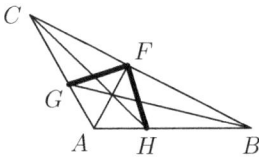

1990.63. In triangle ABC, angle A equals $120°$, and AF, BG, and CH are the triangle's angle bisectors. Prove that angle GFH is right angle.

1990.64.* The kingdom of Olympia has one hundred towns, and every two of them are connected by exactly one road with one-way traffic. There are two towns A and B such that one cannot drive from A to B without breaking the traffic rules. Prove that the king can choose some town and change the direction of all roads connecting it with all other towns so that after this procedure it would be possible to drive from any town in Olympia to any other one obeying the rules of the road.

1990.65. Continuous function $f \colon \mathbb{R} \to \mathbb{R}$ satisfies the equation $f(x + f(x)) = f(x)$ for all real x. Prove that function f is constant.

1990.66.* Several numbers with positive sum are written around the circle. It is allowed to replace any three consecutive numbers x, y, z with $x + y$, $-y$, $z + y$ (in this exact order). Prove that using these operations you can obtain exactly one arrangement consisting of non-negative numbers.

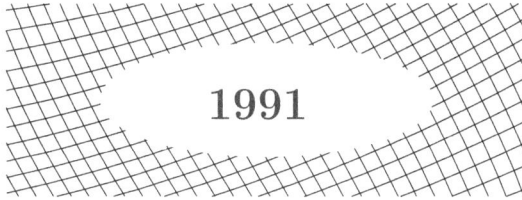

GRADE 6

1991.01. Forty students attend the school rocket-building club. Each one of them has some bolts, screws, and nails. There are fifteen students with number of bolts different from the number of nails, and ten students with number of screws equal to the number of nails. Prove that there are at least fifteen students with the number of screws different from the number of bolts.

1991.02. On the black market in the village of Perestroikino, one can exchange any two food stamps for some three other stamps, and vice versa. Can co-op worker Ivan exchange 100 bread stamps for 100 meat stamps by giving away exactly 1991 stamps in the process?[95]

1991.03. Four cars A, B, C, and D start racing on the closed loop track from the same point; the first two go clockwise, and the other two—counterclockwise. Speeds of all four cars are constant but may be different. It so happens that the first time A meets C is the same as the first time B meets D. Prove that the first time one of the cars A and B catches up to the other is the same as the first time when one of the cars C and D catches up to the other.

1991.04. For many years, baron Munchhausen starts every day by going duck hunting. On August 1, 1991 he said to his cook, "Today I got more ducks than two days ago but less than a week ago". What is the maximum number of days baron can repeat that sentence? (Don't forget that baron Munchhausen never lies!)

1991.05. Three sticks—red, blue, and white—are all one meter long. Nick breaks the first stick into three parts; Vera does the same with the second stick, and finally Nick breaks the third stick in three. Can Nick break his sticks in such a way that regardless of Vera's action, three triangles can be formed from the resulting nine sticks, each triangle with one side red, one side blue, and one side white?

---

[95] This problem is an ironic play on Soviet economic reforms of 1989–1991 which led to rampant inflation and introduction of countrywide system of food stamps which lasted until 1992.

1991.06. Nine teams have played a round-robin volleyball tournament. Is it true that there must be two teams A and B such that every other team has lost to either team A or to team B?

GRADE 7

1991.07. Place numbers 1 through 12 on the twelve segments shown in the figure so that all sums of the numbers on the sides of every small square are the same.

1991.08. Pearl divers have brought to the surface n pearls, $n \leqslant 1000$. They divide the treasure in the following manner. One by one they step to the pile of pearls, and each diver takes either exactly one half or exactly one third of the remaining pearls. After every diver took his or her share, the rest was thrown back into the sea as a sacrifice to Poseidon. What is the largest possible number of divers who took part in this pearl hunt?

1991.09. There are 1991 representatives of four tribes—people, gnomes, elves, and goblins—sitting at the round table. People never sit next to goblins, and elves never sit next to gnomes. Prove that some two neighbors at the table belong to the same tribe.

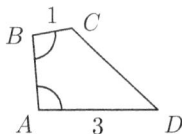

1991.10. In convex quadrilateral $ABCD$, angles A and B are equal. Given that $|BC| = 1$ and $|AD| = 3$, prove that $|CD| > 2$.

1991.11. The statement of a problem in the textbook reads: "Prove that if a collection of n numbers is such that the sum of any ten of them is greater than the sum of all other numbers, then all numbers in the collection are positive." We know that character n in the statement is a typo—some specific natural number not equal to 20 was supposed to be printed there instead. What is that number? (You need to find all possible answers.)

1991.12. In some country, every two towns are connected by exactly one road which is either a highway or a railroad. Prove that you can always select one mode of transportation—roads or railways—so that each town can be reached from any other using only this transportation mode and visiting no more than two other towns on the way.

1991.13. Square 7×7 is dissected into polyominoes ⊞, ⊞, and ⊞. Prove that this dissection uses exactly one tetromino (that is, ⊞ or ⊞).

GRADE 8

1991.14. See Question 9.

1991.15. Natural number x that does not have zeros in its decimal representation, satisfies the equation $x \cdot \bar{x} = 1000 + P(x)$. Here $\bar{x}$ denotes the number which has the same digits as x but in reversed order, while $P(x)$ is the product of all digits of number x. Find all such numbers.

1991.16. See Question 5.

1991.17. Perpendiculars AX and AY are dropped from vertex A of triangle ABC onto angle bisectors of external angles B and C. Prove that the length of segment XY is equal to half-perimeter of triangle ABC.

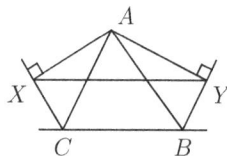

1991.18. Prove the equality
$$\frac{(2^3 - 1)(3^3 - 1)\ldots(100^3 - 1)}{(2^3 + 1)(3^3 + 1)\ldots(100^3 + 1)} = \frac{3367}{5050}.$$

1991.19. Square $(2n - 1) \times (2n - 1)$ is dissected into polyominoes ⊟, ⊞, and ⌗. Prove that this dissection contains at least $4n - 1$ triminoes (that is, figures ⊟).

1991.20. For any collection A of ten different real numbers, one can construct a collection $A^{(5)}$ consisting of all possible sums of five different numbers from A. Do there exist two different collections A and B such that the collections $A^{(5)}$ and $B^{(5)}$ are identical?

GRADE 9

1991.21. For any non-negative real numbers a, b, and c prove the inequality
$$\max(a^2 - b, b^2 - c, c^2 - a) \geqslant \max(a^2 - a, b^2 - b, c^2 - c).$$
Here $\max(x, y, \ldots)$ denotes maximum number in finite collection $\{x, y, \ldots\}$.

1991.22. In an acute-angled triangle ABC, side BC is shorter than side AB. Points X and Y are selected on sides AB and BC, respectively, so that $|AX| = |BY|$. Prove that $|XY| \geqslant |AC|/2$.

1991.23. Lengths x, y and z of the sides of a triangle are integers. Also one of the triangle's altitudes is equal to the sum of two other altitudes. Prove that $x^2 + y^2 + z^2$ is a perfect square.

1991.24. Sequence of natural numbers (a_n) is constructed as follows. Each term with even index a_{2n} is obtained from the previous term a_{2n-1} by subtracting one of its digits, and each term with odd index a_{2n+1} is obtained from the previous term a_{2n} by adding one of its digits. Prove that all terms of this sequence do not exceed $10a_1$.

1991.25. Point P is located outside circumference S centered at point O. Lines ℓ_1 and ℓ_2 pass through point P, ℓ_1 touches S at point A, and ℓ_2 intersects S at points B and C. Lines tangent to S at B and C intersect at point X. Prove that lines AX and PO are perpendicular.

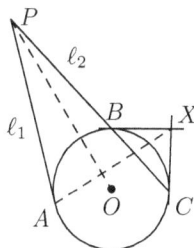

1991.26. At a conference, each delegate knows at least one other attendee, but for every two delegates, there is another one acquainted with neither of these two. Prove that it is possible to split all the delegates into three groups so that each attendee knows at least one person from their group.

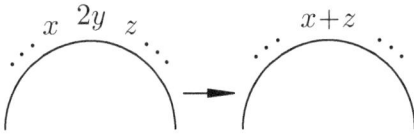

1991.27. Several integers are placed around the circle. It is permitted to do the following. Erase any even number and then replace both its neighbors by one number equal to their sum. These operations are performed until there are no even numbers or until only one or two numbers are left. Prove that the number of integers on the circle at that point does not depend on the choice or the order of operations but only on the original number sequence.

GRADE 10

1991.28. Let a, b, c, and d be four non-negative real numbers. Prove the inequality

$$\max(a^2 - b, b^2 - c, c^2 - d, d^2 - a) \geqslant \max(a^2 - a, b^2 - b, c^2 - c, d^2 - d).$$

1991.29. Two circumferences with centers O_1 and O_2 intersect at points A and B. Circumference (O_1BO_2) intersects the second circumference also at point P.[96] Prove that points O_1, A, and P lie on the same straight line.

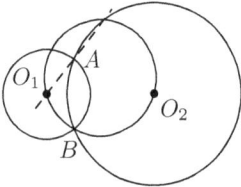

1991.30. At a conference, every delegate is acquainted with at least one other attendee but not with all of them. Prove that it is possible to split all the delegates into two groups so that each attendee knows at least one person from their group.

1991.31. Continuous strictly monotonically increasing function f is such that $f(0) = 0$ and $f(1) = 1$. Prove that

$$f\left(\frac{1}{10}\right) + f\left(\frac{2}{10}\right) + \cdots + f\left(\frac{9}{10}\right) + f^{-1}\left(\frac{1}{10}\right) + f^{-1}\left(\frac{2}{10}\right) + \cdots + f^{-1}\left(\frac{9}{10}\right) \leqslant \frac{99}{10}.$$

1991.32. A **KPK-1991** computer is able to execute the two following operations on the natural numbers:

    a) compute the square of a number;

    b) for any given natural number x, compute number $a + b$, where a is the number formed by the last three digits of x, and b is the number formed by the first $n - 3$ digits of x.

Is it possible to use this computer to obtain number 703 from number 604?

---

[96] In addition to point B.

1991.33. Straight line L, point P, and polygon M with n sides are drawn on the plane. Line L intersects all sides of polygon M in interior points, and each of these points is the foot of the perpendicular dropped from P onto the corresponding side of M. Prove that $n = 4$.

1991.34.* Squares of $n \times n$ board are colored in red, blue, and green in such a way that there is a blue square next to every red square, a green square next to every blue square, and a red square next to every green square (that is, the corresponding squares have a common side). Prove that the number of the red squares R satisfies the inequalities

(a) $R \leqslant \frac{2n^2}{3}$.

(b) $R \geqslant \frac{n^2}{11}$.

GRADE 11

1991.35. See Question 28.

1991.36. Is it possible to split numbers 1 through 100 into three groups so that the sum of the numbers in the first group is divisible by 102, the sum of the numbers in the second group is divisible by 203, and the sum of the numbers in the third group is divisible by 304?

1991.37. See Question 25.

1991.38. Sequence of natural numbers (a_n) is constructed as follows. Each term with even index a_{2n} is obtained from number a_{2n-1} by adding one of its non-zero digits, and each term with odd index a_{2n+1} is obtained from number a_{2n} by subtracting one of its non-zero digits. Prove that all terms of this sequence do not exceed $4a_1 + 44$.

1991.39. Do there exist four different numbers such that any two of them—x and y— satisfy the equation
$$x^{10} + x^9 y + x^8 y^2 + \cdots + xy^9 + y^{10} = 1 \text{ ?}$$

1991.40. Points X and Y are selected on sides AB and BC of triangle ABC, respectively, in such a way that
$$\angle AXY = 2\angle ACB, \quad \angle CYX = 2\angle BAC.$$
Prove the inequality
$$\frac{S_{AXYC}}{S_{ABC}} \leqslant \frac{|AX|^2 + |XY|^2 + |YC|^2}{|AC|^2}.$$

1991.41.* There are 1991 cities on planet Tranai, and each two of them are connected by a two-way highway. Every day Tranai Department of Transportation closes three of the two-way highways for repair, and then introduces one-way traffic on one of the remaining roads. Prove that the Department can choose these roads in such a way that in the end it is possible to drive from each city to any other while obeying the rules of the road.

ELIMINATION ROUND

GRADES 9–10

1991.42. There are seventy different natural numbers not exceeding 200. Prove that some two of them differ either by 4, or by 5, or by 9.

1991.43. Two circumferences with equal radii intersect at points A and B. An arbitrary straight line passing through point B intersects these circumferences also at points X and Y, respectively. Find the locus of the midpoints of segments XY.

1991.44. Natural numbers $a_1, a_2, \ldots, a_n$ are such that, for any natural number $k < n$, the sum of any k numbers from this sequence is at least $k(k-1)$ and the sum of all numbers in the sequence is $n(n-1)$. Prove that a round-robin soccer tournament[97] with n teams can have the results coinciding with the numbers a_i.

1991.45. Find eight natural numbers a_i such that

$$\sqrt{\sqrt{a_1} - \sqrt{a_1 - 1}} + \cdots + \sqrt{\sqrt{a_8} - \sqrt{a_8 - 1}} = 2.$$

1991.46. Does there exist a function $f \colon \mathbb{N} \to \mathbb{N}$ such that

$$\underbrace{f(f(\ldots f(x)\ldots))}_{f(x) \text{ times}} = x + 1$$

for any natural x (note that f is applied $f(x)$ times)?

1991.47.[*] Twenty-six non-zero digits are written in a row. Prove that this row can be split into several contiguous intervals so that the sum of the numbers formed by the digits in these intervals is divisible by 13.

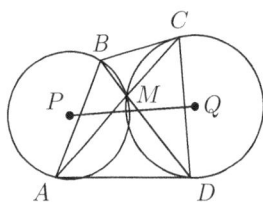

1991.48. Diagonals of convex quadrilateral $ABCD$ intersect at point M. Let P and Q be the circumcenters of triangles ABM and CDM. Prove that $|AB| + |CD| \leqslant 4|PQ|$.

1991.49.[*] We will call a *shuffle* the following procedure performed with a deck of playing cards—the deck is divided into several (arbitrary number of) portions, which are then re-arranged in the reverse order without changing the order of cards within these portions. Prove that any arrangement of a deck with 1000 cards can be transformed into any other arrangement with no more than 56 shuffles.

---

[97] At the time, a team received two points for the win in a soccer match, one point for the draw, and zero points for the loss.

GRADE 11

1991.50. A black checker is placed on the upper left corner of the 8×8 board. It is permitted to put a white checker on any free square of the board and then repaint all the checkers on the neighboring squares (that is, the squares that have a common vertex with it). Can we obtain the board completely filled with white checkers?

1991.51. Let AB be a chord of a circumference, and points M and N—the midpoints of the arcs that A and B divide the circumference into. When rotated by some angle around point A, point B is transformed to point B', and point M—to point M'. Prove that the segments connecting the midpoint of BB' with points M' and N are perpendicular.

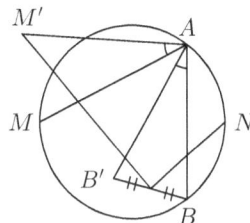

1991.52. Real numbers $x_1, x_2, \ldots, x_n$ lie in $[-1; 1]$, and the sum of their cubes equals zero. Prove that $x_1 + x_2 + \cdots + x_n$ does not exceed $\frac{n}{3}$.

1991.53. Two operations are used to transform a given natural number.

a) Multiplication by any natural number;
b) Removal of some zeros from its decimal representation.

Prove that, using these operations, it is always possible to produce a one-digit number.

1991.54.* For any two continuous functions $f, g \colon [0; 1] \to [0; 1]$ such that f is a monotonically increasing function, prove inequality

$$\int_0^1 f(g(x))\, dx \leqslant \int_0^1 f(x)\, dx + \int_0^1 g(x)\, dx .$$

1991.55.* Let us call a finite sequence $a_1, a_2, \ldots, a_n$ p-balanced if all sums of form

$$a_k + a_{k+p} + a_{k+2p} + \cdots \quad (k = 1, 2, \ldots, p)$$

are equal. Prove that if a sequence with 50 terms is p-balanced for each $p = 3,\ 5,\ 7,\ 11,\ 13,\ 17$, then all of its terms are zeros.

1991.56.* Prove that number $512^3 + 675^3 + 720^3$ is composite.

1991.57.* Let us call an arbitrary set of n vertices of a regular polygon P with $2n$ sides a *pattern*. Is it true that for any pattern, there always exist one hundred rotations of P such that the images of the pattern cover the entire set of all $2n$ vertices of P?

PART 3
Solutions

1961

1961.01. Answer: $\frac{7}{5}$ times as fast.

Indeed, let us say that in one hour they do x, y, and z percent of the work, respectively. Then $y + z = 2x$, $x + z = 3y$. So we have $x = \frac{4}{5}z$, $y = \frac{3}{5}z$, and therefore $x + y = \frac{7}{5}z$.

1961.02. Assuming $d = \gcd(a + b, \operatorname{lcm}(a, b))$ is divisible by a power of some prime number p^k, we have that $\operatorname{lcm}(a, b) \vdots p^k$; that is, one of the numbers a and b is divisible by p^k. Since $a + b$ is also divisible by p^k, both of them are multiples of p^k. It follows then that both a and b are divisible by d, and $\gcd(a, b)$ is also divisible by d. Clearly d is divisible by $\gcd(a, b)$, and therefore $d = \gcd(a, b)$.

1961.03. Let us construct a graph whose vertices correspond to the students, and two vertices are connected with an edge if and only if there is a question solved by both students. The degree of each vertex is two, thus the graph is a disjoint union of several cycles (some of them are possibly only two edges long).[98] Orient these cycles and split each one into pairs: vertex and the edge coming out of it. This will produce the desired presentation.

1961.04. Answer: Claire has to start $\frac{3}{11}$ hours before Abby and Beth.

If Abby and Beth arrived to N simultaneously, then they have walked the same portion of their trip. Therefore, if Abby and Claire met at the distance of s km from N, then Beth rode the bicycle to meet Claire for s km. During that time Abby walked $\frac{2}{5}s$ km, after which the distance between her and Claire started decreasing at the rate of 21 km/h and thus she walked $\frac{3}{5}s \cdot \frac{6}{21} = \frac{6}{35}s$ km more. So we have $15 - s = \frac{2}{5}s + \frac{6}{35}s$, thus $s = \frac{105}{11}$. It follows that Beth reached the point of her meeting with Claire in $\frac{1}{15}s = \frac{7}{11}$ hours, and we know that it took Claire $\frac{1}{6}15 - s = \frac{10}{11}$ hours to walk there. Thus she had to start $\frac{10}{11} - \frac{7}{11} = \frac{3}{11}$ hours before Abby and Beth.

1961.05. Consider one of these people and call him X. It is clear that among the other five, X has either at least three acquaintances or at least three strangers. Without loss of generality, we can assume that X is acquainted with at least three of them—let us call them A, B, and C. If any two of them (say, B and C) are acquainted, then we have a trio of mutually acquainted people X, B, and C. If none of them are, then we have a trio A, B, and C of mutual strangers.

1961.07. Let us translate circle O parallel to line L by the given distance. This can

---

[98] We remind the readers that the degree of a vertex in a graph is the number of edges incident to that vertex, and the cycle is the closed path without repeating edges.

easily be performed using ruler-and-compass operations. Then intersect the resulting circle O' with square K. Any of the intersection points can serve as an endpoint of the segment we need.[99]

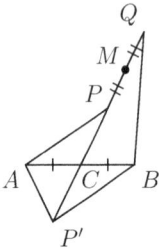

1961.08. $\overline{bca} + \overline{cab} = 110c + 11a + 101b = 37 \times (3a + 3b + 3c) - \overline{abc}$, and that is divisible by 37.

1961.09. Let P' be a point symmetric to P with respect to C. Then $|P'Q| = 2|CM|$, $|P'B| = |AP|$, and the inequality we want to prove is simply the triangle inequality for $\triangle P'QB$. Same solution without any additional construction can be done via vectors, by proving that $2\overrightarrow{CM} = \overrightarrow{AP} + \overrightarrow{BQ}$.

1961.10. For every subset X of a set of $2n+1$ objects, there is its complement Y which consists of all the objects that were not included in X. It is obvious that numbers of objects in sets X and Y are always of opposite parity. This proves that the number of ways to choose a set with even number of objects is equal to the one for the odd number of objects.

1961.11. It is easy to construct the triangle knowing the midpoints of the sides. Thus, to construct the quadrilateral, it is sufficient to have the midpoints of its sides and of one diagonal.

Let us mark two points M and N, at the given distance between them. They will serve as the midpoints of the diagonals of the quadrilateral. Now, we can construct midpoint P of any side since we know the distances $|PN|$ and $|PM|$—they are equal to exactly one half of the corresponding sides of the quadrilateral.

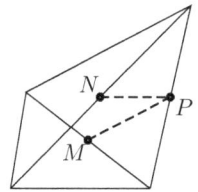

1961.12. Let $\sqrt{a} + \sqrt{b} = r_1$, then $\sqrt{ab} = \frac{1}{2}(r_1^2 - a - b) = r_2$ is rational. Numbers $\sqrt{a}$ and $\sqrt{b}$ are the roots of the quadratic equation

$$t^2 - r_1 t + r_2 = 0,$$

and therefore they are equal to $\frac{1}{2}(r_1 \pm \sqrt{r_1^2 - 4r_2})$. Since

$$4a = \left(2\sqrt{a}\right)^2 = 2r_1^2 - 4r_2 + 2r_1\sqrt{r_1^2 - 4r_2},$$

we obtain that $\sqrt{r_1^2 - 4r_2}$ is rational, and therefore the roots of the equation are rational as well.

1961.13. Answer: $\sqrt[3]{4}$.

Obviously, there are no solutions for $|x| \leqslant 1$. If $x < -1$, then the left-hand side is negative. When $x \geqslant 2$, function $x^3 - x$ is monotonically increasing, and therefore $x^3 - [x] \geqslant x^3 - x \geqslant 2^3 - 2 > 3$. Finally, $[x] = 1$ when $x \in [1; 2)$, so in this case our equation transforms into $x^3 - 1 = 3$.

1961.14. In any triangle, an angle bisector also bisects the angle between the altitude and the corresponding radius of the triangle's circumcircle. Indeed, if in triangle ABC angle

---

[99] The number of such points depends, of course, on the relative positions of O, K, and L.

B is acute, O is the circumcenter, and AH is altitude, then $\angle BAH = 90° - \angle B$. So the central angle $\angle AOC$ equals $2\angle B$. Hence, $\angle CAO = 90° - \angle B = \angle BAH$, and bisector of angle A evenly divides the angle between the altitude and radius AO.

Thus it follows that in this triangle, the circumcenter, point O, lies on the median. This is only possible when either the median is perpendicular to its side (isosceles triangle), or when O is the midpoint of the side (right triangle).

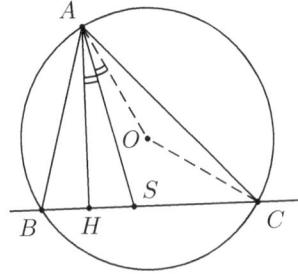

1961.15. If the sum of n numbers, each of them either $+1$ or -1, equals zero, then the number of the positive summands must be equal to the number of the negative summands, that is, $\frac{n}{2} = k$ is an integer. Since the product of all these summands $(x_1 x_2 \ldots x_n)^2$ is obviously positive, the number of the negative summands must be even as well, i.e., k is even, and therefore n is a multiple of 4.

1961.16. Let us use induction on n. The basis is obvious. Now, assuming that all the points but one lie inside arc AB shorter than $120°$ (where A and B are some two of these points). If X lies on the same arc, then the induction step is over. If not, then clearly one of the arcs XA or XB (containing arc AB) subtends an angle less than $120°$.

1961.17. Quick calculation of the angles shows that in any triangle ABC with angle bisectors AA_1, BB_1, and CC_1 intersecting at point I, we have equality $\angle BAA_1 + \angle BIC_1 = \frac{\pi}{2}$.

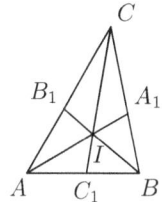

Let us mark the given point as vertex A. The bisector lines meet at angle $\angle BIC_1$. If we draw two lines through point A, forming easily constructible angle $\frac{\pi}{2} - \angle BIC_1$ with the angle A's bisector, we will have two sides of our triangle. Intersecting these lines with two other bisectors gives us the two missing vertices.

1961.19. Consider all possible ways to split these points into n pairs and connect them with segments, and select the one for which the sum of the lengths of the segments is the smallest. If some two segments AB and CD happen to intersect, then we can replace them with AC and BD and obtain the collection of n segments of even lower sum of the lengths which contradicts our choice of the "smallest" split.

1961.20. Let O be the vertex of the given angle. Since quadrilateral $OACB$ is inscribed, we have $\angle BOC = \angle A$, and therefore ray OC does not depend on the position of the triangle.

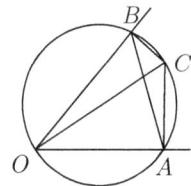

The set of all possible distances $|OC|$ is the same as the set of all distances from C to the points of arc AB on the circumference circumscribed around triangle ABC. That set obviously coincides with interval $[\min\{|AC|, |BC|\}; 2R]$, where R is the circumradius.

Therefore the answer is: the trajectory is the segment on the ray described above, whose ends are positioned at distances $\min\{|AC|, |BC|\}$ and $2R$ from point O. The first end corresponds to the case when either B or C coincide with O, and the second end—when

AC and BC are both perpendicular to the respective sides of the angle.

Note. If configuration is different—that is, points C and O lie to the same side of straight line AB, there is no easy elementary solution. In that case the trajectory is an arc of an ellipse centered at point O.

1961.21. Assume that such a covering is possible and consider the regular chessboard coloring pattern. Since every T-shaped tetromino contains three squares of one color and one square of the other, then taking into account that the total numbers of black and white squares are the same, we conclude that the overall number of the tetrominoes must be even—the number of Ts with three whites and one black square must be equal to the number of Ts with one white and three black squares. But that is impossible since there are 25 tetrominoes, and that number is odd.

1961.22. Since k and n are co-prime, there exist integers a and b such that $ka+mb = -1$. Multiplying this equality by n, we get $kna + mnb = -n$. Hence $m(nb) + n$ is divisible by k.

1961.23. Hint: These numbers are the squares of numbers $333\ldots334$.

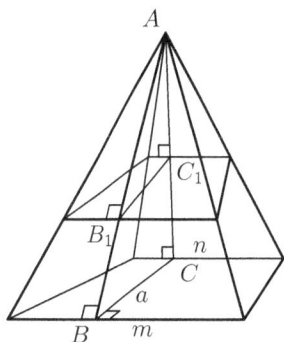

1961.24. It is easy to see that a plane satisfies the "trapezoid section" condition if and only if it separates the vertex of the original pyramid from its base and is parallel to the bases of the original trapezoid.

Let a be the altitude of the original trapezoid, and h—the altitude of the pyramid; a_1 and h_1—similar parameters for the second pyramid. Then

$$V = \frac{1}{6}(m + n)ah, \quad U = \frac{1}{6}(m_1 + n_1)a_1h_1,$$

and the problem is reduced to proving equality

$$\frac{a_1h_1}{ah} = \frac{m_1n_1}{mn}.$$

Let A be the common vertex of the pyramids, B and C are the feet of the altitudes dropped from A onto bases m and n of the first trapezoid (or onto their extensions), B_1 and C_1 are similarly constructed points for the second pyramid.

Note that $B_1 \in AB$, $C_1 \in AC$, and $|AB_1|/|AB| = m_1/m$, $|AC_1|/|AC| = n_1/n$, because all bases of trapezoids are parallel. In addition, $|BC| = a$, $|B_1C_1| = a_1$, h and h_1 are altitudes of triangles ABC and AB_1C_1, dropped from A. Therefore,

$$\frac{m_1n_1}{mn} = \frac{|AB_1| \cdot |AC_1|}{|AB| \cdot |AC|} = \frac{S_{AB_1C_1}}{S_{ABC}} = \frac{a_1h_1}{ah},$$

and the proof is complete.

1961.25. First, let us reformulate the question. Consider the set S of all school subjects as a generic set of $N = 2n$ elements. Each student x then can be represented as a subset $S_x \subset S$—namely, as the collection of all those subjects in which he or she has received grade of **A**. Then the condition that student x is better than student y in all subjects is equivalent to inclusion $S_y \subset S_x$. Now, statement that no student is "better" than any other can be "translated" to say that not one of the subsets S_x contains any other such subset.

Let us denote our subsets simply as $S_1, S_2, \ldots, S_m$, where m is the number of students; also let a_k be the number of elements in S_k $(k = 1, \ldots, m)$. Then for each k from 1 to m, define C_k as set of all subset chains of the form

$$B_1 \subset B_2 \subset \cdots \subset B_N = \{1, 2, \ldots, N\},$$

where each B_i contains exactly i elements, and set B_{a_k} coincides with S_k (the last condition means that the chain must "pass through" the subset S_k). It is obvious that the number of such chains is equal to the product of the number of chains of the form

$$B_1 \subset B_2 \subset \cdots \subset B_{a_k} = S_k,$$

and the number of chains of the form

$$S_k = B_{a_k} \subset B_{a_k+1} \subset \cdots \subset B_N,$$

which are equal to $a_k!$ and $(N - a_k)!$, respectively. Since S_i is never a subset of S_j, then the chain sets C_k are disjoint. Therefore, the number of elements in their union—which is equal to $\sum a_k! (N - a_k)!$—cannot exceed the number of all subset chains, that is, $N!$. Thus

$$\sum_{k=1}^{m} a_k! (N - a_k)! \leqslant N! \quad \Longrightarrow \quad \sum_{k=1}^{m} \frac{1}{\binom{N}{a_k}} \leqslant 1. \qquad (*)$$

Finally, the maximum possible value of binomial coefficient $\binom{N}{a}$ is $\binom{2n}{n}$, so each summand in the right-hand side of $(*)$ is at least $\binom{2n}{n}^{-1}$, implying that the number m of those summands cannot exceed $\binom{2n}{n} = \frac{2n!}{n!n!}$.

1961.26. Answer: $2a$.

Indeed,

$$\frac{a^2 + x^2}{x} = a\left(\frac{a}{x} + \frac{x}{a}\right) \geqslant 2a,$$

where equality is reached if and only if $x = a$.

1961.27. Let $p = \frac{r}{s}$, where r and s are co-prime integers. Evaluating polynomial $f(x)$ at p, we obtain

$$r^n + a_1 r^{n-1} s + \cdots + a_n s^n = 0.$$

Since all the summands after the first one are divisible by s, the first one, r^n, must be as well. However, since r and s are co-prime, that is possible only if $s = 1$. Therefore, p is an integer.

Now, let us divide polynomial $f(x)$ by $x - p$ to obtain $f(x) = g(x)(x-p)+a$, where a is an unknown integer remainder. Evaluating at $x = p$ we have $a = 0$. Hence, $f(x) = g(x)(x-p)$ and $f(m)$ must be divisible by $p - m$ for any integer m.

1961.28. Let us use the following identity

$$\cos 20° \cos 40° \cos 80° = \frac{1}{8},$$

which is easily proved by multiplying both sides by $\sin 20°$. Then,

$$20° + \tan 80° = \frac{\sin 80°}{\cos 20° \cos 80°} = 8 \sin 80° \cos 40°$$

$$= 4\sin 120° + 4\sin 40° = 4\sin 60° + 4\sin 40°.$$

Since $3\sqrt{3} = 6\sin 60°$, it is now left to prove that $4\sin 40° - \tan 40° = 2\sin 60°$, or that $2\sin 80° - \sin 40° = 2\sin 60° \cos 40°$. Finally, we have

$$2\sin 80° - \sin 40° = \sin 100° + (\sin 80° - \sin 40°)$$
$$= \sin 100° + \sin 20° = 2\sin 60° \cos 40°.$$

There is another, more "geometric" solution, that the reader could see in the figure. Triangle ABC has angles 80°, 80°, and 20°; AB and other highlighted segments have length 2; $|CH| = \tan 80°$. Finally, the sum of the projections of segments AF and CG onto the vertical is $3\sqrt{3}$, and the length of FG is equal to $8\sin 10° \sin 20° = \tan 40° - \tan 20°$—to prove that, consider the isosceles triangles DGF and CDG.

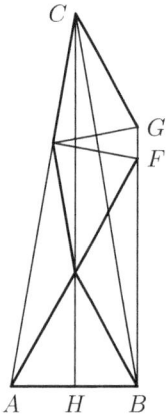

1961.29. If we draw the diagonals that split the faces of the cube, then each vertex of the cube has either one or three incident diagonals—otherwise the sums of the black and the white angles cannot be equal. The cube has 8 vertices, and six diagonals have 12 ends; therefore, there are exactly two vertices with three incident diagonals.

These two vertices cannot belong to the same edge—if they do, then the face containing that edge will have two diagonals which is impossible.

They also cannot be opposite vertices of the same face—then this face will have no diagonals (see the figure). Thus, they must be the two opposite vertices of the cube, and that uniquely determines the diagonals. Quick examination of all possible cases shows that there are only two ways to paint the triangles.

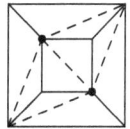

1961.30. We will call the original direction of the snail *horizontal,* and the one perpendicular to it—*vertical.* Then the snail moves horizontally during the first and the third quarters of the hour. So, during its travel, it has to move the same distance to the right as to the left—otherwise its horizontal position will not be the same at the end of the trip. This means that the number of the first and third quarters is even, and therefore, the number of the first quarters equals the number of the third quarters. Same applies, of course, to the numbers of the second and the fourth quarters. Hence, the total number of quarters is twice some even number, which makes it divisible by 4. Therefore the whole snail trip lasted an integer number of hours.

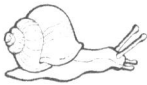

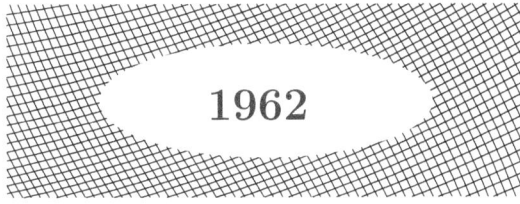

1962

1962.01. Answer: $3\frac{1}{9}$ hours.

Note. It is highly likely that some additional condition was added when the question was presented at the contest—otherwise this problem would be quite difficult to solve even for the high school students.

Let us call the travelers X, Y, and Z. They can act in the following manner. First X and Y ride the motorcycle, while Z goes on foot. After 50 km (traveled in $1\frac{1}{9}$ hour), X and Y stop, Y rides the motorcycle back toward Z, while X walks the remaining 10 kilometers. When Y and Z meet, exactly 24 hours will have passed from the beginning of the trip—that will happen at the distance of 10 km from A; by that time, Z will have walked 10 km, and Y will have ridden $50 + 40 = 90$ km. Finally, after Y collects Z, they will ride the remaining 50 km on the motorcycle in $1\frac{1}{9}$ hr and will arrive at B simultaneously with X.

Now, let us prove that this method is the fastest possible. Suppose that our travelers somehow managed to reach B even faster. We can assume that X was always ahead of Y and Z (or at least one of the leaders)—because if one of the others passed him on the road, we can simply have X right then swap places with that other traveler. Similarly, we can assume that Z was always behind all the others.

Thus it is obvious that each traveler walked no more than 10 km—otherwise he would have spent more than $3\frac{1}{9}$ hours on the road (indeed, $3\frac{1}{9} = \frac{10}{5} + \frac{50}{45}$). Let us paint the stretches of the road where X and Z traveled on foot; then the unpainted portion—where both of them rode the motorcycle—will be at least 40 km long.

But they could not have ridden the motorcycle at the same time! Why? Because if X and Z were simultaneously in the same place, then where was Y supposed to be? X is always in the lead, Z is always the last one—that means that Y must be somewhere between those two. But then he would have to be on the motorcycle with both of them, which is impossible!

That means that every motorcycle rode through every unpainted segment of the road in the direction $A \to B$ at least twice. It follows that it had to also pass through each such segment going in the opposite direction. Thus the motorcycle rode through all those 40 kilometers at least three times, and therefore, its overall travel distance must have been at least $140 = 3 \cdot 40 + 20$. We can safely assume that the motorcycle ended up in B—if the last traveler who rode the bike decided to abandon it somewhere on the road, then riding it instead of walking could not make things worse.

And finally, to travel 140 km, the motorcycle had to ride for at least $140/45 = 3\frac{1}{9}$ hr, proving our claim.

Note. This proof is, obviously, too complicated for the sixth graders. However, it can be significantly simplified if we were allowed to sacrifice some rigor. For instance, we could consider only some, allegedly "best", ways of travelling. It is highly unlikely that the sixth-graders were expected to offer anything better than that. Alas, generally speaking, one should not trust such tricks—formally, they cannot be a part of a rigorous proof. For instance, in a similar problem about two travelers and a one-seat motorcycle, one of the travelers (the leader) can easily leave the motorcycle on the road and continue walking while the other traveler shortly reaches that motorcycle and rides it to the destination, catching up with the leader.

1962.02. Answer: only 1 and 2.

Indeed, any common divisor p of both $a + b$ and $a - b$ has to also divide both $(a + b) + (a - b) = 2a$ and $(a + b) - (a - b) = 2b$. Since a and b are co-prime, the only non-trivial common divisor of $2a$ and $2b$ is 2.

1962.03. Answer: 23 years old.

We can safely assume that the year of birth X begins with $18\ldots$ or $19\ldots$. Thus the sum $S(X)$ of its four digits lies in the range $[9; 28]$. We know that

$$1962 - X = S(S) + 1 \quad \Rightarrow \quad X + S(S) = 1961. \tag{$*$}$$

Now if we use remainders modulo 9 and remember that $S(X) \equiv X \pmod 9$, we obtain $2S(X) \equiv 8 \pmod 9$, or $S(X) \equiv 4 \pmod 9$. It follows that $S(X)$ is either 13 or 22 giving us two cases—$X = 1939$ and $X = 1948$. The second case does not satisfy the equation $(*)$, so only the first one fits.

1962.04. Take any one magazine, and consider the fraction of the area of the table which is covered **only** by this magazine. If it is less than $\frac{1}{15}$, then the other 14 magazines cover more than $\frac{14}{15}$ of the table—split them into two collections of seven magazines, and then obviously one of those two collections must cover at least $\frac{7}{15}$ of the table which would conclude the solution. Thus, we can assume that for every magazine, the area of the portion of the table which is covered by only that one magazine is greater than $\frac{1}{15}$ of the table. But that contradicts the fact that these portions do not intersect each other, yet the sum of their areas is greater than the area of the entire table.

1962.05. Hint: Prove that the knight can traverse the two-squares wide border strip; then, using that fact, use induction to prove that such traversal exists for any board with dimensions $(4n + 1) \times (4n + 1)$.

The figure on the next page shows how the knight starts in the lower left corner of the board $N \times N$, traverses the border strip, then ends up in the lower left corner of the empty $(N - 4) \times (N - 4)$ board. Thus, we can reduce the size of the board from 201 to 197, then to 193, and so on, to 5 and then, finally, to 1.

1962.06. No, that is not possible, because at least one of the two last digits of a perfect square must be even. Suppose that both of them are odd. If $n = x^2$, then x can be represented as $x = 10a + b$, where b is the last digit of x and a is an integer whose last digit is the digit of tens in x. Thus, we have

$$n = x^2 = (10a + b)^2 = 100a^2 + 10(2ab) + b^2.$$

The first summand is divisible by 100, and therefore, does not affect the last two digits of n. The second only affects the digit of tens by increasing it by an even number. Finally, since b is odd, b^2 is one of the numbers 01, 09, 25, 49, 81 (we write them like this to show their digits of tens even if they are absent in usual decimal representation). As you can see, for all of these numbers, their digit of ten is even.

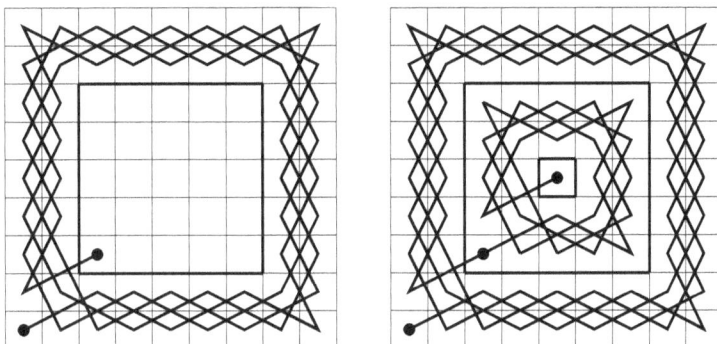

1962.07. Denote the sides of the quadrilateral by a, b, c, d, where a is the shortest and d is the longest side. If not all the sides are equal, then numbers $a - d$, b, c are positive and satisfy the triangle inequality. Let us build triangle ABC such that $|BC| = a - d$, $|AC| = b$, and $|AB| = c$. Then take point D on ray CB so that $|CD| = a$, and construct point E such that $ABDE$ is parallelogram. Then $ACDE$ will be the required trapezoid.

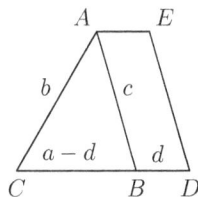

1962.10. Hint: Use that $10^6 - 1$ is divisible by 7.

If the original number is $x = \overline{abcdef}$, then the other number is $y = \overline{fabcde}$. Then the difference $x - 10y$ is

$$(a \cdot 10^5 + b \cdot 10^4 + c \cdot 10^3 + d \cdot 10^2 + e \cdot 10^1 + f \cdot 10^0)$$
$$- 10 \cdot (f \cdot 10^5 + a \cdot 10^4 + b \cdot 10^3 + c \cdot 10^2 + d \cdot 10^1 + e \cdot 10^0) = (1 - 10^6)f,$$

and therefore, divisible by 7. But since $x \vdots 7$, then $10y \vdots 7$, thus y itself is divisible by 7 as well.

1962.11. Let us call the regions' boundaries (except their parts that lie on the circumference—the circle's boundary) *curves*. Curves' endpoints divide the circumference into several arcs. Consider any of these arcs—say, AB—which has color 1. Diametrically opposite arc $A'B'$ can contain 0, 1, or more curve endpoints. If there are more than one of them, then the figure looks like the one on the left. We will slightly change the regions by merging them like it is shown on the right figure—the number of endpoints will decrease by two.

If arc $A'B'$ has no endpoints, then it lies inside arc CD, where C and D are some curves' endpoints. Thus A and B lie inside arc $C'D'$, and we can perform the procedure that we have just described above. Every time we perform this procedure, we decrease the number of endpoints. Of course, this cannot continue indefinitely, and therefore, eventually we will obtain a dissection of the circle into regions such that for every elementary arc AB (A and B

are the neighboring endpoints of the curves), arc $A'B'$, diametrically opposite to it, contains exactly one other endpoint.

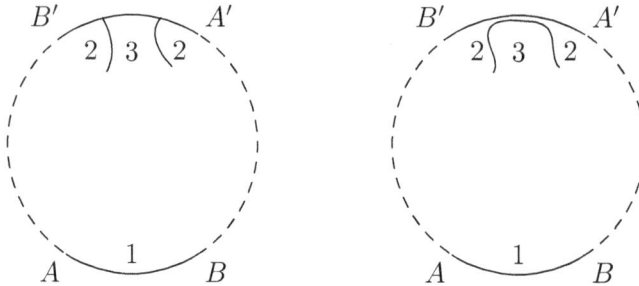

This easily follows that the total number of curve endpoints is odd which is impossible.[100]

1962.12. Since points O_1, M_1, O_2, and M_2 are the midpoints of the sides of quadrilateral $ADLC$, we have that $O_1M_1O_2M_2$ is a parallelogram. It is then enough to prove that segments AL and AD are perpendicular and have equal length. Indeed, they can be obtained from each other by rotating around point B by $90°$.

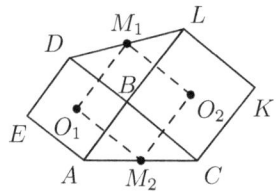

1962.13. Centers of the circumferences—points A, B, C, and D—produce a circumscribed quadrilateral $ABCD$. Let us denote the tangency points by K, L, M, and N—they lie on segments AB, BC, CD, and DA, respectively. Then

$$\angle NKL = 180° - (90° - \tfrac{1}{2}\angle NAK) - (90° - \tfrac{1}{2}\angle KBL)$$
$$= \tfrac{1}{2}(\angle DAB + \angle ABC).$$

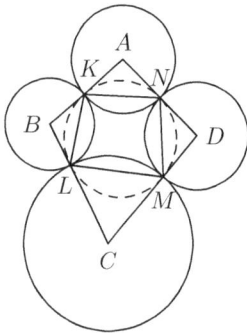

Similarly we get $\angle NML = \tfrac{1}{2}(\angle ADC + \angle DCB)$, and the sum $\angle NKL + \angle NML$ is equal to half-sum of the angles of $ABCD$, i.e., to $180°$. Hence $KLMN$ is inscribed.

1962.14. Hint: Use the identity $(x^2 - 5y^2)(u^2 - 5v^2) = (ux + 5vy)^2 - 5(uy + vx)^2$.

1962.15. Answer: if $d = 0$, then the equation has two roots for $a > 0$, one root for $a = 0$ and no roots for $a < 0$. If $d \neq 0$, then

a) for $a < -d^4$: zero roots;
b) for $a = -d^4$: one root;
c) for $9d^4/16 < a$: two roots;
d) for $a = 9d^4/16$: three roots;
e) for $-d^4 < a < 9d^4/16$: four roots $x = -3d/2 \pm \sqrt{5d^2/4 \pm \sqrt{d^4 + a}}$.

---

[100] This solution was communicated to us by Mikhail Gusarov (1958–1999).

If we define $y = x(x+3d) = x^2 + 3dx$, then for y we have quadratic equation $y(y+2d^2) = a$, or equivalently $(y + d^2)^2 = a + d^4$. From this, assuming that $a \geqslant -d^4$, it follows that

$$y = \pm\sqrt{a + d^4} - d^2 \, .$$

Now, we need to solve another quadratic equation $x^2 + 3dx - y = 0$. Its roots are

$$\frac{-3d}{2} \pm \sqrt{\frac{9d^2}{4} + y} = \frac{-3d}{2} \pm \sqrt{\frac{5d^2}{4} \pm \sqrt{a + d^4}} \, ,$$

which is exactly what was "promised" above. Of course, in some cases we have only two or three of those four possible roots—a more detailed analysis of that we leave to the reader.

1962.16. Solution 1 (vectors): Consider vectors $\vec{v}_1 = (a, b, c)$, $\vec{v}_2 = (m, n, p)$. Their lengths are equal to 1, and thus, the absolute value of their scalar product $am + bn + cp$ does not exceed the product of their lengths which is 1.

Solution 2 (direct computation): Let us prove that $(am + bn + cp)^2 \leqslant (a^2 + b^2 + c^2)(m^2 + n^2 + p^2)$. Expand the expressions to get

$$a^2m^2 + b^2n^2 + c^2p^2 + 2ambn + 2amcp + 2bncp$$
$$\leqslant a^2m^2 + a^2n^2 + a^2p^2 + b^2m^2 + b^2n^2 + b^2p^2 + c^2m^2 + c^2n^2 + c^2p^2 \, .$$

After crossing out the identical terms and carefully grouping the remaining ones, we obtain

$$(a^2n^2 + b^2m^2 - 2ambn) + (a^2p^2 + c^2m^2 - 2amcp) + (b^2p^2 + c^2n^2 - 2bncp) \geqslant 0$$

or

$$(an - bm)^2 + (ap - cm)^2 + (bp - cn)^2 \geqslant 0 \, ,$$

which is obviously true.

1962.17. Let O be the center, and r the radius of the given semicircle. We claim that the largest possible area r^2 is possessed by the triangle with vertices in the ends of the diameter and the midpoint of the semi-circumference. Consider any other triangle inscribed in the semicircle. The case, when two of its vertices lie on the diameter, is obvious. If all the vertices A, B, and C lie on the semi-circumference (in that order), then $|AC|$ does not exceed the diameter, and altitude from vertex B does not exceed radius r. Therefore, its area cannot be greater than r^2. Finally, if exactly one vertex lies on the diameter, then it can be moved to one of its ends increasing the area and reducing the problem to one of the cases we already considered.

1962.18. Let O be the common point of the three circumferences with centers O_1, O_2, and O_3, with radius r, and C, B, and A—the other points of intersection of the first and the second, the first and the third, the second and the third circumferences, respectively. Triangle $O_1O_2O_3$ is inscribed into the circle with center O and radius r, therefore it will suffice to prove the congruence of triangles ABC and $O_1O_2O_3$. Since AO_2OO_3 and BO_1OO_3 are rhombuses, ABO_1O_2 is a parallelogram, hence $|AB| = |O_1O_2|$. Similarly, $|BC| = |O_2O_3|$, $|AC| = |O_1O_3|$, that is, triangles ABC and $O_1O_2O_3$ are congruent (**SSS** rule).

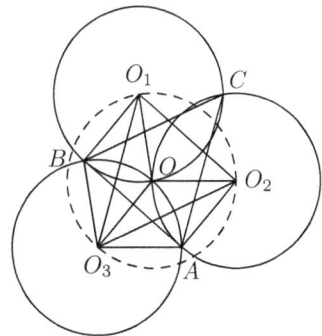

1962.19. Answer: if the triangle is acute, then it is the circumcircle; if it is obtuse, then it is the circle which is built on the largest side of the triangle as diameter.

Let us prove that this circle is the smallest possible. Let O' and R' be the center and the radius of some circle that covers triangle ABC. Let AB be the largest side of the triangle; then by triangle inequality, we have $|AB| \leqslant |O'A| + |O'B| \leqslant 2R'$, and thus $R' \geqslant |AB|/2$, which is enough if ABC is obtuse or right-angled. In the case of an acute triangle, let O and R be the circumcenter and circumradius of $\triangle ABC$; then it is easy to see that if $O \neq O'$, then at least one of the angles $O'OA$, $O'OB$, $O'OC$ is obtuse. Let it be $\angle O'OA > 90°$; then $R' \geqslant |O'A| > |OA| = R$.

1962.20. Denote this polynomial by $f(x)$. Obviously $f(-1) = f(1)$. From $f(x) \vdots (x-1)$ it follows that $f(1) = 0$. Then $f(-1) = 0$, and $f(x)$ is divisible by $x + 1$ and therefore, by $x^2 - 1 = (x - 1)(x + 1)$.

1962.21. Consider numbers 1, 11, 111, ..., 111...11 ($p + 1$ ones). Since there are more than p of them, then some two of them have the same remainders modulo p. The difference of these two numbers is $111...11000...00$. Since p is co-prime with 10, then $111...11000...00$ being divisible by p implies that $111...11$ is divisible by p as well.

1962.24. Let the coordinates of the given point be (m, n). From the four vertices of the square we select one (with coordinates (x, y)) such that both numbers $m - x$, $n - y$ are odd. Then $(m - x)^2 + (n - y)^2$ has remainder 2 modulo 4 and therefore cannot be a square of an integer (and consequently of a rational number).

1962.25. If we subtract twice the sum of all the pairwise products from the square of the sum of all numbers we will obtain the sum of the squares of these numbers. In our particular case, it follows that this sum must be zero, and therefore all ten numbers are zeros.

1962.26. Let a be the smallest side, b and c—the two others. Consider two cases: $b + c \geqslant 3a$ and $b + c < 3a$. In the former case, the perimeter is at least $4a$, and the sum of the distances is less than $3a$, so we are done. In the latter case, the condition that the distances are less than a is unnecessary. We can note that the sum of the distances from a point to the vertices of triangle T (and generally to any finite set of points T on the plane) attains its maximum at one of these points of T—this follows from the convexity of the distance function. Thus the sum of the distances does not exceed $b + c$, but

$$b + c = \frac{3}{4}(b + c) + \frac{1}{4}(b + c) < \frac{3}{4}(b + c) + \frac{1}{4}(3a) = \frac{3}{4}(a + b + c).$$

1962.28. Solution here is nearly identical to the solution to Question **1961.19**.

1962.29. Let us call a box "small" if it weighs 300 kg or less. There are obviously no more than 44 other ("large") boxes, so we can put them into eleven trucks, four boxes per truck. Now, we will prove that a small box (with weight $w \leqslant 300$ kg) can always be added onto one of the trucks. Assuming the opposite, we obtain that the cargo in each truck weighs more than $1500 - w$ kg. Adding up these eleven inequalities, we obtain that the total weight of the cargo on the trucks is greater than $16500 - 11w$. Since at the same time that weight cannot exceed $13500 - w$, we have $10w > 3000$, bringing about the contradiction.

1962.30. Let the vertices of the quadrilateral split the sides of the square into segments $\frac{1}{2} \pm x$, $\frac{1}{2} \pm y$, $\frac{1}{2} \pm z$ and $\frac{1}{2} \pm t$. If all the sides of the quadrilateral are less than $1/\sqrt{2}$, then the sum of their squares is less than 2. From the Pythagoras' Theorem, the sum of these squares equals the sum of the squares of all the eight segments forming the sides of the square, that is,

$$\left(\frac{1}{2} - x\right)^2 + \left(\frac{1}{2} + x\right)^2 + \left(\frac{1}{2} - y\right)^2 + \left(\frac{1}{2} + y\right)^2$$
$$+ \left(\frac{1}{2} - z\right)^2 + \left(\frac{1}{2} + z\right)^2 + \left(\frac{1}{2} - t\right)^2 + \left(\frac{1}{2} + t\right)^2$$
$$= 2 + 2x^2 + 2y^2 + 2z^2 + 2t^2 \geqslant 2.$$

1962.31. See solution to Question **1962.25**.

1962.32. First, we use the formula

$$x^3 + y^3 + z^3 - 3xyz = (x + y + z)(x^2 + y^2 + z^2 - xy - yz - xz),$$

applied to numbers $x = \sqrt[3]{a}$, $y = \sqrt[3]{b}$, and $z = \sqrt[3]{c}$, and then use the formula

$$s^3 - t^3 = (s - t)(s^2 + st + t^2),$$

applied to $t = 3xyz$ and $s = x^3 + y^3 + z^3$.
   Thus the original fraction is equal to p/q, where

$$p = (x^3 + y^3 + z^3)^2 - 3xyz(x^3 + y^3 + z^3) + 9x^2y^2z^2)(x^2 + y^2 + z^2 - xy - xz - yz),$$
$$q = (x^3 + y^3 + z^3)^3 - 27x^3y^3z^3,$$

with denominator q being equal to $(a + b + c)^3 - 27abc$.

1962.33. Answer: $x = \pm 1$, $x = (36 \pm \sqrt{671})/25$.
   Indeed, we have

$$(3x + 2)^4 - (2x + 3)^4 = (4x - 2)^4 - (2x - 4)^4.$$

Factoring the sides of the equation, it is easy to see that we can divide both by $(x+1)(x-1)$, after which we are left with a quadratic equation $25x^2 - 72x + 25 = 0$.

1962.34. Let A be the vertex of the given angle. Let us measure the given segments off A along the sides of the angle, and denote the endpoints by B and C. Notice that the areas will not change if we change the given segments to AB and AC. For any point M inside the angle, $S_{MAB} + S_{MAC} = S_{ABC} \pm S_{MBC}$, where plus or minus is selected based on whether M is outside or inside triangle ABC.
   Since S_{ABC} and BC are constant, the sum of the areas depends only on the distance from M to line BC and on which side of BC point M is. Thus the answer is the segment with its endpoints on the sides of the angle which (a) passes through point O and (b) is parallel to BC.

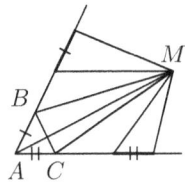

1962.35. Answer: this sum is zero when $n > 1$, and (-1) when $n = 1$. **Hint:** Use equality $k\binom{n}{k} = n\binom{n-1}{k-1}$ for $n \geq k \geq 1$.

1962.36. Answer: $x_1 = x_2 = 1$, $x_k = k$ for $k > 2$.

Since the number on the left-hand side of the equation is greater than $\frac{1}{2}$, we have $x_1 = 1$. Further, let $a = 2 + 1/(3 + 1/(4 + \cdots))$, $x = x_2 + 1/(x_3 + 1/(x_4 + \cdots))$. Simplifying the original equation, we get $\frac{a-1}{a} = \frac{x}{x+1}$. Thus, $x = a - 1 = 1 + 1/(3 + 1/(4 + \cdots))$ giving us the desired answer.

1962.37. Let us assume that such a coloring is possible. Without loss of generality, we can say that the "special" face (mentioned in the problem's statement) is colored white. Let us count the number of the sides of the black faces—on one hand it has to be the number of polyhedron's edges, and on the other hand, some natural number divisible by n. Similar count for the white faces will give us the same number of edges of the polyhedron, but the sum cannot be divisible by n. This contradiction proves the impossibility of such coloring.

1962.38. Answer: there are $\binom{n}{4}$ points inside and $2\binom{n}{4}$ points outside of the polygon.

Obviously the intersections inside can come only from the lines that contain the polygon's diagonals. Every quartet of the polygon's vertices uniquely determines a pair of intersecting diagonals. Thus, the number of interior intersection equals $\binom{n}{4}$. The overall number of intersection points is $\binom{n(n-1)/2}{2}$, of which $\frac{1}{2}n(n-1)(n-2)$ points lie on the polygon's boundary because each of the n vertices is the point of intersection of $\frac{1}{2}(n-1)(n-2)$ pairs of lines. Thus, the number of exterior intersections is

$$\binom{\frac{1}{2}n(n-1)}{2} - \binom{n}{4} - \frac{n(n-1)(n-2)}{2} = 2\binom{n}{4}.$$

1962.40. Clearly $f(x) = g(x^k)$ for some polynomial g. Since $f(x) \vdots (x - 1)$, then $f(1) = 0$. Therefore $g(1) = f(1) = 0$, hence $g(x) \vdots (x - 1)$, and $g(x^k) \vdots (x^k - 1)$.

1962.42. Answer: this sum is zero when $n > 2$; (-1) when $n = 1$; and 2 when $n = 2$. **Hint:** Use equality $k\binom{n}{k} = n\binom{n-1}{k-1}$ for $n \geq k \geq 1$.

1962.47. Assume that is not so. Let us draw lines parallel to the given side of the square passing through all the vertices of the polygon. They will dissect the polygon into triangles and trapezoids with bases not exceeding $\frac{1}{2}$. Since the sum of altitudes in these triangles and trapezoids does not exceed 1, then by adding up all the areas of these parts, we will have the area of the polygon not exceeding $\frac{1}{2}$.

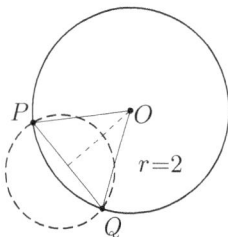

1962.48. Divide the circumference M (boundary of our given circle S) with radius 2 by six points A, B, C, D, E, and F into six equal arcs and construct six circles of radius 1 on diameters AB, BC, CD, DE, EF, and FA. Add one more identical circle centered at the center of S, and we will have seven circles covering S.

Why is it impossible to cover a larger circle S' with radius $r > 2$? One of the unit circles has to cover the center of S' and therefore cannot intersect its boundary, circumference M'. Hence, M' has to be covered by six circles. Consider an arc that is the intersection of M' and any of these circles—we claim that its angular measure α is less than $60°$.

First, we can assume that $\alpha \leqslant 180°$.

Second, if $\alpha \geqslant 60°$, then the distance between the arc's endpoints P and Q is $2r \sin \alpha/2 > 4 \sin 30° = 2$. Thus, points P and Q cannot belong to a circle of unit radius, a contradiction. Finally, it suffices to note that six arcs, each measuring less than $60°$, cannot cover entire circumference M'.

1962.49. This follows from inequality

$$5 \geqslant S_1 + \cdots + S_9 - S_{12} - S_{23} - \cdots - S_{89},$$

where S_i denotes the area of ith polygon, and S_{ij}—the area of intersection of ith and jth polygons.

1962.50. It is not quite clear whether the regions were assumed to be connected. Here we will present the solution that works even for non-connected regions consisting from some finite number of connected sub-regions. We will assume that the boundary of each region is the union of a finite number of simple closed curves, lest we take a deep plunge into some complicated areas of low-dimensional topology.

Now if we assume the opposite, then no point is colored into all three colors—otherwise that point's diameter would serve. Let us find the boundary curve L which encircles the sub-region A of smallest possible area; it lies inside one of our three regions, and we can assume it has color 1, while the area on the other side of that curve has color 2. Now, consider set A' symmetric to A with respect to center of the sphere. The boundary of A' must be entirely colored with color 3, or we would immediately find the required diameter with endpoints of the same color. Since sub-region A is the minimal one (area-wise), it is clear that A' must lie entirely inside some sub-region (which obviously has color 3). Next, repaint A to color 2 and erase curve L that used to separate it from color 2. It is obvious that this operation cannot generate any same-colored diameters, but the number of the sub-regions will decrease by one. Continuing in the same manner, we will eventually obtain the coloring with only one sub-region, meaning that the entire sphere has only one color, making it quite easy to find the required diameter.

1962.51. This problem has defied numerous and quite serious attempts to solve it; certainly, an extremely hard nut to crack. However, we hope that the readers will still try. Remember that it is quite possible that the statement is wrong—try looking for a counterexample.

It should be mentioned here that this problem was subjected to computer analysis using various Python and C++ programs. All known natural numbers (see [19], Problem **F24**) whose squares are written by no more than two digits (excluding trivial cases of one-digit numbers and numbers of form $n = m10^k$, $m = 1, 2, 3$) are listed below.

$$11, 12, 15, 21, 22, 26, 38, 88, 109, 173, 212, 235, 264, 3114, 81619.$$

For instance, $81619^2 = 6661661161$.

1962.52. Let us denote these numbers by x_1, x_2, $\ldots$, x_k and consider the sums x_1, $x_1 + x_2$, $x_1 + x_2 + x_3$, $\ldots$, $x_1 + x_2 + \cdots + x_k$. If none of them is divisible by k, then some two must have the same remainder modulo k, and so their difference is divisible by k. But any such difference represents a sum of several numbers from our collection.

1962.53. It is quite likely that this problem's statement was written down incorrectly. Otherwise the problem is just too simple for the elimination round.

Answer: $x \neq 0$, 1 for $n = 1$; $x \neq 1$ for $n = 2$; $x = 0$, -1 for $n \geqslant 3$.

Indeed, $x = 0$ is one of the roots for $n \geqslant 2$. Dividing by $x^{[n/2]}$, we obtain $x^{n(n-1)/2-[n/2]} = 1$ and $x \neq 1$. If $n = 2k$, then $x^{k(2k-2)} = 1$ which is true for $x = -1$ for any k and for any $x \neq 0$ if $k = 1$. If $n = 2k+1$, then $x^{k \cdot 2k} = 1$ which is true for $x = -1$ for any k and for any $x \neq 0$ if $k = 0$.

1962.54. Answer: n sequences.

For example, let the first k terms in all the sequences be equal to 1, and all the other terms of the mth sequence be equal m $(m = 1, \ldots, n)$.

Suppose that we have $n + 1$ sequences satisfying the condition. For each two of them we will count the pairs of all the identical terms. Obviously there will be $k\binom{n+1}{2}$ such pairs in total. On the other hand, for each natural m among the terms of all $n + 1$ sequences with index m, there exist two equal ones (since we have $n + 1$ natural numbers not exceeding n), and so they produce the pair of identical terms at the mth position of the sequences. Therefore, the number of these pairs must be infinite, giving us the contradiction.

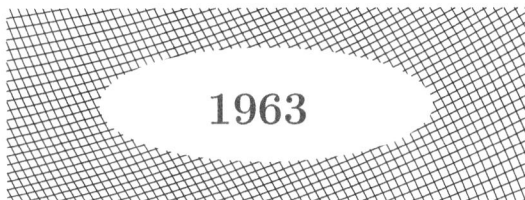

1963.01. Answer: $|AB| = 10$ km.

Indeed, if we denote that distance (in kilometers) by x, then we have equation

$$\frac{x}{5} - 1 = \frac{x-6}{4}.$$

Solving it, we get $x = 10$.

1963.02. Answer: the bus velocity is 30 km/h.

Let us denote that velocity by v (in km/h). Then the "meetings" of some fixed bus A with the buses that go in the opposite direction occur with intervals of 5 minutes, and thus the distance between two consecutive buses is $\frac{v}{6}$ km. Indeed, the speed of A relative to the bus B that goes towards it is $2v$, and so in 5 minutes they cover $2v \cdot \frac{5}{60}$ km. If we now consider the buses going in the same direction as the pedestrian, their velocity relative to the pedestrian is $(v-5)$ km/h. Therefore, the interval between the times when such a bus catches up to the pedestrian is $\frac{v}{6(v-5)}$ hours, and so in two hours we will have $\frac{12(v-5)}{v}$ of these "catch-up" events. Similarly, the number of "pass" events (for the buses which go in the opposite direction) is $\frac{12(v+5)}{v}$. Thus, we have $12(v-5) + 4v = 12(v+5)$, and solving this for v we get $v = 30$ km/h.

1963.03. Number 43^4 ends with 1, and therefore number $x = 43^{43} = (43^4)^{10} \cdot 43^3$ ends with 7. Similarly, 17^4 ends with 1, and $y = 17^{17} = (17^4)^4 \cdot 17$ ends with 7. Hence, difference $x - y$ ends with zero, i.e., is divisible by 10.

1963.04. See solution to Question **1963.13**.

1963.05. Answer: $|EC| = 150$ km.

From triangle inequality, it follows that $|CD| \leqslant |CB| + |BA| + |AE| + |ED|$, and the equality is attained only if points B, A, and E lie inside segment CD in exactly that order. Therefore $|EC| = 80 + 30 + 40$ km.

1963.06. Answer: no, that is impossible.

If we had such an arrangement, then out of any three consecutive numbers only one could be even. We can now split all 1963 numbers except for one into 654 such trios and conclude that there can be no more than 655 even numbers, while in fact there must be 981 of them.

1963.07. Answer: the measure of that angle is $30°$.

Indeed, the area of the quadrilateral equals half of the product of diagonals multiplied by the sine of the angle between them. From this formula we can compute the sine; we obtain $2 \cdot 3/(2 \cdot 6) = \frac{1}{2}$, and thus the angle itself is $\arcsin(\frac{1}{2}) = 30°$.

1963.08. This number is divisible by 3. Since square of 2 is the same as 1 modulo 3, we have that any even power of 2 is 1 modulo 3. Thus, any odd power of 2 is 2 modulo 3, and the proof follows.

1963.09. Hint: Only two outcomes are possible. The contestant in the last place gained either 0 or 0.5 points, and the winner then gained 10.5 or 10, respectively, with all the other players getting exactly 10 points.

Indeed, each one of the first eighteen players accumulated at least 10 points which, together with the 19th place's result, add up to $180 + 9.5 = 189.5$ points. At the same time, the overall number of matches played is $20 \cdot 19/2 = 190$, that number being only 0.5 points above the previous estimate. The remaining half-point has to go either to the last place participant, or to the winner. For all other players their result is fixed—9.5 points for the 19th place contestant, 10 points for players who placed 2 through 18.

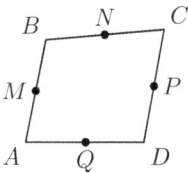

1963.10. Let $ABCD$ be the given quadrilateral and points M, N, P, and Q—the midpoints of sides AB, BC, CD, and DA, respectively. Since we have vector equalities $\overrightarrow{AB} + \overrightarrow{DC} = 2\overrightarrow{QN}$ and $\overrightarrow{BC} + \overrightarrow{AD} = 2\overrightarrow{MP}$, the equality for the lengths $|AB| + |BC| + |CD| + |DA| = 2(|QN| + |MP|)$ can be true only if $\overrightarrow{AB} \parallel \overrightarrow{DC}$ and $\overrightarrow{BC} \parallel \overrightarrow{AD}$; that is, $ABCD$ is a parallelogram.

1963.11. Obviously, after all the fares are paid, every passenger will be in possession of at least one coin in exchange for his or her fare. On the other hand, the driver is paid 2 rubles and so he received at least ten coins. Therefore, at least fifty coins are required, and this proves the statement.

Exercise. Can you find an example when fifty coins is enough?

1963.12. Answer: $a = 10$.

When divided by $a - 1$, $a - 2,\ldots$, $[\frac{1}{2}a] + 1$, number a produces remainders 1, 2, ..., $a - [\frac{1}{2}a] - 1$, respectively. These represent all possible non-zero remainders we can get, because when divided by 1, 2,..., $[\frac{1}{2}a]$, number a will produce remainders not exceeding $[\frac{1}{2}a] - 1 \leqslant a - [\frac{1}{2}a] - 1$. Thus, number a must satisfy equation

$$1 + 2 + \cdots + \left(a - \left[\frac{a}{2}\right] - 1\right) = a.$$

When a is even, we obtain $\frac{1}{2} \cdot (\frac{1}{2}a - 1) \cdot \frac{1}{2}a = a$, and therefore $a = 10$; when a is odd, we obtain $\frac{1}{2} \cdot \frac{1}{2}(a - 1) \cdot \frac{1}{2}(a + 1) = a$, and it is easy to see that such a does not exist.

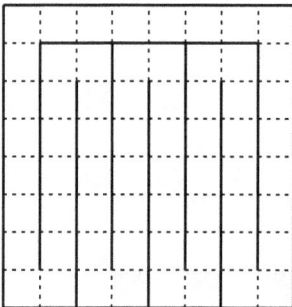

1963.13. If the two cut squares have the same color (remember that the chessboard has standard coloring in black and white), then the answer is negative. Indeed, each domino always covers two squares, one of them white and one—black. Hence the number of covered black squares must be the same as the number of covered white squares, and if the cut squares are the same color, the number of remaining white squares is different from the number of remaining black squares.

But if the cut squares have different color, then the answer is positive: such tiling by dominoes is possible. Consider the "corridor" shown in the figure. Those two squares divide the corridor into two parts, each one consisting of even number of squares, and therefore each of these two parts can be covered by dominoes.

1963.14. Answer: $ys/(x+y)$, $xs/(x+y)$.

Let us denote the distances from point A to the sides by h_1 and h_2. Also we will draw the segments from A to the endpoints of the base. The areas of the two new triangles are $xh_1/2$ and $yh_2/2$. But since A lies on the median, these areas must be equal, and so we have $xh_1 = yh_2$. Now, using equality $h_1 + h_2 = s$ gets us the required answer.

1963.15. Consider all pairs (x, y), where x and y—two consecutive digits from the given decimal representation. Since there are infinitely many pairs, we can find two identical ones. Now, moving back along the sequence, it is easy to see that without loss of generality we can assume that the earlier pair is (a, b). Since we have pair (a, b) at some other position, it is obvious that beginning at this position the sequence will start repeating itself, and therefore the fraction is periodic.

1963.16. Answer: $2n + 2$. **Hint:** Find an estimate for the number of intersection points.

Let us assume that all the intersection points are simple intersections, that is, the sides intersect at interior points (we can always achieve that by a tiny shift of one of the polygons without changing the number of regions). If $k > 0$ is the number of intersection points, then the number of regions equals $k + 2$. Indeed, one polygon separates the plane into two regions, and every part of the other polygon's contour between two intersection points adds another region.

From convexity of the first polygon (the one with m sides) it follows that each side of the second one intersects it no more than twice and therefore the number of intersection points is at most $2n$, and therefore the number of regions is at most $2n + 2$. To provide an example, consider two identical regular polygons with n sides and common center (for instance, one of them is obtained from the other by rotation around its center by $\frac{\pi}{n}$). Then we can turn one of them into an m-sided polygon by inserting $m - n$ very short edges next to one of the vertices.

1963.17. Perfect squares, when divided by 9, produce only the following remainders: 0, 1, 4, and 7. Obviously, the sum of any three different remainders from that list cannot be divisible by 9.

1963.18. Consider the following pairs of remainders modulo $2k$:

$$(1, 2k-1), \quad (2, 2k-2), \quad \ldots, \quad (k-1, k+1),$$

and also singular remainders 0 and k. Altogether we have $k + 1$ sets of remainders. Since we have $k + 2$ numbers, some two of them have remainders modulo $2k$ which belong to the same set. Those two numbers are the ones we need.

1963.19. Answer: let O be the center of the given circle S, R—its radius, C—vertex of the angle, and M—the midpoint of OC. Then the locus is the circumference centered at M with diameter $\sqrt{2R^2 - |OC|^2}$.

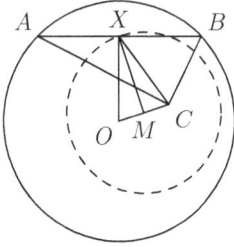

Indeed, consider an arbitrary point X inside S. There is a unique chord AB such that X is its midpoint, and since $AB \perp OX$, we have $|OX|^2 + |AX|^2 = R^2$. Point X belongs to the locus if and only if $\angle ACB = 90°$. But this is equivalent to $|CX| = |AX|$, and therefore, to $|OX|^2 + |CX|^2 = R^2$. Since XM is a median in triangle OXC, we can use a well-known formula expressing the length of that median through the lengths of the sides—$|XM| = \frac{1}{2}\sqrt{2|OX|^2 + 2|CX|^2 - |OC|^2}$—to rewrite this equality as $|XM| = \frac{1}{2}\sqrt{2R^2 - |OC|^2}$.

1963.20. Since $\overline{bcdea} = 10\overline{abcde} + a - 100000a = 10\overline{abcde} - 99999a$, and 99999 is divisible by 41, number $\overline{bcdea}$ is also divisible by 41. Continuing to permute the digits of the number in cyclic order, we will prove the divisibility condition for all the other numbers that can be obtained by such permutations of digits.

1963.21. For any indexes i and j, the following inequality is true

$$a_i b_i + a_j b_j \geqslant a_i b_j + a_j b_i,$$

because $(a_i - a_j)(b_i - b_j) \geqslant 0$. Adding up these inequalities for all pairs of indexes (i, j) where $i + j = n + 1$, we obtain the required result.

1963.22. Let $ABCD$ be that trapezoid with sides $|AB| = a$, $|CD| = b$ and diagonals $|BD| = u$, $|AC| = v$; let $x = |AD|$. (Triangle inequality requires $u + v > a + b$.) Since AD is parallel to BC, we have $S_{ABD} = S_{ACD}$. Using Heron's formula, we obtain

$$(x + a + u)(x - a - u)(x - a + u)(x + a - u) = (x + b + v)(x - b - v)(x - b + v)(x + b - v)$$

or, after some transformations,

$$2x^2(a^2 + u^2 - b^2 - v^2) = (u^2 - a^2 + v^2 - b^2)(u^2 - a^2 - v^2 + b^2),$$

and therefore

$$x = \sqrt{\frac{(u^2 - a^2 + v^2 - b^2)(u^2 - a^2 - v^2 + b^2)}{a^2 + u^2 - b^2 - v^2}},$$

if the denominator is not zero. Since all these operations, including square root, are constructible (using compasses and straightedge), from that formula we can construct segment $x = |AD|$, after which points B and C can be built uniquely up to line symmetry with respect to AD.

If the denominator of the fraction above is zero, then the numerator must be zero as well, and therefore $a = b$ and $u = v$. In this case there are infinitely many isosceles trapezoids that satisfy the conditions of the problem.

Note. Currently, we do not know any purely geometric solution to this problem.

1963.23. Let us assume that our two circumferences coincide, and instead of "superim-positions", we will simply consider rotations of the circumference. Angles of these rotations will be represented by the numbers from $[0; 1]$ (rotation by angle α will be represented by number $a = \alpha/2\pi$.)

Then for every selected point we will mark the set of all numbers in $[0; 1]$ such that corresponding rotation will map that point inside one of the given arcs. Such set will consist of several segments whose total length is less than $\frac{1}{20}$. Since we have 20 such sets, their union cannot cover the entire segment $[0; 1]$; hence, there exists number a such that rotation by angle $2\pi a$ does not map any of the selected points inside one of the arcs.

1963.25. Answer: $x = 1$, $y = 2$, $z = 2$.

Remainder modulo z for the left-hand side is 1, and for the right-hand side, it is -1. Thus $z = 1$ or $z = 2$. For $z = 1$, we have $2^x = 0$, which is not possible. For $z = 2$, we have $3^x + 1 = 2^y$. Since $3^x + 1$ cannot be divisible by 8, then we have only two cases: $y = 1$ and $y = 2$.

1963.26. Hint: Two irrational numbers of the form $a_1 + \sqrt{b_1}$ and $a_2 + \sqrt{b_2}$ with rational a_1, a_2, b_1, b_2 coincide if and only if $a_1 = a_2$ and $b_1 = b_2$.

Since both polynomials have leading coefficient equal to 1, their rational roots must be integers. Thus, if a root is not an integer, it must be irrational. On the other hand, it is the root of their difference, so it must be a solution for equation $(p_1 - p_2)x = (q_1 - q_2)$. Such a solution must be rational except for the case of identity, that is, $q_1 = q_2$, $p_1 = p_2$.

1963.27. Answer: consider plane parallel to face $F = BOC$, with its distance to F equal to one third of the distance between point A and F. The locus will then coincide with the part of that plane that lies inside the given trihedral angle.

Let O be the vertex of that angle. Notice that the center of mass of triangle ABC can be obtained from the midpoint of BC by homothety with center A and coefficient $\frac{2}{3}$. The locus of the midpoints of segments BC is the entire interior of face F.

1963.28. Without loss of generality, we can assume that $a_1 \neq 0$. Then it is clear that for each n-tuple $(x_1, x_2, \ldots, x_n)$ representing a solution of the given equation, n-tuple $(1 - x_1, x_2, \ldots, x_n)$ is not a solution. Thus, the number of solutions cannot exceed half of the number of all n-tuples.

1963.29. Consider the longest edge of the pyramid. On any face that contains it, the angle opposite to it is the largest angle of that triangular face. Hence, the angles adjacent to that edge must be acute.

1963.30. Let us prove that the indexes p and q can be selected even with an additional condition $p \geqslant q$.

Lemma. *From any infinite sequence (x_n) of natural numbers one can always select a non-decreasing subsequence.*

Indeed, if for any number x_n there is a subsequent term which is greater than or equal to it, then the statement is obvious. If we assume the opposite, then starting at some index, the sequence is bounded from above, and a constant infinite subsequence can be found. □

Now we complete the solution using induction on k. Basis is trivial. For the induction step from $k-1$ to k, we choose a non-decreasing subsequence $(x_{n_m}^{(k)})$ from sequence $(x_n^{(k)})$, and then apply induction hypothesis to the sequences $(x_{n_m}^{(i)})$, $1 \leqslant i \leqslant k-1$. Then indexes p and q $(p \geqslant q)$ for them will serve for the original sequences as well.

1963.31. We can assume that the squares have unit sides. Now, let us assume the opposite—that some square K touches nine other squares. Consider square K' with side 2 obtained from K by homothety with respect to the center of K. Then the following is true.

Lemma. *Interior of any unit square that touches K externally, when intersected with the contour of K' produces broken line whose length is at least 1.*

From this lemma the contradiction will immediately follow as there are nine squares while perimeter of K' is 8—therefore, some two squares must have common interior points.

The rest of the solution consists of proving the lemma.

Let X be the unit square that touches K. Case when sides of X are parallel to the sides of K is trivial. So, we will assume from now on that these sides are not parallel. That means, in particular, that the intersection of contours of K and X consists of the finite number of points. We will denote by P the intersection of entire square X with the boundary of K'. It obviously consists of several segments with endpoints on the vertices of K' and points of intersection of contours of K' and X. Our objective is to prove that the sum of the lengths of these segments is at least 1.

First, we start with some ideas which allow us to reduce the number of possible cases. If we assume that the angles between the sides of these two squares are fixed, and the combinatorial structure of that intersection is fixed—by the combinatorial structure we mean the list of data of the following type "K's side number a intersects X's side number b"—then the length of P, expressed through the coordinates of the center of X, is a linear function f.

If one of the squares K and X touches some side of the other square, then it can be translated along this side into such direction that function f is not decreasing until either these two squares stop touching, or the combinatorial structure of P changes. Thus, it is enough to consider only the "boundary" cases:

a) K and X have a common vertex;
b) one of the vertices of X lies on the boundary of K';
c) one of the vertices of K' lies on the boundary of X.

We can assume without loss of generality that K's sides are parallel to the coordinate axes. Now, consider the intersection P. The following cases are possible.

Case 1. Contour of K' intersects two opposite sides of X. The configuration shown in the figure is impossible because in that case, squares K and X would have common interior points. Then P is a segment or a broken line which goes from one side of X to the opposite side, and the length of that intersection is then obviously at least the distance between those sides, that is, 1.

Case 2. Contour of K' intersects two neighboring sides of X, and P contains a segment between these two sides (figure (2) shows one of the possibilities). Then the leg of the right triangle formed by this segment and those two sides contains at least one point from K, while its hypotenuse is a part of a side of K'. Therefore, the altitude of this triangle from the vertex of the right angle is at least $\frac{1}{2}$, and thus the length of the median from the same vertex is at least $\frac{1}{2}$ implying that the hypotenuse cannot be shorter than 1.

Case 3. Contour of K' intersects two neighboring sides of X, and P is a connected broken line. This case can be a boundary case only if the common vertex of these two sides of X is a vertex of K. Thus, the broken line is a union of horizontal segment of length $\frac{1}{2}(1 - \tan \alpha)$ and vertical segment of length $\frac{1}{2}(1 + \tan \alpha)$ (where α is the angle between the sides of squares K and X), and therefore its length is 1.

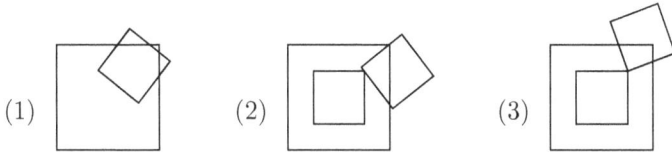

1963.32. The sum $(5 + \sqrt{26})^{1963} + (5 - \sqrt{26})^{1963}$ is an integer (this can be proved by using Newton's binomial theorem). From $-0.1 < 5 - \sqrt{26} < 0$ it follows that $-10^{-1963} < (5 - \sqrt{26})^{1963} < 0$. Thus, the first 1963 digits of number $(\sqrt{26} + 5)^{1963}$ after the decimal point are zeros.

1963.33. Answer: this box has to be positioned in such a way that the plane passing through the second endpoints of the edges coming out of one vertex of the box is parallel to the given plane.

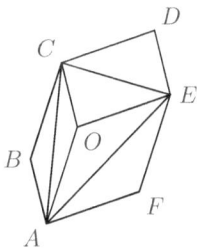

Projection of a parallelepiped must be a convex (possibly degenerate) hexagon—we will denote it by $ABCDEF$—which consists of three parallelograms $ABCO$, $CDEO$, and $EFAO$, each one of them being a projection of one of the parallelepiped's faces. Segments AC, CE and EA divide each parallelogram into two equal parts, and thus $S_{ABCDEF} = 2S_{ACE}$.

Triangle ACE is obviously a projection of a triangle whose sides are the diagonals of the parallelepiped's faces. Since during projection the area of a triangle can only decrease, the area of ACE does not exceed the area of the diagonal triangle, and the equality can only be reached if that triangle is parallel to the projection plane.

1963.34. Let B_i be the given circles, O_i—their centers, r_i—their radii ($i = 1, 2, 3, 4, 5$). If any of the three centers lie on the same straight line, then the corresponding circles have a common point on that line. Indeed, each circle cuts a segment on that line, we have that any two of these segments intersect, and therefore all these segments must have a common point.

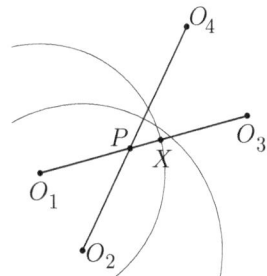

Assume that no three centers are collinear. Then it is again easy to show that some four of them lie in the vertices of a convex quadrilateral.

Let $O_1 O_2 O_3 O_4$ be that convex quadrilateral, and P—the point where its diagonals intersect. Since circles B_1 and B_3 intersect, their union contains segment O_1O_3, and P belongs to at least one of these two circles. Same is true for circles B_2 and B_4. Without loss of generality, assume that P belongs to B_1 and B_2, $|PO_1| = r_1 - d_1$, $|PO_2| = r_2 - d_2$, $d_1 \leqslant d_2$. Let us find point X on segment PO_3 such that $|PX| = d_1$; if X does not exist (that is, $|PO_3| < d_1$), set $X = O_3$. Then X belongs to circles B_1, B_2, and B_3: $|O_1X| \leqslant |PO_1| + d_1 = r_1$, $|O_2X| \leqslant |PO_2| + d_1 \leqslant |PO_2| + d_2 = r_2$ by triangle inequality. Distance $|O_3X|$ equals either 0 or $|O_3P| - d_1 = |O_1O_3| - r_1$, thus not exceeding r_3 (since B_1 and B_3 intersect).

1963.35. Consider all summands taking part in representation of $2a$. Since $a \geqslant \text{lcm}(1, 2, \ldots, 9) = 2050$, at least one of the numbers 1, 2, ..., 9 is included in that collection at least 9 times—otherwise $2a < 9(1 + 2 + \cdots + 9) < 405$). Let k be that number; then we temporarily remove 9 copies of k from the given representation. If we can select several of the remaining numbers so that their sum is divisible by k and lies between $a - 9k$ and a, then we can complement them with a few (no more than 9) copies of k to obtain sum a.

Let us start selecting these numbers, taking k equal numbers at a time. Obviously either at some point the sum of the selected numbers will be in the required range or it will always be below $a - 9k$ (or, adding the temporarily removed numbers, below a). At the moment when we cannot find k equal numbers among the not-yet-selected, the sum of the non-selected numbers will be less than $k(9 + 8 + \cdots + 1) \leqslant 45 \cdot 9 < a$. Thus, the sum of all numbers in the original collection must be less than $2a$, a contradiction.

1963.36. Let us denote the four given numbers by $a \leqslant b \leqslant c \leqslant d$.

Lemma. *Areas of the faces of tetrahedron relate as $a : b : c : d$ if and only if the linear combination of the vectors normal to the faces with coefficients a, b, c, and d equals zero-vector $\vec{0}$. That is, if we denote those normal vectors by $\vec{v_i}$ $(i = 1, 2, 3, 4)$, then*

$$a\vec{v_1} + b\vec{v_2} + c\vec{v_3} + d\vec{v_4} = \vec{0}.$$

Now, it would suffice to construct in space four vectors in general position, whose sum is zero while their lengths are equal to a, b, c, and d. Indeed, after that we can take any tetrahedron bounded by planes perpendicular to vectors v_i, and transform it using some appropriate homothety.

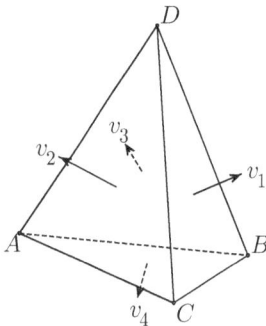

To construct these vectors we will take number e which is slightly less than $a + b$ (so that numbers c, d, and e satisfy the triangle's inequality) and construct in space triangle T from three vectors with lengths c, d, and e. Then we represent vector with length e as a sum of two vectors of length a and b, respectively, that do not lie in the plane of triangle T (this can always be done if $|a - b| < e < a + b$).

Thus we have constructed a non-planar quadrilateral whose sides have lengths a, b, c, and d; their sum (as vectors) is, obviously, zero. Applying the lemma completes the proof.

1963.37. Answer: $x = y = z = t = 0$.

To prove this, let us assume that some of the values are non-zero. Dividing them all by the maximum possible power of two, we can also assume that at least one of them is odd. Find the first odd number in the sequence x, y, z, t. If x is odd, then expression $S = x^4 - 2y^4 - 4z^4 - 8t^4$ is odd. Else, if y is odd, the same expression is not divisible by 4. Else, if z is odd, then S is not divisible by 8, and finally, if only t is odd, then S is not divisible by 16. In any case S cannot be zero, a contradiction.

1963.38. Consider two points A and B on the broken line L such that the length of L between those points equals $\frac{1}{2}$. Then circle S of radius $\frac{1}{4}$, centered at the midpoint of segment AB, covers L entirely. Indeed, if L intersects boundary of S at some point C, then we have $|AC| + |BC| > |AB| = \frac{1}{2}$, and this contradicts the fact that the length of both "halves" of L between A and B is $\frac{1}{2}$.

1963.39. Answer: yes, it does. Obviously, it is enough to construct such a triangle with rational lengths of its sides, altitudes, and medians.

Consider triangle ABC with angles α, β, and γ such that $\sin\frac{\alpha}{2}$, $\cos\frac{\alpha}{2}$, $\sin\frac{\beta}{2}$, $\cos\frac{\beta}{2}$ are all rational (for instance, we can choose $\alpha = \beta = \arcsin\frac{3}{5}$). Then $\sin\frac{\gamma}{2}$, $\sin\alpha$, $\sin\beta$, and $\sin\gamma$ are also rational. In addition, we can have one of the triangle's sides rational. Then it follows from the law of sines that all the sides are rational. Altitude from vertex A is equal to $AB\sin\gamma$ and is, therefore, rational; same applies to the other two altitudes. Finally, for angle bisector AL from the law of sines for triangle ABL we have $|AL| = |AB|\sin\gamma/\sin\frac{\alpha}{2}$, and so it is rational as well. Similar formula proves it for all other angle bisectors.

1963.40. Answer: only for $n = 3$. Indeed, if $2^n + 1 = a^p$, then

$$2^n = a^p - 1 = (a - 1)(a^{p-1} + a^{p-2} + \cdots + a + 1).$$

Then $a^{p-1} + a^{p-2} + \cdots + a + 1$ is a power of 2 different from 1. Since a is odd but this sum is even, the number of summands must be even. Thus $p = 2q$, and

$$2^n = a^{2q} - 1 = (a^q - 1)(a^q + 1),$$

that is, both $a^q - 1$ and $a^q + 1$ are powers of 2 which differ by 2. This is only possible when $a^q = 3$ and so $2^n = 8$.

1963.41. The case when the lines are vertical is trivial. So we can denote the slope of these lines by $k \in \mathbb{R}$. Moving the origin of the coordinate system we can achieve the situation when one line's equation is $y = kx$, and the other's—$y = kx - a$, $a > 0$. If $k = \frac{m}{n}$ is rational, then all points of form (an, am) lie inside the strip because they belong to the first line. If k is irrational, then consider numbers $b_n = kn - [kn]$, $n \in \mathbb{N}$. They all belong to segment $[0; 1]$, and there are infinitely many of them. Hence there are two of them, say, b_n and b_m, such that $0 < b_n - b_m < a$. Thus we have $0 < kn - [kn] - km + [km] < a$, and for $y = [kn] - [km]$ we obtain that $y < k(n - m) < y + a$ which means that the point with coordinates $(n - m, y)$ lies inside the strip. Then we can find another pair of terms of sequence (b_n) which are even closer to each other and so on, thus, producing infinitely many nodes inside the strip.

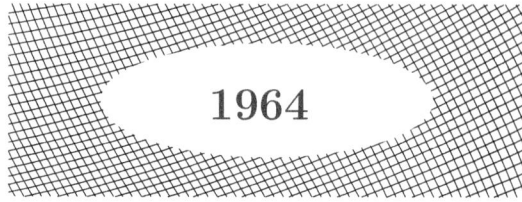

1964

1964.01. Answer:

- Adam: $25 + 20 + 20 + 3 + 2 + 1$
- Bob: $50 + 10 + 5 + 3 + 2 + 1$
- Charles: $25 + 20 + 10 + 10 + 5 + 1$

Since the total sum of points is 213, each shooter scored 71 points. The first three shots by Adam that sum up to 43 can only be 20, 20, and 3. Now, who could have hit 50? That shooter could not have also shot 25 or 20; he could not have made two shots of 10, but he had to make one 10 point shot because otherwise the most he could have scored would be $50 + 5 + 5 + 3 + 3 + 2 = 68 < 71$. His remaining four shots amount to 11 points, and the only possible way to do that is $5 + 3 + 2 + 1$. Then the three remaining shots of Adam (worth 28 points) must be $25 + 2 + 1$. All the other shots were made by the third shooter who therefore did not have a 3 point hit. Hence, it was Charles.

1964.02. See solution to Question **1961.21**.

1964.03. Consider the largest number on the chessboard. Obviously, all the neighboring numbers must be the same. The ones neighboring these numbers must be equal to them as well, and so on. Therefore all the numbers on the board are the same, and we know that the sum of some four of them is 16. That means that they all are equal to 4, so the number in square $e2$ equals 4 as well.

1964.04. Answer: 4 letters.

Since in any 2×2 squares all the letters must be different, it is impossible to do that with less than four letters. To fill the table with four letters, tile it with 2500 squares 2×2, and in each one of them, write the same four letters in exactly the same order.

1964.05. Answer: John Smith is taller.

Consider Smith's column and Brown's row. We can assume that the scout who stands at their intersection is neither Smith nor Brown (if he is then they share a column or a row, and therefore Smith is taller). This scout is shorter than Smith (since they stand in the same row) and he is taller than Brown (they are in the same column). Thus Smith is taller than Brown.

1964.06. Answer: their product is equal to 0.

Let us denote these numbers by x, y and z. Then we have

$$xy + yz + xz = \frac{1}{2}\big((x + y + z)^2 - x^2 - y^2 - z^2\big) = 0\,,$$

$$1 - 3xyz = x^3 + y^3 + z^3 - 3xyz = (x + y + z)(x^2 + y^2 + z^2 - xy - xz - yz) = 1\,,$$

and therefore $xyz=0$. But, actually, deriving such a long equation can be avoided.

Indeed, if there are no zeros among these numbers, then for each of these numbers we have $0 < x^2 < 1$, and therefore $|x^3| < x^2$. So the absolute value of the sum of the cubes is strictly less than 1, a contradiction.

1964.07. Since all the angles are obtuse, n cannot be less than 5. Consider now all diagonals that connect vertices separated by exactly one other vertex. They form an n-pointed star, the contour of which obviously has length greater than the polygon's perimeter. This proves our inequality.

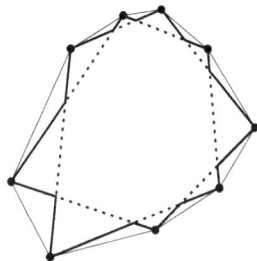

1964.08. Since
$$x^4 + 4y^4 = (x^2 + 2xy + 2y^2)(x^2 - 2xy + 2y^2) = ((x + y)^2 + y^2)((x - y)^2 + y^2),$$
this number is prime only if $x = \pm y = \pm 1$. Since all these cases do indeed produce a prime number (namely, 5), those four pairs of integers constitute the answer.

1964.09. Let us build triangle LKX (with segment LK as a side), obtained from ABC by a parallel translation. Then triangle XMF is the one we need. Indeed, $XM = KN$ since $MNKX$ is a parallelogram (MN is mapped into XK by a parallel translation); similarly, $|FX| = |GL|$ since $FGLX$ is a parallelogram.

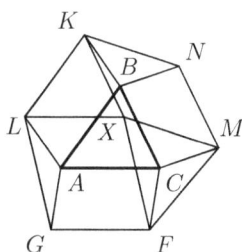

1964.11. Answer: 16 numbers.

Example: 1 and all primes less than 50. If some number in the collection is not prime, replace it with all of its prime factors, and the resulting collection will have more elements but still possess the same mutual "co-primeness" property.

1964.12. Since median ME in triangle MBC is longer than half of side BC, angle BMC is acute. Therefore, angle BMA is obtuse, and in triangle ABM, median MD is shorter than half of side AB.

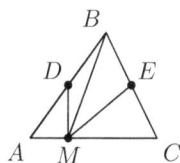

1964.13. Answer: $\{p, q, r\}$ equals $\{7, 5, 2\}$.

Since product pqr is divisible by 5, one of the prime factors, say, p, is 5. Then $qr = 5 + q + r$ or $(q - 1)(r - 1) = 6$. So (up to some permutation) $q = 3$, $r = 4$ or $q = 2$, $r = 7$. First case does not fit since 4 is composite.

1964.14. We can rewrite the condition as $(10a+b)c = (10b+c)a$, that is, $9ac+bc = 10ab$. The equality we need to prove is equivalent to
$$\left(10^n a + \frac{10^n - 1}{9} b\right) c = \left(10b \cdot \frac{10^n - 1}{9} + c\right) a,$$
which after dividing by $\frac{1}{9}(10^n - 1)$ turns into the first equality.

1964.15. Let us start building triangle ABC whose angle CAB coincides with the given angle XAY. Vertex B is the intersection point of ray AY with the line parallel to AX at distance equal to the given altitude.

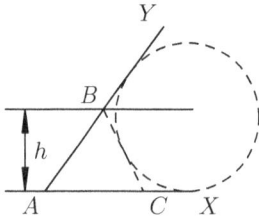

Consider the externally inscribed circumference S touching side BC of triangle ABC—S touches rays $AB = AY$ and $AC = AX$ at the distance from A equal to half the triangle's perimeter. Since we know that number, we can construct S, and after that we construct line BC by drawing tangent line from point B to S.

1964.16. Denoting the sum of the squares by A we get

$$A = a_1^2 + a_2^2 + \cdots + a_n^2 .$$

Consider now the following trivial identity

$$(a_1^2 + a_2^2 + \cdots + a_n^2)^2 = (a_1^2 - a_2^2 - \cdots - a_n^2)^2 + (2a_1 a_2)^2 + \cdots + (2a_1 a_n)^2 .$$

The first square on the right-hand side could be zero; however, if we choose a_1 as the smallest of numbers $\{a_i\}$, we can guarantee that it will not be. Thus, we obtain the required representation for A^2.

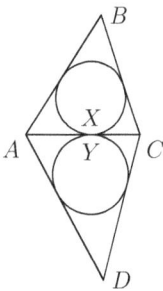

1964.17. Let circumference inscribed in triangle ABC touch diagonal AC at point X, and the one inscribed in ACD—at point Y. In the given case, $X = Y$, and so we have $|AX| = |AB| + |AC| - |BC|$, $|AY| = |AC| + |AD| - |CD|$. From this we obtain $|AB| + |CD| = |AD| + |BC|$, that is, quadrilateral $ABCD$ is circumscribed. Now, performing the same computations in reverse order for the second diagonal BD, we come to the required conclusion.

1964.18. Let q be such a natural number that $q - 1 < \sqrt{n} \leqslant q$.

There are $q^2 \geqslant n$ pairs (x, y) in which $1 \leqslant x, y \leqslant q$. If for none of these pairs number $ax - y$ is divisible by n, then there must be two different pairs (x_1, y_1) and (x_2, y_2) such that $a(x_1 - x_2) - (y_1 - y_2) = (ax_1 - y_1) - (ax_2 - y_2) \vdots n$. Note that $|x_1 - x_2|, |y_1 - y_2| \leqslant q - 1 < \sqrt{n}$. If $x_1 - x_2 = 0$, then $y_1 - y_2 \vdots n$, and therefore $y_1 - y_2 = 0$. And if $y_1 - y_2 = 0$, then $a(x_1 - x_2) \vdots n$; and since $(a, n) = 1$, then $x_1 - x_2 \vdots n$, and so we have $x_1 - x_2 = 0$. In both cases we end up with $(x_1, y_1) = (x_2, y_2)$, which is impossible.

1964.20. Let us split the square into 25 small squares with side length $\frac{1}{5}$. By pigeonhole principle, one of these small squares contains at least three points. Since each small square can be covered by a circle with radius $\frac{1}{7}$, the proof is complete.

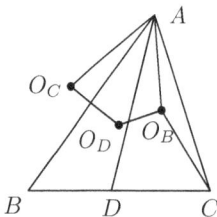

1964.21. Reducing both sides by $[n/1] + [n/n]$ and $2 + [(n-1)/1]$ we can conclude that equality $[n/k] = [(n-1)/k]$ must hold true for any k from 2 to $n - 1$. But if n is a composite number, divisible by $p \geqslant 2$, then $[n/p] > [(n-1)/p]$—a contradiction. Thus, n must be prime.

1964.23. Let us use here the *oriented* angles between the lines. Denote by O_B, O_C, O_D the centers of circumferences circumscribed around triangles ACD, ABD, ABC, respectively.

Then we have $\angle(AO_B, O_BO_D) = \angle(AO_B, O_BC)/2 = \angle(AD, DC)$ (the first equality follows from O_BO_D being perpendicular bisector to AC, and the second is simply a relation between inscribed and center angles). Similarly, $\angle(AO_C, O_CO_D) = \angle(AD, DB)$. Thus, $\angle(AO_B, O_BO_D) = \angle(AD, BC) = \angle(AO_C, O_CO_D)$, and therefore points A, O_B, O_C, O_D lie on the same circumference.

All these circumferences contain chord AO_D which does not depend on the location of point D. Radius of the circumference is minimal when this chord is its diameter, that is, $\angle(AO_B, O_BO_D) = \angle(AD, BC) = \frac{\pi}{2}$. The last equality means that D is the foot of altitude dropped from A onto BC.

1964.25. See solution to Question **1975.30**.

1964.26. For any "integer" point on the plane $A = (a, b)$, $a^2 + b^2 < n^2$, consider a square centered at this point and with unit sides parallel to the coordinate axes. Obviously, these N squares are mutually disjoint. Let us prove that they completely cover circle S defined by inequality $x^2 + y^2 \leqslant (n-1)^2$. Thus the sum of the areas of these squares is greater than or equal to the area of S. Thus, $N \geqslant \pi(n-1)^2 \geqslant 3(n-1)^2$.

To prove that these squares indeed cover S, consider any point $X = (x, y)$ in this circle. Denote by a and b the integers nearest to x and y, respectively. Then $|x - a| \leqslant \frac{1}{2}$, $|y - b| \leqslant \frac{1}{2}$, and the distance between points X and $A = (a, b)$ does not exceed $\frac{\sqrt{2}}{2} < 1$. Hence, $|OA| \leqslant |OX| + |XA| < (n-1) + 1 = n$ (where O is the origin of the coordinate system), and therefore one of our small squares is centered at point A. That is precisely the square that contains point X.

1964.27. Let us represent these points with vectors going to them from the center of the circumference. Let $\vec{a}$, $\vec{b}$, $\vec{c}$, $\vec{d}$ be these points (vectors). It is easy to see that the point of intersection of circumferences passing through $\vec{a}$ and $\vec{b}$ and through $\vec{b}$ and $\vec{c}$, is vector $\vec{a} + \vec{b} + \vec{c}$. Thus we only need to show that points $\vec{a} + \vec{b} + \vec{c}$, $\vec{b} + \vec{c} + \vec{d}$, $\vec{c} + \vec{d} + \vec{a}$, and $\vec{d} + \vec{a} + \vec{b}$ lie on the same circumference. Translating these points by adding vector $-(\vec{a} + \vec{b} + \vec{c} + \vec{d})$, we get points $-\vec{a}$, $-\vec{b}$, $-\vec{c}$, $-\vec{d}$, that indeed lie on the same unit circumference.

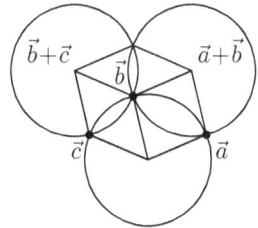

1964.28. Let $\alpha_0 = 0$, $S_0 = 0$. Adding S_{2^n}, we transform collection $(S_0, S_1, \ldots, S_{2^n-1})$ into collection $(S_{2^n}, \ldots, S_{2^{n+1}-1})$ (modulo 2), since $\alpha_k = \alpha_{k+2^n}$ $(0 < k < 2^n)$. Now, using induction, we obtain that collection $(S_0, \ldots, S_{2^n-1})$ has equal number of odd and even elements.

1964.29. As it is well known, locus of points X such that $|AX| + |BX| = d$ is an ellipse whose axes a and b can be easily constructed. The one that lies on line AB has the length of $\frac{1}{2}(d - |AB|)$. The ends of the other are the points of intersection of circumferences with radius $\frac{d}{2}$ centered at A and B.[101]

---

[101] It is assumed here that d is positive and sufficiently large. For values of d less than a certain number, this locus is empty.

Now, consider the transformation that stretches the plane away from line AB with coefficient $\frac{a}{b}$. It maps the ellipse to circumference S with diameter a; and maps L to easily constructible line L'. Find the points where L' intersects S, then use the stretch with reciprocal coefficient $\frac{b}{a}$ which will produce all possible locations of point E.

1964.30. Denoting the sum of all numbers in S by N, we have that if $a \in S$, then aN and N have the same remainders modulo p (because multiplication by a simply permutes the remainders of numbers in S modulo p). Now, choose $a \neq 1$, and since $(a - 1)N \vdots p$, we have $N \vdots p$.

1964.31. Denote the bases of these triangles by a and b. For each triangle, consider the rectangle which is "circumscribed" around it so that one of its sides coincides with the base of the triangle lying on the contour of the square. Then intersection of the triangles lies inside the intersection of these rectangles, which is itself a rectangle with sides a and b. Thus, the area of the intersection does not exceed ab.

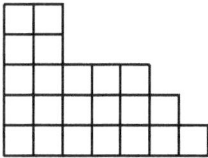

1964.32. Let us construct a simple "polyomino" histogram (a so-called Young diagram) with numbers $a_1 \geqslant a_2 \geqslant \cdots \geqslant a_n$ representing the heights of the diagram's columns (an example for $\{a_k\} = \{5, 5, 3, 3, 3, 2, 1\}$ is shown in the figure). Then $a_1 + a_2 + \cdots + a_n$ is, obviously, the total number of the squares in the diagram, and $f_1, f_2, \ldots$ are the widths of the diagram's rows, and their sum, of course, equals the very same number.

1964.33. Let us split the square into the parts which are the smallest non-empty intersections of these figures. Then its area $S = 1$ is equal to $s_1 + s_2 + \cdots + s_n$, where s_i denotes the area of the ith part. Since each part is covered by at least q figures, the sum of all areas must be at least $qs_1 + qs_2 + \cdots + qs_n = qS$. Therefore, one of the figures' area is greater than or equal to $qS/n = q/n$.

1964.34. First, we need to prove this fact for $r = 1$. This is relatively easy. Consider two different numbers x and y that belong to the same group (e.g., group #1). If even one of the numbers $\frac{1}{2}(x+y)$, $2y - x$, and $2x - y$ belongs to the same group, then that number forms the required trio with x and y. If all three belong to group #2, then $\{x - 2y, y - 2x, \frac{1}{2}(x+y)\}$ form such a trio.

Now, let us consider any positive r. As we have just proved, there exist different numbers a, b, and c belonging to the same group (say, group #1), and such that b is the arithmetic mean of a and c; we can, without loss of generality, assume that these numbers are 0, 1, and 2. Then numbers $r + 1$, $r + 2$, $2r + 2$ either all belong to group #2, or one of them is in group #1. In the former case, they form the required trio, and in the latter case, this member of group #1 forms such a trio with two of the numbers 0, 1, and 2.

1964.35. Let $A_0 B_0 C_0$ be the given regular triangle with unit side, α, β, γ being the angles of triangle ABC we are trying to construct. Without loss of generality, we can assume that $\alpha \leqslant \beta \leqslant \gamma$.

Lemma. *Maximum area of triangle ABC is attained in one of the three following cases.*

1) $A = A_0$, $B = B_0$;

2) $A = A_0$, point B lies on side $A_0 B_0$, point C lies on side $B_0 C_0$;

3) $A = A_0$, points B and C both belong to segment $B_0 C_0$.

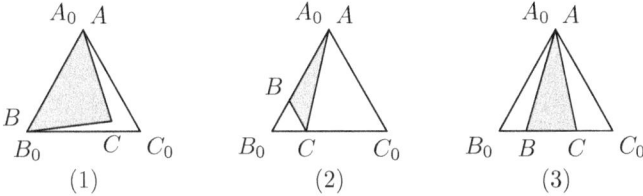

(1)   (2)   (3)

This can be proved in a rather tedious way by going through all possible cases of how the smaller triangle can fit inside the larger one. However, we will offer a different proof using convexity-based approach.

Start with introducing the coordinate system. Let (x_1, y_1), (x_2, y_2) and (x_3, y_3) be coordinates of vertices A, B, and C, respectively. Fix orientation of triangle ABC, and then since the angles are fixed, point C is uniquely determined by points A and B; moreover, its coordinates x_3 and y_3 are linear functions in x_1, y_1, x_2, and y_2 (that is very easy to verify using complex numbers).

The condition that a point with coordinates (x, y) lies within triangle $A_0 B_0 C_0$ can be written as a system of three linear inequalities in x and y, with each inequality expressing the fact the point belongs to a specific half-plane defined by one of the triangle's sides. Writing these inequalities for points A, B, and C, then substituting coordinates of C via coordinates of A and B, we will obtain system of nine linear inequalities in four independent variables x_1, y_1, x_2, and y_2.

Area of triangle ABC is proportional to the square of side AB, thus it is a convex function in the same variables. As it is well known, at the point of maximum of a convex function of n variables which must satisfy the system of linear inequalities, at least n of the inequalities must turn into equalities—using geometric language, maximum of a convex function defined on a (possibly infinite) polyhedron in $\mathbb{R}^n$ is attained at some vertex of the polyhedron. In our case, it means that the point of maximum satisfies four conditions of form "one of the vertices A, B, and C lies on one of the straight lines $A_0 B_0$, $B_0 C_0$, $A_0 C_0$". Hence, either some two of the vertices of ABC coincide with some vertices of $A_0 B_0 C_0$ (Case 1), or one of the points A, B, and C coincides with a vertex of $A_0 B_0 C_0$, and two others lie on its sides (Case 2 or Case 3).

It is left to prove that vertex A is the one which coincides with a vertex of $A_0 B_0 C_0$, and in Case 2, it will be vertex B on one of the adjacent sides of the same triangle. If both angles α and β do not exceed 60°, then Case 1 is possible, and it obviously delivers the maximum area, since side AB equals the maximum possible distance between the points of $A_0 B_0 C_0$. And if $\beta > 60°$, then only vertex A will "fit" into an angle of equilateral triangle.

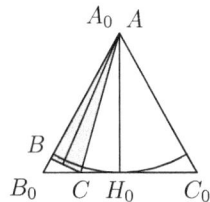

Assume that $A = A_0$, $\gamma > \beta$, and C lies on the side adjacent to A_0. Then, reflecting ABC with respect to the angle bisector of $\angle A$, we obtain congruent triangle, one of whose vertices lies strictly inside $A_0B_0C_0$. That means this is not the case of maximum area, which leaves us with Cases 2 and 3. This concludes the proof of the lemma. $\square$

As it was already mentioned, if Case 1 is possible, then it realizes the maximum area. To compare areas for Cases 2 and 3, consider altitude AH in triangle ABC. In Case 3, it coincides with altitude A_0H_0 in triangle $A_0B_0C_0$. In Case 2, the inequality $|AH| \geqslant |A_0H_0|$ holds true if and only if $\angle ACB > \angle ACC_0$, that is, $\gamma \geqslant \alpha + 60°$.

Thus, if $\beta \leqslant 60°$, then the maximum area is attained in Case 1. If $\beta > 60°$, then for $\gamma \geqslant \alpha + 60°$, it is attained in Case 2, and for $\gamma \leqslant \alpha + 60°$—in Case 3.

1964.36. First of all, in order to find the weight of one genuine coin, we will use our first weighing. Without loss of generality, we can assume that weight to be 1.

Then we take one coin from each of the coin makers and determine their total weight S. If $S = n$, then all the coins are genuine. If $S \neq n$, then the difference between the weight of a counterfeit and the genuine coin belongs to the finite set of numbers $\{d_k = (n - S)/k\}$, $k = 1, \ldots, n$.

Let x be some natural number. For one remaining weighing, we will take 1 coin from coin maker #1, x coins from coin maker #2, x^2 coins from coin maker #3, and so on, until the last coin maker #n, from whom we will take x^{n-1} coins. Then we will find their total weight. Obviously, it will be different from $(1 + x + \cdots + x^{n-1})$, and the difference (let us denote it by δ) will be equal to $d_r(x^{m_1} + x^{m_2} + \cdots + x^{m_r})$, where r is the number of the coin makers who make counterfeit money, and $m_1, \ldots, m_r$ are indexes of these counterfeiters. All expressions of this type are different polynomials in x, and so they can coincide only in a finite number of points. Thus, x can be chosen in such a way that all these polynomials have different values at x, and knowing δ, we will be able to determine all the counterfeiters.

1965

1965.01. Answer: 49 books.

After each book's binding is done, the difference between the numbers of sheets of white and red paper stays the same. Since that difference originally equals $135 - 92 = 43$, that means that we currently have 43 white sheets and 86 red sheets, and thus the number of bound books is $92 - 43 = 49$.

1965.02. Out of the multiplied numbers, 393 are divisible by 5, while 78 of them are divisible by 5^2, fifteen—by 5^3, and three—by 5^4. In addition, 982 of these numbers are even. Thus, 1965! is divisible by $2 \cdot 10^{489}$ but is not divisible by 10^{490}, completing the proof.

1965.03. Answer: after 9375 km.

Assume the automobile was driven for x km, after which the tires were rotated, and then it was driven y km more. We obtain the equation $\frac{x}{25000} + \frac{y}{15000} = \frac{x}{15000} + \frac{y}{25000} = 1$, and therefore $x = y = 9375$.

1965.04. Answer: 1318 parts.

Indeed, inside the rectangle, the diagonal intersects 18 vertical and 64 horizontal lines. Thus, it intersects the interior grid lines in $18 + 64$ points, and all of these 82 points are different, meaning that the diagonal intersects interior of 83 squares adding to the already existing $19 \cdot 65$ parts 83 more. Hence, we obtain $(18 + 1) \cdot (64 + 1) + 18 + 64 + 1 = 1318$ parts.

1965.05. Answer: $110\,768 : 112 = 989$.

Likely, there is something missing in this question's statement. As it is currently presented, there are multiple possible solutions—the one shown above is just one of the many. E.g., one other such solution is $103\,886 : 127 = 818$. The original answer is taken from the jury's records; it is not really clear what kind of additional condition could have been added in order to ensure the uniqueness of the answer. If the original question was about finding just any solution, then it would simply not be a very good problem (one of its most obvious drawbacks is that such an answer/solution can be easily copied from another participant).[102]

1965.06. Answer: regardless of how we choose the numbers, their sum will be 125.

Since all the numbers lie in different rows, their digits of tens are 0, 1, 2, 3, and 4, each one occurring exactly once. Similarly, their digits of ones are 1, 3, 5, 7, and 9. Thus, the sum of the numbers is $(10 + 20 + 30 + 40) + (1 + 3 + 5 + 7 + 9) = 125$.

---

[102] However, this version of the problem is still viable.

C

E   K

O

A   B

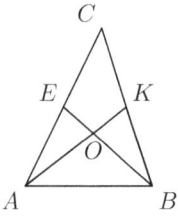

1965.07. Let us split all the divisors of natural number n into pairs $(d, n/d)$. Since the number of divisors is odd, one of them must form such a pair with itself.

1965.08. Answer: $\frac{S}{3}$.

Since $S_{KCA} = \frac{S}{2} = S_{BEA}$, the area of the part of triangle ABC covered twice by these two triangles is the same as the area of the part of ABC which is not covered at all; that is, $S_{CKOE} = S_{BOA}$. Then $S_{BOA} = \frac{S}{3}$, because triangles AOB and ACB have the same base, and the altitude from vertex O is three times shorter than the altitude from vertex C (remember, point O divides median from C in the ratio $1 : 2$).

1965.09. Answer: after 9375 km.

Assume that the automobile drove for x km, after which the tires were rotated, and then it drove for y km more. Then we have $\frac{x}{25000} + \frac{y}{15000} \leqslant 1$ and $\frac{x}{15000} + \frac{y}{25000} \leqslant 1$. Adding these inequalities, we obtain $(x + y)(\frac{1}{15000} + \frac{1}{25000}) \leqslant 2$, that is, $x + y \leqslant 18750$. Equality is attained when $\frac{x}{25000} + \frac{y}{15000} = 1$ and $\frac{x}{15000} + \frac{y}{25000} = 1$, implying $x = y = 9375$.

1965.10. Answer: $24 \cdot 60 + 72 = 1512$.

See solution to Question **1965.04**.

Indeed, inside the rectangle the diagonal intersects 23 vertical and 59 horizontal lines, passing through 11 grid nodes. Thus, it intersects the interior grid lines in $23 + 59 - 11 = 71$ points, meaning that it intersects the interior of 72 squares, adding to the already existing $24 \cdot 60 = 1440$ parts 72 more.

1965.11. Answer: $x = \frac{7}{15}$ or $x = \frac{4}{5}$.

Denoting left and right-hand sides of the equality by n, we have $n \leqslant \frac{1}{8}(5 + 6x) < n + 1$ and $n = \frac{1}{5}(15x - 7)$. Expressing x via the second equation and then substituting it into the first inequality, we obtain $n \leqslant \frac{1}{8}(2n + \frac{39}{5}) < n + 1$. It is easy to see that only $n = 0$ and $n = 1$ satisfy this system.

1965.12. Consider the vertices of an equilateral triangle with side length of 1965 meters. Some two of its vertices must be colored the same.

1965.13. If you draw a line very close to the diagonal you will be able to add 83 more parts. See solution to Question **1965.21**.

1965.14. Answer: the number of the lawyers is always three.

Label all the people by their first initials. From the second part of the condition (C said that D was a lawyer), it follows that these two people must have different occupations. So there are only two options for pair (C, D): LS and SL. Since F and G are both scientists, the statement "F denies that G is a scientist" is false. Hence, the number of lawyers between A and F (not including F but including A) must be odd. Since A is a lawyer, and there is exactly one lawyer among C and D, pair (B, E) also must contain exactly one lawyer. Summing it up: A is a lawyer, F and G are not lawyers, pair (B, E) contains exactly one lawyer, and pair (C, D) contains exactly one lawyer. So we have three lawyers. Possible results for (A, B, C, D, E, F, G) are $LSLSLSS$, $LSSLLSS$, $LLLSSSS$, $LLSLSSS$.

1965.15. The answer is shown in the figure. Lines AB, AF, DC, and DE are tangent to the circumference, angles BAF and CDE have measure of 60°, and the length of AD is 12 km.

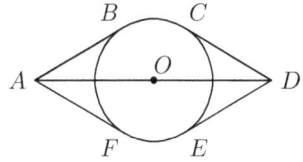

1965.17. Let us prove it using induction on the number of towns. The basis is obvious. Now if we choose one town A (out of n towns) and separate it from the rest, then the remaining $n-1$ towns can be all visited via some route $B_1 \to B_2 \to \cdots \to B_{n-1}$. If there exists road $A \to B_1$ or $B_{n-1} \to A$, then the induction step, clearly, follows from that. If both these roads are oriented in the opposite direction, then there must exist index k such that we have roads $B_k \to A$ and $A \to B_{k+1}$. Then we can visit all n towns using the route

$$B_1 \to B_2 \to \cdots \to B_k \to A \to B_{k+1} \to \cdots \to B_{n-1}.$$

1965.18. Answer: there is only one such octuplet consisting of eight numbers 2.

Indeed, let us denote these numbers by p_i ($i = 1, \ldots, 8$); we know that $p_1^2 + \cdots + p_8^2 + 992 = 4p_1 p_2 \ldots p_8$. For any odd p, its square p^2 has remainder 1 modulo 8. Thus, if none of the numbers p_i equals 2, then the left-hand side is divisible by 8 and the right-hand side is not. Now, let $k \geqslant 1$ be the number of twos among these eight primes. Then the left-hand side modulo 8 is $4k + (8 - k) = 3k + 8$, and the right-hand side is divisible by 8. It follows that $k = 8$, i.e., all the numbers are equal to 2.

1965.19. Answer: maximum is reached when the ends of the segment are equidistant from the angle's vertex.

First, we will prove the following preposition.

Lemma. *Given the angle and the opposite side, the triangle with the maximum possible area is an isosceles triangle with the given angle at the vertex.*

Indeed, if we fix the given side XY, then the vertex of the given angle sweeps one of the arcs of the circumference passing through the endpoints of XY. Obviously, the maximum distance to the side is reached at the midpoint of this arc—that is, the maximum length of the altitude (and thus, the maximum area) is attained for the isosceles triangle. □

Now, back to the solution. It will suffice to show that if we replace an arbitrary collection of segments with the one given in the answer above, the area will not decrease. Let O be the vertex of the angle, AB and BC—unit segments, where A and C lie on the sides of the angle, while B lies inside of it.

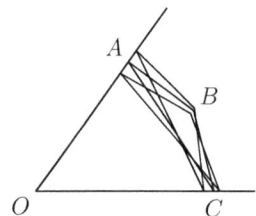

Move triangle ABC (preserving its shape) so that A and C end up equidistant from point O. Area of $OABC$ will not decrease because S_{ABC} stays the same, and it follows from the lemma that S_{OAC} will not decrease as well. After the move, B lies on the bisector of angle O, and we can apply the lemma to the half-angles AOB and BOC.

1965.21. Answer: $mn + m + n - 1$.

Each line intersects at most $(m-1)+(n-1)$ interior lines parallel to the sides of the parallelogram. Thus, it intersects no more than $m+n-1$ squares, and therefore adds no more than $m+n-1$ to the existing mn parts. To obtain exactly $mn+m+n-1$ parts, choose the line which is parallel and very close to the diagonal of this parallelogram so that it intersects precisely the same interior grid lines (except two of the boundary segments) but does not pass through any grid nodes. (Compare this to Questions 4 and 10.)

1965.22. Start by multiplying three inequalities

$$|AC'| \cdot |C'B| \leqslant \tfrac{1}{4}(|AC'| + |C'B|)^2 = \tfrac{1}{4}|AB|^2 \,,$$

$$|BA'| \cdot |A'C| \leqslant \tfrac{1}{4}(|BA'| + |A'C|)^2 = \tfrac{1}{4}|BC|^2 \,,$$

$$|CB'| \cdot |B'A| \leqslant \tfrac{1}{4}(|CB'| + |B'A|)^2 = \tfrac{1}{4}|CA|^2 \,.$$

Therefore, the left-hand side of the inequality we are trying to prove does not exceed

$$\tfrac{1}{64}(|AB| \cdot |BC| \cdot |AC|)^2 < |AB| \cdot |BC| \cdot |AC| \,.$$

1965.23. If all the remainders are different, then their sum is $0 + 1 + \cdots + (n-1) = \tfrac{1}{2}n(n-1)$, and that means that the sum is not divisible by n. On the other hand, the sum of these people is $2(1 + 2 + \cdots + n) = n(n+1)$, and that number is divisible by n.

1965.24. Let us enlarge all the radii of the circles by the factor of three, keeping their centers fixed. Then if some two of the original circles intersected, we remove the one with the smaller radius (together with the corresponding "triple" circle). This removal satisfies the following property: the "triple" circles still cover the entire set covered by the original circles.

To prove that, we need to note that if two original circles with radii r_1 and r_2 have a common point, then the larger one of their "triple" circles completely covers the smaller original circle.

After as many such removals as possible, we will end up with the collection of mutually disjoint circles whose "triples" cover the figure of area 1. Therefore, the sum of their areas is at least $\tfrac{1}{3^2} = \tfrac{1}{9}$.

1965.25. Answer: $x = 3$, $y = 14$ and $x = -24$, $y = -2$.

Adding 360 to both sides, transform them to $(2x+3)(3y-2) = 360 = 3^2 \cdot 5 \cdot 2^3$. Since $2x+3$ is odd and must be divisible by 3^2 (since $3y-2$ is not divisible by 3), we are left with only four possibilities $2x+3 = \pm 3^2, \pm 3^2 \cdot 5$.

1965.26. Answer: $\frac{\sin n\alpha}{\sin \alpha \cos n\alpha} - \frac{1}{\cos \alpha}$.

To prove that, simply rewrite each summand as follows

$$\frac{1}{\cos(k-1)\alpha \cos k\alpha} = -\frac{1}{\sin \alpha}\left(\frac{\sin(k-1)\alpha}{\cos(k-1)\alpha} - \frac{\sin k\alpha}{\cos k\alpha}\right).$$

1965.28.

Grade 10 version. Let us denote the sides of the triangle by a, b, c and its angles by α, β, γ; also $x = |AM_1|$. Then one by one we compute $|AM_2| = x$, $|CM_3| = |CM_2| = b - x$, $|BM_4| = |BM_3| = a - (b - x) = a - b + x$, $|AM_5| = |AM_4| = c - (a - b + x) = c - a + b - x$, $|CM_6| = |CM_5| = b - (c - a + b - x) = a - c + x$, $|BM_7| = |BM_6| = a - (a - c + x) = c - x$, and finally, $AM_7 = x$. Therefore, $M_7 = M_1$, that is, the closed curve we obtain will have at most six arcs with the total length

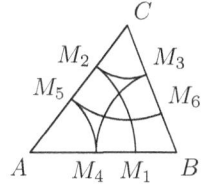

$$L = x\alpha + (b - x)\gamma + (a - b + x)\beta + (c - a + b - x)\alpha$$
$$+ (a - c + x)\gamma + (c - x)\beta = (a + b + c)\pi - 2a\alpha - 2b\beta - 2c\gamma.$$

This curve can contain less than six arcs (e.g., three) only if $M_4 = M_1$, i.e., if $c - a + b - x = x$ or equivalently $x = \frac{1}{2}(b + c - a) = p - a$. Then M_1, M_2, M_3 are the points where the sides touch the inscribed circle. In this case, the length of the curve would be equal to $L/2$.

Grade 11 version. We will prove it using induction on n. Basis $n = 1$ is obvious. Now for the step, we have

$$(a + b)^{n+1} \leqslant (a + b)2^{n-1}(a^n + b^n)$$
$$= 2^{n-1}(a^{n+1} + b^{n+1} + ab^n + ba^n).$$

Since $ab^n + ba^n \leqslant a^{n+1} + b^{n+1}$ (because $(a^n - b^n)(a - b) \geqslant 0$), it follows that

$$2^{n-1}(a^{n+1} + b^{n+1} + ab^n + ba^n) \leqslant 2^n(a^{n+1} + b^{n+1}).$$

1965.29. Answer: 216 parts.

Consider the coordinate system with origin in one of the cube's vertices. Then plane $x + y + z = 18$ intersects a unit cube with the corner at (a, b, c) if and only if $15 < a + b + c < 18$, that is $a + b + c = 16$ or 17 (by the corner we mean that vertex of the unit cube which is the closest one to the origin). It is left to find the number of solutions for the equation $a + b + c = 16$ or 17 in non-negative integers not exceeding 11.

Each solution of the equation $a + b + c = 16$ in non-negative integers not exceeding 11 corresponds to a solution of the equation $A + B + C = 19$ in natural numbers not exceeding 12. The latter equation has $\binom{18}{2} = \frac{18 \cdot 17}{2} = 153$ solutions in natural numbers without the condition of not exceeding 12 (let us write 19 ones in a row, then write 18 pluses between them, after which we need to choose two pluses which will be removed; another way of thinking about it is to imagine a strip of paper 1×19, and then choose two out of eighteen possible places to make a cut). Which solutions do not fit our original setup? There are $3 = 3 \cdot 1$ solutions in which one number is 17; $6 = 3 \cdot 2$ solutions in which one of the numbers is 16; $9 = 3 \cdot 3$ solutions with 15; 12 and 15 solutions with 14 and 13, respectively. Thus we have $153 - 3(1 + 2 + 3 + 4 + 5) = 108$ solutions.

Replacing numbers a, b, c by $11 - a$, $11 - b$, $11 - c$, we obtain a solution of the equation $a + b + c = 17$ in non-negative integers not exceeding 11. Thus, the number of such solutions is also 108, and the total is 216.

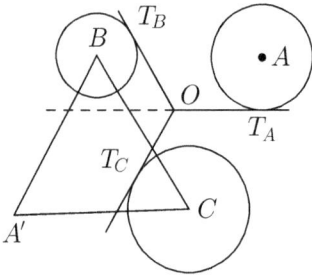

1965.30. Let A, B, and C be the centers of the given circumferences, and R_A, R_B, and R_C—their radii. Let us assume the required point O is already known, and OT_A, OT_B, OT_C are the "right" tangents mentioned in the problem. We will also assume that ray OT_A is directed horizontally and to the right, rays OT_B and OT_C are directed into the upper and the lower half-plane, respectively, forming $120°$ angles with OT_A (see the figure).

Construct equilateral triangle BCA' (vertices are listed clockwise). Let us prove that point A' lies at distance $R_B + R_C$ from line OT_A; also that it lies below that line and to the left of point O. Temporarily, we can forget that points B and C are already given to us. Consider them variable with restrictions $BT_B \perp OT_B$ and $CT_C \perp OT_C$ (points O, T_B, and T_C are fixed), and treat point A' as a function of points B and C.

If $B = T_B$ and $C = T_C$, then $OBA'C$ is inscribed, and so $\angle A'OB = \angle A'CB$, that is, A' lies on the ray complementary to ray OT_A. If we shift point B down-and-left by some vector $\vec{v}$ perpendicular to OT_B, then point A' (which is the image of B under rotation by $60°$ counterclockwise around point C) will shift by the image of $\vec{v}$ under that rotation, that is, by the vector of the same length pointed straight downward. The same happens if we shift point C down-and-right away from line OT_C. Hence, if distances of B and C from the corresponding lines are equal to R_B and R_C, respectively, then A' lies at distance $R_B + R_C$ down from line OT_A.

This means that the circumference with radius $R_B + R_C$ and center A' touches line OT_A. It follows that the way to construct point O is to construct point A', then construct circumference S with radius $R_B + R_C$ and center A', and then line OT_A as a common interior tangent line for S and the circle centered at A. Similarly, we can construct line OT_B and then find O as the intersection of these two straight lines. Since triangle $T_A T_B T_C$ can be oriented in two different ways, two locations of O are possible.

1965.31. Consider the row or column which has the minimum possible sum of numbers—let us assume it is a row with the sum equal to S. If $S \geqslant n/2$, then the statement is obvious. So we will assume that $S < n/2$. That row has $k \geqslant n - S$ zeros, and in each column that contains one of them, the sum of the numbers is at least $n - S$. In each of the other $n - k$ columns, the sum must be at least S. Therefore, the sum of all numbers in the table is greater than or equal to

$$k(n - S) + (n - k)S = k(n - 2S) + Sn$$

$$\geqslant (n - S)(n - 2S) + Sn = \frac{n^2}{2} + \frac{(n - 2S)^2}{2} \geqslant \frac{n^2}{2}.$$

1965.32. Let $f(t) = (t - a)(t - b)(t - c)$ and $g(t) = (t - x)(t - y)(t - z)$. Then

$$f(t) = t^3 - t^2(a + b + c) + t(ab + ac + bc) - abc,$$
$$g(t) = t^3 - t^2(x + y + z) + t(xy + yz + xz) - xyz,$$

and therefore $f(t) - g(t) = \lambda t$, where $\lambda = (ab + bc + ac) - (xy + yz + zt)$. If $\lambda > 0$, then $f(c) = g(c) + \lambda c > 0$, since $g(t) \geqslant 0$ when $t \geqslant z$, which is impossible, since c is a root of

$f(t)$. If $\lambda < 0$, then $f(a) = g(a) + \lambda a < 0$ because $g(t) \leqslant 0$ when $t \leqslant x$, and that is, again, impossible. Thus, $\lambda = 0$ and therefore $f = g$ and trios $\{a, b, c\}$ and $\{x, y, z\}$ coincide.

1965.33. Answer: $x^2 - \frac{1}{2}$.

Maximum of the absolute value of this polynomial on $[-1; 1]$ equals $\frac{1}{2}$. If for polynomial $f(x) = x^2 + ax + b$ maximum of its absolute value is less than $\frac{1}{2}$, then $|b| = |f(0)| < \frac{1}{2}$. Therefore, $|f(1)| = |1 + b + a| \geqslant 1 - |b| + |a| > \frac{1}{2}$ for $a \geqslant 0$, and $|f(-1)| = |1 + b - a| \geqslant 1 - |b| + |a| > \frac{1}{2}$ for $a \leqslant 0$, a contradiction.

1965.34. Let us prove this not just for a square, but for any arbitrary rectangle inside the triangle.

First, we will reduce the question to the case when two of the triangle's sides are parallel to the sides of the rectangle. Let us assume that the rectangle itself is positioned "horizontally", that is, its sides are parallel to the coordinate axes. If the triangle has no vertical sides, then one of the lines passing through its vertices lies between the two others. Cut our figures along this line, and the rectangle is now split into two rectangles (it is convenient for us here to consider an empty set as a rectangle as well), and the triangle—into two triangles with each one of them possessing one vertical side. So, it is enough for us to prove the inequality separately for one of these triangles and the corresponding rectangle. Using the same reasoning with horizontal lines, we can reduce the problem to the case when triangle T has one vertical and one horizontal side.

To prove the inequality for that case, clearly, we can assume that one of the rectangle's corners coincides with the right angle of T while the other corner lies on the hypotenuse of T. Then the set difference of T and the rectangle consists of two triangles similar to T with ratios k_1 and k_2, where $k_1 + k_2 = 1$ because their hypotenuses constitute the hypotenuse of T. That implies the inequality $k_1^2 + k_2^2 \geqslant \frac{1}{2}$, and therefore the sum of these triangles' areas is at least one half of the area of T. Hence, the area of the rectangle does not exceed the other half.

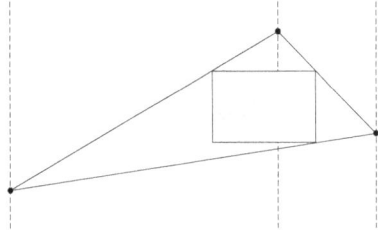

1965.35. Denote the figure by F. Now instead of moving it, we will move the entire grid. Let us select one the unit squares T in the grid. We will move all the unit squares which intersect with F, making them coincide with T (together with the corresponding parts of F). Since the total area of F is less than 1, there exists point A in T which is not covered by any of these images. Thus, in every unit square, the corresponding point ("copy" of A) is not covered by figure F. Now we simply translate the grid so that its nodes map into point A and its "copies" in all other unit squares.

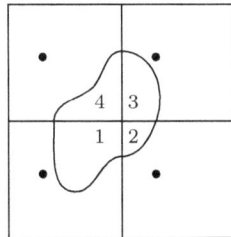

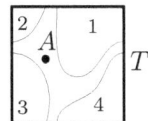

Note. This can be generalized to the following interesting fact.

Blichfeldt's Lemma. *A figure of area S is drawn on the plane with unit grid. Prove that it can be translated so that the image contains at least $\lceil S \rceil$ grid nodes.*

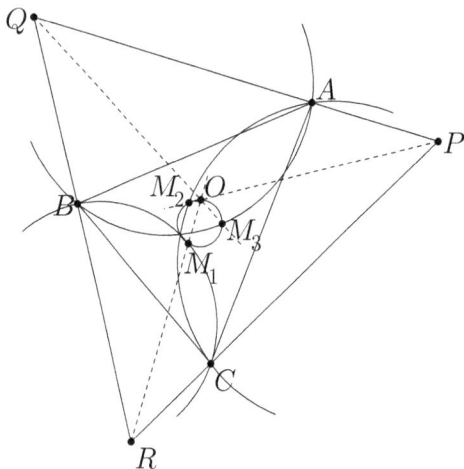

1965.36. Answer: draw three circumferences, with sides of the given triangle $T = \triangle ABC$ as their chords, such that their arcs inside the triangle have measure of 120°. Let M_1, M_2, and M_3 be the midpoints of these arcs. Then the required locus is the arc of the circumference S which passes through points M_1, M_2, and M_3 with its endpoints corresponding to the boundary cases (when one of the sides of a regular triangle contains some side of T).

To prove this, notice that the vertices of any such equilateral triangle PQR lie on the exterior arcs of these circumferences; the center O of a regular triangle is the point of intersection of its angle bisectors; the angle bisectors pass through points M_1, M_2, and M_3 because the bisector of an angle inscribed into circumference divides in half the arc subtended by that angle.

Since angle bisectors in equilateral triangle form 120° angles with each other, angles M_1OM_2, M_2OM_3, and M_3OM_1 (or their supplementary angles) are all equal to 120°. Thus, if we construct circumference S_1 with chord M_2M_3 which subtends angle of 120° in this circumference, point O must lie on it. Same is true for similar circumferences S_2 with chord M_1M_3 and S_3 with chord M_1M_2. If two of these circumferences coincide, then obviously all three do. Therefore, we either have just one point of intersection $O = S_1 \cap S_2 \cap S_3$ (if T is equilateral), or we have an arc of circumference $S = (M_1M_2M_3)$ as the required locus.

1966

1966.01. Answer: the first number is greater than the second one.
To prove that, multiply both numbers by 10. The results are equal to

$$1 + \frac{9}{\underbrace{1000\ldots001}_{1966 \text{ zeros}}} \quad \text{and} \quad 1 + \frac{9}{\underbrace{1000\ldots001}_{1967 \text{ zeros}}},$$

respectively. Obviously, the second number is smaller.

1966.02. There are 30 possible numbers of matches played by each specific team—from 0 to 29. Obviously, there cannot be two teams which have played 0 and 29 matches, respectively. Thus we have 30 teams and no more than 29 possible outcomes, and it follows that some two teams have played the same number of matches.

1966.03. Let us track the sum of the numbers on the blackboard. Originally it is equal to $1967 \cdot 983$, and thus it is odd. Now, notice that when we replace pair (a, b) with difference $a - b$, the parity of the sum does not change. Hence, the sum of the numbers is always odd and therefore it is not possible to end up with all zeros.

1966.04. Consider two points of the same (say, white) color lying on the x-axis; assume they represent numbers 0 and 1. If one of the points 2, $\frac{1}{2}$, and -1 is white, then it forms the required trio with the original two points. If they are all black, then these three points form that trio.

1966.05. Answer: 14 chess players.
Denote the number of participants by n, and by k—the number of times that each player played every one of the other $n - 1$ players.

If n is even, then each participant played in every round; thus, each one played 26 games. Obviously, we have $26 = k(n - 1)$, and so $n - 1$ divides 26.

If n is odd, then each round was skipped by exactly one player—therefore, $26 = kn$, so n divides 26.

Since $n > 3$, the only possible cases are $n = 13$ and $n = 14$. If $n = 13$, then in each round of the tournament, exactly 6 points were distributed to the players and the sum of the points after 13 rounds is 78, which cannot be represented as a sum of one odd number and several even numbers. Now, all we need is to present an example for the latter case ($n = 14$). Since the first 13 rounds constitute one full round-robin tournament between 14 players, we need to define the result of every possible match between any two participants. Namely, we choose one player A and have him defeat all the others, thus gaining 13 points (13 is odd). Everyone else has one loss (from their match with A) with all other twelve games drawn; hence, each player except A has $12/2 = 6$ points (6 is even).

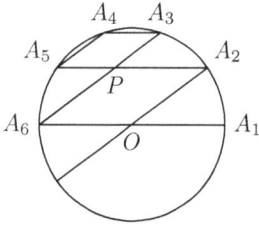

1966.07. Denote the six consecutive points from the ten that divide the circumference in ten equal arcs by A_1, A_2, A_3, A_4, A_5, and A_6. Then line A_2A_5 is parallel to diameter A_1A_6 and to line A_3A_4, and line A_3A_6 is parallel to A_4A_5. Let P be the intersection of A_2A_5 and A_3A_6. Then $PA_3A_4A_5$ is a parallelogram, and it suffices to show that the length of A_2P equals the circumference's radius. Since A_2OA_6P is also a parallelogram, and O is the circumference's center, then $|A_2P| = |OA_6|$.

1966.08. Let $p < 1966$ be a prime number. We will prove that p divides our product $A = n(2n+1)(3n+1)\ldots(1966n+1)$. If $n \vdots p$, then $A \vdots n \vdots p$. If p does not divide n, then consider the remainders modulo p for numbers n, $2n$, $\ldots$, pn. No two of these p remainders can be equal. Indeed, if $1 \leqslant a < b \leqslant p$ and $an \equiv bn \pmod{p}$, then $(b-a)n \vdots p$; but neither n nor $b - a$ is divisible by p.

Hence, one of these remainders is $p - 1$; so for some a we have $an \equiv p - 1 \pmod{p}$. If $a \geqslant 2$, then $A \vdots (an+1) \vdots p$; if $a = 1$, then $A \vdots ((p+1)n+1) \vdots p$.

1966.09. Answer: 2.

Let us denote that number by k. If $k = 1$, then n could be any number greater than 1. If $k \geqslant 3$, the number n of lines could be $k + 1$ (k parallel lines and one line that intersects them all), or it could be k (a "sheaf" of $k - 1$ lines passing through one point A and one more line that intersects them all and does not pass through A). Therefore, when $k \neq 2$ there is no unique solution.

For $k = 2$, the problem has a unique solution $n = 3$. Indeed, if there is less than three lines, the number of their intersection points is at most one. Assume we have four or more lines and they intersect at more than one point—then we either have three pairwise non-parallel lines which are not concurrent, or the lines can be split into two groups of parallel lines.

In the former case, we have at least three intersection points. In the latter case, we have xy intersections, where x and y are numbers of lines in these two groups. Since $x + y \geqslant 4$, we have $xy \geqslant 3$.

1966.11. Consider triangle ABC which has the largest area S of all triangles whose vertices lie on the given points. Then it is obvious that each of the given n points lies inside triangle $A'B'C'$, where $C \in A'B' \parallel AB$, $A \in B'C' \parallel BC$, $B \in A'C' \parallel AC$. Since the area of triangle $A'B'C'$ is exactly $4S < 4$, then $A'B'C'$ is the required triangle.

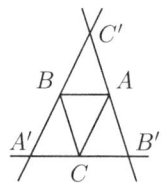

1966.15. Let $n = 2^k m$, where m is an odd number. Split all the odd divisors of n^2 (except for m) into pairs $(d, m^2/d)$.

Sum of the divisors in each pair is even, and therefore the sum of all odd divisors (including m) must be odd. Sum of all even divisors of n^2 is, of course, even, and therefore the total sum is odd.

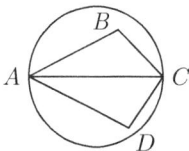

1966.16. Denote the quadrilateral by $ABCD$ with A being its acute angle. Consider circumference S which has AC as a diameter.

Since angles ABC and ADC are obtuse, points B and D lie inside S, and therefore distance between them is less than the diameter of S; that is, $|BD| < |AC|$.

1966.17. Since $x^4 + 1 > x$ for any real x, we have $(x^4 + 1)/5x \geqslant 1/5$ for $x > 0$. Since function $x^4 - 10x + 1$ is convex, it attains maximum value $[1/5; 2]$ at one of the segment's ends. At points $\frac{1}{5}$ and 2, this function is negative. Thus, if $\frac{1}{5} \leqslant x \leqslant 2$, then $(x^4 + 1)/5x \leqslant 2$.

1966.18. Let $a \leqslant b$ be the sides of the rectangle, vertical side being the shorter one. Let ℓ be the straight line which forms angle ϕ with the horizontal and angle $\frac{\pi}{2} - \phi$ with the vertical. Then the lengths of its intersections with the horizontal strip of width a and with the vertical strip of width b are, respectively, $\frac{a}{\sin \phi}$ and $\frac{b}{\cos \phi}$. Thus, the length of intersection of this line with the rectangle does not exceed $F(\phi) = \min\left(\frac{a}{\sin \phi}, \frac{b}{\cos \phi}\right)$.

Hence, the doubled area of the rhombus inside the rectangle (equal to the product of the rhombus' diagonals) cannot exceed $F(\varphi)F(\frac{\pi}{2} - \varphi)$, where φ is the angle formed by one of its diagonals with the horizontal. Without loss of generality, $0 \leqslant \varphi \leqslant \frac{\pi}{4}$.

From definition of $F(\varphi)$, we have

$$F(\varphi)F\left(\frac{\pi}{2} - \varphi\right) \leqslant \frac{b}{\cos \varphi} \cdot \frac{a}{\sin(\frac{\pi}{2} - \varphi)} = \frac{ab}{\cos^2 \varphi},$$

$$F(\varphi)F\left(\frac{\pi}{2} - \varphi\right) \leqslant \frac{a}{\sin \varphi} \cdot \frac{a}{\sin(\frac{\pi}{2} - \varphi)} = \frac{a^2}{\sin \varphi \cos \varphi}.$$

Obviously, if $\varphi \in (0; \frac{\pi}{4})$, then the right-hand side of the first inequality is the increasing function of φ, and the right-hand side of the second inequality—the decreasing function. Let $\varphi_0 = \arctan \frac{a}{b}$. Then if $\varphi \leqslant \varphi_0$, then from the first inequality we get $a(\varphi) \leqslant ab\left(1 + \frac{a^2}{b^2}\right)$; if $\varphi \geqslant \varphi_0$, then we get the same from the second inequality. It is left to note that the rhombus, one of whose diagonals coincides with the rectangle's diagonal, has area equal to $\frac{a(b^2 + a^2)}{2b}$.

1966.19. Answer: 15 solutions.

Rewrite the equation as $\sqrt{y} = 14\sqrt{10} - \sqrt{x}$, and square it to obtain $y = 1960 + x - 28\sqrt{10x}$. Then number $\sqrt{10x}$ must be rational and therefore it is an integer.[103] Hence $x = 10a^2$ for some non-negative integer a. Similarly $y = 10b^2$ (b is non-negative integer). The original equation can be then rewritten as $a\sqrt{10} + b\sqrt{10} = 14\sqrt{10}$, that is, $a + b = 14$. This equation, obviously, has 15 solutions in non-negative integers.

1966.20. Hint: Tile the plane with squares with sides equal to 1300. Let us index them all with pairs of integers according to the coordinates of their left lower corners. We will color squares indexed (a, b) and (c, d) into the same color if both $a - c$ and $b - d$ are divisible by 3. Then any two points inside one square lie at the distance less than $1300 \cdot \sqrt{2} < 1966$, and any two points in different squares of the same color have one coordinate difference at least $1300 \cdot 2 > 1966$; thus, the distance between them is at least 1966.

7	8	9	7	8	9
4	5	6	4	5	6
1	2	3	1	2	3
7	8	9	7	8	9
4	5	6	4	5	6
1	2	3	1	2	3

---

[103] A very useful and very simple fact which we leave as an exercise for the reader.

1966.21. $q^3 - 1 = (q-1)(q^2 + q + 1)$ is divisible by p. Since $p > q$, then $q^2 + q + 1$ is divisible by $p = kq + 1$. Let $(kq+1)m = q^2 + q + 1$, that is, $m = q^2 + q - kmq + 1$. Then either $m = 1$ or $m \geqslant q + 1$. In the latter case, $q^2 + q + 1 \geqslant (q+1)^2$, which is absurd. Thus, $m = 1$.

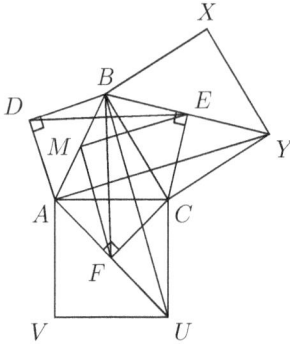

1966.22. Consider point M—the middle of segment AB. It is easy to prove that segments MF and ME are equal by length and perpendicular to each other. It follows from the fact that triangles CYA and CBU are obtained from each other by rotation by angle $90°$ around point C (here Y and U are vertices of squares $CBXY$ and $ACUV$ centered at E and F). Therefore, this rotation by $90°$ around point M maps B to D, and F—to E. Hence, segments BF and DE are equal and perpendicular.

1966.23. Let us denote that number by x_n. It is easy to see that $x_1 = 0$, $x_2 = k(k-1)$, $x_3 = (k-1)^3 - (k-1)$, $x_4 = (k-1)^4 + (k-1)$. Also for any n we have

$$x_n = (k-2)x_{n-1} + (k-1)x_{n-2}. \qquad (*)$$

Indeed, say, we have colored the sides from the first through the $(n-2)$th. Number of the colorings for which the color of the $(n-2)$th side is different from the color of the first side equals x_{n-1}; then you can color the $(n-1)$th side into one of $k-2$ other colors. The number of colorings, where the color of the $(n-2)$th side is the same as the color of the first side, is x_{n-2}—then we can color the $(n-1)$th side into one of the remaining $k-1$ colors. This proves $(*)$.

Now, by induction we can prove that

$$x_n = (k-1)^n + (k-1)(-1)^n.$$

Indeed, the basis (for $n = 1, 2$) is trivial. Then, using $(*)$ and the induction hypothesis, we can write

$$
\begin{aligned}
x_n &= (k-2)x_{n-1} + (k-1)x_{n-2} \\
&= (k-2)\left((k-1)^{n-1} + (k-1)(-1)^{n-1}\right) \\
&\qquad + (k-1)\left((k-1)^{n-2} + (k-1)(-1)^{n-2}\right) \\
&= (k-1)^{n-2}\left((k-2)(k-1) + (k-1)\right) \\
&\qquad + (-1)^n\left((k-1)^2 - (k-2)(k-1)\right) \\
&= (k-1)^n + (-1)^n(k-1).
\end{aligned}
$$

1966.26. Add a few more colors into the solution for **1966.20**.

1966.28. The statement is somewhat ambiguous. Here is how the more formal statement of this problem should begin (at least, that is how we choose to interpret it).

"Given natural number n, for which values of $\varepsilon > 0$ it is possible to dissect...".

Answer: if $n = 1$, 2, then such ε does not exist; if $n = 2$, then for all $\varepsilon > 0$; if $n = 3$, then for $\varepsilon \leqslant \frac{1}{3}a$; and if $n > 3$, then for $\varepsilon < \frac{1}{3}a$.

Case of $n \leqslant 2$ is trivial. Now we need examples for $n \geqslant 3$. For $n = 3$ we split the segment into three equal parts, for $n > 3$—into three segments that have the length almost equal to $\frac{2}{3}a$ and several segments of almost zero length ("almost equal to x" here means "differs from x by less than $\frac{1}{n}|\varepsilon - \frac{1}{3}a|$").

Any segment that can be constructed by putting together several segments of this dissection has length very close to one of the following numbers 0, $\frac{2}{3}a$, $\frac{4}{3}a$, $2a$ ("very close" means here "differing by less than $|\varepsilon - \frac{1}{3}a|$"); it follows that the dissection satisfies the required condition.

Now it is left to prove that the values $\varepsilon > \frac{1}{3}a$ (and $\varepsilon = \frac{1}{3}a$ for $n > 3$) do not fit; that is, for every dissection one can always combine several segments to form a segment with length inside interval $[a - \frac{1}{3}a, a + \frac{1}{3}a]$ (and for $n > 3$—strictly inside that interval).

If at least one segment in the dissection is longer than $\frac{2}{3}a$, then this segment is, just by itself, the one we need. If all segments do not exceed $\frac{2}{3}a$, then we start with the longest one, and gradually add others one by one until their total length exceeds a. Since the last segment we added is not longer than $\frac{2}{3}a$, then either before or after we added it the total length belongs to the required interval. In addition, the equality is reached only if the first two segments had length equal to $\frac{2}{3}a$; then for $n > 3$ the inequality can be made strict by replacing the second segment by one of the remaining ones.

1966.29. Answer: all x_i equal 1.

Set $S_k = x_1^k + \cdots + x_n^k$, $(0 \leqslant k \leqslant n)$. Let $\sigma_1, \sigma_2, \ldots, \sigma_n$ be the elementary symmetric polynomials in variables $x_1, \ldots, x_n$ ($\sigma_1 = \sum x_i$, $\sigma_2 = \sum x_i x_j$, ..., $\sigma_n = x_1 x_2 \ldots x_n$). A well-known theorem on symmetric polynomials tells us that $\sigma_1, \sigma_2, \ldots, \sigma_n$ can be represented as functions (actually, polynomials) of $S_1, S_2, \ldots, S_n$. To prove that, use Newton's Formula

$$S_k = S_{k-1}\sigma_1 - S_{k-2}\sigma_2 + S_{k-3}\sigma_3 - \cdots + (-1)^{k-1}S_0\sigma_k,$$

which can be easily checked by direct computation. Note that $\sigma_1 = S_1$, and therefore σ_1 has the required form. Newton's Formula shows that if $\sigma_1, \sigma_2, \ldots, \sigma_{k-1}$ are all expressible as polynomials in S_i, then σ_k is also a polynomial in S_i.

Thus, we have proved that all solutions $\{x_i\}$ of the original system $S_1 = S_2 = \cdots = S_n = n$ produce the identical sets of values $\sigma_1, \sigma_2, \ldots, \sigma_n$. Hence, by Vieta Theorem, all such collections $\{x_i\}$ are the sets (more precisely, multisets) of the roots of one fixed polynomial of nth degree, and thus they differ only by permutation of their elements. It is left to note that one of such collections is $x_1 = x_2 = \cdots = x_n = 1$.

1966.30. Assume that there exists a prime number p such that both $2^n + 1$ and $2^m - 1$ are divisible by p; in other words, that 2^n and 2^m have remainders -1 and 1 modulo p. Let d be the greatest common divisor of n and m. Since there exist natural numbers a and b such that $an - bm = d$, then 2^d has remainder 1 or -1 modulo p. But if $2^d \equiv 1 \pmod{p}$, then $2^n \equiv 1 \pmod{p}$. Hence $2^d \equiv -1 \pmod{p}$. But since m is odd, $k = m/d$ is odd, and $2^m = (2^d)^k = (-1)^k \equiv -1 \pmod{p}$, a contradiction.

1966.31. Let M_1 and M_2 be $(n-1)$-gon and $(n+1)$-gon with maximum area, inscribed into the given polygon M. We will assume that their vertices are numbered counterclockwise, and let $\{A_1, A_2, \ldots, A_{2n}\}$ be the union of sets of vertices of M_1 and M_2, listed in the order of counterclockwise traversal of the boundary of M. If some vertices of M_1 and M_2 coincide, we will still consider them as different and choose their order randomly. The further reasoning assumes that all vertices are different.

We will call points A_1, A_2, ..., A_{2n} *vertices*, oriented (directed) segments connecting them—*chords*, regions of M situated to the right of any such chord—*segments*. Area of any polygon P inscribed into M equals area of M minus the sum of the areas of the segments cut off by the sides of P, oriented in the standard counterclockwise order. For any chord, its *length* will be the number of vertices in the segment it defines, increased by 1 (thus the length of chord $A_i A_j$ equals either $j - i$ or $j - i + 2n$ depending on the sign of $j - i$).

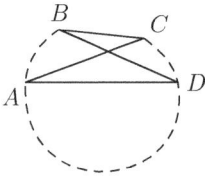

Note that each chord $A_i A_j$, which is the side of either M_1 or M_2, possesses the following property: there exists triangle $A_i A_j X$ which contains the entire chord's segment—otherwise it is easy to immediately increase the area of one of these polygons. We will call such chords *short*.

Sides of M_1 and M_2 form a system of $2n$ chords. Let us transform that system so that the sum of the segments' areas does not increase and the following conditions are kept:

a) for each vertex there is exactly one chord coming out of it and exactly one chord that goes into it;

b) all chords are short;

c) the sum of the chords' lengths is $4n$.

The last condition is equivalent to the fact that each point on the contour of M (except for vertices A_i) belongs to exactly two of the segments. Clearly, at the beginning all these conditions are met.

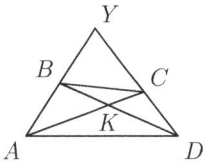

Suppose that in the system of chords that complies with these conditions there exists chord BC of length 1. Find chord AD for which BC lies inside the chord's segment. Replace chords AD and BC with chords AC and BD. Clearly, all the conditions are still met. Let us verify that the sum of the segments' areas does not increase. It is easy to see that the replacement operation adds to this sum the difference $S_{BCK} - S_{ADK}$, where K is the point of intersection of AC and BD. Thus, it is sufficient to prove that $S_{BCK} < S_{ADK}$. Since chord AD is short, the sum of angles A and D in quadrilateral $ABCD$ is less than $180°$, and therefore rays AB and DC intersect at some point Y. Ratio of altitudes CL and DN of triangles ABC and ABD, respectively, is equal to $CL/DN < 1$. Thus this particular altitude of ABC is shorter than the corresponding altitude of triangle ABD, and therefore $S_{ABC} < S_{ABD}$. Subtracting area of triangle ABK, we get the required inequality $S_{BCK} < S_{ADK}$.

Our operation decreases the number of chords of length 1. By repeating it, sooner or later we will obtain the system where all the chords have lengths of 2 or more; but the sum of their lengths is still $4n$ and therefore all the lengths are exactly 2. Therefore, these chords form two n-gons M_1' and M_2', whose vertices alternate on the contour of M. Since the sum of the segments' areas did not increase, we have $S(M_1') + S(M_2') \geqslant S(M_1) + S(M_2)$, and thus,

$$T_n \geqslant \max\{S(M_1'), S(M_2')\} \geqslant \frac{S(M_1') + S(M_2')}{2} \geqslant \frac{S(M_1) + S(M_2)}{2} = \frac{T_{n-1} + T_{n+1}}{2}.$$

1966.32. Assume the opposite and split all numbers 1 through 2^n into several groups: we assign two numbers to the same group, if they have the same largest odd divisor. If some odd number k satisfies conditions $2^s < k < 2^{s+1}$ (there are exactly 2^{s-1} of such numbers), then its group contains $n - s$ elements: $k, 2k, \ldots, 2^{n-s-1}k$; also the group containing 1 comprises $n + 1$ elements. It is obvious that in a group that contains t elements, at least $\left[\frac{t}{2}\right]$ numbers had to be removed—otherwise we can find the required pair. Altogether, at least $\sum_{s=1}^{n} 2^{s-1}\left[\frac{n-s}{2}\right] + \left[\frac{n+1}{2}\right]$ numbers were removed. But

$$\sum_{s=1}^{n} 2^{s-1}\left[\frac{n-s}{2}\right] + \left[\frac{n+1}{2}\right] > \left[\frac{2^n - 2}{3}\right],$$

which can be easily proved by induction. Basis $n = 1$ and $n = 2$ is obvious. For the induction step from n to $n + 2$, both sides of the inequality are increased by the same number $(1 + 2 + 2^2 + \cdots + 2^{n-1}) + 1 = 2^n$. Thus, we prove the inequality obtaining the desired contradiction.

1966.33. Obviously, we have to consider only non-trivial tilings, excluding the one-rectangle tiling, the square itself. Indeed, any tiling can be obtained by subdividing the trivial tiling. Thus, we exclude case $n = 1$ from any consideration.

Answer: for $n = 2$, $n = 5$, and $n \geqslant 7$. The figure below shows you examples of tiling with two rectangles, five rectangles, and then (easily adjusted for general case) examples with any odd and any even number of rectangles.

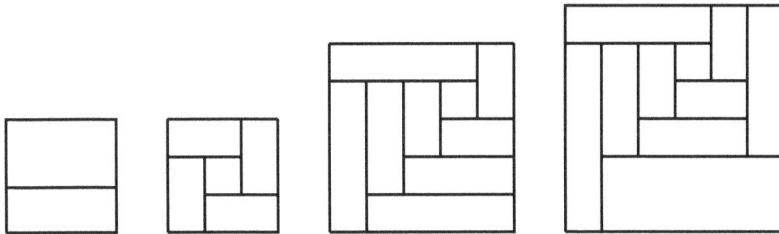

Proving the non-existence of primitive tilings for $2 < n < 5$ or $n = 6$ is a rather tedious case-by-case analysis. We will sketch it here, skipping some of the less interesting details.

Case $n = 3$ is pretty obvious. Still, we will prove that by using one simple technique which will be used for greater values of n. Mark all four corners of the square, and then we can reason as follows. A rectangle cannot cover two opposite marks lest it covers the entire square. A rectangle also cannot cover two marks in neighboring corners because that will mean there is a vertical or horizontal line dividing the square which means that the tiling is not primitive as it can be obtained by subdivision from a tiling into two rectangles.

For $n = 4$, none of the rectangles can cover two corners for the same reason. Thus, we have one rectangle per each corner; these rectangles cannot overlap, and they must cover the entire square. The figure must look like the one shown below (otherwise we would have a non-primitive tiling). The marks show the three corners that must be covered by the two remaining rectangles, so one of them must cover two marks. In any such case, it is clear that such tiling is not primitive.

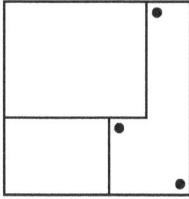

Now, to the most complicated case of $n = 6$. We, of course, start with assuming the opposite—existence of a primitive tiling with 6 rectangles. As we already know, each corner mark is covered by a separate rectangle. Consider two cases: (a) one of the sides (S) is covered by more than one rectangle; (b) each side is covered by exactly two rectangles. Case (b) is quite easy and we will leave it to the reader (you either get a non-primitive tiling or a tiling with exactly five rectangles).

Case (a) is somewhat more complicated. We will split it into two subcases depending on where the "shortest" one of those rectangles is located. Assume that side S is horizontal and find the rectangle with the smallest height that has common points with S. Case (a1): that rectangle also touches one of the vertical sides. Case (a2): it does not. The left figure below explains the rest of the proof which proceeds more or less as follows.

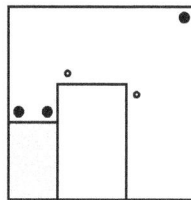 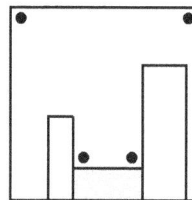

(a1)   (a2)

In Case (a1), we have three "major" corner marks, which are not covered by any of the rectangles touching S, and each one of them requires a separate rectangle (to keep this tiling primitive). Also there are two additional "minor" marks such that if one of them is covered by any of the already listed six rectangles, then the other one is not.

In Case (a2), there are four marks, and one of the remaining rectangles has to cover two of them. That is either impossible, or it would result in a non-primitive tiling.

1966.34. The fact we are about to prove is called Cayley's Formula. Let us reformulate this in the graph theory terminology, with points serving as vertices and segments—as edges. Index the vertices by numbers 1 through n. We need to prove that there are n^{n-2} trees with vertices in the given n indexed points (isomorphic trees which differ only by indexing are counted as different trees; see the figure).

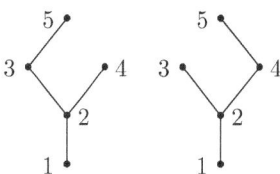

Let us build a one-to-one correspondence between the set $\mathcal{T}$ of all these trees and set $\mathcal{M}$ of all sequences consisting of $n - 2$ natural numbers 1 through n.

Consider an arbitrary tree T with the vertices on our n points, and find pendant vertex v_1 with the minimal index. Then we write down index of v_1 only neighbor as the first term of the sequence.

Remove v_1 from the tree to obtain tree T_1 with $n-1$ vertices. Then we repeat the previous step, finding vertex v_2 and writing down index of v_2 only neighbor as the second term in the sequence. And so on.

After $n-2$ steps, we will have the entire sequence (of course, the resulting tree T_{n-2} will consist of two vertices and one edge). Therefore, we have mapping $\mathcal{F} : \mathcal{T} \to \mathcal{M}$.

Inversely, let us consider sequence $A = \{a_1, \ldots, a_{n-2}\} \in \mathcal{M}$ and construct a tree using A. Denote by I the set of all natural numbers 1 through n. Define v_1 as the minimum number in I not belonging to A. Now, connect vertex with index v_1 with the vertex with index a_1. After this, we remove a_1 from A as well as number v_1 from I.

Repeat the previous step for new sets A and I, and so on. After $n-2$ steps, we will have exhausted sequence A, while I at that moment will consist of only two numbers. Finally, we connect the two vertices indexed by these two remaining numbers.

It is easy to see that the graph we have constructed has n vertices, $n-1$ edges and no cycles. Thus we have constructed a tree, and so have defined mapping $\mathcal{G} : \mathcal{M} \to \mathcal{T}$. It is also obvious that mapping $\mathcal{F}$ is the inverse of $\mathcal{G}$ and vice versa.

1966.35. Select midpoints of the polygon's sides. The circles C_i centered at these points, with respective radii $\frac{1}{6}na_i$ cover the entire polygon.

To prove that, consider an arbitrary point X inside the circle, and verify that X is covered by one of these circles. If X lies outside the polygon, between the circle's boundary S and side $AB = a_k$, then it is covered by circle C_k. Indeed, $\angle AXB > 90°$ since the center of S lies inside the polygon—therefore X belongs to the circle C_{AB} with diameter AB, and that circle is completely covered by C_k since $\frac{1}{6}na_k \geqslant \frac{1}{2}a_k$.

Now if X is inside the polygon, then we can find side AB with $|AB| = a_k$ seen from X at an angle which is greater than or equal to $\frac{2\pi}{n}$. We will prove that X is covered by C_k, or equivalently, distance from X to the midpoint M of side AB does not exceed $\frac{1}{6}na_k$. For $n = 3$, this is the same as a trivial statement that for $\angle AXB \geqslant 120°$, point X lies inside circle C_{AB}. Assume $n \geqslant 4$. Then the locus of the points which lie in the "interior" half-plane defined by AB, from which AB is seen at angle $\frac{2\pi}{n}$, is the arc of measure $2\pi - \frac{4\pi}{n}$ with endpoints A and B. Condition $\angle AXB \geqslant \frac{2\pi}{n}$ means that X lies in the region bounded by that arc and segment AB. Clearly, in this region, the point which is farthest away from M is simply the midpoint X_0 of that arc. Then AX_0B is an isosceles triangle with angle $\frac{2\pi}{n}$ at vertex X_0. Hence,

$$|MX| \leqslant |MX_0| = \frac{a_k}{2\tan(\frac{\pi}{n})} < \frac{na_k}{2\pi} < \frac{na_k}{6}$$

(using inequalities $\tan x > x$ and $\pi > 3$).

1967

1967.01. Answer: 350, 700, and 1050 liters.

If the first jar originally had x liters of water, then the second one had $2x$ and the third one—$3x$ liters. After the first round of transfers, the levels became equal, thus the volumes of liquid were proportional to the areas of the bases, that is, $1 : 4 : 9$. Thus, the jars had $\frac{3}{7}x$, $\frac{12}{7}x$ and $\frac{27}{7}x$ liters, respectively. After the next two pourings, the first jar ended up with $(x - 100)$ liters, and the second one with twice as much, since the water level is half of the first jar's, but the area of the base is 4 times greater. At the same time the total amount of water stayed the same. Hence, $3(x - 100) = \frac{15}{7}x$, and we have $x = 350$.

1967.02. Answer: that happens twice, at midnight and at noon. Let us assume that the smallest ("seconds") hand moves every second on the second, and let us call one sixtieth of the full $360°$ angle a *tick*. If all hands point in the same direction at x hours y minutes z seconds ($0 \leqslant x < 24, 0 \leqslant y, z < 60$), then we have the seconds hand deviating from zero by z ticks and the minutes hand—by $y + \frac{1}{60}z$. From this, we have $y + \frac{1}{60}z = z$, or $60y = 59z$. That is possible only if $y = z = 0$, but then, obviously, $x = 0$ or $x = 12$.

Note. This approach (and the answer) is valid even if all the hands are moving continuously.

1967.03. Denote Leningrad's population size by n. Then each person can have from 0 to $n - 1$ acquaintances—n possibilities overall. If all n Leningradians have different number of acquaintances, then obviously one of them (we will call him A) has 0 acquaintances, and some other—$(n - 1)$ acquaintances (we will call him B; that guy knows everyone). But do A and B know each other? It is obvious they cannot coexist.

1967.04. There are 14 possible positive differences for two unequal natural numbers between 1 and 15. Our eight numbers have $8 \cdot \frac{7}{2} = 28$ positive differences. Obviously, number 14 can be presented as such a difference only once, as $14 = 15 - 1$. Thus, the remaining 13 possibilities are "covered" by the remaining 27 pairwise differences, and if each value is covered no more than twice, there can be no more than 26 differences, a contradiction.

1967.05. Answer: 70 km.

The bee will fly for exactly $2 = \frac{100}{20+30}$ hours, thus travelling $2 \cdot 50 = 100$ kilometers. In the end, it has to arrive at the point where the riders meet, 40 km away from A. That means it has flown 70 kilometers in direction $A \to B$ and 30 kilometers in direction $B \to A$.

1967.06. We need trapezoid $ABCD$ with bases $|AD| = a$, $|BC| = b$ and sides $|AB| = c$, $|CD| = d$. If $a = b$, then there are infinitely many solutions for $c = d$ (case of parallelogram), and no solutions for $c \neq d$; for $a \neq b$, there is only one solution if numbers $|a - b|$, c, and d

satisfy the triangle inequality, and no solutions if they do not.

Let $a > b$ and trapezoid $ABCD$ be such that we have bases $|AD| = a$, $|BC| = b$ and sides $|AB| = c$, $|CD| = d$. Take point E on segment AD such that $|DE| = b$, and then $BEDC$ is a parallelogram, and sides of triangle ABE are equal to $a - b$, c, and d, respectively. Therefore, numbers $a - b$, c, d must satisfy the triangle inequality, and triangle ABE is uniquely defined. In particular, numbers a, b, c, and d determine the measure of $\angle A$ and similarly, of $\angle D$. Thus, there is only one such trapezoid.

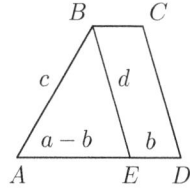

If $a > b$, then construct triangle ABE with sides $|AE| = a - b$, $|AB| = c$, $|BE| = d$. Take point D on ray $|AE|$ such that $|AD| = a$, and then find point C such that $BEDC$ is a parallelogram. Then trapezoid $ABCD$ is the required one.

1967.07. For $x \leqslant 0$, the inequality is obvious because

$$-102x^{101} \geqslant 0, \quad 1 + x + x^2 + \cdots + x^{100} = \frac{1 - x^{101}}{1 - x} \geqslant 0.$$

For $x > 0$, after expanding, we will get the sum of all powers of x from 1 to x^{202} except for x^{101}. Let us write out inequalities $1 + x^{202} \geqslant 2x^{101}$; $x + x^{201} \geqslant 2x^{101}$, and so on; they are equivalent to the obviously true inequalities $(1 - x^{101})^2 \geqslant 0$, $x(1 - x^{100})^2 \geqslant 0$, etc. Adding them up gives us the required result.

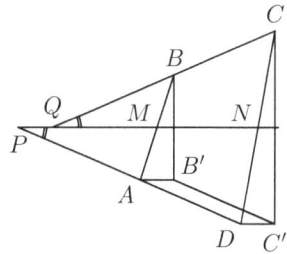

1967.08. Let us reflect vertices B and C about line MN to obtain points B' and C', respectively. Lines AD and $B'C'$ must be parallel, but since both lines AB' and DC' are parallel to MN, we have that $AB'C'D$ is a parallelogram, and therefore $|AD| = |B'C'| = |BC|$.

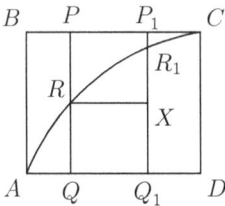

1967.10. Answer: point R is the intersection of arc AC and rectangle's midline parallel to BC.

Suppose we drew a different line P_1Q_1 intersecting arc AC at point R_1. Let X be midpoint of P_1Q_1, and assume that P_1 lies on segment PC. Then

$$S_{P_1PRR_1} < S_{P_1PRX} = S_{Q_1QRX} < S_{Q_1QRR_1}.$$

Therefore,

$$S_{AQ_1R_1} + S_{CP_1R_1} < S_{AQR} + S_{CPR}.$$

1967.12. Since $x + y = c$ and $xy = -c$, we have that $x^3 + y^3 + (xy)^3 = (x + y)^3 - 3xy(x + y) + (xy)^3 = c^3 + 3c^2 - c^3 = 3c^2 \geqslant 0$.

1967.13. Draw a line tangent to both circumferences at point A. Denote the point of intersection of that line and line CD by E.

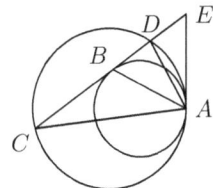

It is clear that $\angle EDA = \angle CAE$ and $\angle ECA = \angle DAE$, since triangles CEA and DEA are similar. Further, triangle BEA is isosceles, and therefore

$$2\angle BAE = 180° - \angle BEA = \angle DAE + \angle EDA = \angle DAE + \angle CAE.$$

1967.14. It would suffice to prove that $2^{3^k} + 1$ is divisible by 3^{k+1}. This is done using induction on k. Basis $k = 1$ is obvious. For the induction step, we use identity $2^{3x} + 1 = (2^x + 1)(2^{2x} - 2^x + 1)$; the second factor on the right-hand side is divisible by 3, when x is odd.

1967.16. Consider a graph, whose vertices represent the people in the group. We will connect two vertices with a red edge if these people are enemies and with a green edge if they are friends. Degree of each vertex in this graph is 2, and therefore it is a union of several cycles. Each cycle has an even number of vertices because edge colors in it alternate; thus, we can mark half of the cycle vertices as "odd", and the other half as "even". All "odd" vertices will then constitute one subgroup, and all "even" vertices—the other.

1967.17. If sequence $\{a_i\}$ contains a negative number, then the smallest number in the sequence—let it be a_k—cannot be the first one or the last one. Since $a_k \geq \frac{1}{2}(a_{k+1} + a_{k-1})$, both numbers a_{k+1} and a_{k-1} must coincide with a_k, otherwise we get a contradiction with a_k being the smallest. Continuing to $k - 1$ and $k + 1$ and so forth, we will obtain that all the numbers are equal and therefore non-negative.

1967.18. Let $q = 2k + 1$, $p = 2k - 1$. Then

$$p^p + q^q \equiv (-q)^p + q^q = q^{2k-1}(q^2 - 1) \pmod{(p+q)},$$

and

$$q^{2k-1}(q-1)(q+1) = q^{2k-1}2k(2k+2) = q^{2k-1}4k(k+1),$$

which concludes the solution since the right side is divisible by $4k = p + q$.

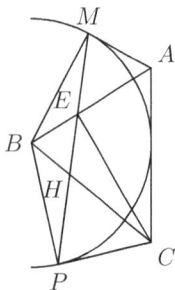

1967.19. Denote angles of triangle ABC by α, β, and γ. Then in isosceles triangle BMP we have $\angle MBP = 2\angle ABC = 2\beta$, $\angle BMP = 90° - \beta$, and $\angle AME = 90° - \angle BMP = \beta$. In addition, $\angle MAE = \alpha$; hence, triangle AME is similar to triangle ABC and so $|AE|/|AC| = |AM|/|AB| = \cos\alpha$. Therefore, $|AE|/|AC| = \cos\angle EAC$, meaning that CE is the altitude. Proof for AH is the same.

1967.20. Let t_2 be the moment when this operation was performed for the first time with the second rightmost number. Obviously, this operation will not be performed again, lest the sequence repeats. If the operation is never performed for the second rightmost number, then we set $t_2 = 0$. Further, let t_3 be the moment when the operation is performed with the third rightmost number for the first time after t_2 (if that never happened, then we simply set $t_3 = t_2$). Similarly, we will define $t_4, t_5, \ldots,$ and t_k. After t_k, the next operation will inevitably result in repeating the sequence.

1967.21. Let us draw the maximum possible number of non-intersecting segments connecting these 106 points. The square is then dissected into triangles with vertices on the given points (if some region is a polygon with $k > 3$ sides, then we can always add one more segment, not intersecting any of the existing segments, inside that polygon). Sum of the square's angles is 2π, same as the sum of the angles around any of the 102 interior points. Hence the sum of all angles in all triangles equals 206π, and therefore the number of the dissection triangles is 206. Since they do not have common interior points, at least 107 of them must have area less than 0.01.

1967.22. Consider expression

$$S = \sum_{k=1}^{n-1}(a_k - a_{k+1})(b_1 + \cdots + b_k).$$

Expanding this, it is easy to see that this is the same as $a_1b_1 + a_2b_2 + \cdots + a_nb_n$. On the other hand, since $a_k - a_{k+1}$ is non-negative, and any expression of form $b_1 + b_2 + \cdots + b_k$ has absolute value not exceeding B, we obtain

$$|S| \leqslant B((a_1 - a_2) + (a_2 - a_3) + \cdots + (a_{n-1} - a_n)) \leqslant Ba_1.$$

1967.23. It is enough to travel along half-circumference of radius $\frac{1}{\pi}2507$ meters. Indeed, if we assume that during this walk the tourist is always inside the forest, then this convex forest must contain the corresponding semicircle. The area of such semicircle equals $\frac{1}{2\pi}2.507^2 \approx 1.0002$, thus exceeding the area of the polygon (forest), a contradiction.

1967.24. See solution to **1967.18**.

1967.25. This follows from the equalities

$$|AB_1|^2 + |BC_1|^2 + |CA_1|^2 = |AC_1|^2 + |BA_1|^2 + |CB_1|^2,$$
$$|AB_1| + |BC_1| + |CA_1| = |AC_1| + |BA_1| + |CB_1|.$$

Indeed, square the second equality and subtract the first one.

The first equality is true because by Pythagoras' Theorem both of these sums are equal to $|AM|^2 + |BM|^2 + |CM|^2 - |MA_1|^2 - |MB_1|^2 - |MC_1|^2$. To prove the second one, note that it is true if $M = A$, and furthermore, if we move point M in direction parallel to one of the sides, then both of the sums in the equality do not change. For instance, if we move M by distance a in the direction of vector $\vec{AC}$, then $|AB_1|$ increases by a, while $|BC_1|$ and $|CA_1|$ diminish by $\frac{a}{2}$ each. Hence, the sum on the left does not change; same is true for the sum on the right.

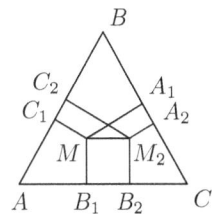

1967.29. Let us prove that for any k there exists a set of points (let us denote it by S_k) such that for each point $X \in S_k$ there are at least k points in S_k at unit distance from X. We do that using induction on k. Basis $k = 1$ is trivial. To prove the induction step from $k = n - 1$ to $k = n$, consider set S_{n-1} and translate it by unit distance in such a direction that none of the new points coincide with any of the "old" ones. The new set S_n

is the union of S_{n-1} and $T(S_{n-1})$, where T is the translation we just described. It is quite obvious that for any point $X \in S_n$, you can find at least n points in this new set at unit distance from X.

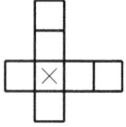

1967.30. Answer: this figure consists of the seven squares shown in the picture. The first player will place an **X** onto the marked square. Regardless of the second player's move, there will be a line (row or column) with no **O**'s in it. The next move of the first player will be placing an **X** onto the other central square of that line creating the winning "fork".

Now, if a figure has only six squares, then the first player must end the game on his third move. That means that after his second move, there must be a fork of two **X**'s which means they must be standing right next to each other, and therefore there must be free squares on both sides of these two squares. Thus, the figure must contain a 1×4 polyomino (the case of diagonal fork is not possible with less than seven squares), and since we only have six squares, there can be only one such "line" L that contains this subpolyomino. Moreover, none of these four squares can contain an **O** after the first player's second move.

But that means that all the second player needs to do in order to prevent this from happening is to put an **O** into one of the central squares of L. dividing it into pieces shorter than four squares each.

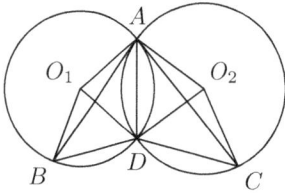

1967.31. Let O_1 and O_2 be the centers of these circumferences. Then it is easy to see that triangles O_1BD and O_2AD are similar; and so are triangles O_2CD and O_1AD. Thus, we have

$$\frac{R}{|AD|} = \frac{r}{|BD|}, \quad \frac{R}{|CD|} = \frac{r}{|AD|},$$

where r and R are the radii of the circumferences. Multiplying these equalities and using $r^2/R^2 = |AB|^2/|AC|^2$, we obtain the desired result.

1967.32. Denote these polynomials by F and G; we can assume that for some x_0 polynomials F and εG (where $\varepsilon = 1$ or -1) are both increasing on interval $[x_0; \infty)$.

Let us suppose that $F - \varepsilon G$ is not a constant, $F(x_1) = a$, $G(x_1) = b$, where $x_1 > x_0$, for some integers a, b. Define $H(x) = F(x) - \varepsilon G(x) - (a - \varepsilon b)$. Since F is increasing, we can find $x_2 > x_1$ such that $F(x_2) = a+1$. Then, obviously, $\varepsilon G(x_2) = \varepsilon b+1$, that is, $H(x_2) = 0$. Similarly, if $x_3 > x_2$ is such that $F(x_3) = a + 2$, then $\varepsilon G(x_3) = \varepsilon b + 2$ and $H(x_3) = 0$, and so on. Thus, $H(x)$ has infinitely many roots—hence, $H(x) \equiv 0$.

1967.33. Adding the perimeters of the corner triangles and subtracting the perimeter of the middle one, we will, on one hand, obtain the perimeter of the large triangle, and on the other hand—double the perimeter of the small one. Therefore, perimeter of any small triangle equals half of the perimeter of the large one.

Now, let ABC be the original large triangle with angles α, β, and γ, and A_1, B_1, C_1—selected points on sides BC, AC, and AB, respectively. Also, let A_0, B_0, C_0 be the midpoints of the corresponding sides. Suppose that $A_1 \neq A_0$, and assume, without loss of generality, that A_1 lies on segment A_0C. Then B_1 lies on segment B_0A—otherwise triangle A_1CB_1 is entirely contained inside triangle A_0CB_0, and therefore, has a smaller perimeter. Similarly, C_1 lies inside C_0B.

Now, consider triangles A_0CB_0 and A_1CB_1. Since they have equal perimeters, they possess common escribed circumference S inside angle C (it follows from the fact that the distance from triangle's vertex to the points of tangency of the escribed circumference equals half-perimeter). Let O and r be the center and the radius of S. Then it is easy to see that $\angle A_0OB_0 = 90° - \frac{\gamma}{2} = \angle A_1OB_1$, and therefore $\angle A_0OA_1 = \angle B_0OB_1$. Denoting that angle by δ, we have $\angle OA_0B = 90° - \frac{\beta}{2}$, $\angle OA_1B = 90° - \frac{\beta}{2} - \delta$. Computing angles in the right triangle OA_0X (X is the point where S touches line BC), we get

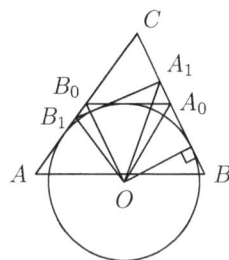

$$|OA_0| = \frac{r}{\sin \angle OA_0B} = \frac{r}{\cos \frac{\beta}{2}}.$$

From the sine theorem for triangle OA_0A_1, we obtain

$$|A_0A_1| = \frac{|OA_0| \sin \angle A_0OA_1}{\sin \angle OA_1B} = \frac{r \sin \delta}{\cos \frac{\beta}{2} \cos(\frac{\beta}{2} + \delta)} > \frac{r \sin \delta}{\cos^2 \frac{\beta}{2}}.$$

Similarly,

$$B_0B_1 = \frac{r \sin \delta}{\cos \frac{\alpha}{2} \cos(\frac{\alpha}{2} - \delta)} < \frac{r \sin \delta}{\cos^2 \frac{\alpha}{2}}.$$

Hence, $\dfrac{|A_0A_1|}{|B_0B_1|} > \dfrac{\cos^2 \frac{\alpha}{2}}{\cos^2 \frac{\beta}{2}}$, and therefore $\dfrac{|B_0B_1|}{|C_0C_1|} > \dfrac{\cos^2 \frac{\beta}{2}}{\cos^2 \frac{\gamma}{2}}$ and $\dfrac{|C_0C_1|}{|A_0A_1|} > \dfrac{\cos^2 \frac{\gamma}{2}}{\cos^2 \frac{\alpha}{2}}$. Multiplying these three inequalities results in

$$\frac{|A_0A_1|}{|B_0B_1|} \cdot \frac{|B_0B_1|}{|C_0C_1|} \cdot \frac{|C_0C_1|}{|A_0A_1|} > \frac{\cos^2 \frac{\alpha}{2}}{\cos^2 \frac{\beta}{2}} \cdot \frac{\cos^2 \frac{\beta}{2}}{\cos^2 \frac{\gamma}{2}} \cdot \frac{\cos^2 \frac{\gamma}{2}}{\cos^2 \frac{\alpha}{2}}.$$

Both left-hand and right-hand sides are equal to 1, and we have the desired contradiction.

1967.34. Each person at the party either has nine or more acquaintances or nine or more non-acquaintances. Without loss of generality, we can assume that some person A is acquainted with at least nine attendees. Take any one of them—we will call her B. If B has six (or more) non-acquaintances among the other eight people, then we can reduce our problem to **1961.05**. Thus, she has at least three acquaintances there. As a matter of fact, she should be acquainted with exactly three—otherwise we will have four persons such that if any two of them are acquainted, then these two together with A and B will form the quadruplet of pairwise acquainted party-goers. And if no two of those four are acquainted, then we have a quadruplet of pairwise non-acquainted people.

Now, let us count the number x of all acquaintances inside the nine. Each one of them is acquainted with exactly three others. Adding them up, we get $2x = 9 \cdot 3 = 27$ (because in this sum each acquaintance is counted exactly twice), and therefore $x = 13.5$, which is absurd.

1967.35. Let us construct a bipartite graph G, with n vertices in each of its two parts. The vertices of the first part will represent the columns of the table, and the vertices of the second part—the rows of the table. A vertex in the first part is connected to a vertex in the second part if and only if the number on the intersection of the corresponding column and the corresponding row is positive. Then the conditions of the Hall's Marriage Theorem are met—indeed, if we select any k columns, then the sum of the numbers in them equals k, and therefore the positive numbers in these k columns cannot be covered by less than k rows. Thus, there exists a perfect matching between columns and rows, and that implies exactly what we need to prove—n positive numbers, no two of which belong to the same row or to the same column.

1967.36. It is clear that $a_0 < 1$—otherwise a_1 is not defined or equal to 0. Then $a_0 = \sin x$ for some $0 < x \leqslant \frac{1}{2}\pi$. Then $a_n = |\sin 2^n x|$—that can be proved by induction using the formula

$$|\sin 2z| = 2|\sin z \cdot \cos z| = 2|\sin z| \cdot \sqrt{1 - \sin^2 z}\,.$$

The condition of monotonicity $a_i < a_{i+1}$ for $i = 0, 1, \ldots, 25$ now implies that $a_{n+1}/a_n = 2|\cos 2^n x| > 1$ for $n = 0, 1, \ldots, 25$.

Consider the smallest non-negative integer k for which $2^k x > \frac{1}{3}\pi$. Then, obviously, $2^k x \leqslant \frac{2}{3}\pi$ and therefore, $|\cos 2^k x| \leqslant \frac{1}{2}$. This is possible only if $k \geqslant 26$ and consequently $2^{25} x < \frac{1}{3}\pi$; that is, $x \leqslant 3^{-1} 2^{-25}\pi$. It follows that

$$x \leqslant \frac{\pi}{3 \cdot 32} \cdot 2^{-20} < \frac{\pi}{96} \cdot 10^{-6} < 7 \cdot 10^{-8},$$

and then $a_0 = \sin x < x < 7 \cdot 10^{-8}$.

1968

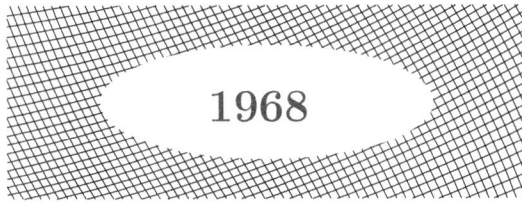

1968.01. Answer: 7 rubles.

If the prices for the backpack, the pen, and the book are x, y, and z, respectively, then we have

$$\frac{x}{5} + \frac{y}{2} + \frac{2z}{5} = 2 \quad \text{and} \quad \frac{x}{2} + \frac{y}{4} + \frac{z}{3} = 3.$$

Thus, $x = 5 - \frac{1}{3}z$ and $y = 2 - \frac{2}{3}z$, and $x + y + z = 7$.

1968.02. Answer: the number on the right is greater than the one on the left by $444\ldots44$ (nineteen digits 4).

Indeed,

$$88\ldots8 \times 33\ldots3 = 2 \times 44\ldots4 \times 33\ldots3$$
$$= 44\ldots4 \times 66\ldots6 = 44\ldots4 \times 66\ldots7 - 44\ldots4.$$

1968.03. Answer: 335 km. **Hint**: Use the triangle's inequality.

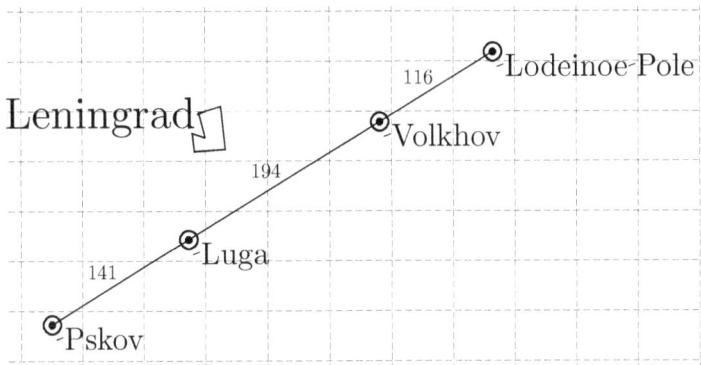

If points A, B, C, D represent Pskov, Luga, Volkhov, and Lodeinoe Pole, respectively, then we have $|AD| = |AB| + |BC| + |CD|$. That is only possible if these four points lie on the same straight line in exactly the specified order. It follows that $|AC| = |AB| + |BC| = 141 + 194 = 335$.

Note. Using Google Maps or some other online map tool, you can easily check that the actual distances between these four cities are very close to the numbers given in this problem's statement. They do, indeed, lie on one straight line (well, almost).

1968.04. Let us denote the weights of the four objects (A, B, C, and D) by a, b, c, and d. Start with comparing the first three objects, every one with each other. Without any loss of generality, we can assume that $a > b > c$. Then compare D and B. If D is heavier, weigh it against A; if lighter, then against C. Thus, we determine D's place in the lineup.

1968.05. Assume that team T defeated N teams. If some team X is not among their number, and also it has not been defeated by any of them, then obviously it gained at least $N+1$ points, contradicting T being the winner. Thus, X either lost to T or to one of the teams that T defeated.

1968.06. It immediately follows from the solution of **1968.01** that $x > y$; that is, the backpack costs more than the pen.

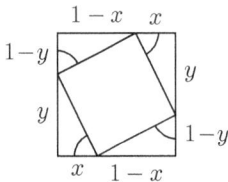

1968.07. Hint: Show that two parallel chords of the square have the same length only if they are symmetric with respect to the center, or if they connect opposite sides of the square.

Let us assume that the side of the square has length 1. Sides of the quadrilateral cut four right triangles off the square, and these triangles are similar (because their hypotenuses are parallel or perpendicular). Since the hypotenuses of the opposite triangles are equal, the triangles themselves are congruent. Let us denote the legs of these triangles by x and y. Then the legs of the neighboring triangle are $1-y$ and $1-x$. From the triangles' similarity follows that $x(1-x) = y(1-y)$, or $(x-y)(x+y-1) = 0$. Since $x + y \neq 1$ (the rectangle is not a square), we must have $x = y$. Therefore, the sides of the rectangle are parallel to the square's diagonals, and its half-perimeter is equal to $x\sqrt{2} + (1-x)\sqrt{2} = \sqrt{2}$.

1968.08. Answer: -1, 1, 3, 6, 11.

All pairwise sums are different, and therefore the numbers must be different. Denote them by $a < b < c < d < e$. Then the smallest sum is $a + b$, and the second smallest is $a + c$; the largest one is $d + e$ and the second largest is $c + e$. The sum of all pairwise sums is $4(a + b + c + d + e)$, and so we have the system of equations

$$a + b = 0, \ a + c = 2, \ c + e = 14, \ d + e = 17, \ 4(a + b + c + d + e) = 80,$$

which only has one solution, presented above.

1968.09. Perfect squares, when divided by 9, only have remainders 0, 1, 4, and 7. The sum of the digits of the given number equals $5 \cdot 999 + d$, where d is its only non-five digit. Thus, if this number is a perfect square, then d must be equal to 0, 1, 4, or 7. Since a square cannot end with two odd digits (numbers like that have remainder 3 modulo 4, so they cannot be perfect squares), our number must end in 50, 54, 05, or 45.

In the first two cases, this number has remainder 2 modulo 4 (also impossible for a perfect square), and in the last two cases, the number is divisible by 5 but not by 25.

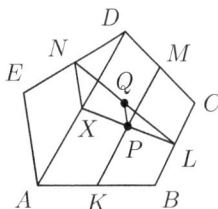

1968.11. Let X be the midpoint of segment AD. Then segment XN is the midline of triangle ADE parallel to side AE— hence, its length equals one half of $|AE|$. In quadrilateral $ABCD$, the midlines KM and XL are divided in half by the point of their intersection—therefore, P is the midpoint of segment XL. Thus, segment PQ is the midline of triangle XLN, and so it is parallel to XN and is equal to $\frac{1}{2}|XN| = \frac{1}{4}|AE|$.

1968.12. Let us project all the circles onto some arbitrary diameter D of the large circle. The sum of the lengths of the projections is clearly the same as the sum of the diameters of the small circles, that is, 50. Since the length of D is 6, we can conclude that if every point of D is covered by no more than eight projections, then the sum of their lengths would not exceed $6 \cdot 8 = 48 < 50$. Thus, there must be a point on D covered by at least nine projections. The line perpendicular to D, passing through that point, is the one we need.

1968.13. We will prove that line BD touches the circumcircle of triangle ABM. To do that, it will suffice to check that angle $\angle ABD$ equals the angle measure of arc $\overset{\frown}{AMB}$ of this circumference. Note that $\overset{\frown}{AMB} = 180° - \angle AMB = \angle BMC$. Furthermore, $\angle BMC = \angle BDC$ since the quadrilateral $BCDM$ is inscribed. Then we have $\angle BDC = \angle ABD$ because $ABCD$ is a parallelogram, and $\overset{\frown}{AMB} = \angle ABD$, as desired. Similarly, we prove that BD touches the circumcircle of $\triangle ADM$.

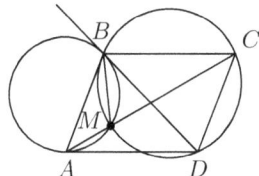

1968.14. Using Vieta's Theorem, we can easily calculate that $s = x^4 + y^4 = a^4 + 4a^2 + 2$, and $t = x^5 + y^5 = -a^5 - 4a^3 - 3a$. Since $as + t = -a$, we can see that $\gcd(s, t) = \gcd(a, s) = \gcd(a, 2) = 1$ because a is odd.

1968.15. Let us color the plane into two colors as it is shown in the figure. Then with every reflection, the color of the triangle changes. Since at the end, the color is the same as in the beginning, the number of reflections had to be even.

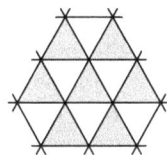

1968.17. The sequence

$$99\ 98\ 88\ 87\ 77\ \ldots\ 22\ 21\ 11\ 12\ 23\ 34 \ldots\ 89\,,$$

comprises all 2-digit numbers which end with any non-zero digit. We will split all the numbers, that do not belong to this sequence, into pairs of form $(ab, \overline{ba})$ and then insert these pairs into any gap in this sequence which separates two digits a. This proves the first part of the problem.

Now, consider transformation

$$F : a_1 a_2 \ldots a_n \mapsto \overline{(10 - a_1)(10 - a_2) \ldots (10 - a_n)}.$$

If some 162-digit number A satisfies the problem's conditions, then number $F(A)$ does too. Furthermore, inequality $A > B$ implies inequality $F(A) < F(B)$. Thus, the largest and the smallest of such numbers are obtained one from the other by transformation F. Therefore, their sum is equal to $10 + 100 + 1000 + \cdots + 10^{162} = 111 \ldots 110$.

1968.18. Hint: Consider numbers

$$1222222222,\ 2122222222,\ \ldots,\ 2222222221\,,$$

and prove that the signature for one of them equals 1. Let it be number 1222222222. Then, obviously, $2333333333 \to 2$ and $3111111111 \to 3$ (this is how we will denote the signature mapping). Prove that the signature always coincides with the first digit of the number.

Lemma. *This is true for all numbers written with only two digits.*

Indeed, if only two digits, say, 1 and 2, were used, and the first digit is, say, 1, then obviously the signature of this number is not 3, and since $2333333333 \to 2$, then that digit is not 2 either.

Now, let us take any number A that begins, say, with 2. Consider numbers $\overline{1ab \ldots c}$ and $\overline{3xy \ldots z}$, both written only with ones and threes so that they are different from A in all places. It follows from the lemma that the signatures of these two numbers are 1 and 3, and therefore signature of A must be 2.

1968.19. Obviously, the roots of quadratic equation $x = 1 - 1968x^2$ also serve as two solutions of the original equation; the other two solutions are the roots of the "remaining" quadratic polynomial $1968x^2 - 1968x - 1967$.

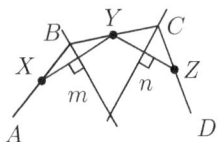

1968.20. Let $ABCD$ be the desired quadrilateral, X, Y, and Z—midpoints of its three equal sides AB, BC, and CD, respectively. Vertex B lies on the perpendicular bisector m for segment XY, and vertex C lies on the perpendicular bisector n for segment YZ. Since B and C are symmetric with respect to point Y, point B can be constructed as intersection of line m and line symmetric to n with respect to Y. Other vertices are then constructed in a rather obvious manner.

1968.21. Answer: 8.

We will solve this problem in its more general form. Namely, we will prove that for n colors, the smallest possible number of pairs of neighbor colors is $n - 1$. An example for $n - 1$ pairs is very simple: color exactly one square in color 1, one square in color 2, $\ldots$, one square in color $n - 1$, so that no two of these squares are adjacent; then color all other squares in color n.

Construct graph G whose vertices represent colors, and two of them are connected by an edge if and only if the two corresponding colors are neighbors. Consider any two colors i and j. Let us find square A of color i and square B of color j.

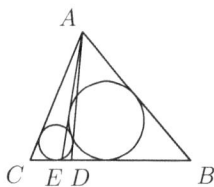

Since we can travel from A to B by stepping to the adjacent squares, there is a path in graph G which connects vertex-color i and vertex-color j. Thus, graph G is a connected graph, and therefore it has no less than $n - 1$ edges.

1968.22. From vertex A, draw segments AD and AE tangent to the first and the second circumference, respectively; thus, D and E lie on BC. Obviously, triangles ACD and AEB cover triangle ABC, and therefore the sum of the radii of the circumferences equals

$$\frac{2\,S(ACD)}{p(ACD)} + \frac{2\,S(ABE)}{p(ABE)} > \frac{2\,\big(S(ACD) + S(ABE)\big)}{p(ABC)} \geqslant \frac{2\,S(ABC)}{p(ABC)},$$

where the last expression is exactly the formula for the inradius of triangle ABC.

1968.23. Checking that this inequality is true for $n = 2$ and $n = 3$ is quite easy. Now, we assume that $n \geqslant 4$. Then, for $2 \leqslant k \leqslant n - 2$, the inequality $(n - k)!k! \leqslant \frac{2}{3}(n - 1)!$

holds true. Indeed, it is sufficient to prove it for $k = 2$ (trivial), and then note that $(n - k - 1)!(k + 1)! < (n - k)!k!$ for $k < n/2$ (for $n/2 \leqslant k \leqslant n - 2$, this inequality is then true by virtue of symmetry).

Therefore, the left-hand side of the original inequality does not exceed $(n - 1)!(1 + \frac{2}{3} + \frac{2}{3} + \cdots + \frac{2}{3} + 1) = (n - 1)!(2 + \frac{2}{3}(n - 3)) = \frac{2}{3}n!$.

1968.24. Let us assume that such a dissection is possible. We will classify the endpoints of all interior sides of the dissection triangles into three types. Type 1 are the points on the side of the large triangle (four internal sides of the dissection triangles meet there); Type 2 are the points inside the large triangle where six triangles (with twelve sides altogether) meet; and Type 3 are the points inside the large triangle lying inside one of the interior sides (three triangles and six of their sides meeting there).

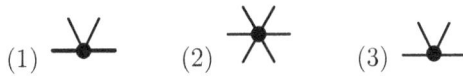

Denote the number of points of these three types by a_1, a_2, a_3, respectively. Then the number of interior sides equals $(4a_1 + 12a_2 + 6a_3)/2 = 2a_1 + 6a_2 + 3a_3$. At the same time, each point of Type 3 corresponds to three interior sides—the one on which it lies, and the two sides that end at it. It is clear that each interior side of the dissection corresponds to one of the points of Type 3—otherwise some two dissection triangles are congruent. Therefore, the number of interior sides is at most $3a_3$. Consequently, $a_1 = a_2 = 0$, that is, the dissection triangles have no vertices on the sides of the large triangle; hence, the entire dissection must consist of just one large triangle.

1968.25. Heron's formula gives us $S = \sqrt{p(p - a)(p - b)(p - c)}$; S is the triangle's area, p—its half-perimeter, and a, b, and c are the triangle's sides. Using obvious notation, we obtain

$$p = p_1 + p_2 \geqslant 2\sqrt{p_1 p_2},$$
$$p - a = (p_1 - a_1) + (p_2 - a_2) \geqslant 2\sqrt{(p_1 - a_1)(p_2 - a_2)},$$

and so on. Finally, multiplying four inequalities obtained in this manner, we arrive at $S^2 \geqslant 16 S_1 S_2$.

1968.26. Answer: there is only one solution.

It is clear that $x_i \in [-1, 1]$. Then $x_{i+1} = \cos(x_i) > 0$, and therefore all x_i are positive. Note that $\cos(\cos(\ldots(\cos(x_i))\ldots)) = x_i$. Consider function $f(x) = \cos(\cos(\ldots(\cos(x))\ldots)) - x$. It is easy to see that $f'(x) < 0$ when $x \in [-1; 1]$; that is, on that interval, f is decreasing.

At the same time, $f(0) > 0$ and $f(1) < 0$, and so $f(x)$ has exactly one root—let us call it a—in $[-1; 1]$. Therefore, $x_i = a$ for every i, and the system has one solution.

Note. As a matter of fact, number a is actually the root of a simpler equation $\cos(x) = x$.

1968.27. Answer: $\sqrt{a^2 + b^2 + 2ab \cos(\alpha^* + \frac{2}{3}\pi)}$, where α^* is the oriented angle between vectors $\overrightarrow{AA_1}$ and $\overrightarrow{BB_1}$ (thus the regular angle can be equal to α, $-\alpha$, $\pi - \alpha$, or $\pi + \alpha$).

To prove this, we will use complex numbers (assuming that point A corresponds to complex zero). If triangle ABC is oriented counterclockwise, then vector $\overrightarrow{AC}$ is obtained from $\overrightarrow{AB}$ via rotation by oriented angle $\frac{1}{3}\pi$, which is equivalent to the multiplication by $\xi = \frac{1}{2} + \frac{\sqrt{3}}{2}i$ (its absolute value is 1, and its argument is $\frac{1}{3}\pi$). Thus, we have

$$C - A = \xi(B - A), \text{ or } C = (1 - \xi)A + \xi, \ B = \bar{\xi}A + \xi B.$$

Note that for complex numbers, overline means complex conjugation.

Similarly, $C_1 = \bar{\xi}A_1 + \xi B_1$. Subtracting, we get $C_1 - C = \bar{\xi}(A_1 - A) + \xi(B_1 - B)$, or $c = \bar{\xi}a + \xi b$, where a, b, and c are complex numbers corresponding to vectors $\overrightarrow{AA_1}$, $\overrightarrow{BB_1}$, and $\overrightarrow{CC_1}$. Thus, vector c is uniquely determined by vectors a and b. It is left to find its length. Using formula $|z|^2 = z\bar{z}$, we obtain

$$|c|^2 = c\bar{c} = (\bar{\xi}a + \xi b)(\xi\bar{a} + \bar{\xi}\bar{b}) = \xi\bar{\xi}a\bar{a} + \xi\bar{\xi}b\bar{b} + \bar{\xi}^2\bar{a}b + \xi^2 a\bar{b} = |a|^2 + |b|^2 + \xi^2\bar{a}b + \bar{\xi}^2 a\bar{b}.$$

The last two summands in the right-hand side are conjugated to each other, and therefore their sum is equal to

$$\xi^2\bar{a}b + \bar{\xi}^2 a\bar{b} = 2\mathrm{Re}(\xi^2\bar{a}b) = 2|\xi^2\bar{a}b| \cdot \cos\mathrm{Arg}(\xi^2\bar{a}b)$$

$$= 2|a| \cdot |b| \cdot \cos(2\mathrm{Arg}(\xi) - \mathrm{Arg}(a) + \mathrm{Arg}(b)) = 2|a| \cdot |b| \cdot \cos(\alpha^* + \frac{2}{3}\pi).$$

Rewriting this in the original notation gives us the desired result.

1968.28. Answer: 9. See solution to Question **1968.21**.

1968.29. Assume that such a tiling exists. Clearly, there is an edge of the dissection that lies on the face of the tetrahedron but not on its edge. That dissection edge, obviously, contains a point which is not a vertex of the dissection. Dissection in the neighborhood of that point looks like dissection of half-space into dihedral angles with a common edge; measure of each of these angles must be equal to α—the measure of dihedral angle of a regular tetrahedron. But that is impossible because π is not an integer multiple of α. Indeed, it is quite easy to check that $\alpha = \arccos(\frac{1}{3})$ and $\frac{1}{3}\pi < \alpha < \frac{1}{2}\pi$.

1968.30. Denote $\sqrt[k]{2}$ by p. Then we can rewrite this inequality as

$$(p - 1)\left(\frac{b_1}{p} + \frac{b_2}{p^2} + \cdots + \frac{b_n}{p^n}\right) \leqslant \sqrt[k]{b_1^k + \cdots + b_n^k},$$

where $b_i = a_i p^i$. Substituting $t = 1/p$, we obtain

$$b_1 + b_2 t + \cdots + b_n t^{n-1} \leqslant \frac{1}{1 - t}\sqrt[k]{b_1^k + \cdots + b_n^k}.$$

Since $\sqrt[k]{b_1^k + \cdots + b_n^k} \geqslant B = \max b_i$, we then have

$$b_1 + b_2 t + \cdots + b_n t^{n-1} \leqslant B(1 + t + \cdots + t^{n-1}) = \frac{B(1 - t^n)}{(1 - t)}$$

$$\leqslant \frac{B}{(1 - t)} \leqslant \frac{1}{1 - t}\sqrt[k]{b_1^k + \cdots + b_n^k}.$$

1968.31. Denote by H the foot of the perpendicular dropped from M. Assume that $|AC| > |AB|$ and select point C' on the extension of segment AC so that $|AB| = |AC'|$. It is easy to verify, by counting the angles, that AM is the bisector of angle BAC'. Since triangle BAC' is isosceles, then $MA \perp BC'$, $|MB| = |MC'|$, that is, $|MC| = |MC'|$. Therefore, MH is a median in triangle MCC', and H is the midpoint of CC'.

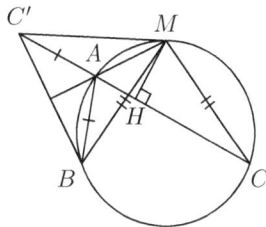

1968.32. Let us denote the original point by X_1, and the end of the first segment (first move of the first player) by X_2. Also define set of points $S = \{X_1, X_2\}$.

After the first move of the second player (consisting of drawing segment X_2—X_3), the first player will draw segment X_3—X_1, "returning" to point X_1. We will also expand set S by adding point X_3 to it.

Obviously, the second player now cannot make a move to any of the points X_1, X_2, or X_3. He will have to draw a segment X_1—X_4, where X_4 is different from any of the points in $S = \{X_1, X_2, X_3\}$. Then the first player draws segment X_4—X_2, "returning" to X_2.

Clearly, every move by second player must be made to point X_k outside of S because both points X_1 and X_2 are already connected to all points in S. The first player's response will consist in "returning" to point X_1 or X_2 alternately (by connecting X_k with it). Then we expand S by adding point X_k.

Sooner or later, there are no more points outside S and the second player loses the game.

1968.33. Let us prove this by induction on the number of points. The basis ($m = 3$) is obvious.

For the step of induction, assume that the statement is already proved for any set of m points if not all of them lie on the same line.

Consider some set S of $m + 1$ points that do not all belong to the same straight line. Let us select point $A \in S$ in such a way that not all other points lie on the same line. By induction hypothesis, there exist at least $\frac{1}{2}(m-1)(m-2)$ triangles with vertices on the points of S different from A. It is left to find $m-1$ more triangles with point A as a vertex. Consider line L passing through A and containing $t \geqslant 1$ other points of S. As we know, $t < m$ and $m - t$ points of S lie outside L.

Therefore, there exist at least $t(m-t) \geqslant m-1$ triangles with vertices on the points of S, one of which coincides with A.

1968.34. Let us assume that $M = a_n \geqslant a_{n-1} \geqslant \cdots \geqslant a_1 = m$. Then

$$S = \sum_{i<j} |a_i - a_j| = (n-1)a_n + (n-3)a_{n-1} + \cdots - (n-3)a_2 - (n-1)a_1.$$

On one hand $S = (n-1)(a_n - a_1) + (n-3)(a_{n-1} - a_2) + \cdots \leqslant Q(a_n - a_1) = Q(M - m)$, where $Q = (n-1) + (n-3) + \cdots + 1 = \frac{1}{4}n^2$ for even n, and $Q = (n-1) + (n-3) + \cdots + 2 = \frac{1}{4}(n^2 - 1)$ for odd n. On the other hand, $S \geqslant (n-1)(a_n - a_1) = (n-1)(M - m)$.

1968.35. Actually, the stronger statement is true: this sum does not exceed the sum of any three edges coming out of any vertex of the tetrahedron. In other words, for points X

inside the tetrahedron the function $f(X)$, computing the sum of the distances from point X to the vertices of the tetrahedron, reaches its maximum at one of the vertices. That immediately follows from the property of convexity that the distance function possesses; that, in turn, is easily deduced from the triangle's inequality.

1968.36. See solution to Question **1981.36**.

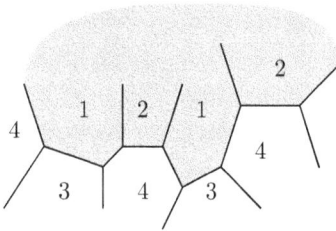

1968.37. Let us mentally re-color the faces of the first and the second color into the common "fifth" color. Consider regions of the "fifth" color on the surface of the polyhedron. Moving along the contour of one such region, we see that on the other side, the colors of the adjacent regions (they must be colored into colors three and four) alternate. It follows that the contour of any such region R has an even number of vertices if we exclude the common vertices of the first and the second colors. Furthermore, we see that if we count the number of vertices of the faces of the first color on R's boundary, and then count the similar number for the second color, then these two numbers must have the same parity.

Adding these numbers over all regions of the "fifth" color, we obtain that the parity of the number of vertices of the faces of the first color (that is, the number of the faces of the first color with an odd number of sides) must be the same as the similar parity for the second color.

1968.38. Let us call word W "proper" if:
(a) it has the same number of letters A and letters B, and
(b) for any end-segment of W (that is, several consecutive letters at the end of W), the number of letters A in it does not exceed the number of letters B in it.

We will prove that the set of words that can be obtained solely by using operation #1, or solely by using operation #2, is the set of all "proper" words.

First of all, it is easy to check that all the words that can be obtained by our operations are proper. Second, consider any proper word W. It is then possible, using only operations inverse to #1 (or only operations inverse to #2), to obtain word AB.

For operation #1, word W, of course, has a pair of neighboring letters AB. Removing this pair, we obtain proper word W' with fewer letters. Continue doing that until we have a word with only two letters—that word must be AB.

For operation #2 consider the last letter of W—obviously, it must be B. Now, find the shortest end-segment S of W containing the same number of As and Bs (S can coincide with entire W). Clearly, S begins with A. By removing this letter A and the final letter B, we obtain proper word W' with fewer letters. Continue doing that until we have the word with only two letters—that word must be AB.

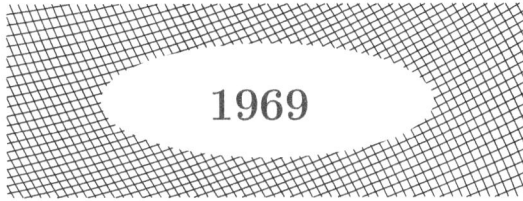

1969.01. A square on the chessboard is white if and only if the sum of the indexes of its column and its row is odd. Since the sum of all column and row indexes is $2(1+\cdots+8) = 72$ (note that it is even), the number of the squares that contribute an odd summand must be even. Therefore, the number of the rooks on the white squares is even, as well as the number of the rooks standing on the black squares.

1969.02. Answer: 14.

The sum of the numbers in the table is 110 which has remainder 2 modulo 4. The sum of all crossed-out numbers is divisible by 4, and thus the remaining number must also have remainder 2. There is only one such number in the table.

1969.03. Answer: at 3 pm.

Set distance between A and B at 1, and denote the speeds of Alex, Mike, and Johnny by v_1, v_2, and v_3, respectively. Then the first condition can be written as $2v_1 = v_2 + 1 - v_3$, and the second—as $2 - 3v_3 = \frac{3}{2}(v_1 + v_2)$. From these two equalities, we obtain $2 \cdot 3v_2 = 1 - 3v_3 + 3v_1$; that is, Mike will be in the middle between Alex and Johnny exactly in three hours after they all started their rides.

1969.04. Let us add one nut to all piles with indexes 1, 2, ..., 23; then add one nut to all piles with indexes 24, 25, ..., 35, 1, 2, ..., 11, and finally to the piles with indexes 12, 13, ..., 34. This triple operation results in increasing all the piles by two nuts except for pile #35 which has increased by only one nut. So, relatively to the level of all other piles, this triple operation is equivalent to removing one nut from pile #35.

It is obvious now that using such triple operations we can "remove" any number of nuts from any of the piles. Doing that first for the largest pile, then for the second largest and so on, we will make all the piles equal.

1969.05. Let us prove by induction that it is possible to place all n-digit numbers written by digits 1 and 2 around the circle so that they satisfy the problem's conditions. Basis $n = 1$ is obvious. Now the induction step. It is easy to see that if $\overline{a_1}, \overline{a_2}, \ldots, \overline{a_{2^n}}$ is the required sequence of n-digit numbers, then $\overline{1a_1}, \overline{1a_2}, \ldots, \overline{1a_{2^n}}, \overline{2a_{2^n}}, \ldots, \overline{2a_2}, \overline{2a_1}$ is the desired sequence of $(n+1)$-digit numbers.

1969.06. Let us assume that n questions were asked, and all the answers were negative. We will prove, using induction on n, that each mathematician can be absolutely certain that his opponent's number is at least $n+1$. Basis $n = 0$ is trivial. Now, for the induction step. After $n - 1$ such questions (all answered in the negative), each mathematician knows that the opponent's number is at least n. If mathematician A's number equals n, then, knowing that opponent's number cannot be $n-1$, to the nth question A would answer, "Yes, I know your number. It's $n + 1$." Thus, after n negatively answered questions both scientists know that their numbers are greater than or equal to $n + 1$.

It also follows that if A's secret number is k, then he knows that his opponent's number is at most $k + 1$, and therefore he cannot give more than k negative answers.

1969.08. Answer: 64.

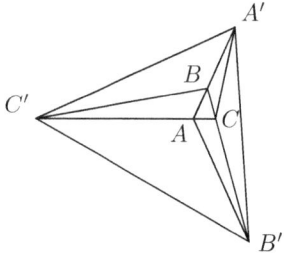

For the triangles with the same altitude, the ratio of their areas equals the ratio of their bases, and therefore

$$S_{C'A'A} = 3\,S_{C'BA} = 21\,S_{ABC},$$
$$S_{A'B'B} = 5\,S_{A'CB} = 10\,S_{ABC},$$
$$S_{B'C'C} = 8\,S_{B'AC} = 32\,S_{ABC}.$$

Hence, $S_{A'B'C'} = S_{C'A'A} + S_{A'B'B} + S_{B'C'C} + S_{ABC} = 64\,S_{ABC}$.

1969.09. The left expression can be rewritten as

$$\frac{1}{x-1} - \frac{1}{x+1} + \frac{1}{x-2} - \frac{1}{x+2} + \cdots + \frac{1}{x-10} - \frac{1}{x+10}.$$

Let us regroup these summands. Specifically, group term #1 with term #20, term #3 with term #18, and so on, and finally, term #19 with term #2. Then we obtain

$$\left[\frac{1}{x-1} - \frac{1}{x+10}\right] + \left[\frac{1}{x-2} - \frac{1}{x+9}\right] + \cdots + \left[\frac{1}{x-10} - \frac{1}{x+1}\right]$$

$$= 11\left[\frac{1}{(x-1)(x+10)} + \frac{1}{(x-2)(x+9)} + \cdots + \frac{1}{(x-10)(x+1)}\right].$$

1969.10. Through point W, representing the wolf's position, draw two lines parallel to the square's diagonals. The dogs' strategy is to always be in the points of intersections of these lines with the square's sides. First, it is obvious that this strategy guarantees that every time the wolf crosses the boundary, it will be met there by at least two hounds. Second, in this strategy, the speed of the dogs at any moment is at most $\sqrt{2}$ greater than the speed of the wolf (note that $\sqrt{2} = 1.4142\ldots < 1.5$). Indeed, look at the figure below.

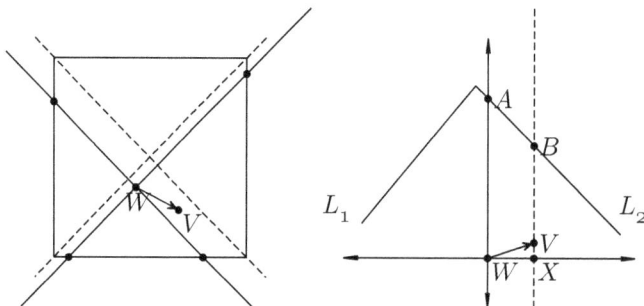

On the left is the original position W of the wolf together with its movement vector $\overrightarrow{WV}$; on the right, the result of $45°$ rotation together with the coordinate system centered at W

and the axes parallel to the original square's diagonals. Let lines L_1 and L_2 be the two sides of the square which contain the current positions of the dogs. Obviously, to obtain the hounds' positions, we can simply "project" vector WV onto the lines L_1 and L_2 by drawing lines parallel to the axes through point V and then intersecting them with L_1 and L_2. So, for instance, as the wolf moves from W to V, one of the hounds runs from A to B.

Using notation of the figure, it is evident that

$$|AB| = \sqrt{2}|WX| \leqslant \sqrt{2}|WV|.$$

1969.11. Answer: yes, they can.

Consider square $ABCD$ formed by the four villages. Find points P and Q such that ABP and CDQ are isosceles triangles with obtuse angle of $120°$ at vertices P and Q, respectively. Then the system of roads AP, BP, PQ, CQ, and DQ has total length of $10 \cdot (1 + \sqrt{3})$, which is approximately equal to 27 km 321 m. That is less than 28 kilometers, and so the answer is positive.

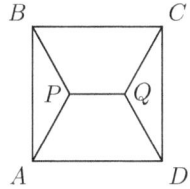

1969.13. Let X be a point on segment AD such that $CX \parallel AB$. Then obviously $p(ABX) = p(BCX)$, where $p(T)$ denotes the perimeter of triangle T. The same is true for triangles ABE and BCE, and therefore $p(ABX) - p(BCX) = |XE| \pm ||CE| - |CX||$; it follows from the triangle inequality that this difference of the perimeters is greater than $|XE| - |XE| = 0$, a contradiction. Therefore, $X = E$. Similarly, $BX \parallel CD$. Hence, E is the midpoint of AD and $|AE| = |BC|$.

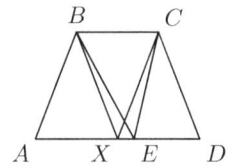

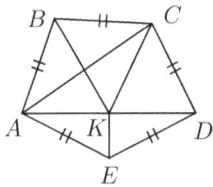

1969.14. Consider the midpoint K of the largest diagonal of our pentagon. It is easy to see that $\angle AKE = \angle EKD = 90°$. Since triangles ABC and ADE are isosceles and $|AC| < |AD|$, we have $\angle EAD < \angle CAB < \angle KAB$. Hence, points A and B lie to the same side of line EK. Therefore, $\angle AKB < 90°$. Similarly, $\angle DKC < 90°$.

Finally, if $\angle BKC > 90°$, then BC is the largest side in triangle BKC. Then AB is the largest side in AKB and CD—in CKD. Then $\angle AKB > 60°$, $\angle DKC > 60°$, and $\angle AKB + \angle BKC + \angle CKD > 180°$, which is not possible.

1969.15. Answer: 10 cities.

Consider any city A. It is connected with no more than three others, and each one of them is connected with no more than two other cities, different from A. For any city which is different from all these (no more than ten) cities, the route that connects it to A has more than one stopover. Thus, the country cannot have more than ten cities. An example with exactly ten cities is shown in the figure (this is a so-called Petersen graph).

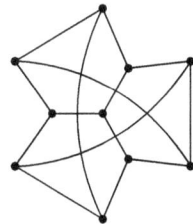

1969.17. Answer: 108, 117, 135, 180.

Let $a < b < c < d$ be our unknown numbers and S—their sum. The conditions imply that $d - a < 100 \leqslant a$ and therefore, $d < 2a$. Thus, $S < 3d + a < 7a$ (that is, $a > \frac{S}{7}$), and

$S > 3a + d > 3 \cdot \frac{d}{2} + d > 2d$, or $d < \frac{S}{2}$. Hence a, b, c, and d lie in the interval $(\frac{S}{7}, \frac{S}{2})$. This segment contains only four numbers—$\frac{S}{6}$, $\frac{S}{5}$, $\frac{S}{4}$, and $\frac{S}{3}$—that can divide S. We know that three of them belong to collection $T = \{a, b, c, d\}$. Note that $\frac{S}{6}$ and $\frac{S}{3}$ cannot be included in T simultaneously lest inequality $d < 2a$ is false. Therefore collection T contains either numbers $\frac{S}{6}$, $\frac{S}{5}$, and $\frac{S}{4}$ or numbers $\frac{S}{5}$, $\frac{S}{4}$, and $\frac{S}{3}$.

In the first case, the fourth number in the collection must be equal to $S(1 - \frac{1}{6} - \frac{1}{5} - \frac{1}{4}) = \frac{23}{60} S$, contradicting the inequality $d < 2a$ since $\frac{23}{60} > 2 \cdot \frac{1}{6}$. Thus, it has to be the second case, and then the fourth number equals $S(1 - \frac{1}{5} - \frac{1}{4} - \frac{1}{3}) = \frac{13}{60} S$. Hence, S must be divisible by 60 since all these numbers are integers. If $S = 60k$, then $a = 12k$, $b = 13k$, $c = 15k$, $d = 20k$. For $k = 9$ we have our answer. All other values of k do not fit: if $k \leqslant 8$, then $a \leqslant 12 \cdot 8 < 100$; if $10 \leqslant k \leqslant 14$, then $d \geqslant 200$ but $a \leqslant 196$; and if $k \geqslant 15$, then $d - a = 8k \geqslant 8 \cdot 15 > 100$.

1969.18. Somewhere in the sequence, there must be quintuples $abcd0$ and $abcd1$, where $abcd$ are the last four digits of the sequence. Thus, quadruple $abcd$ occurs three times—but there is only two digits that can stand in front of it: 0 and 1. To avoid the repeat of a quintuple, one of the quadruples $abcd$ must be preceded by nothing, that is, it has to be the first one.

1969.19. Triangle ABC divides the circumference into four parts. Consider the one that contains point P. If P lies inside segment defined by chord BC, then $\angle PBA > \angle CBA = \angle CAB > \angle PAB$ and therefore, $|AP| > |PB|$ (in triangle APB, the greater side lies opposite the greater angle), a contradiction.

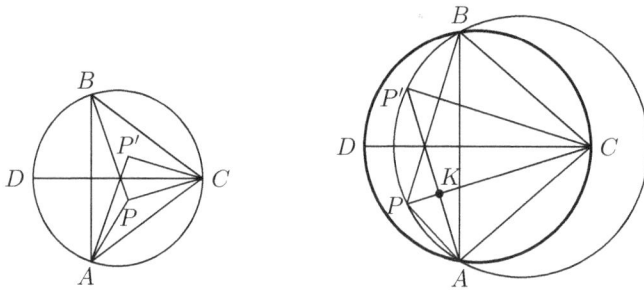

If P lies within the segment defined by chord AC, then $\angle APC = \angle APB + \angle BPC > \angle BPC$.

To analyze the remaining cases, denote by P' the point symmetric to P about diameter CD.

If P lies inside triangle ABC, then it follows from $|AP| < |BP|$ that point P lies inside $AP'C$ and $\angle APC > \angle AP'C = \angle BPC$.

And if P lies on the segment defined by chord AB, then define point K as intersection of segments AP' and PC. Considering triangles APK and $P'KC$, we can see that the required inequality $\angle APC > \angle AP'C$ is equivalent to the inequality $\angle PAP' < \angle PCP'$. Finally, the latter inequality follows from the fact that point C lies inside circumference $APP'B$.

1969.20. Answer: there are 6700 of such tickets.

Let $a_1, \ldots, a_6$ be the ticket's digits, and S—half of their sum (which must be an even integer).

It follows from equalities $a_1 + a_2 + a_3 = S = a_1 + a_3 + a_5$ that $a_2 = a_5$. Thus, in order for the ticket to be double-lucky, it is necessary and sufficient to satisfy equation $a_1 + a_3 = a_4 + a_6$. There is only one pair of digits with sum 0, two pairs with sum 1, ..., 10 pairs with sum 9, 9 pairs with sum 10, ..., one pair with sum 18. Thus, the number of all possible ways to select digits a_1, a_3, a_4, a_6 is equal to $1^2 + 2^2 + \cdots + 10^2 + 9^2 + \cdots + 1^2 = 670$. Finally, for each one of these selections, we have ten ways to choose digit a_2.

1969.21. Indeed,

$$\sqrt{2} - \frac{m}{n} = \frac{(2 - m^2/n^2)}{(\sqrt{2} + m/n)} > \frac{(2n^2 - m^2)}{2\sqrt{2}n^2} \geqslant \frac{1}{2\sqrt{2}n^2}.$$

1969.22. Answer: the locus is an arc of the circumference which has as a diameter the segment connecting midpoints of the quadrilateral's diagonals; the ends of the arc correspond to the boundary positions of the rectangle when some of its sides lie on the sides of the given quadrilateral.

Hint: Note that the center of any rectangle is the point of intersection of its midlines which are perpendicular and pass through the midpoints of the diagonals.

1969.23. If $x_{n-1} \equiv -a \pmod{x_n}$, then we obviously have

$$x_n \equiv 0 \pmod{x_n},$$
$$x_{n+1} = kx_n - x_{n-1} \equiv a \pmod{x_n},$$
$$x_{n+2} = kx_{n+1} - x_n \equiv ka \pmod{x_n},$$
$$\ldots,$$

and we can see that sequence $x_{n+1}, x_{n+2}, x_{n+3}, \ldots$ is congruent to $ax_1, ax_2, ax_3, \ldots$ modulo x_n. Thus, for any $m > 0$, we proved congruence $x_{n+m} \equiv ax_m \pmod{x_n}$. Therefore, if we choose $m = n$, then x_{2n} is divisible by x_n.

1969.24. Let us index all the tourists by numbers 0 through 59. On the first day, the manager will sit them in the following order (counting clockwise from the manager's place):

$$0,\ (0 - 1),\ (0 - 1 + 2),\ (0 - 1 + 2 - 3), \ldots,\ (0 - 1 + 2 - \cdots + 58 - 59) \pmod{60},$$

and on the $(k + 1)$th day the manager will use the sequence

$$k,\ (k - 1),\ (k - 1 + 2),\ \ldots,\ (k - 1 + 2 - 3 + 4 - \cdots + 58 - 59) \pmod{60}.$$

1969.25. Denote vectors $\overrightarrow{AB}$, $\overrightarrow{AC}$, and $\overrightarrow{AD}$ by $\overrightarrow{b}$, $\overrightarrow{c}$, and $\overrightarrow{d}$, respectively. From perpendicularity $\overrightarrow{AB} \perp \overrightarrow{CD}$ we get equality $\overrightarrow{b} \cdot \overrightarrow{c} = \overrightarrow{b} \cdot \overrightarrow{d}$ for the scalar products. If we denote vector to point O by $\overrightarrow{r}$, then the required equality can be rewritten as

$$(\overrightarrow{r} - \overrightarrow{d}/2)^2 + (\overrightarrow{r} - (\overrightarrow{b} + \overrightarrow{c})/2)^2 = (\overrightarrow{r} - \overrightarrow{c}/2)^2 + (\overrightarrow{r} - (\overrightarrow{b} + \overrightarrow{d})/2)^2,$$

and after subtracting the identical terms, we obtain $\overrightarrow{b} \cdot \overrightarrow{c} = \overrightarrow{b} \cdot \overrightarrow{d}$.

1969.28. See solution to Question **1969.23**.

1969.29. Let us construct this sequence S step by step. At step $\#\,n$, we obtain sequence S_n of $2n$ numbers such that the collection of all its pairwise differences contains numbers $1, 2, \ldots, n$, and in addition every number in that collection occurs there exactly once. At every step the next constructed sequence S_{n+1} extends the previous sequence S_n (that is, $S_n \in S_{n+1}$).

At first, we set our sequence to be $S_1 = \{1, 2\}$. On the $(n+1)$th step, we find the minimum natural number x not yet present among all the pairwise differences between the terms of our current sequence S_n. Now, add two numbers N and $N+x$ with $N > 2M$, where M is the largest number in S_n—that is, $S_{n+1} = S_n \cup \{N, N+x\}$. Notice that x is now present among the pairwise differences of the terms of S_{n+1}. Also, any such pairwise difference that involves one of the two new numbers cannot be equal to any pairwise difference between terms of S_n. Same is true for the difference $N - a$ and $(N+x) - b$, where a and b belong to S_n.

For example, the first three steps of this process will produce the following sequences:

$$S_1 = \{1, 2\}\,,$$
$$S_2 = \{1, 2, 5, 7\}\,,$$
$$S_3 = \{1, 2, 5, 7, 15, 22\}\,,$$
$$S_4 = \{1, 2, 5, 7, 15, 22, 45, 54\}\,.$$

Continuing this process indefinitely, we obtain the infinite sequence $S = \bigcup_{k=1}^{\infty} S_k$, satisfying the desired condition.

1969.30. Expressing areas of triangles via a, b, c and angles α, β, γ ($\angle BAC$, $\angle ABC$, and $\angle ACB$), we rewrite the left-hand side of the equality as

$$\frac{1}{2}\, abc \sin \alpha \sin \beta \sin \gamma \, (a \cos \beta \cos \gamma + b \cos \alpha \cos \gamma + c \cos \alpha \cos \beta)\,.$$

Similarly, using $R = abc/4S$ (*), the right-hand side transforms into

$$\frac{(abc \sin \alpha \sin \beta \sin \gamma)^2}{4S}\,.$$

Dividing by $abc \sin \alpha \sin \beta \sin \gamma$ and once again using (*), we obtain the following equality to prove

$$a \cos \beta \cos \gamma + b \cos \alpha \cos \gamma + c \cos \alpha \cos \beta = 2R \sin \alpha \sin \beta \sin \gamma\,.$$

Since

$$a = 2R \sin \alpha, \quad b = 2R \sin \beta, \quad c = 2R \sin \gamma\,,$$

we can see that the last equality is equivalent to the obviously true identity $\sin(\alpha + \beta + \gamma) = 0$.

1969.31. Answer: 5.

We will prove that a circle with radius $\frac{3}{2}$ cannot be covered by four circles with radius 1. Assume the opposite and denote the centers of the four unit circles by A_1, A_2, A_3, and A_4 so that rays OA_1, OA_2, OA_3, and OA_4 are listed in the clockwise order. If A_i coincides with point O, draw ray OA_i arbitrarily. Without loss of generality, we can assume that $\angle A_1 O A_2 = \alpha \geqslant \frac{\pi}{2}$. Let B be the intersection of the bisector of that angle with the given circumference (with radius $\frac{3}{2}$). If $\alpha \geqslant \pi$, then it is easy to see that all distances BA_i are at least $\frac{3}{2}$, which is impossible.

And if $\alpha < \pi$, then all distances BA_i cannot be lower than the distance from B to lines OA_1 and OA_2. That is,

$$BA_i \geqslant \frac{3}{2} \sin \frac{\alpha}{2} \geqslant \frac{3}{2} \sin \frac{\pi}{4} > 1 \,.$$

Thus, in this case, point B is not covered as well. The contradiction proves that at least five circles are needed.

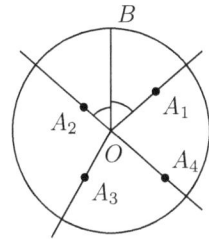

To construct a covering with five circles, let us dissect circle with radius $\frac{3}{2}$ and center O into five equal sectors. Select point D on the angle bisector of some sector so that $|OD| = 1$. It is not too difficult to verify that the circle with radius 1 centered at D covers that sector.

1969.32. Answer: the minimum number of the casualties is 10.

Here is an example with exactly ten casualties. Consider a grid made of regular triangles with unit sides. Select on this grid two adjacent nodes A and B; there are other eight nodes surrounding these two, situated at unit distance from one of the two "center" nodes. Put ten gangsters into these nodes. It is quite easy to "organize" the shootout in such a way that only A and B will be killed. Now "multiply" this construction by five, and we will have a shootout between $50 = 5 \cdot 10$ gangsters with $5 \cdot 2 = 10$ casualties.

We will prove that on average there must be at most five shots per casualty.

Consider any gangster O, who was killed by more than five shots, and connect him (by straight line segments) with all the gangsters who shot at him. Order these gangsters by the angle of direction from O: $A_1, \ldots, A_m$. Since any two "neighbors" A_i and A_{i+1} shot at O, not at each other, then side $A_i A_{i+1}$ must be the largest one in triangle $OA_i A_{i+1}$. Therefore, each of the angles $\angle A_1 O A_2$, $\angle A_2 O A_3$, $\ldots$, $\angle A_m O A_1$ is greater than or equal to $60°$. Therefore, condition $m > 5$ can be true only if $m = 6$, and all triangles $A_1 O A_2$, $A_2 O A_3$, $\ldots$, $A_6 O A_1$ are equilateral.

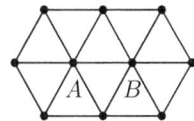

Now consider circle ω with center O and radius OA_1. It is clear that ω (boundary included) cannot contain any other gangsters except A_1, A_2, $\ldots$, A_6 and O. Gangster O himself shot at one of these six—e.g., A_2. Then A_2 was shot at by at most three gangsters save O. Indeed, as we know, A_1 and A_3 did not shoot at A_2. Since the reasoning from the previous paragraph can be applied to A_2, we have at most $3 = 6 - 3$ gangsters, different from O, who shot at A_2.

Thus, for each gangster who was killed by six shots, there is a "neighbor" who was killed by no more than four shots. It is also obvious that if we have two different gangsters killed by six shots, then the corresponding "neighbor casualties" are also different. The only exception would be the situation when we have two "six-shot casualties" O_1 and O_2 who shot at the same gangster A; in that case, gangster A was killed by exactly two shots—therefore, we have three casualties per thirteen shooters.

Thus, on average we have no more than five shooters per casualty—this proves that there are at least ten casualties. (This is still somewhat informal—the reader needs to make sure that we can split all gangsters into disjoint groups we have described above.)

1969.33. Using inequality

$$\frac{a_i a_k}{a_i + a_k} \leqslant \frac{a_i + a_k}{4} \,,$$

we obtain

$$\sum_{i<k} \frac{a_i a_k}{a_i + a_k} \leqslant \frac{1}{4} \sum_{i<k} (a_i + a_k) = \frac{n-1}{4},$$

because in the sum $\sum_{i<k} (a_i + a_k)$ each summand a_j occurs exactly $n-1$ times.

1969.34. Suppose that the plane is colored with red, white, and blue, and a triangle with vertices of one color does not exist. Consider two points A and B of the same (say, red) color at distance d. Draw lines L_1 and L_2, parallel to AB and at the distance of $\frac{2}{d}$ from it. Clearly, both these lines have no red points.

Now, let us prove that there exists entire line consisting of points of one color. If L_1 and L_2 are not such lines, then every pair of points at distance of $\frac{d}{2}$ on either one of them must be colored differently, and therefore any points at the distance of d have the same color. That means that line AB cannot contain points of white and blue colors, and therefore the entire line is red.

Thus, there exists a one-colored line L (say, red). Obviously, there can be no more red points on the plane. Let us take any two points X and Y of the same color such that XY is parallel to L. Similarly, we obtain line M, parallel to XY and to L, and colored into one (say, blue) color. That means that the rest of the plane (outside of lines L and M) is colored white, and then finding a one-color triangle of area 1 is trivial.

1969.35. We can assume that the square was cut in the lower left corner. We introduce the coordinate system with its origin in that same corner and adjust the scale so that the side of the small square is 1, and the side of the large square is 10^9. Now suppose that in every one of our ten rectangles the ratio of the sides does not exceed 9.

Let P be one of the rectangles whose upper right corner A lies strictly inside the square and has coordinates (x_0, y_0). Considering all possible positions for the tessellation rectangles in the neighborhood of A, it is easy to see that there exists rectangle P_1 for which A is either the upper left or the lower right corner. In the former case, height of P_1 does not exceed y_0; its width, therefore, does not exceed $9y_0$ and coordinates (x_1, y_1) of its upper right corner must satisfy conditions $x_0 < x_1 \leqslant x_0 + 9y_0$ and $y_1 = y_0$. In the former case, we obtain $x_1 = x_0$ and $y_0 < y_1 \leqslant y_0 + 9x_0$. In both cases, we have $\max\{x_1, y_1\} \leqslant 10 \max\{x_0, y_0\}$ with equality attained only for $x_0 = y_0$.

If instead of P we use that corner square, treating it as one more tessellation rectangle, we obtain that our tessellation must contain rectangle P_1 with $\max\{x_1, y_1\} \leqslant 10$, where (x_1, y_1) are the coordinates of its upper right corner. Applying the same reasoning to P_1, instead of P, we can find rectangle P_2 such that the coordinates of its upper right corner are less than 10^2 (inequality is strict, since $x_1 \neq y_1$). Continuing in the same manner, we build a chain of rectangles $P_1, P_2, \ldots, P_9$, such that the coordinates of the upper right corner of P_k are less than 10^k for $k = 2, \ldots, 9$. Note that these nine rectangles are pairwise different and do not contain any of the corners of the original square. Therefore, the remaining tessellation rectangle must contain three of those corners which is not possible.

1969.36. Let r be the radius of these circumferences, a—the side of the square formed by their centers. Note that these four circumferences are all contained within the circle with radius $r + a/\sqrt{2}$ centered at the center of the square; they touch that circle's boundary at four respective points, forming a square.

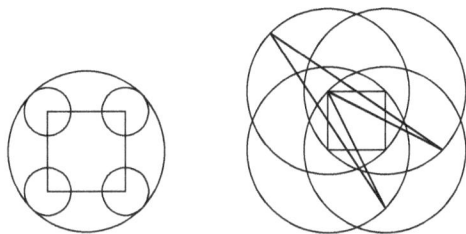

This square is the sought after convex quadrilateral with the maximum perimeter. That immediately follows from a more general and well-known fact: among all convex quadrilaterals inside the given circle, the inscribed square has the largest possible perimeter.

Note. If ratio a/r is sufficiently large, then any quadrilateral with vertices on these circumferences is convex, and therefore the square we described above is the one with the maximum perimeter. However, if ratio a/r is sufficiently small, then the "maximum" quadrilateral is not convex. The maximum quadrilateral is very close to the four copies of the circumferences' diameters, and therefore its perimeter is approximately equal to $8r$ (while the perimeter of the square above is $4a + 4\sqrt{2}r$). It is quite likely that the conditions of the problem included some restriction which eliminated the case of non-convex quadrilateral.

1969.37. Let x_0 be an integer such that

$$y_1 = |F_1(x_0)| > 1, \ldots, \quad y_n = |F_n(x_0)| > 1.$$

Consider number $a = x_0 + ky_1y_2 \ldots y_n$. From a well-known property of polynomials with integer coefficients, we have $F(p) - F(q) \vdots p - q$ for any integers p and q. Therefore, $F_1(a) - F_1(x_0)$ is divisible by $ky_1y_2 \ldots y_n$, and so $F_1(a)$ is divisible by y_1. Similarly, $F_2(a)$ is divisible by y_2 and so on. Choosing k sufficiently large so that all numbers $|F_i(a)|$ are greater than $\max(y_1, y_2, \ldots, y_n)$, we get the value of a we are looking for.

1969.38. Let us divide all terms on the left-hand side into two groups—with $x_k > n^2$ and with $x_k \leqslant n^2$. For any term from the first group

$$\frac{\sqrt{x_k - x_{k-1}}}{x_k} < \frac{\sqrt{x_k}}{x_k} = \frac{1}{\sqrt{x_k}} < \frac{1}{n},$$

and therefore the sum of the terms in the first group does not exceed 1. For any term in the second group, we have

$$\frac{\sqrt{x_k - x_{k-1}}}{x_k} \leqslant \frac{x_k - x_{k-1}}{x_k} \leqslant \frac{1}{x_{k-1} + 1} + \cdots + \frac{1}{x_k}.$$

Adding all these inequalities proves that the sum of the terms in the second group does not exceed

$$\frac{1}{2} + \frac{1}{3} + \cdots + \frac{1}{n^2}.$$

Add 1 (the first group sum) to the last expression to obtain the desired upper bound.

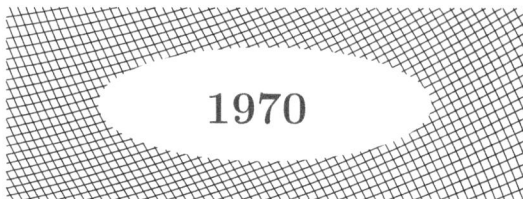

1970

1970.01. Since $123456789 \cdot 8 = 987654312$, the only pair with the required property is $(123456789, 987654312)$. Any attempt to increase the smaller number in such a pair—for instance, from 123456789 to 123456798—will result in the larger number being greater than 987654321. But that is the largest possible nine-digit number with different non-zero digits.

1970.02. Answer: 2, 3, and 4.

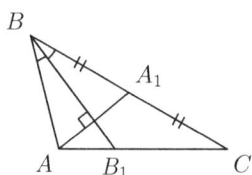

Let median AA_1 and angle bisector BB_1 (which are perpendicular) intersect at point O. Triangle ABA_1 is isosceles, since $|AB| = |BA_1| = |BC|/2$. Thus either $|AB| = 1$ and $|BC| = 2$, or $|AB| = 2$ and $|BC| = 4$. In the former case, $|AC| = 3$ contradicting the triangle inequality. In the latter case we have $|AC| = 3$ which gives us the only solution.

1970.03. Answer: no, that is not possible.

Indeed, each exchange involves exactly three coins, and therefore the overall number of coins used in all the exchanges must be divisible by three. On the other hand, if every villager gave exactly ten coins, then the same number would have been equal to 1970×10 which is not divisible by 3.

1970.04. Before each procedure, change the last digit of the number to zero. Obviously, that does not affect the result. So we obtain a monotonically decreasing sequence of three-digit numbers ending in 0 (the only non-negative number which does not decrease after such procedure is zero). Clearly, it cannot contain more than 99 different numbers; therefore, the 100th number must be equal to zero.

1970.05. Solution uses induction on the number of the cities n. The basis is obvious. To do the induction step, assume that we have $n + 1$ cities, and select an arbitrary one of them, say, A. Choose a type of transportation (air or water)—we will call it $\mathcal{T}$—which connects all the cities except A. If A is connected by $\mathcal{T}$ with any of the other n cities, then, clearly, $\mathcal{T}$ connects all $n + 1$ cities. If not, then the second type of transportation connects A with every other city on the country, and thus it connects all $n + 1$ cities together.

1970.06. Assume that is not true. Then for any three teams A, B, and C we know that if A defeated B and B defeated C, then A defeated C. It follows quite easily that team X that gained the maximum number of points defeated every other team. Further, team Y that took the second place defeated every other team except X, and so on. Thus, in such tournament, if team M got more points than N, then M defeated N. It means that the team that took the kth place has $12 - k$ points, and so the team that took the fifth place gained 7 points.

1970.07. Answer: $\angle B = 30°$.

Let perpendicular bisector for segment AC intersect ray AB at point B'. Then $|AB'| = |CB'|$, $\angle ACB' = 75°$, $\angle AB'C = 180° - \angle B'AC - \angle ACB' = 30°$. In right triangle $CB'H$, leg CH lies opposite angle of $30°$, and therefore $|CH| = |CB'|/2$. Hence, $|AB'| = |CB'| = 2|CH| = |AB|$, points B and B' coincide, and $\angle ABC = 30°$.

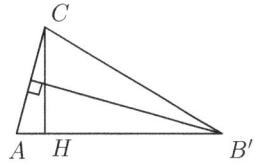

1970.09. Find on side AB point E such that BE is equal to base BC. Then $\angle CEB = \angle ECB = 50°$, $\angle ACE = 30°$, and since in any triangle the greater side lies opposite the greater angle, we have $|AE| > |CE|$ and therefore $|CE| > |CB|$. So we have $|AB| = |AE| + |BE| > 2|CB|$.

To prove item (b), we will "triple" our triangle by reflecting it twice across its side (see the figure). Since the length of broken line $CBXY$ must be greater than the length of segment CY, we are done.

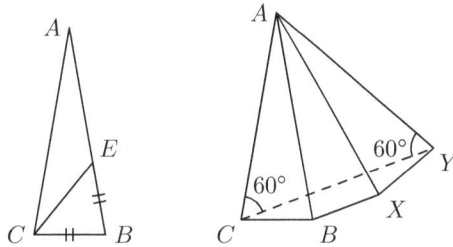

1970.11. Answer: $x = \pm 1$, $y = 1$.

Substituting $x^2 = 2y - 1$ into the second equation, we get $y^4 + 4y^2 - 4y - 1 = 0$. Dividing by $y - 1$, we obtain

$$y^3 + y^2 + 5y + 1 = 0.$$

Since $2y - 1 = x^2 \geqslant 0$, number y is positive while the equation above clearly has no positive solutions. Therefore, $y = 1$ is the only solution, and then $x = \pm 1$.

1970.12. There is at most one match left for each team to play. Let us split the teams into 18 pairs in such a way that every team has played every other team except maybe the team it is paired with—first, pair every team who still has a match to play with its last opponent, then arbitrarily split the remaining teams into pairs. Let these pairs be (A_1, A_{19}), (A_2, A_{20}), ..., (A_{18}, A_{36}).

Then we can divide the teams into three groups as follows: $A_1, A_2, \ldots, A_{12}$—the first group, $A_{13}, A_{14}, \ldots, A_{24}$—the second, and $A_{25}, A_{26}, \ldots, A_{36}$—the third.

1970.13. Let us denote the other points of tangency that lie on sides AB and BC by D and E, respectively; the second points of intersection of the first and the second circumference with segment AC by X and Y, respectively. Then from the tangent-secant theorem, we

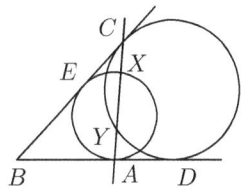

have $|CX| \cdot |CA| = |CE|^2$, $|AY| \cdot |AC| = |AD|^2$. Also from $|CE| = |AD|$, it follows that $|CX| = |AY|$, or $|AX| = |CY|$.

1970.14. Answer: $(x, y, z) = (1, 2, 4)$, $(-1, -2, -4)$, $(\sqrt{7}, 0, -2\sqrt{7})$, $(-\sqrt{7}, 0, 2\sqrt{7})$.

Subtracting the first equation from the second, we get $(z - y)(x + y + z) = 14$. Doing the same with the third and the second equations, we obtain $(y - x)(x + y + z) = 7$. Then $z - y = 2(y - x)$, $z = 3y - 2x$. Substituting value z into the second derived equality, we obtain $(y - x)(4y - x) = 7$, that is, $x^2 - 5xy + 4y^2 = 7$. Subtracting the first equation of the system, we arrive at $3y^2 - 6xy = 0$. If $y = 0$, then from the first equation we have $x = \pm\sqrt{7}$, and from the last—$z = \pm 2\sqrt{7}$; the second equation implies that the signs of x and z are opposite. And if y is not zero, then $y = 2x$ and $z = y - 2x = 4x$. Substitute these values into the first equation to obtain $x = \pm 1$.

1970.15. Answer: lengths of both segments are equal to $\frac{1}{2}$.

Denote by a_i the length of the ith side ($i = 1, \ldots, 5$). Using the "equality of tangents" rule, it is easy to see that $a_1 + a_4 - a_3 - a_5$ is equal to the difference of the segments into which the second side is divided by its tangency point. Hence, this difference is an integer. But both these segments are shorter than 1 (since they are equal in length to portions of the first and of the third side), and therefore they must be equal. Finally, their sum is an integer as well, and therefore they must both be equal to $\frac{1}{2}$.

1970.16. Suppose this is not true. Consider the rightmost column and the bottom row of the table. The number of painted squares in them must be at least eight. Indeed, if it is less than eight, then the remaining part of the table, which is a 4×4 square, has at least nine painted squares. Dissecting it into four 2×2 disjoint subtables, we can conclude, using pigeonhole principle, that one of them must have at least three painted squares, which proves our claim.

Now, obviously, if our column-and-row selection contains nine painted squares, then the bottom-right 2×2 subtable is the one we are looking for. So there must be exactly eight painted squares in that selection; moreover, the "missing" square must be inside the bottom-right subtable. It follows that all the squares marked by stars (see the figure) are unpainted. Thus, the upper-left 3×3 subtable contains eight painted squares, immediately giving us a 2×2 subtable with at least three painted squares in it.

1970.17. Let a be the side of the square. The circumference s inscribed into the square has radius $\frac{a}{2}$. Since the triangle contains s, it follows that this radius is less than the inradius of the triangle.

Now consider circumference S circumscribed around the square; its radius equals $\frac{a}{\sqrt{2}}$. S intersects or touches all sides of the triangle—therefore, its radius is greater than the inradius of the triangle (this follows rather easily from consideration of the areas of the corresponding triangles). Thus, the circumradius of the triangle lies between numbers $\frac{a}{2}$ and $\frac{a}{\sqrt{2}}$.

1970.18. Suppose that none of these segments intersect. Consider a planar graph whose vertices are the given points, and edges are the segments. This graph splits the plane into several connected regions. If boundary of some region consists of four or more vertices, then some two of them are not directly connected. Let us connect them by an edge that does not have to be a straight segment (it can be a curve). Let us draw these edges until all the regions are "triangular". The number of edges E will only increase; hence, $E \geqslant 100$. Since each edge separates two regions, and each region has exactly three edges on its boundary, the number of regions R equals $\frac{2E}{3}$. The Euler Formula for planar graphs states that $2 = 35 - E + R$; hence, $33 = \frac{E}{3}$ or $E = 99$, a contradiction.

1970.19. Ratio AC/AM equals the ratio of the homothety which maps circle O_1 to O; that is, equals the ratio of their radii. Similarly, ratio BD/BM equals the ratio of these radii as well. Thus, from *intercept theorem* (also called Thales' Theorem) follows $AB \parallel CD$.

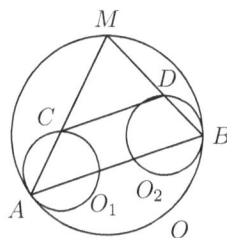

1970.21. Answer: both lengths are equal to $\frac{1}{2}$.

Same as Solution **1970.15**, using expression $a_1 + a_4 + a_6 + a_8 - a_3 - a_5 - a_7 - a_9$ instead of expression $a_1 + a_4 - a_3 - a_5$.

1970.22. Let $p = a^2 + b^2 = c^2 + d^2$, $q = a^3 + b^3 = c^3 + d^3$. Note that
$$p^3 - q^2 = a^2 b^2 (3p - 2ab) = c^2 d^2 (3p - 2cd),$$
and therefore both numbers ab and cd are the roots of polynomial $f(x) = 2x^3 - 3px^2 + (p^3 - q^2)$. Since numbers a and b are positive, we have
$$0 \leqslant ab \leqslant \frac{a^2 + b^2}{2} = \frac{p}{2} \quad \text{and} \quad 0 < cd \leqslant \frac{c^2 + d^2}{2} = \frac{p}{2}.$$
Since derivative $f'(x) = 6x^2 - 6px = 6x(x - p)$ is negative on interval $(0; \frac{p}{2}]$, polynomial $f(x)$ has no more than one root on that interval. Therefore, $ab = cd$.

1970.23. Answer: $x = 1$, $y = 3$, $z = 2$, $t = 4$.

Rewrite the equation as $31(zt + 1) = (40 - 31x)(yzt + y + t)$. The right-hand side is non-negative, and therefore $x = 1$. Then we have $9t = (zt + 1)(31 - 9y) > (31 - 9y)t$, hence $0 < 31 - 9y < 9$. Therefore, $y = 3$ and $9t = 4zt + 4$, then $4 = t(9 - 4z)$, and the answer immediately follows.

1970.24. See solution to Question **1970.30**.

1970.25. Each face of the tetrahedron contains at least one vertex of the cube. Indeed, if all vertices of the tetrahedron lie on the union of three planes with each of them not intersecting tetrahedron's interior, then either two of the planes are parallel or all three are parallel to one straight line—obviously such planes cannot contain faces of a tetrahedron. Now, same as in Solution **1970.17**, we prove that our edge ratio lies between $\frac{2}{\sqrt{3}}$ and 2 (radius of a ball circumscribed around cube with edge length a is equal to $\frac{1}{2}a\sqrt{3}$). Finally, use trivial inequalities $1 < \frac{2}{\sqrt{3}} < 2 < 1 + \sqrt{2}$.

1970.26. Let us set $x = \sin 1$, $y = \cos 1$. Then we need to prove that
$$2xy + 1 - 2x^2 + 2(x - y) \geqslant 1.$$
This can be rewritten as $(x - 1)(y - x) \geqslant 0$, and that is true because $x < 1$ and $y < x$.

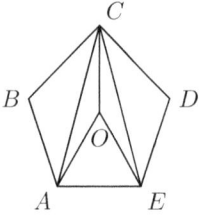

1970.27. Assume the opposite—the lengths of all sides in pentagon $ABCDE$ equal 1, all angles are below 120° and $\angle C \leqslant 90°$.

Since angles B and D are less than 120°, angles $\angle BCA$, $\angle BAC$ and $\angle ECD$ are greater than 30°. Therefore, $\angle ACE = \angle C - \angle BCA - \angle DCE < 30°$, $\angle CAE = \angle BAE - \angle BAC < 90°$, and it follows that triangle ACE is acute. Then its circumcenter O lies inside the triangle. Since $\angle AOE = 2\angle ACE < 60°$, we have $|AO| = |EO| = |CO| > 1$. Subsequently, $\angle AOC < \angle ABC < 120°$; similarly, $\angle COE < 120°$, and therefore $\angle AOC + \angle COE + \angle EOA < 300°$, which is impossible.

1970.29. Solving the system

$$
\begin{cases}
f(x) + f\left(1 - \frac{1}{x}\right) = \frac{1}{x} \\[2mm]
f\left(1 - \frac{1}{x}\right) + f\left(\frac{1}{1-x}\right) = \frac{x}{x-1} \\[2mm]
f\left(\frac{1}{1-x}\right) + f(x) = 1 - x
\end{cases}
$$

we obtain

$$
f(x) = \frac{(x^3 - x^2 + 1)}{2x(1 - x)}.
$$

1970.30. It is enough to show that none of the numbers from the first row occur on the same positions in rows 2, 3, $\ldots$, n. Clearly, after every procedure, when we "transition" from one row to the next, the index of each number increases by either $m - n$, or $n - k$, or $m - k$. Suppose that after several procedures some number "visited" the left group a times, the middle group—b times, and the right group—c times. Then altogether its index has increased by $x = a(n - k) + b(m - k) + c(m - n)$. If it has returned to its original location, then $x = 0$, while $0 \leqslant a \leqslant k$, $0 \leqslant b \leqslant n - m - k$, $0 \leqslant c \leqslant m$. It is easy to check that each vector (a, b, c) which provides a solution for equation $x = 0$ must be an integer-valued linear combination of vectors $\vec{v}_1 = (0, n - m, m - k)$ and $\vec{v}_2 = (1, -1, 1)$. Finally, it will suffice to prove that vector $\vec{v} = y_1\vec{v}_1 + y_2\vec{v}_2$ (y_1 and y_2 are integers; so coordinates of $\vec{v}$ are also integers) cannot have coordinates satisfying the inequalities above.

1970.31. Consider any two adjacent digits a and b of our number, and then two numbers we obtain after removing one of them. These numbers differ only in one position (say, 10^k), and therefore their difference is equal to $(a - b) 10^k$. Since both numbers are divisible by 7, we have that $a - b$ is divisible by 7, and therefore a and b have the same remainders modulo 7. Therefore, all digits have equal remainders modulo 7—it follows that if even one of the digits is 4, then all of them must have remainder 4 modulo 7, and therefore be equal to 4.

1970.32. Suppose that $t_k(t_{k+1} + 1) < 2$ for any k. Multiplying these inequalities, we obtain

$$
(t_1 + 1)(t_2 + 1) \cdots (t_n + 1) < 2^n.
$$

Since $1 + t \geqslant 2\sqrt{t}$ for any $t > 0$, we have

$$(t_1 + 1)(t_2 + 1) \cdots (t_n + 1) \geqslant 2^n \sqrt{t_1 t_2 \ldots t_n} = 2^n,$$

which contradicts the previous inequality.

1970.33. Assume that we have two such polygons M_1 and M_2. Let us mark on each of them the points that correspond to the vertices of the other polygon.

Polygon M_1 is split by the marked points into segments $X_1 X_2$, $X_2 X_3$, ..., $X_{k-1} X_k$, $X_k X_{k+1}$ ($X_{k+1} = X_1$), and polygon M_2 is similarly split into segments $Y_1 Y_2$, $Y_2 Y_3$, ..., $Y_{k-1} Y_k$, $Y_k Y_{k+1}$ ($Y_{k+1} = Y_1$), where the corresponding segments $X_i X_{i+1}$ and $Y_i Y_{i+1}$ have equal length. By condition, $Y_i Y_{i+2} \leqslant X_i X_{i+2}$, and therefore $\angle Y_i Y_{i+1} Y_{i+2} \leqslant \angle X_i X_{i+1} X_{i+2}$ for any i. If we sum up all these inequalities, both the left-hand side and the right-hand side add up to $\pi(k-2)$; hence, $\angle Y_i Y_{i+1} Y_{i+2} = \angle X_i X_{i+1} X_{i+2}$ for all i. The congruence of the polygons immediately follows from the equality of the corresponding segments and angles between them.

1970.34. Among numbers a_1, a_2, ..., a_m there exists either a number divisible by m or two numbers with equal remainders modulo m. Since $a_m < 2m$, then a_k (as well as $a_k - a_l$) can be divisible by m only if it equals m.

1970.35. Consider circumference passing through points A, D, N and intersecting the given circumference S again at some point E. Line NE intersects S again at point $\tilde{X}$, and line AX—at point O. Then by the well-known property of inscribed angles, triangles ODE and ONA are similar, as well as triangles ODE and $O\tilde{X}X$. Hence,

$$\angle(NE, NA) = \angle(DE, DA) = \angle(DE, DX) = \angle(\tilde{X}E, \tilde{X}X).$$

Thus, triangles ONA and $O\tilde{X}X$ are similar, and it follows that $NA \parallel \tilde{X}X$. Draw line AE which intersects S once more at point $\tilde{Y}$. Similarly, we have $\tilde{Y}Y \parallel NA$. Therefore, lines $A\tilde{Y}$ and BY are symmetric with respect to the diameter of S perpendicular to AB. It follows that points C and E are symmetric about that same diameter, and since lines $E\tilde{X}$ and CX are symmetric, so are points N and M.

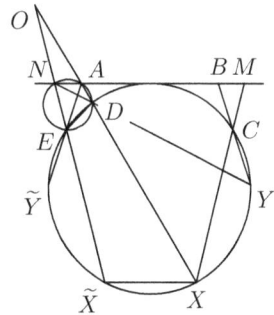

1970.36. Let us call a number any finite sequence of digits (beginning possibly with a few zeros). Set $N = \max\{n_i\}$. For each $i = 1, 2, \ldots, k$, consider all N-digit numbers, for which their n_i-digit head coincides with number a_i (there are 10^{N-n_i} such numbers). Together with every n_i-digit number, we will consider all of its cyclic shifts to the left by $1, 2, \ldots, n_i - 1$ positions. Altogether we have $n_1 10^{N-n_1} + \cdots + n_k 10^{N-n_k}$ of these numbers. They all have the form $\overline{c_i x_i b_i}$, where b_i and c_i are contiguous parts of number a_i, $a_i = \overline{b_i c_i}$ (and c_i must have at least one digit while b_i could be empty), and x_i is an arbitrary $(N - n_i)$-digit number.

Suppose some two of these N-digit numbers coincide: $\overline{c_i x_i b_i} = \overline{c_j x_j b_j}$. Then one of numbers c_i and c_j is a head of the other, and one of numbers b_i and b_j is a tail of the other.

It is easy to see that either one of the numbers $a_i = \overline{b_i c_i}$ and $a_j = \overline{b_j c_j}$ is a part of the other, or a head of one of them coincides with a tail of the other. By definition, it is possible only if $a_i = a_j$, $b_i = b_j$, and $c_i = c_j$. But then we have $x_i = x_j$.

Thus, all these N-digit numbers are different. In addition, number $00\ldots0$ is not among them. Hence, $\sum n_i 10^{N-n_i} < 10^N$, or after division by 10^N,

$$\frac{n_1}{10^{n_1}} + \frac{n_2}{10^{n_2}} + \cdots + \frac{n_k}{10^{n_k}} < 1.$$

Note. We did not use the fact that $n_i < 2n_j$ for every i and j. It seems likely that the original (possibly, simpler) solution somehow used that inequality.

1970.37. Let us prove that the sum of the numbers in any rectangle cannot exceed 4. Similarly, we can prove that the sum is always greater than or equal to (-4).

Assume that the sum of the numbers in some rectangle $a \times b$ equals $4 + x$, where $x > 0$. Then solution to **1970.38** (see below) allows us to obtain a rectangle with dimensions $|2a - b| \times |a - 2b|$ with the sum of the numbers at least $4 + 3x$ (obviously ratio of the sides cannot be equal to 2, because such a rectangle can be dissected into two disjoint squares, and therefore the sum of the numbers in the rectangle would not exceed 2). Applying this procedure again and again, we will get the sequence of rectangles with the sums of the numbers increasing to infinity.

Note that for $b < a < 2b$, rectangle $a \times b$ is followed by a rectangle with dimensions $(2a - b) \times (2b - a)$ that has the same perimeter but triple the difference of the sides. However, the difference of the sides cannot increase indefinitely when the perimeter is fixed—therefore, eventually we will encounter rectangle $p \times q$ with the ratio of the sides exceeding 2.

It is easy to see that two steps in our sequence result in $3p \times 3q$ rectangle, and $2k$ steps—in $3^k p \times 3^k q$ rectangle. For any k, this latter rectangle can be split into pq disjoint squares with dimensions $3^k \times 3^k$, and therefore the sum of the numbers in it never exceeds pq, contradicting the fact that this sequence tends to infinity.

1970.38. Let us assume that the sum of the numbers in some rectangle $ABCD$ equals $4 + x$, with $x > 0$. Consider four squares such that three sides of each one of them lie on some three sides of the rectangle (see the figure below). Their fourth sides define another rectangle $PQRS$. Let us prove that the sum of the numbers in $PQRS$ is at least $4 + 3x$. Then, if $ABCD$ is the rectangle with the largest sum, we would obtain the contradiction that proves the desired statement.

Without loss of generality, we can assume that $|AB| > |AD|$. There can be two significantly different situations, represented by two figures below. The first case is $|AB|/|AD| < 2$.

Let the sum of the numbers in square $AFGD$ be a and in $BCHE$—b. Since the sum of the numbers in $ABCD$ is $4 + x$, we can express the sums in rectangles $BCGF$, $AEHD$, and $EFGH$. Furthermore, if the sums in the corner squares are m, v, t, and r, and in squares with sides AB and CD—φ and ψ, respectively, then we can find the sums in rectangles $PQFE$ and $RSHG$. They are $\psi - m - r - 4 - x$ and $\varphi - t - v - 4 - x$. Thus, the absolute value of the sum of the numbers in rectangle $PQRS$ equals

$$\left| (\psi - m - r - 4 - x) + (-x - 4 + a + b) + (\varphi - t - v - 4 - x) \right|$$
$$= |\varphi + \psi - m - r + a + b - t - v - 12 - 3x| \geqslant 4 + 3x,$$

since all the numbers a, b, m, r, t, v, φ, and ψ do not exceed 1 by absolute value.

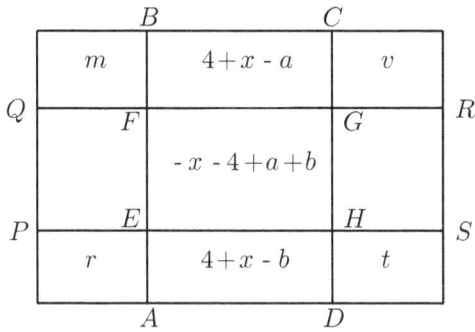

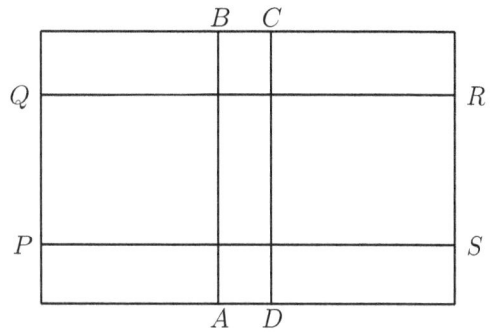

For the second case, we have $|AB|/|AD| > 2$, and the reasoning is virtually the same. If $|AB|/|AD| = 2$, then rectangle $PQRS$ becomes degenerate. In that case, rectangle $ABCD$ can be split into two squares, and the sum of the numbers in it does not exceed 2.

1971

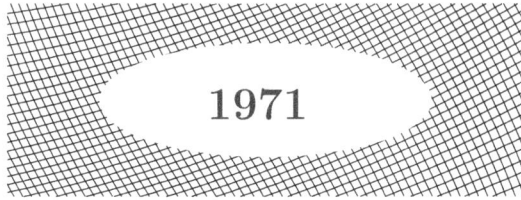

1971.01. Answer: meeting between John and Johnson will be closer to A. Indeed, let us denote Peter's velocity by x and Peterson's velocity by y. Then John's velocity is $2x$, and Johnson's velocity is $3y$. We know that the velocity at which John and Peterson ride toward each other is the same as the one for Johnson and Peter. Thus, $2x + y = x + 3y$. Solving this equation, we obtain $x = 2y$, and therefore ratio of velocities of John and Johnson is $4 : 3$, ratio for Peter and Peterson is $2 : 1$. Thus, by the time of the meeting, John will have ridden $\frac{4}{7}$ of the distance between the towns and Peter—$\frac{2}{3}$ of that distance. It is sufficient now to note that $\frac{4}{7} < \frac{2}{3}$.

1971.02. Answer: the equilateral triangle.

In any triangle, the sum of the altitudes does not exceed the sum of the medians because the length of any median is greater than the length of the corresponding altitude. The equality is attained only when all altitudes coincide with medians, that is, in the equilateral triangle.

1971.03. Answer: Nadia.

Altogether the kids gained 40 points. Thus, $40 = Sn$ where S is the total number of points awarded for one exam, and n is the number of exams. Furthermore, $S \geqslant 1+2+3 = 6$ and $n \geqslant 3$, since one cannot get 22 points for just two exams (obviously, the first place gets you no more than eight points; look at Eugene!). Hence, there are only two cases.

Case 1. $n = 4$, $S = 10$. Nick gained 22 points, and so the first place is worth at least six points. Thus, all possible cases of one exam point distribution are: $6 + 3 + 1$, $6 + 4 + 0$, $7 + 3 + 0$, and $7 + 2 + 1$. In the first three cases, Nick would not be able to get 22 points, and in the fourth case, Eugene would gain at least $10 = 7 + 1 + 1 + 1$ points.

Case 2. $n = 5$, $S = 8$. Then the first place is worth at least five points, and we have the following possible cases of one exam point distribution: $S = 5+3+0$ and $S = 5+2+1$. In the first case, it is impossible to get 22 points after five exams. In the second case, Nick can gain 22 points only as $5 + 5 + 5 + 5 + 2$; Eugene obtains 5 points for algebra, and therefore, his nine points can be expressed only as $5 + 1 + 1 + 1 + 1$. Obviously, Nick was first in all subjects except algebra, where he took the second, and in all those subjects, Eugene was third. Therefore, Nadia took the second place in physics.

1971.04. Answer: 18.

Let us denote the unknown integer by a. Since the sum of the digits of a^6 is divisible by 9, then obviously a is divisible by 3. Furthermore,

$$12^6 = 2\,985\,984 < 10\,000\,000 < a^6 < 64\,000\,000 = 20^6,$$

and therefore, $12 < a < 20$. Thus, the only viable cases are $a = 15$ (does not fit because 15^6 ends with 5) and $a = 18$ (fits because $18^6 = 34\,012\,224$).

1971.05. If the number on the blackboard is less than 998, the first player skips the move. For 998, he subtracts the number on the die, and for 999, he adds the number on the die. It is obvious that this strategy guarantees that the first player does not lose.

1971.06. Answer: yes, that is possible.
The example is shown in the figure.

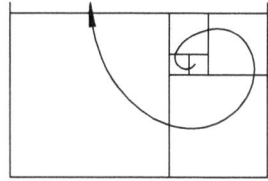

1971.07. Answer: $x = -2$, $y = -\frac{1}{2}$.
From the first equation, subtract the second one multiplied by $y + 1$. We obtain $x + 2 = 0$, and substituting $x = -2$ into the second equation gives us $2y + 1 = 0$.

1971.08. Answer: the equilateral triangle.
Solution here is absolutely identical to Solution **1971.02** with word "median" changed to "angle bisector".

1971.09. Consider rotation by angle $90°$ which maps the square onto itself. Then lines AK, BK, CK, and DK will be transformed into the perpendiculars mentioned in the problem, and point K—into the common point of all these perpendiculars.

1971.11. First, consider the cases when our polygon P has a side not parallel to any other side of P. Let it be side AB. Point K is then the vertex which lies farther away from line AB than any other vertex of P, while points L and M are the vertices adjacent to K. Then rays KL and KM intersect AB, and P lies inside the resulting triangle.

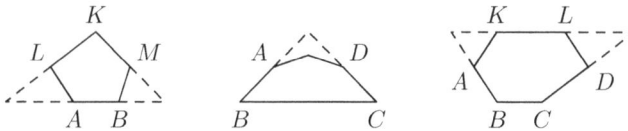

Now consider the cases when for each side of P, there exists another side of P parallel to it. Since P is not a parallelogram, there exists a side of P such that the sides adjacent to it are not parallel. Let A, B, C, D be consecutive vertices of P such that $AB \parallel CD$, and KL is the side of P parallel to BC. Now, if $\angle ABC + \angle BCD < 180°$, then the required sides are AB, BC, and CD; if $\angle ABC + \angle BCD > 180°$, then the required sides are AB, CD, and KL.

1971.12. Considering the bottom of the box tessellated into unit squares, we will mark some of them as it is shown in the figure. Then each 2×2 tile covers exactly one marked square, and each 1×4 tile covers either 0 or 2 marked squares. Thus, the parity of the number of 2×2 tiles must coincide with the parity of the number of the marked squares. Clearly this parity changes when a 2×2 tile is replaced by a 1×4 tile.

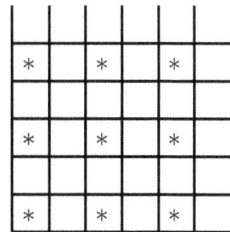

1971.13. Let points A and B move along lines L_A and L_B, respectively, and assume that at the start their positions were A_0 and B_0. Consider rotation of the plane that maps L_A to L_B, and point A_0 to point B_0.

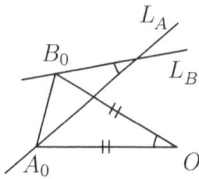

Center O of this rotation is obviously the desired point. It is situated in the vertex of isosceles triangle A_0OB_0 with the angle at the vertex equal to the angle between lines L_A and L_B.

1971.14. Answer: 1972 digits.

Digital representation of 5^{1971} has k digits, of 2^{1971}—m digits. Then

$$10^{k-1} \cdot 10^{m-1} < 2^{1971} \cdot 5^{1971} < 10^k \cdot 10^m,$$

therefore, $k + m - 2 < 1971 < k + m$, and $k + l = 1972$.

1971.15. Answer: 4951.

Obviously, if we have two chains complementary to each other, precisely one of them has the positive sum of the numbers. Indeed, if the sum of the numbers in one chain of a pair is x, and in the other one—y, then $x + y = 1$; since these numbers are integers, exactly one of them is positive.

Split the set of all chains (except the one that has all one hundred numbers) into pairs of complementary chains. Then exactly half of the chains has the positive sum of the numbers. The total number of chains is $100 \cdot 99$, giving us 4950 chains with a positive sum. Adding the chain that covers all one hundred numbers, we get the final answer.

1971.16. Answer: yes, the first player always wins.

The winning strategy is as follows. With every move, the first player divides a pile with even number of matches into two piles with odd numbers of matches (and, of course, gets rid of the other pile). To prove that the first player always has a move, we must show that after every move of the second player, at least one of the piles has an even number of matches. Indeed, after the first player's move, both piles have odd number of matches. One of them is then removed, and the other one is split into two piles, with one of these new piles, obviously, containing an even number of matches.

1971.17. Since $n + 1 \vdots 24$, then $n \equiv 3 \pmod 4$, and therefore n is not a perfect square. Thus, we can split all of its divisors into pairs $(d, n/d)$. Let us prove that the sum of the numbers in each such pair, that is, $(n + d^2)/d$, is divisible by 24. We know that $\gcd(d, 24) = 1$, and so it will suffice to prove that $(n + d^2) \vdots 24$. But $d^2 \equiv 1 \pmod 3$ and $d^2 \equiv 1 \pmod 8$ (because $\gcd(d, 24) = 1$), and $n \equiv -1 \pmod{24}$, so it follows that $n + d^2$ is divisible by both 3 and 8, and thus by 24 as well.

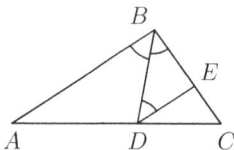

1971.18. Select point E on side BC such that $BE = ED$. Then lines AB and DE are parallel, triangles ABC and DEC are similar, and therefore $|BE| = |ED| = 6$. From the triangle's inequality, $|BD| < |BE| + |ED| = 12$.

1971.19. Answer: there are 202 solutions. They are: $x_0 = \frac{1}{2}$, $x_k = \pm\frac{1}{2}$, $x_j = 0$ for $j \neq 0$, k (200 solutions); and $x_1 = x_2 = \cdots = x_{100} = 0$, $x_0 = 0$ or 1 (two more solutions).

If x_0 is different from $\frac{1}{2}$, then, using in reverse order all equations from the last one to the second one, we obtain $x_{100} = x_{99} = \cdots = x_1 = 0$, and from the first equations we get $x_0 = 0$ or $x_0 = 1$.

If $x_0 = \frac{1}{2}$, we have the system

$$
\begin{cases}
x_1^2 + \cdots + x_{100}^2 = \frac{1}{4} \\
x_1 x_2 + x_2 x_3 + \cdots + x_{99} x_{100} = 0 \\
\cdots \\
x_1 x_{100} = 0 \\
0 = 0 .
\end{cases}
$$

If not all of the variables vanish, then consider the minimum index $i \geqslant 1$ and maximum index $j \leqslant 100$, for which $x_i \neq 0$, $x_j \neq 0$. If $i \neq j$, then equality

$$0 = x_1 x_{j-i+1} + x_2 x_{j-i+2} + \cdots + \underline{x_i x_j} + \cdots + x_{100-j+i} x_{100}$$

is false, since its right-hand side has exactly one non-zero summand (underlined). Thus, only one of x_i (say, x_k) differs from zero, and from the first equation $x_k = \pm\frac{1}{2}$.

1971.21. Suppose that $b^2 - 4ac = n^2$. Then we have

$$4a(ax^2 + bx + c) = (2ax + b + n)(2ax + b - n) .$$

From this, substituting $x = \underline{10}$ gives us that either $20a + b + n$ or $20a + b - n$ is divisible by prime number $\overline{abc}$. But then, obviously, $20a + b + n \geqslant 100a + 10b + c$, which is impossible since $n < b$.

1971.22. Answer: (to the additional question) The first player cannot win only if both piles have odd number of matches.

Indeed, after any move by the first player, the two resulting piles have different parity, and the second player can use the strategy already described in Solution **1971.16**.

1971.23. Since $\angle EDK = \frac{1}{2}\angle ADC + \frac{1}{2}\angle BDC = 90° = \angle ECK$, quadrilateral $CEDK$ is inscribed. Thus $\angle BKD = \angle DEC$ and $\angle DEK = \angle DCK = \angle DCE = \angle DKE$; from this we have $|DE| = |DK|$. Rotate triangle BKD by $90°$ around point D so that segment DK maps into segment DE. Then from $\angle BKD = \angle DEC$ we have that point B maps into point B' which belongs to ray EC. Then Pythagoras' Theorem for right triangle ADB' produces the required equality.

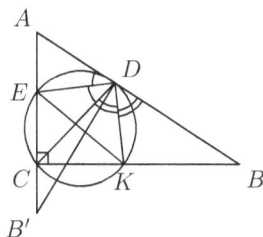

1971.24. Answer: yes, that is possible.

Let us start with the tessellation of 12×21 rectangle T shown below; let us denote the ratio $\frac{21}{12} = \frac{7}{4}$ of its sides by q. The idea is to cover the plane with homothetic copies of this rectangle (and its tessellation), with rational ratios of homothety.

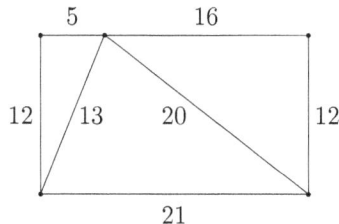

We will construct these copies by induction, where at each step the area covered by the first n rectangles $T_1, \ldots,$ T_n is also a rectangle—we will denote it by R_n. The first such rectangle R_1 coincides with T_1. The homothety ratio is, of course, $r_1 = 1$.

Now we construct T_2 as follows: its upper shorter side will coincide with the lower (longer) side of $R_1 = T_1$. And, of course, it must be homothetic to R_1. Clearly, we have homothety ratio $r_2 = q$. Together, T_2 and R_1 form rectangle R_2. It is a "vertical" rectangle, by which we mean that its vertical side is longer than its horizontal (R_1 was "horizontal").

Then we construct T_3, using a similar construction, with one change which is caused by "verticality" of R_2: we will attach T_3 to the right vertical side of R_2. Together, T_3 and R_2 will form rectangle R_3. Homothety ratio r_3 can be computed as $r_3 = 1 + q^2 = \frac{65}{16}$.

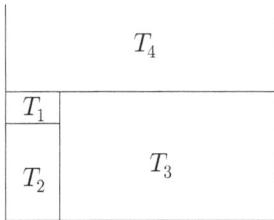

Then on to T_4 and R_3—this time we attach the next rectangle T_4 on top of R_3, continuing to build the spiral of rectangles. It is easy to see that ratio r_4 is equal to $1 + q + q^3$.

At each step k, rectangle T_k must be homothetic to R_1, so the ratio of its sides is always $\frac{7}{4}$. Obviously, all the sides' length of rectangles T_k and R_k are rational numbers and so the homothety ratio is also always rational, guaranteeing us that the sides of the tessellation triangles inside T_k are rational.

It is also quite obvious that no two of the tessellation triangles are congruent. That immediately follows from the fact that the homothety ratios r_k are increasing.

1971.26. Let $ABCD$ be the given tetrahedron, its altitude AH_a intersecting altitudes BH_b and CH_c. Note that the projection of any altitude of a tetrahedron on any of its faces (except the one it is dropped on) coincides with the altitude of that face. Project AH_a, BH_b, and CH_c onto face ABC—the three resulting altitudes must intersect at one point, and therefore the points of intersection of AH_a with BH_b and with CH_c coincide.

Now consider projection onto face BCD. Altitude AH_a maps into one point H_a, and altitudes BH_b and CH_c—into altitudes of triangle BCD. Therefore, H_a is orthocenter of BCD, and it follows that the fourth altitude DH_d must also intersect AH_a (because its projection also passes through same orthocenter). Similarly, we prove that DH_d passes through the point of intersection of the three other altitudes.

1971.27. Take two arbitrary moments of time t_1 and t_2. Let X_1 and Y_1 be the locations of the flies at t_1, and X_2 and Y_2—at t_2. Now find in space point O equidistant from X_1 and Y_1 as well as from X_2 and Y_2. This point exists (actually, there are infinitely many such points) because the corresponding loci of equidistance are the planes (perpendicular bisectors for segments X_1Y_1 and X_2Y_2) which are not parallel (otherwise points X_1, Y_1, X_2, and Y_2 would be coplanar and that would contradict the problem's condition). Now we will prove that point O is the one we need. Triangles OX_1X_2 and OY_1Y_2 are congruent (**SSS** rule), and therefore there exists a motion keeping point O in place while mapping X_1 to Y_1, and X_2 to Y_2. Obviously, this motion maps the trajectory of the first fly to the trajectory of the second fly—therefore, at any moment the flies are equidistant from point O.

Note. Another way is to use vector representation of uniform point motion in space.

1971.28. Let

$$x = \sum_{i=1}^{201} \frac{1}{2^{201-i}} \cdot \frac{1}{i} \underset{k:=i-1}{=} \frac{1}{2^{200}} + \sum_{k=1}^{200} \frac{1}{2^{200-k}} \cdot \frac{1}{k+1}$$

be the expression from the left-hand side of the inequality. Then

$$2x = \sum_{i=1}^{201} \frac{1}{2^{200-i}} \cdot \frac{1}{i} = \frac{2}{201} + \sum_{k=1}^{200} \frac{1}{2^{200-k}} \cdot \frac{1}{k}.$$

Hence,

$$x = 2x - x = \sum_{k=1}^{200} \frac{1}{2^{200-k}} \left(\frac{1}{k} - \frac{1}{k+1} \right) + \frac{2}{201} - \frac{1}{2^{200}}$$

$$< \frac{2}{201} + \sum_{k=1}^{200} \frac{1}{2^{200-k}} \frac{1}{k(k+1)} = \frac{2}{201} + \sum_{k=1}^{180} + \sum_{k=181}^{200}$$

$$\leqslant \frac{2}{201} + \left(\frac{1}{2^{199}} + \cdots + \frac{1}{2^{20}} \right) + \frac{\left(1 + \frac{1}{2} + \cdots + \frac{1}{2^{19}} \right)}{180 \cdot 181}$$

$$\leqslant \frac{2}{201} + \frac{1}{2^{19}} + \frac{1}{90 \cdot 181} < \frac{1}{90}.$$

1971.29. Suppose that this number is rational, and our decimal representation is periodic. The period contains ℓ digits, forming ℓ-digit number A, and starts at the kth place. Consider a large enough power of 2, $n = 2^N$, such that it covers at least four periods. We can assume that the last digit of n is the last digit of period A. If that is not so, we can simply cycle-shift the period's digits. Since n contains at least 4ℓ digits, we have $N \geqslant 4\ell$ and $n \vdots 2^{4\ell}$.

Then $n = x \cdot 10^{4\ell} + A \cdot 10^{3\ell} + A \cdot 10^{2\ell} + A \cdot 10^{\ell} + A = x \cdot 10^{4\ell} + A \cdot (10^{3\ell} + 10^{2\ell} + 10^{\ell} + 1)$, and therefore A is divisible by $2^{4\ell}$. But $0 < A < 10^{\ell} < 2^{4\ell}$, a contradiction.

1971.30. Let us cover all the points where the lines meet as well as point P with a circle C of radius r and center O. Now consider number $t < 1$ such that cosines of all angles between the given lines are less than t. Then all the projection points will always be covered by the circle centered at O with radius $2r/(1-t)$.

Indeed, take point A lying on line L and project it onto line M, obtaining point B. Lines L and M intersect at point K inside C. Then $|BK| < t \cdot |AK|$. Furthermore,

$$|BO| < r + |BK| < R + t|AK| < r + t(|AO| + r) < 2r + t|AO| < \frac{2r}{1-t}.$$

The last inequality follows from $|AO| < 2r/(1-t)$.

1971.31. Rewriting the equation as

$$x^n + a_1 x^{n-1} = a_2 x^{n-2} + \cdots + a_n,$$

and dividing it by x^{n-1}, we obtain

$$x + a_1 = \frac{a_2}{x} + \cdots + \frac{a_n}{x^n}.$$

Since on the left we have a monotonically increasing function and on the right we have a function that monotonically decreases on interval $[0; \infty]$, this equation cannot have more than one positive solution.

1971.32. Denote by a the arithmetic mean of numbers a_1, a_2, ..., a_n. Assume that the statement is false, and so for any $k \leqslant n$ we have $A_k > a$. Let us prove by induction on k that $a_1 + a_2 + \cdots + a_k > ka_k$ for all $k \leqslant n$. Then for $k = n$, this inequality would contradict the definition of a.

The basis follows from inequality $a_1 = A_1 > a$. Now, assuming that the statement is true for all $k \leqslant m$, we need to show that it is true for $k = m + 1$ as well. Since $A_{m+1} > a$, there exists index i for which $(a_i + a_{i+1} + \cdots + a_{m+1})/(m + 2 - i) > a$, that is, $a_i + a_{i+1} + \cdots + a_{m+1} > (m + 2 - i)a$. Furthermore, by induction hypothesis, we know that $a_1 + a_2 + \cdots + a_{i-1} > (i - 1)a$. Adding the last two inequalities, we obtain $a_1 + a_2 + \cdots + a_{m+1} > (m + 1)a$. The induction proof is complete, providing us with the desired contradiction.

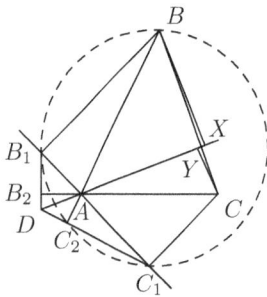

1971.33. Lines B_1B_2 and C_1C_2 meet at point D. Since $\angle BB_1C_1 = \angle BC_2C_1 = 90°$, points B, B_1, C_1, and C_2 lie on the same circumference. Therefore, $|AB| \cdot |AC_2| = |AB_1| \cdot |AC_1|$ (both products are equal to the power of point A with respect to this circumference). Similarly, $|AC| \cdot |AB_2| = |AB_1| \cdot |AC_1|$. Thus, $|AB| \cdot |AC_2| = |AC| \cdot |AB_2|$. Drop perpendiculars BX and CY from points B and C onto line AD. Since $\angle BXD = \angle BC_2D = 90°$, points A, X, C_2, and D lie on the same circumference, and therefore $|AX| \cdot |AD| = |AB| \cdot |AC_2|$. Similarly, $|AY| \cdot |AD| = |AC| \cdot |AB_2|$. Finally, we have $|AX| \cdot |AD| = |AY| \cdot |AD|$, that is, $X = Y$, and it follows that side BC is perpendicular to AD.

1971.34. We will say that index p *succeeds* index q (and write that as $p \succ q$) if the sum $b_p + b_{p+1} + \cdots + b_{q-1}$ is positive. It is obvious that for any two indexes p and q either p succeeds q, or q succeeds p (but not simultaneously!).

Also it is easy to show that for any three indexes p, q, and r, the following is true: if $p \succ q$ and $q \succ r$, then $p \succ r$. Now it suffices to note that number N_k is simply the number of indexes j such that $k \succ j$. Therefore, for any indexes p and q, if $p \succ q$, then $N_p > N_q$.

1971.35. Color all nodes of square S in two colors—black and white—defined by the parity of the sum of the node's coordinates. Obviously, for even n, the number of black nodes equals the numbers of white nodes, and for odd n, these two numbers differ by one. Also for any 2×2 square, all four of its vertices and its center are colored the same.

Now, assume that $a \neq b$. Then the number of nodes on the border of K (i.e., its perimeter $4n - 4$) is less than the number of nodes on the border of P (equal to $2(a+b) - 4$). Indeed, for $a \neq b$ we have $n = \sqrt{ab} < \frac{1}{2}(a + b)$.

Therefore, there exists an interior node X in S that is mapped to a border node of P lying, say, on side s with length a. If we now consider 2×2 square K centered at X, then obviously all of its vertices are mapped into the nodes lying on the same side s—otherwise one of the vertices of the corresponding parallelogram will lie outside of P. Assume that node X is black. We have proved that all four black nodes which are neighbors of X (via diagonal) are also mapped into nodes of side s. The same is true for any of these four neighboring nodes, provided it is also an interior node. Continuing this reasoning, we can see that all black nodes in S must be mapped inside segment s. But the number of the black nodes is at least $(n^2 - 1)/2 = (ab - 1)/2 \geqslant (3a - 1)/2 > a$, contradicting the requirement that different nodes of S must correspond to different nodes of R.[104]

Note. Without restriction $a, b > 2$, the statement is false. Indeed, let $n = 4$, $a = 2$, $b = 8$. Then the counterexample is shown below—the corresponding nodes are labeled by identical symbols (you will note that black nodes are marked by letters, white—by digits).

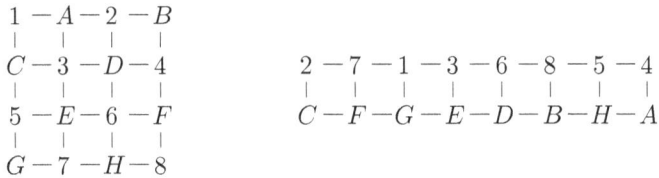

$$
\begin{array}{cccc}
1 - A - 2 - B \\
| \quad | \quad | \quad | \\
C - 3 - D - 4 \\
| \quad | \quad | \quad | \\
5 - E - 6 - F \\
| \quad | \quad | \quad | \\
G - 7 - H - 8
\end{array}
\qquad\qquad
\begin{array}{cccccccc}
2 - 7 - 1 - 3 - 6 - 8 - 5 - 4 \\
| \quad | \quad | \quad | \quad | \quad | \quad | \quad | \\
C - F - G - E - D - B - H - A
\end{array}
$$

1971.36. Induction on n. The basis is obvious. To prove the induction step from n to $n + 1$, in complete graph G_{2n+1}, select set X of n vertices which are connected by edges of, say, color # 1 (here we use induction hypothesis, which refers to finding such n-element set in complete graph G_{2n-1}, but it is, obviously, true for G_{2n+1} as well). Remove—temporarily—one of these vertices, and in the remaining G_{2n} subgraph, select one more n-tuple Y, whose vertices are also connected by edges of one color.

Case 1. This color is again # 1. If the two selected sets of vertices X and Y intersect, then their union is the required subgraph in G_{2n+1}. If they do not intersect, then the only remaining vertex A is connected with these $2n$ vertices by the edges colored only in # 2 and # 3. Then there are at least n edges of the same color, and the corresponding n vertices together with A form the required subgraph with $n + 1$ vertices.

Case 2. This color is # 2 (or # 3, but the proof is the same). Let us split all $2n + 1$ vertices into two subsets: the first one will contain the vertices from $X \cap Y$, and the second one—all the others. It is easy to see that, excluding one trivial subcase, each of these subsets is split into two parts so that the vertices from different parts can be connected only by edges of color # 3—otherwise we immediately have the required subgraph. One of these two subsets must contain at least $n + 1$ vertices connected by edges of color # 3.

1971.37. For any integer t, trio $x = 1 + 6t^3$, $y = 1 - 6t^3$, $z = -6t^2$ is a solution of the given equation.

1971.38. Let us attach another plane P to the boat and swim it through the river again, repeating the trip; as it swims, we will rotate and translate the plane P as it moves and turns together with the boat (you can imagine the plane as being rigidly nailed to the boat). Denote the complement of the river by X, and at every moment of the trip, remove from P its intersection with X. That part of P that remains attached to the boat at the end of the trip is the raft we are looking for.

---

[104] This solution was communicated to the authors by Sergey Berlov.

1972.01. Let us prove that the first plane will arrive first. Remove identical segments from both routes. The first route will then be reduced to twice the segments AC and BD, and the second route to twice the segments AB, BC, CD, and DA. From triangle inequality, we have $|AB| + |BC| \geqslant |AC|$, $|BC| + |CD| \geqslant |BD|$, $|CD| + |DA| \geqslant |AC|$, $|DA| + |AB| \geqslant |BD|$. The first and the third inequalities turn into equality only if points B and D lie inside line segment AC, and the second and the fourth—if A and C lie inside line segment BD; obviously, these two conditions cannot both be true at the same time. Adding up these inequalities, we obtain that the second route is always longer.

1972.02. Let a, b, and c be the numbers of the men who took part in **only** the first outing, **only** the second outing, and **both** the first and the second outings, respectively. Then the number of women in the first outing was two thirds of the number of men, that is, $\frac{2}{3}(a+c)$. The number of women in the second outing was one third of the number of men, that is, $\frac{1}{3}(b+c)$. Therefore, the number of women in the meeting cannot exceed $\frac{2}{3}a + \frac{1}{3}b + c$, which is less than or equal to total number of men $a + b + c$.

1972.03. One of such examples is the five vertices and the center of a regular pentagon.

1972.04. Answer: the last digit is even. It is easy to see that 5 goes into prime factorization of 100! with exponent of 24, and 2—with the exponent that is at least 25. Thus 100! is divisible by $2 \cdot 10^{24}$, but is not divisible by 10^{25}. This implies that 100! ends in 24 zeros with some even digit preceding those zeros.

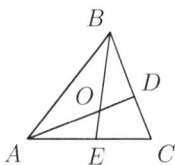

1972.05. Same as the solution to Question **1986.13**.

1972.06. That is not possible. If it were, and we added up all these vertex sums, the result S must be divisible by 8, since all the sums are the same. On the other hand, each number 1 through 12 in this sum of sums is counted exactly twice, and therefore $S = 12 \cdot 13 = 156$.

1972.07. For equality $S(AOB) = S(BOD) = S(AOE)$ to be true, it is necessary to have $|AO| = |OD|$ and $|BO| = |OE|$. But that would mean that O belongs to two midlines of triangle ABC, which is impossible.

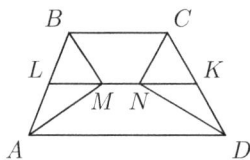

1972.09. Answer: $|MN| = \big||BC| + |AD| - |AB| - |CD|\big|/2$.

Let L and K be midpoints of sides AB and CD, respectively. Point M is equidistant from lines AD and AB, as well as from lines AB and BC. Hence, it is equidistant from BC and AD, and therefore lies on midline LK. Similarly, N lies on LK, and so $|NM| = \big||LK| - |ML| - |NK|\big|$. Since $\angle MBA + \angle MAB = 90°$, we have $\angle AMB = 90°$, thus, $|ML| = |AB|/2$. Similarly, $|NK| = |CD|/2$, and finally, $|LK| = (|BC| + |AD|)/2$.

1972.10. Let us index the tokens by numbers 1 through 12, and assume that such three tokens do not exist. So the tokens 1, 5, and 9 cannot be colored the same. Due to symmetry, we can assume, without loss of generality, that tokens 1 and 5 are black, and token 9 is white. Triangle formed by tokens 1, 3, and 5 is isosceles—therefore, token 3 is white. Both triangles formed by tokens 3, 6, 9 and 3, 9, 12 are isosceles, showing that tokens 6 and 12 must be black. Triangles formed by tokens 1, 11, 12 and 5, 6, 7 are isosceles, and so tokens 11 and 7 are white. Then white tokens 7, 9, and 11 form isosceles triangle, a contradiction.

1972.12. Denote our even number by $2n$. Since $2n = x^2 + y^2$, where x and y are integers of the same parity, the equality

$$n = \left(\frac{x+y}{2}\right)^2 + \left(\frac{x-y}{2}\right)^2$$

provides us with the desired representation.

1972.13. Since quadrilateral $AKCM$ is inscribed, $\angle KAM = 180° - \angle KCM = \angle MCD$. Then from $\angle MAB = \angle KAM$ it follows that $\angle MAB = \angle MCD$. These points could have a different order on the circumference, but the nearly identical reasoning will apply as well.

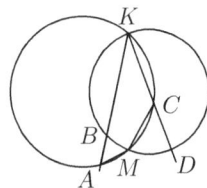

1972.14. Answer: yes, such a number exists.
Consider $n = 10^k - 9$. Then

$$n^2 = 10^{2k} - 18 \cdot 10^k + 81 = 10^k \cdot (10^k - 18) + 81 = 99\ldots98200\ldots081.$$

Sum of the digits is $9(k-2) + 8 + 2 + 8 + 1 = 9k + 1$, so for $k = 219$, the sum of the digits is exactly what we need.

1972.15. Let the diagonals of quadrilateral $ABCD$ intersect at point O. We can assume that $|AO| \leqslant |CO|$ and $|BO| \leqslant |DO|$. On segments CO and DO, find points A_1 and B_1 such that $|A_1O| = |AO|$ and $|B_1O| = |BO|$. Then ABA_1B_1 is a parallelogram and $P(COD) = P(AOB) = P(A_1OB_1)$ (P here denotes the perimeter function). On the other hand, triangle inequality implies $P(A_1OB_1) \leqslant P(A_1OD) \leqslant P(COD)$, and therefore, $P(A_1OB_1) = P(COD)$; that is, $A_1 = C$ and $B_1 = D$. This means that the original quadrilateral is also a parallelogram and it follows that it is, in fact, a rhombus.

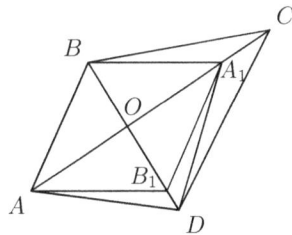

1972.16. Let us mentally draw the curves which connect precisely those circles which are **not** connected by the segments. On each curve, we will write a different prime number (since the time of Euclid, we know there are infinitely many of them). After that, label each circle with the product of numbers written on the curves connecting this circle with all the others.

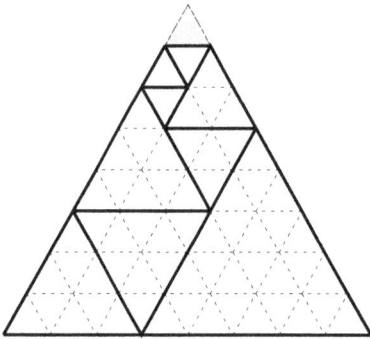

1972.17. Start with examining the dissection triangles that lie along the left side of the large triangle T (let us call them "left-side triangles"), and their vertices on that side (in the top-to-bottom direction). Construct the sequence $\alpha = (a_1, a_2, \ldots)$ of their distances from the top vertex of T. Obviously, $a_1 = 1$ and $a_1 < a_2 < a_3 < \cdots$. Similarly we define "right-side triangles" and sequence $\beta = (b_1, b_2, \ldots)$ for the dissection vertices on the right side of T.

In the example shown here we have $\alpha = (1, 2, 5, 8)$ and $\beta = (1, 3, 8)$.

Now, obviously, these numbers must satisfy some conditions so that the corresponding triangles do not overlap. Let us find out what these conditions are. Consider any left-side triangle A (whose left side's ends represent numbers a_i and a_{i+1}), and any right-side triangle B (whose right side's ends represent numbers b_j and b_{j+1}).

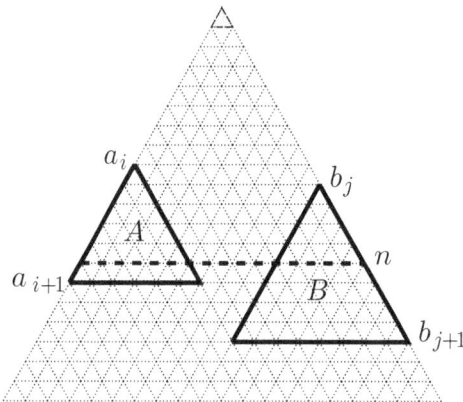

Take any natural number n which lies between a_i and a_{i+1} as well as between b_j and b_{j+1}, and draw the line parallel to the base of T which represents the lower boundary of the top n rows of T. This line intersects both A and B.

Since these two triangles cannot overlap, it follows that the sum of the lengths of the segments formed by intersection of this line with A and B cannot exceed n. Thus, we have $(n - a_i) + (n - b_j) \leqslant n$ or, equivalently, $a_i + b_j \geqslant n$.

From this we conclude that

$$a_i + b_j \geqslant \min(a_{i+1}, b_{j+1}). \qquad (*)$$

As a matter of fact, a somewhat stronger fact is true.

Lemma 1. *Form a monotonically non-decreasing sequence* $\gamma = (c_k)$ *by merging sequences* α *and* β. *Then* $c_{k+2} \leqslant c_{k+1} + c_k$ *for any* $k > 0$.

In the example shown above, we have $\gamma = (1, 1, 2, 3, 5, 8, 8)$ with $c_1 = 1$, $c_2 = 1$, $c_3 = 2$ and so on. Obviously, the last two terms are always the same, and one of them can be discarded without any effect on the validity of **Lemma 1**.

To prove this Lemma, we have to consider three cases.

1) Numbers c_k and c_{k+1} belong to different sequences (one—to α, the other—to β).
2) c_k and c_{k+1} belong to the same sequence, but c_{k+2} belongs to another (e.g., the first two belong to α and the third one—to β.
3) All three numbers c_k, c_{k+1} and c_{k+2} belong to the same sequence.

Case 1 is already proved—see inequality $(*)$ above.

In Case 2, for some indexes i and j, numbers c_k, c_{k+1}, and c_{k+2} are a_i, a_{i+1}, and b_{j+1}. Since they are the consecutive terms of γ, we have $b_{j+1} \geqslant a_{i+1} > a_i \geqslant b_j$. If we assume that the Lemma's statement is false—namely, that $b_{j+1} > a_i + a_{i+1}$—then we have $b_{j+1} - b_j \geqslant b_{j+1} - a_i > a_{i+1}$. That means that the next left-side triangle (with its top vertex at the lower left vertex of A) cannot protrude below the base of B, and therefore its size cannot be greater than b_j. From that, it follows that the next term of α does not exceed $a_{i+1} + b_j$. Hence,

$$a_{i+2} \leqslant a_{i+1} + b_j \leqslant a_{i+1} + a_i < b_{j+1},$$

which means that b_{j+1} cannot immediately follow a_{i+1} in γ.

Proof for Case 3 is nearly identical to Case 1. Assume that a_i, a_{i+1}, and a_{i+2} are three consecutive terms in γ, while for some index j we have $b_j \leqslant a_i$ and $a_{i+2} \leqslant b_{j+1}$. Then inequality proved in Case 1 for $n = a_{i+2}$ provides us with a contradiction, finalizing the proof of **Lemma 1**. $\square$

Using sequence γ, we can prove the following inequality for the number of triangles in a dissection.

Lemma 2. *If the size of triangle T is N, then the number of triangles in the dissection cannot be less than $p + q$, where p is the number of terms of sequence γ below N, and q is the number of different terms of γ below N.*

Indeed, let us assume that

$$\alpha = (a_1, a_2, \ldots, a_{m+1} = N),$$
$$\beta = (b_1, b_3, \ldots, b_{n+1} = N).$$

Let us discard the last terms of both α and β since they are equal to N and do not play any role in **Lemma 2** (we will, however, remember that the sum of the last two terms of γ obtained by merging these two truncated sequences has to be greater than or equal to N). Sequence γ has $m+n$ terms some of which could be equal (but obviously each number can occur no more than twice)—we will call them *duplicates*.

It is easy to see that we have exactly $p = m + n$ left- and right-side triangles. All of them are upward-oriented—their vertex is pointing up. Then there are m downward-oriented triangles whose left vertex coincides with one of the m top vertices of the left-side triangles, and n downward-oriented triangles whose right vertex coincides with one of the n top vertices of the right-side triangles (in the figure they are colored light gray). However,

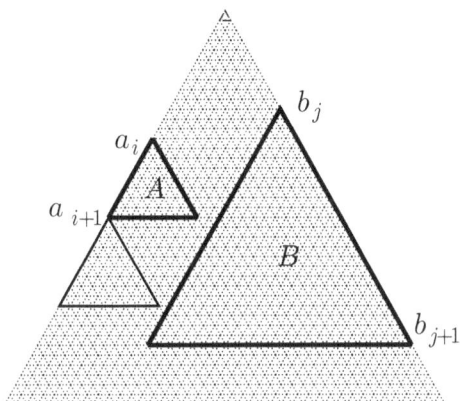

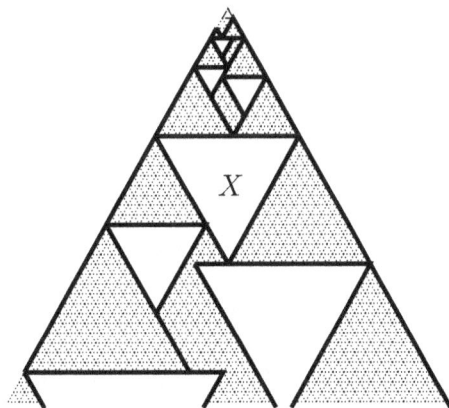

some of these triangles could be the same. Clearly, that can happen only if such a triangle has vertices on both sides of T (see triangle X in the figure)—these vertices correspond to two equal terms of sequences α and β, or, equivalently, a duplicate term of γ. If d is the number of duplicates in γ, then we have at least $m + n - d = p - d$ downward-oriented triangles. Since $q = p - d$, **Lemma 2** is proved. $\square$

Now our main result will be a corollary of **Lemma 2** if we prove inequality $p + q \geqslant 15$ for $N = 32$. From **Lemma 1**, sequence γ cannot grow faster than Fibonacci series

$$1, 1, 2, 3, 5, 8, 13, 21, 34, \ldots,$$

and therefore $p \geqslant 8$. Thus, we can split the remainder of the proof into three cases: $p = 8$, $p = 9$, and $p \geqslant 10$.

First, it is easy to see that $q \geqslant \frac{1}{2}p$ (and therefore $p + q \geqslant \frac{3}{2}p$). Indeed, if we have p numbers, no three of which are equal, then obviously there are at least $\frac{1}{2}p$ different numbers. It immediately follows that if $p \geqslant 10$, then $p + q \geqslant \frac{3}{2}p \geqslant 15$.

In case $p = 9$, we have $p + q \geqslant \frac{3}{2}p = 13\frac{1}{2}$, and thus $p + q \geqslant 14$. So the only chance for $p + q$ to be less than 15 is to have $q = 5$, and so γ must have four pairs of duplicates. If we compute eight ratios $r_k = c_{k+1}/c_k$ with $k = 1, \ldots, 8$, then four of these numbers are equal to 1, and no two of these 1's can be consecutive (otherwise γ would have more than two equal consecutive terms). This, with $r_1 = 1$, implies that sequence (r_k) must look like this: $1, r_2, 1, r_4, 1, r_6, 1, r_8$. Obviously, the terms of this sequence distinct from 1 cannot exceed 2, and, therefore, we have $c_8 \leqslant 8$ and $c_9 \leqslant 16$. Hence, the sum of the last two terms of γ does not exceed 24, and so it is less than 32, contradicting the **Lemma 1** inequality.

Finally, if $p = 8$, then in order for the sum $p + q$ to be 14 or less, q must be at most 6, implying that that γ has at least one more duplicate in addition to $c_1 = c_2 = 1$. It follows that for all $1 \leqslant k \leqslant 8$, we have $c_k \leqslant p_k$, where p_k is the sequence satisfying Fibonacci formula $p_{k+2} = p_{k+1} + p_k$ for every value of k except one index $\ell > 2$ such that $p_{\ell+1} = p_\ell$. First eight terms of the regular Fibonacci series are

$$1, 1, 2, 3, 5, 8, 13, 21,$$

forcing (p_k) to be one of the following sequences.

$$\ell = 7 : \ 1, 1, 2, 3, 5, 8, 13, 13;$$
$$\ell = 6 : \ 1, 1, 2, 3, 5, 8, 8, 16;$$
$$\ell = 5 : \ 1, 1, 2, 3, 5, 5, 10, 15;$$
$$\ell = 4 : \ 1, 1, 2, 3, 3, 6, 9, 15;$$
$$\ell = 3 : \ 1, 1, 2, 2, 4, 6, 10, 16.$$

In all of these cases, the sum of the last two terms of γ is at most 26, and therefore certainly less than 32, as desired.

Note. Two different examples of dissection with 15 triangles are shown below. They correspond to the following sequences.

$$\gamma = (1, 1, 2, 2, 4, 4, 8, 8, 16, 16), \quad p = 10, \ q = 5;$$
$$\gamma = (1, 1, 2, 2, 4, 4, 8, 12, 20), \quad\quad p = 9, \ q = 6.$$

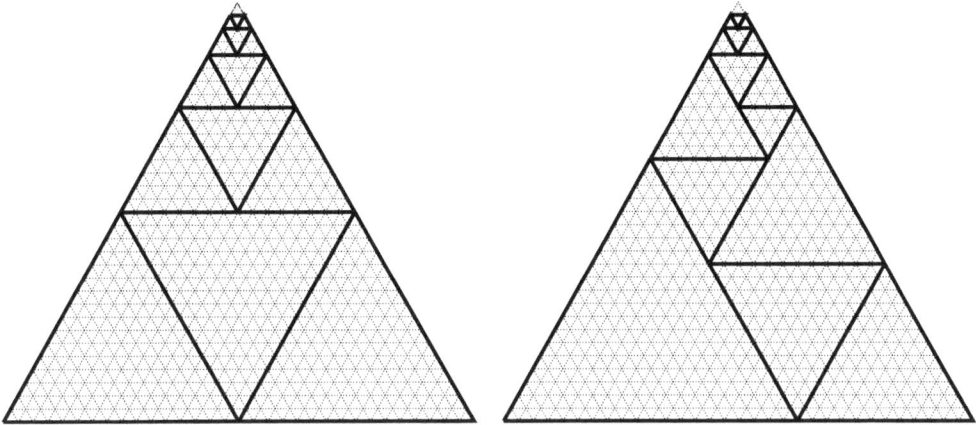

Exercise. Prove that there is no dissection into 15 triangles such that $p = 8$, $q = 7$. **Hint**: Such a dissection must have no "interior" triangles (that is, the triangles touching neither the left nor the right side of T).

Note. It follows from the proof above that when $n = \phi_k$ (kth Fibonacci number), the minimum number of dissection triangles equals $2k - 3$. However, it is not clear whether this lower bound is even asymptotically true for all values of n.

1972.18. Follow the following chain of equalities and inequalities.

$$b = b(kn - ml) = bkn - bml = bkn - anl + anl - bml$$
$$= n(bk - al) + l(an - bm) \geqslant n \cdot 1 + l \cdot 1 = n + l.$$

1972.20. Let X be the intersection point of lines tangent to the circle at vertices A and C. Then triangle ADX is similar to triangle BAX (since $\angle DAX = \angle ABX$ and angle X is common), and therefore, $|AD|/|AB| = |AX|/|BX|$. Similarly, $|DC|/|BC| = |CX|/|BX|$. It is left to notice that $|AX| = |CX|$, since these segments are tangents to the circle drawn from point X.

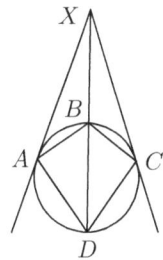

1972.22. Triangles A_1AA_2 and BAC_1 are isosceles; thus, $A_1A_2 \parallel BC_1$. Since $|AB| = |AC_1|$ and $|AA_2| = |AA_1|$, we have $|A_2B| = |A_1C_1|$, and therefore $A_1A_2BC_1$ is an isosceles trapezoid. Hence, points A_1, A_2, B, and C_1 lie on the same circumference.

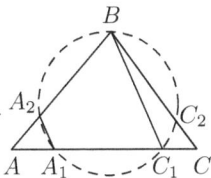

Similarly, points C_1, C_2, B, and A_1 lie on one circumference as well. It follows that all the points lie on the same circumference which incidentally also contains point B.

1972.23. We have $\sqrt[n]{m + 1} \leqslant 1 + m/n$, because $m + 1 \leqslant (1 + m/n)^n$ (use Newton's binomial theorem). Similarly, $\sqrt[m]{n + 1} \leqslant 1 + n/m$, and therefore

$$\frac{1}{\sqrt[n]{m + 1}} + \frac{1}{\sqrt[m]{n + 1}} \geqslant \frac{1}{1 + m/n} + \frac{1}{1 + n/m} = 1.$$

1972.24. Let n be the length of the original sequence. It is obvious that, starting from the third sequence, each number does not exceed the one under which it is written. Any non-increasing sequence of positive integers eventually becomes constant. So after some moment t_1, the first number in the sequence will always be the same; after some moment t_2, the second number in the sequence will always be the same, and so on. Thus, after the moment which is the maximum of all t_k $(k = 1, \ldots, n)$, all the numbers in the sequence will have "stabilized", and therefore the sequence must repeat.

1972.25. Since $b + c \geqslant 2bc$, we have $(b + c)(1 - a) \geqslant 2bc(1 - a)$, that is, $b + c + 2abc \geqslant ab + ac + 2bc$. Thus,

$$a + b + c + 2abc \geqslant a + ab + ac + 2bc = ab + bc + ca + (a + bc)$$
$$\geqslant ab + bc + ca + 2\sqrt{abc}.$$

1972.26. Let us build a broken line, where all the angles between any two consecutive segments are $90°$. If we already have such a broken line with n segments, where some three consecutive segments AB, BC, and CD are co-planar (lie inside the same plane), then we can build a broken line with $n + 2$ segments as follows. Remove segment BC and insert instead segments BX, XY, and YC such that BX and YC are perpendicular to plane ABC. Notice that in this new broken line segments BX, XY and YC are co-planar, and therefore this construction can be iterated as many times as we need. All that is left is to construct two broken lines of this type for $n = 6$ and $n = 7$—we will leave this exercise to the reader.

1972.27. If $p = 3k + 1$, then $4p^2 + 1 = (4k + 2)^2 + (4k + 1)^2 + (2k)^2$; if $p = 3k + 2$, then

$$4p^2 + 1 = (4k + 3)^2 + (4k + 2)^2 + (2k + 2)^2.$$

1972.29. The sides of this regular triangle ABC have length greater than 1. Select on side BC point M_a that is distinct from projections of the vertices of the heptagon onto BC and also relatively close to the midpoint of BC (namely, with distances from this point to B and C greater than $\frac{1}{2}$). Now, draw ray from point M_a perpendicular to BC and pointing outward relative to the triangle ABC. Then do the same for the other two sides of the triangle. These three rays split the outside of the triangle into three parts. By pigeonhole principle, one of these parts contains at least three vertices of the heptagon. Let it be part U_c, that is, the one adjacent to vertex C.

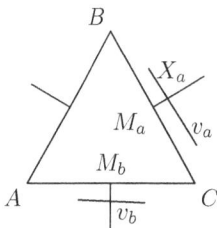

Let X_a be the intersection point of ray issued from M_a with the contour of the heptagon, and v_a be the edge of the heptagon which contains X_a. Triangle BCX_a lies inside the heptagon but outside the triangle; therefore, its area does not exceed 10^{-7}, and its altitude M_aX_a does not exceed $2 \cdot 10^{-7}$. It follows that the line containing v_a forms with line BC angle α not exceeding $4 \cdot 10^{-7}$ (radian). Indeed, if $\alpha > 4 \cdot 10^{-7}$, then line containing v_a intersects segment BC (distance from the intersection point to M_a equals $\frac{|M_aX_a|}{\tan \alpha} < \frac{2 \cdot 10^{-7}}{\alpha} < \frac{1}{2}$). Since the heptagon is convex, it lies to one side of this line, and therefore cannot contain both points B and C, a contradiction.

Similarly, edge v_b, intersecting the ray issued from M_b, forms with AC angle not exceeding $4 \cdot 10^{-7}$. Thus, the angle between oriented edges v_a and v_b differs from angle between $\overrightarrow{BC}$ and $\overrightarrow{CA}$—that is, from $\frac{2}{3}\pi$—by no more than $8 \cdot 10^{-7}$ (the edges are oriented in the direction defined by the order of the heptagon's vertices.)

But this angle equals the sum of the external angles of the heptagon at the vertices lying between edges v_a and v_b, that is, the ones inside area U_c. Since there are three or more of those angles, it follows that at least one of them does not exceed $\frac{1}{3}(\frac{2}{3}\pi + 8 \cdot 10^{-7}) = \frac{2}{9}\pi + \frac{8}{3} \cdot 10^{-7}$. Corresponding internal angle of the heptagon is, therefore, greater than or equal to

$$\frac{7\pi}{9} - \frac{8}{3} \cdot 10^{-7} > \frac{7\pi}{9} - \frac{\pi}{180} = 139° .$$

1972.30. Let's index the teams and denote them by $C_1, C_2, \ldots, C_n$.

Build bipartite graph G in which for every team C_i we have two vertices A_i and B_i, and vertices A_i and B_j are connected if and only if team C_i defeated team C_j. It is easy to see that for any $m \leq n$ and any set of m vertices $A' \subset \{A_1, \ldots, A_n\}$, there exist at least m vertices of the other part of G (that is, $B = \{B_i\}$) adjacent to at least one of the vertices of A'.

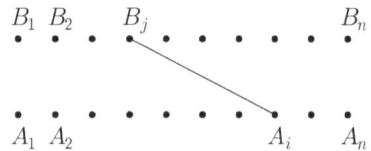

Thus, graph G satisfies conditions of Hall's Marriage Theorem (see Solution to **1967.35**), and there exists a perfect matching $A_1 \to B_{i_1}, \ldots, A_n \to B_{i_n}$. Therefore, in matches $C_1 - C_{i_1}, \ldots, C_n - C_{i_n}$ each team had exactly one win and exactly one defeat.

1972.31. Answer: $k = 201$.

Let us determine for which values of k the difference $\Delta_k = k^2(1.01)^{-k} - (k+1)^2 1.01^{-(k+1)}$ is positive and for which—negative. It is clear that $\Delta_k > 0$ only if

$$1.01 > \left(\frac{k+1}{k}\right)^2 = \left(1 + \frac{1}{k}\right)^2 = 1 + \frac{2}{k} + \frac{1}{k^2} .$$

Thus $\Delta_k > 0$ for $k \geq 201$ and $\Delta_k < 0$ for $k \leq 200$. Further, $a_1 < a_2 < \cdots < a_{201} > a_{202} > a_{203} > \cdots$, where $a_k = k^2(1.01)^{-k}$, and, therefore, a_{201} is indeed the largest number in sequence $\{a_k\}$.

1972.32. Let points X and Y be intersections of the given line with sides BC and AC of triangle ABC, respectively. Denote $|AB| = c$, $|AC| = b$, $|BC| = a$, $|XC| = x$, and $|YC| = y$, and assume that $a \geq b$. We need to prove that

$$\frac{1}{4}(a+b+c) \leq x+y \leq \frac{3}{4}(a+b+c) .$$

Since the given line divides the triangle's area in half, it follows that $xy = \frac{1}{2}ab$. From $x \leq a$, we have $y \geq \frac{1}{2}b$; similarly, $y \leq b$ implies $x \geq \frac{1}{2}a$. Therefore $x + y \geq \frac{1}{2}(a+b) \geq \frac{1}{4}(a+b+c)$. On the other hand, if the sum of two numbers x and y is fixed, their sum increases as the numbers move away from each other, and thus $x + y \leq \max(a + \frac{1}{2}b, b + \frac{1}{2}a) = a + \frac{1}{2}b$. It is left to prove inequality $a + \frac{1}{2}b \leq \frac{3}{4}(a+b+c)$, and it immediately follows from the triangle's inequality $a < b + c$.

1972.33. Let us denote number $999\ldots99$ (99 *nines*) by X. We need to show that there exists number n such that $X{\cdot}10^{100} \leqslant n^2 < X{\cdot}10^{100}+10^{100}$. Assuming otherwise implies that there exists natural number k such that $k^2 < X\cdot10^{100}$ and $(k+1)^2 \geqslant X\cdot10^{100}+10^{100}$. But then $(k+1)^2 - k^2 > 10^{100}$, that is, $2k+1 > 10^{100}$, and therefore, $k \geqslant \frac{1}{2}10^{100}$, contradicting the inequality $k^2 < X\cdot10^{100}$.

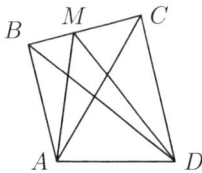

1972.34. Assume that all three radii are equal to r. Let $|BM|:|MC| = \alpha:\beta$, with $\alpha+\beta=1$. Then

$$S(AMD) = \alpha S(ACD) + \beta S(ABD),$$

since all three triangles share common base AD, and therefore their altitudes' ratios are the same as their areas'. After we divide that equality by $\frac{r}{2}$, we obtain $P(AMD) = \alpha P(ACD) + \beta P(ABD)$. Subtracting $|AD|$, we get

$$|AM| + |MD| = \alpha(|AC| + |CD|) + \beta(|AB| + |BD|).$$

On the other hand, $\overrightarrow{AM} = \alpha\overrightarrow{AC} + \beta\overrightarrow{AB}$, while vectors $\overrightarrow{AC}$ and $\overrightarrow{AB}$ are not parallel. Therefore, $|AM| < \alpha|AC| + \beta|AB|$, and similarly $|MD| < \alpha|CD| + \beta|BD|$; it follows that $|AM| + |MD| < \alpha(|AC| + |CD|) + \beta(|AB| + |BD|)$, a contradiction.

1972.35. Assuming the opposite, we obtain that there exists number N such that all the terms of the sequence greater than N are odd primes. Every such number must be equal to the sum of some two previous terms of the sequence, and therefore one of these two terms must be even, and consequently it cannot exceed N. Thus, after some index the difference between two consecutive terms of the sequence does not exceed N. But in the set of natural numbers, there are infinitely many intervals of N consecutive composite numbers—for instance, interval between $m(N+1)!+2$ and $m(N+1)!+(N+1)$, where m is any natural number (indeed, $m(N+1)!+k$ is divisible by k for any $2 \leqslant k \leqslant N+1$). Since our sequence cannot "jump" over such an interval, it cannot be a strictly monotonically increasing infinite sequence—a contradiction.

1972.36. Answer: 21.

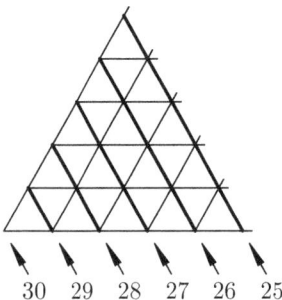

30   29   28   27   26   25

Indeed, select n such grid points, and then for every point determine three of its "coordinates": for each of the three sides of the triangle, index the lines parallel to that side starting at zero and then going in ascending order towards the opposite vertex of the triangle. Obviously, the sum of these coordinates is always 30. Since for any two selected grid points their first (as well as second or third) coordinates are different, their sum must be at least $0+1+2+\cdots+(n-1) = \frac{1}{2}n(n-1)$, and so the sum of all coordinates is at least $\frac{3}{2}n(n-1)$. On the other hand, this sum equals $30n$. Therefore, $30n \geqslant \frac{3}{2}n(n-1)$, or $n-1 \leqslant 20$. We will leave to the reader, as a relatively simple exercise, the task of finding an example with 21 grid points.

1972.37. For this proof, we will need one genuinely difficult lemma (so-called Menger's Theorem), which can be proved by induction.

Lemma. *Any two subway stations in Metropolis are connected by at least ten non-intersecting paths.*

Now consider two stations A and B such that you can get from A to B with no more than 99 transfers. Let $A - X_1 - X_2 - \cdots - X_{99} - B$, $A - Y_1 - Y_2 - \cdots - Y_{99} - B$, ..., $A - Z_1 - Z_2 - \cdots - Z_{99} - B$ be the 10 non-intersecting paths whose existence is guaranteed by the lemma above (that their lengths are exactly 100 is not important). Consider other 980 stations not covered by any of these paths; take any one of them and name it C. It cannot be connected directly to both X_1 and X_{99}; therefore, station C, with one of these two stations, say, X_1, forms a group $\{C, X_1\}$ in which the stations are not connected with each other.

Then take another one of the "not-yet-covered" stations; that station D cannot be connected to both X_2 and X_{99}. Thus, D belongs to group, say, $\{D, X_{99}\}$, where stations are not connected, and so on. In this way, we can form 96 such pair-groups. In the end there are three stations left on the first path—X_k, X_{k+1} and X_{k+2}; hence, we can add another pair $\{X_k, X_{k+2}\}$—this is group number 97. Doing that for every one of the ten paths, we construct 970 of these groups. Two more groups can be formed from A and X_{k+1}, and from B and $Y_{\ell+1}$ (the remaining station from the second path). Remaining $1972 - 2 \cdot 972 = 28$ stations can be considered as 28 separate groups. Thus, we have constructed 1000 groups—972 pairs and 28 "singles"—which comply with the requirements.

1972.38. Let us assume the opposite, and introduce some terminology. We will call an intersection a "source", if all of its streets point away from it; "sink", if it is impossible to leave it (that type was called a "dead-end" in the problem's statement), and a "pass", if it is possible to drive through it.

Two intersections of the same type cannot be located next to each other—that is obvious for the *sources* and the *sinks*; if two *passes* are located right next to each other, then there is a route at least 1500 meters long that can be driven in compliance with the traffic rules.

Now consider the figure shown here (we will call it a *rhomboid*), containing eight intersections. Analyzing all possibilities, it is easy to show that one of the intersections has to be a *sink*. That is true even if some part of the rhomboid lies outside the map—the only thing we require is that the center of the rhomboid belongs to the city map. Obviously, that center must be a midpoint of some street on the map.

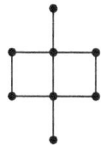

Each rhomboid is uniquely determined by its center and its orientation (horizontal or vertical). Let us draw contours of the squares centered at the map's center, with sides 100, 96,..., 12, 8, and 4 blocks long.

Consider the four sides of any such contour C. To every other midpoint (midpoint of a street, or, in other words, the middle of a block) M on contour C assign a rhomboid whose center is M and whose orientation is equal to that of the side of C containing M. The overall number of these rhomboids is $R = 200 + 192 + \cdots + 8$, so $R = 8(1 + 2 + 3 + \cdots + 25) = 4 \cdot 25 \cdot 26 = 2600$.

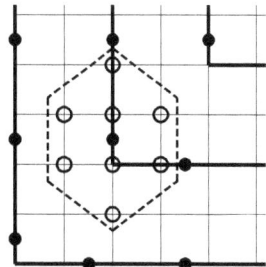

As we know, each rhomboid must contain a sink. Furthermore, it is not difficult to see that every street intersection belongs to no more than two rhomboids.

Therefore, the map must contain at least $\frac{R}{2} = 1300$ sinks.

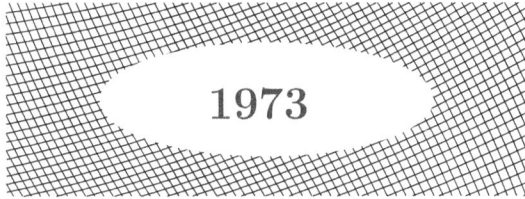

1973.01. Answer: originally the first store had 658 textbooks, the second—615, and the third—700.

The first number must be divisible by 47, by 2, and by 7, that is, by $2 \cdot 7 \cdot 47 = 658$. Similarly, the second number is divisible by 615, and the third—by 100. Thus, we have equation $658x + 615y + 100z = 1973$, and obviously, it has one and only solution in natural numbers $x = y = 1$, $z = 7$. That is quite easy to verify, since, clearly, x, $y \leqslant 3$.

1973.02. Answer: four breaks.

Consider one of the four central pieces and its sides. A break along each one of these sides is needed to separate this piece from all the others. It is obvious that one break cannot "serve" two of those sides—thus, four breaks is the minimum. We leave it to the reader to provide an example of how it can be done in exactly four breaks.

1973.03. Assume the opposite. Perfect square must end with an even number of zeros. Discard them and we will have a perfect square ending with 06 or 66. This is not possible since such a number has remainder 2 modulo 4, and a perfect square can only have remainders 0 and 1 (mod 4).

1973.04. An explicit example is shown in the figure. The large squares here have the side length of 492, and the side of the small squares is 1.

Another construction can be designed in steps. First, prove that each square can be dissected into 4, as well as into 6, smaller squares—these dissections add 3 and 5 squares respectively. Then start with any tessellation (say, into 4 squares), and then add the necessary number of 3's and 5's to increase the number of squares to 1973.

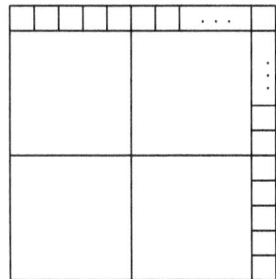

1973.05. Any number from the fifth column is either present in the third column and absent in the fourth (then it occurs even number of times in the first two columns) or absent in the third and present in the fourth (occurs exactly once in the first two columns). Therefore, the fifth column contains exactly the numbers occurring odd number of times in the first three columns. The contents of the seventh column can be computed the same way, and it turns out to be the same.

1973.06. An example is shown in the figure. We took a regular pentagon and then extended three of its sides to their corresponding intersection points.

1973.09. It would be sufficient to prove that $|PD|/|HE| = |MC|/|ED|$. Segment KH is the midline triangle BCP, $|KH| = |CP|/2$. If $|KH| \leqslant |HE|$ (the other case is similar), then point P lies on segment CD, and point M—on KC. Then

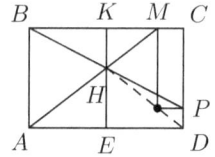

$$\frac{|PD|}{|HE|} = \frac{|CD| - |CP|}{|HE|} = \frac{|KE| - 2|KH|}{|HE|}$$

$$= \frac{|HE| - |KH|}{|HE|} = 1 - \frac{|KH|}{|HE|},$$

$$\frac{|MC|}{|ED|} = \frac{|KC| - |KM|}{|AE|} = 1 - \frac{|KM|}{|AE|} = 1 - \frac{|KH|}{|HE|}.$$

1973.10. Transform the given expression, factoring it as follows.
$$2^{10} + 5^{12} = (2^5 + 5^6)^2 - 2^6 5^6$$
$$= (2^5 + 5^6 - 2^3 5^3)(2^5 + 5^6 + 2^3 5^3) = 14657 \cdot 16657.$$

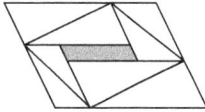

1973.11. The triangles outside of the original quadrilateral Q must cover the half of the area of the parallelogram. Now, consider the parallelograms that contain these triangles as their halves. Their total area must be equal to the area of the original parallelogram. Then the quadrilateral formed by their sides (in the figure it is shown shaded) has zero area. That is equivalent to one of the diagonals of Q containing one of the sides of the parallelograms. Therefore, that diagonal is parallel to a side of the large parallelogram.

1973.12. Factoring the expression $d = 2(a^8 + b^8 + c^8) - (a^4 + b^4 + c^4)^2$, we obtain
$$(a^2 + b^2 + c^2)(a^2 + b^2 - c^2)(a^2 - b^2 + c^2)(b^2 + c^2 - a^2).$$

Since we know that $d = 0$, one of the factors must be equal to zero. The rest follows from the Pythagoras' Theorem.

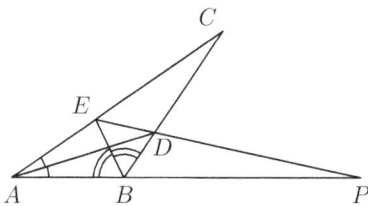

1973.13. Since
$$\frac{|AE|}{|EC|} = \frac{|AB|}{|BC|} > \frac{|AB|}{|AC|} = \frac{|BD|}{|DC|},$$
point E lies farther away from line AB than point D. Therefore, point $P = AB \cap DE$ lies on line AB, and so B is located between A and P. Hence, $\angle ABE > \angle BEP$; that is, $\angle DBE > \angle BED$, and it follows that $DE > BD$. Similarly, we have $\angle EDA > \angle PAD$, and therefore $\angle DAE < \angle EDA$, and $|DE| < |AE|$.

1973.14. Answer: the sum is 7 or 16.

It is clear that the given number is the sum of four powers of ten:
$$a = 10^{k_1} + 10^{k_2} + 10^{k_3} + 10^{k_4},$$

where k_i are non-negative integers. Therefore,

$$a^2 = a \cdot 10^{k_1} + a \cdot 10^{k_2} + a \cdot 10^{k_3} + a \cdot 10^{k_4}.$$

The sum of the digits of each of these four summands is 4, and therefore the sum of the digits of a^2 does not exceed 16. Since a has remainder 4 modulo 9, a^2 must have remainder 7. Hence, its sum of the digits equals 7 or 16. The first is attained when $a = 4 \cdot 10^9$, and the second—when $a = 3 \cdot 10^9 + 1$.

1973.15. Answer: no, they cannot.

Introduce the orthogonal coordinates so that the three given vertices have coordinates $(0,0)$, $(0,1)$, and $(1,0)$. Then operation $*$ applied to these vertices and their images and so on can only produce points with integer coordinates at least one of which is even.

Indeed, all three given points satisfy that condition. Let us prove that if points A and B satisfy the conditions above, then point $C = A * B$ satisfies them as well. Let $A = (x_a, y_a)$ and $B = (x_b, y_b)$. Then $C = (x_c, y_c)$, where $x_c = 2x_b - x_a$ and $y_c = 2y_b - y_a$. Obviously, numbers x_c and y_c are integers; also they have the same parity as x_a and y_a, respectively, proving our "lemma" and finalizing the proof, since point $(1,1)$ does not satisfy the parity conditions.

1973.16. Let us assume that there is no triangle, and that altogether we have n polygons. Then if all the numbers of sides are different, there must exist polygon P with at least $n + 3$ sides. At most three of those sides could lie on the boundary (the sides of the big triangle), and all the others must be common with the remaining $n - 1$ polygons. Then two of those sides of P must also belong to the same polygon Q (different from P), but two convex non-overlapping polygons P and Q cannot have more than one common side.

1973.17. If not all the numbers on the circle are the same, then after performing one operation, the sum of all numbers will decrease. That follows from an obvious fact that G.C.D. of two numbers cannot exceed either one of them. Since this sum cannot decrease indefinitely, eventually all numbers will become the same.

1973.18. Yes, it is always possible. First, mentally remove the maximum possible number of segments so that the connectivity still holds. After that, find a point with only one incident segment—the existence of such can be proved by considering the longest broken line formed by the segments.

Any end of this broken line will serve. Now, we can remove this point with all of its incident segments.

1973.19. We have that $S_{ABM} = S_{CBM}$ because diagonal BM of quadrilateral $ABCM$ divides diagonal AC in half. Then $|AB| \cdot |AM| = |BC| \cdot |CM|$, since $\angle BAM = \pi - \angle BCM$. This implies that $|AM|/|MC| = |BC|/|AB| = |CD|/|DA|$, and therefore triangles AMC and CDA are similar (**SAS** similarity rule). But they have common side AC opposite equal angles $\angle AMC = \angle CDA$, and therefore these triangles are congruent. Thus, $|AM| = |CD|$.

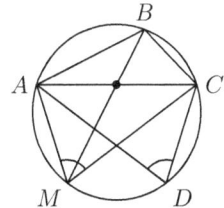

1973.20. Answer: the sum must be equal to 9, 18, or 27.

Clearly, this number can be written as $a = 10^{k_1} + 10^{k_2} + 10^{k_3}$ for some non-negative integers k_i. Then $a^2 = a \cdot 10^{k_1} + a \cdot 10^{k_2} + a \cdot 10^{k_3}$. The sum of the digits of each of these three summands is equal to 3, and therefore the sum of the digits of a^2 does not exceed 9. Further, $a^3 = a^2 \cdot 10^{k_1} + a^2 \cdot 10^{k_2} + a^2 \cdot 10^{k_3}$; the sum of the digits of each summand does not exceed 9, and therefore the sum S of the digits of a^3 does not exceed 27.

Now, since the sum of the digits of a is 3, this number is divisible by 3. Therefore, a^3 is divisible by 9, and so is S. Hence, S equals either 9, or 18 or 27.

All three possibilities can be realized. Namely, $a = 3 \cdot 10^8$, $a = 12 \cdot 10^7$, and $a = 111 \cdot 10^6$ produce sums $S = 9$, $S = 18$, and $S = 27$, respectively.

1973.21. Answer: take any points K and M so that KM is parallel to BC.

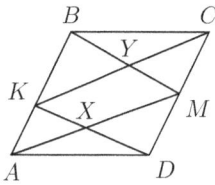

Let segments KD and AM intersect at X, and segments KC and BM—at Y. Set $k = |AK|/|DM| = |KX|/|XD| = |AX|/|XM|$. Then

$$k = \frac{|AK|}{|DM|} = \frac{|KX|}{|XD|} = \frac{|AX|}{|XM|}$$

implies that $S(AXD) = S(KXM)$, $S(AKX) = kS(KXM)$, $S(DXM) = S(KXM)/k$. Therefore,

$$S(AKMD) = \left(2 + k + \frac{1}{k}\right) S(KXM) \geqslant 4\,S(KXM),$$

and the equality can be attained only if $k = 1$, that is, when $KM \parallel AD$. Similarly, $S(KBCM) \geqslant 4\,S(KYM)$ with equality possible only if KM is parallel to BC. Adding these two inequalities, we obtain $S(ABCD) \geqslant 4\,S(KXMY)$. Thus, the maximum possible area of quadrilateral $KXMY$ is equal to $S(ABCD)/4$ and can be attained when $KM \parallel AD$.

1973.24. Answer: eleven arrangements.

Let us index the positions around the circle with numbers 0 through 29 in clockwise order.

Since we can only swap the tokens standing on the positions of equal parity, the number of white tokens on the odd-numbered positions does not change. Thus, we have at least eleven pairwise non-equivalent arrangements.

Let us prove that the equivalency class (the maximal set of pairwise equivalent arrangements) is ultimately defined by the number k of white tokens on the odd-numbered positions—let us call that number a *signature* of the arrangement; it is an integer between 0 and 10.

It would suffice to show that any arrangement can be transformed to have all white tokens on odd-numbered positions move to positions $1, 3, \ldots, 2k - 1$ (where k is the arrangement's signature).

During the operation, we swap tokens standing on positions n and $n + 4$ (modulo 30). Repeating that operation, we can move a token from position n to any position of the same parity. Suppose that one of the positions with indexes $1, 3, \ldots, 2k - 1$—say, position with index p—is not yet occupied by a white token.

Then one of the white tokens stands on an odd-numbered position with a different index q. Using swap $n \leftrightarrow n + 4$ for this token, we can move it to position p (which was, obviously, occupied by a black token before that). Now, applying $n \leftrightarrow n - 4$ swap for that black token, we can move it to position q. After these swaps, all the other tokens return to their original places. Thus, we can gradually move all the white tokens to the assigned positions.

The same, of course, can be done with $(10 - k)$ white tokens standing on the even-numbered positions. Thus, we have proved that any two arrangements with the signature are, in fact, equivalent—they can both be transformed to the same token arrangement.

1973.26. Let $A + kd$, $A + md$, $A + nd$ be any three consecutive terms of this geometric progression with quotient q. Then $(A + md)/(A + kd) = q$, and denoting A/d by t, we have

$$q = \frac{t + m}{t + k}, \qquad t = \frac{qk - m}{1 - q}.$$

But since $(qk - m)/(1 - q) = (qm - n)/(1 - q)$ (second expression is also equal to t), we obtain that $qk - m = qm - n$, and thus $q = (n - m)/(m - k)$ is rational, and so is number t.

1973.27. Each summand can be rewritten as

$$2^k \cos\left(\frac{\pi}{2^{k+2}}\right) - 2^k \cos^2\left(\frac{\pi}{2^{k+2}}\right) = 2^k \cos\left(\frac{\pi}{2^{k+2}}\right) - 2^{k-1} \cos\left(\frac{\pi}{2^{k+1}}\right) - 2^{k-1}.$$

Adding all these expressions for $k = 0, 1, \ldots, 8$, we obtain

$$256 \cos\left(\frac{\pi}{1024}\right) - \frac{1}{2} \cos\left(\frac{\pi}{2}\right) - \frac{1}{2} - 1 - 2 - \cdots - 128 = 256 \cos\left(\frac{\pi}{1024}\right) - 255\frac{1}{2} < \frac{1}{2}.$$

1973.28. If a polyhedron has n faces, then each face can have from 3 to $n - 1$ sides—no more than $n - 3$ possibilities. Thus, some two faces must have the same number of sides.

1973.29. We will prove that our polygon P is mapped onto itself when rotated around point O by angle α. The orbit of any vertex under multiple superpositions of a rotation is either the set of vertices of a regular polygon or infinite. If it is finite, then the number of the sides of P depends only on angle α. In our case, since 1973 is prime, there must be only one orbit, and therefore P is regular.

Now, let us draw two arbitrary rays starting at O and forming angle with measure α, intersecting the boundary of P at points A and B. Showing that $|OA| = |OB|$ would be then enough to prove our lemma—because it would imply that point A maps to point B when rotated by α around point O. Since the rays are arbitrary, we will have the entire polygon P mapping onto itself.

Assume the opposite; say, $|OA| > |OB|$. Rotate rays OA and OB by some angle φ. Let A' and B' be the points where the new rays intersect the boundary. If angle φ is small enough, then segments AA' and BB' lie on the sides of P. Subsequently, $S(OAA') = S(OBB')$, or $|OA| \cdot |OA'| \cdot \sin\varphi = |OB| \cdot |OB'| \cdot \sin\varphi$, and so it follows that $|OA| \cdot |OA'| = |OB| \cdot |OB'|$ and therefore, $|OB'| > |OA'|$. But for small values of φ, we will still have $|OA'| > |OB'|$, a contradiction.

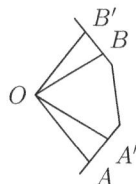

1973.31. Answer: the minimum value is $\frac{1}{4}$.

Arrange these numbers in descending order: $a \geqslant b \geqslant c \geqslant d \geqslant e$. Then, from problem's conditions, we have $4a + 2b - 2d - 4e = 1$, that is, $4(a+b+c+d+e) - 2b - 4c - 6d - 8e = 1$, and therefore, $a + b + c + d + e \geqslant \frac{1}{4}$. This value can be reached when $a = \frac{1}{4}$, $b = c = d = e = 0$.

1973.32. Consider two arbitrary vertices A and B of the convex hull[105] of the given point set. Then all the other points lie on one side of line AB. Denote them by $C_1, \ldots, C_{2k+1}$, listing them in descending order (by angle):

$$\angle AC_1B > \angle AC_2B > \cdots > \angle AC_{2k+1}B,$$

since we know that all these angles must be different. Then the circumference that passes through points A, B, and C_{k+1} encircles exactly k points: namely, points $C_1, \ldots, C_k$.

Note. It is interesting that regardless of the original configuration, this circumference can be selected in exactly $(n+1)^2$ ways.

1973.33. Since polynomial $P(x)$ has integer coefficients, $P(a) - P(b)$ is divisible by $a - b$ for any two integers a and b. Let $P(x_1) = 1$, $P(x_2) = 2$, $P(x_3) = 3$. Then $1 = P(x_2) - P(x_1)$ must be divisible by $x_2 - x_1$—hence, numbers x_1 and x_2 must differ by 1. Similarly, x_2 and x_3 differ by 1.

Suppose that $P(a) = P(b) = 5$ ($a \neq b$). Then $5 - 3 = 2 \vdots (a - x_3)$, that is, $|a - x_3| = 1$ or 2, and $5 - 2 = 3 \vdots (a - x_2)$, that is, $|a - x_2| = 1$ or 3. It is easy to see that these two conditions are not compatible.

1973.34. Answer: seven teams.

First of all, if there is a team that did not lose to any other team, then we can simply discard it and consider only the matches between all other teams. Then the condition of the problem is still true with one team less than before.

Thus, we can assume that each team has lost at least once. But then it follows from the condition that it lost to at least two teams. Furthermore, it has to be at least three teams. Indeed, take any team A, and suppose that there are exactly two teams, B_1 and B_2, it has lost to. Without loss of generality, we can assume that B_1 defeated B_2. Then the pair B_2 and A does not comply with the condition of the problem.

In exactly the same way, we prove that each team defeated at least three other teams. Therefore, there must be a minimum of seven teams.

Let us construct an example of a tournament with seven teams. Denote these teams by A_0, A_1, $\ldots$, A_6. Then for every index i, team A_i defeated three teams A_{i+1}, A_{i+2}, and A_{i+4} (if index exceeds 6, we reduce it modulo 7) and lost to the three others. It is quite easy to verify that (a) this definition is not self-contradictory, and (b) that the problem's condition is met.

1973.35. Denote our inscribed quadrilateral by $ABCD$. It can be split into several parts: the internal small square, and two pairs of opposite triangles outside of that square. Thus, the area of $ABCD$ equals the area of the square (that is, $\frac{1}{4}$) and the sum of the areas of these triangles (for every pair this sum equals $\frac{1}{8}$). Thus, the area of $ABCD$ is equal to $\frac{1}{2}$.

---

[105] The convex hull of a planar point set S is the smallest possible convex set that contains S. Also, it can be defined as the intersection of all half-planes containing set S.

Now, if we prove that one of the diagonals is parallel to the side of the unit square, then we are done. Assuming that AC is not parallel to the square's side, draw segment AC' parallel to that side. Obviously, $S(ABC'D) = \frac{1}{2}$ and therefore $S(ABC'D) = S(ABCD)$. Hence, $S(BCD) = S(BC'D)$, and so $BD \parallel CC'$, that is, diagonal BD is parallel to the side of the unit square.

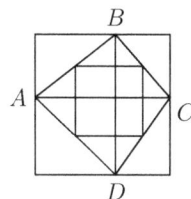

1973.36. Denote the lengths of the segments-intersections by $a_1, a_2, \ldots, a_{10}$. Then we have two inequalities

$$\frac{a_1 + a_2}{2} + \frac{a_2 + a_3}{2} + \cdots + \frac{a_9 + a_{10}}{2} \leqslant 9,$$

$$9\left(\frac{a_1 + a_{10}}{2}\right) \leqslant 9,$$

which follow from comparing the areas of the trapezoids defined by the nine lines with the area of the polygon. Adding these inequalities, we obtain $a_1 + a_2 + \cdots + a_{10} \leqslant 10$.

1973.37. Answer: 10.

Let us prove that nine questions are not enough to determine the number x even if it is written with ones and twos only. Without loss of generality, we can assume that the second player never asks a question that involves several (more than one) digits when some of them are already guaranteed (by the answers to the previous questions) to be identical—he can simply include only one of them in the question.

Now, imagine that in every answer the first player also divulges a bit of additional information; specifically, he names some two places (among those chosen for the question) with different digits. The rest of his answer is, obviously, "1 and 2". Then we can simply assume that the question was asked about those two places only. Indeed, from such an answer, no additional information can be logically deduced about digits in all the other places—they can be chosen absolutely arbitrarily, and the answer would still be the same.

After 9 questions of that form, we will have $k \leqslant 9$ places with the known digits (results of one-place questions) and $9 - k$ pairs of places (results of two-place questions)—it is known that in these pairs digits are different. Let us construct graph G with ten vertices, corresponding to the ten decimal places of our secret number, and $9-k$ edges, corresponding to those pairs. Also, mark k vertices that correspond to the decimal places with the known digits. In order to uniquely determine the secret number, each component of connectedness of G must have at least one marked vertex. Indeed, if there is a component without marked vertices, then all digits in it can be changed—ones to twos and vice versa—and the new number would fit the same set of answers. But if a component with n vertices contains a marked vertex, then there were at least n questions involving this component's places—at least one one-place question and at least $n - 1$ two-place questions (connected graph with n vertices must contain at least $n - 1$ edges). Therefore, at least 10 questions are required to "cover" all 10 vertices.

1973.38. Consider all vertices covered by the square, and denote their convex hull by F. If its area is S and its perimeter P, then $S \leqslant a^2$ and $P \leqslant 4a$. Say, x vertices lie on the boundary of F, and y vertices lie inside of it. Then $x + y = (\frac{x}{2} + y - 1) + \frac{x}{2} + 1$, and by the well-known Pick's formula, $S = \frac{x}{2} + y - 1$. Therefore, we have

$$x + y = S + \frac{x}{2} + 1 \leqslant S + 2a + 1 \leqslant (a + 1)^2 \,.$$

1973.39. See solution to Question **1980.40**.

1973.40. Let us call these figures F_k ($k = 1, \ldots, 1973$)—then consider their complements in the square. Sum of their areas equals

$$\left(1 - S(F_1)\right) + \left(1 - S(F_2)\right) + \cdots + \left(1 - S(F_{1973})\right)$$
$$= 1973 - \left(S(F_1) + S(F_2) + \cdots + S(F_{1973})\right),$$

and therefore this sum is less than 1. Hence, there must be a point in the square that does not belong to any of the complements. This point then, obviously, belongs to all figures F_k.

1974

1974.01. Answer: 134, 144, 150, 288, 294. Since

$$100a + 10b + c = 2(10a + b + 10b + c + 10a + c),$$

it follows that $20a = 4b + c$. Since $4b + c \leqslant 4 \cdot 9 + 9 = 45$, we have $a = 1$ or $a = 2$. The rest is just a quick direct examination of all possible cases.

1974.02. Answer: no, such a polygon does not exist.

If a convex polygon has n vertices, then there are $\frac{1}{2}n(n-3)$ diagonals. However, 1974 cannot be represented as $\frac{1}{2}n(n-3)$.

1974.03. Answer: no, that is not possible.

The sum of all numbers is equal to $(1+2+3)k = 6k$, where k is the number of triangles, and $55 \cdot 3 = 165$ is odd.

1974.04. Answer: yes, they could.

Take eleven triangles $1 - 2 - 3$ and eleven triangles $3 - 2 - 1$ (vertices are listed counterclockwise)—these twenty-two triangles will give us the sum of 44 along each edge. Then add one triangle of each of the three types $1 - 2 - 3$, $2 - 3 - 1$, and $3 - 1 - 2$ (they add sum of 6 along every edge of the prism).

1974.05. Answer: B came first, then A, while C was the third.

With every passing, one of the runners moves from the first (or from the third) place to the second, while the other runner moves from the second to the first (or the third). A was originally in the lead, and therefore after five passes, he has to be in the second place. Since B has finished ahead of A, he is the winner. And then C, of course, is the last one.

1974.06. Consider the set S of all the cities that can be reached from the capital (with stopovers, if necessary; the capital itself is included, of course). If Fartown is not in S, then for each city in S write down the number of air routes from this city and then add all these numbers. This sum equals $101 + 20n$, where n is the number of all cities in S except the capital, and so it is odd. On the other hand, each air route in this sum is counted exactly twice, and so this sum must be even. Therefore, Fartown must belong to S.

1974.07. Answer: $\frac{19}{10}$.

$$\frac{a}{b+c} + \frac{b}{a+c} + \frac{c}{a+b} = \frac{a+b+c}{b+c} + \frac{a+b+c}{a+c} + \frac{a+b+c}{a+b} - 3$$

$$= (a+b+c)\left(\frac{1}{a+b} + \frac{1}{b+c} + \frac{1}{a+c}\right) - 3 = 7 \cdot \frac{7}{10} - 3 = \frac{19}{10}.$$

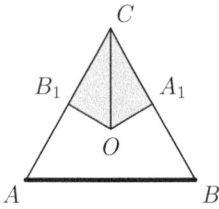

1974.08. Answer: this locus is the union of segment AB and quadrilateral OA_1CB_1, where A_1 and B_1 are the midpoints of sides BC and AC.

Let us prove that each point X from the quadrilateral above belongs to the locus. If, for instance, X belongs to triangle OCB_1, and we have line ℓ that passes through it without intersecting segment OC, then ℓ intersects sides CB_1 and B_1O of triangle OCB_1. Therefore, ℓ does not intersect BC because it does not intersect sides OC and OB of triangle OBC. And since line ℓ cannot intersect only one side AC of triangle ABC, it must intersect AB.

Now, we will prove that any point X belonging to neither quadrilateral OA_1CB_1 nor segment AB does not satisfy the conditions. If X lies outside of triangle ABC, then we can draw a line through it which has no common points with ABC. If X lies inside triangle OAB, then the line that passes through it and is parallel to AB, does not intersect segments AB and OC. And finally, the case when X lies inside triangle AOB_1. Draw line KY through some arbitrary interior point K of segment OB. It is easy to see that it does intersect neither of segments AB and OC.

1974.09. Answer: 1010, 1221, 1452, 1703, 1974.

Since $\overline{abcd} < 1000(a+1)$ and $\overline{ad} \cdot \overline{ada} \geqslant \overline{a0} \cdot \overline{a00} = 1000a^2$, we have $a + 1 > a^2$, that is, $a = 1$. Thus $\overline{ad} \cdot \overline{ada} < 2000$, and therefore $d \leqslant 4$. For $d = 0, 1, 2, 3, 4$ we can easily find all possible answers.

1974.10. See solution to Question **1968.31**.

1974.12. Answer: $36°$, $36°$, and $108°$.

Let $T = ABC$ be the given triangle ($AB = BC$). Denote by L the endpoint of the bisector of angle A, while point M is symmetric to B across line AC. Then $ABCM$ is a rhombus and $|BM| = |AL|$. Diagonals in trapezoid $ABLM$ are equal, and therefore it is isosceles with $\angle LBM = \angle LAM$. So we have $90° - \angle BAC = \frac{3}{2}\angle BAC$, and therefore, $\angle BAC = 36°$.

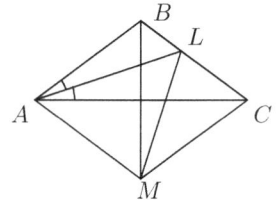

1974.13. Answer: yes, such a point exists. For example, see point M in the figure.

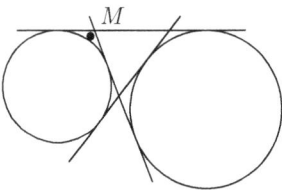

1974.14. Answer: $x = 2$, $y = 3$.

Note that $y^x < 19$, otherwise the right-hand side is negative. Cases where $x = 1$ or $y = 1$ can be checked directly. If $x, y \geqslant 2$, then we only have the following cases: $(x, y) = (2, 2), (3, 2), (4, 2), (2, 3), (2, 4)$. The right-hand side is always even—that rules out case $(3, 2)$. If x and y are even, then $y^{x^y} \vdots 4$ and $x^{x^{x^x}} \vdots 4$ while 74 is not divisible by 4. Hence, the last case $(3, 2)$ is the only answer.

1974.15. Answer: no, that is not possible.

Indeed, consider the sides of the triangles adjacent to the sides of the corner square.

Clearly, the length of one of them does not exceed 1. Since the altitude in the corresponding triangle, dropped onto that side, cannot be longer than 7, the area of that triangle does not exceed $7 \cdot \frac{1}{2} < \frac{63}{17}$.

1974.16. The lines perpendicular to diagonal $a1$–$h8$ are parallel to diagonal $a8$–$h1$. Let us index them by numbers 0 through 14 (index 0 is assigned to the one-square diagonal $a1$, index 14—to the one-square diagonal $h8$). Originally the difference between the indexes of diagonals is 14, at the end it equals (-14). With every move, this difference either decreased by 1 or did not change at all.

Obviously, at some point it was equal to zero. That is the moment when the pawns were standing on one diagonal perpendicular to $a1$–$h8$.

1974.17. Assume the opposite—that such set S exists. Consider triangle ABC with vertices in S possessing the maximum possible area. Then we have point D in S such that $ABDC$ is a parallelogram. It is obvious that all points of S lie inside triangle DEF (it is obtained by drawing lines parallel to the sides of ABC through the opposite vertices of this triangle). Now, recall that set S has more than four points. If some other point $X \in S$ lies inside AEC (other cases are similar), then XBC can be complemented (to a parallelogram) only by point Y, and YBD—by point Z (see figure). It is left to note that triple XYZ can be complemented only by a point that lies outside triangle DEF.

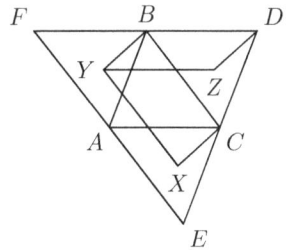

1974.18. Let X_1 be the number of all descendants of the first bacterium, that is, $X_1 = 1000$. One of its "daughters" produced at least $X_1/2$ of them—let us denote this number by X_2. Then we define $X_3 \geqslant X_2/2$ in the same manner, and so on. Consider the first index k such that $X_k \leqslant 667$. Clearly, $X_k \geqslant 334$, because $X_k \leqslant 333$ implies $X_{k-1} \leqslant 666$, contradicting the definition of k.

1974.22. Since circumference S and circumcenter of triangle ABC are homothetic with ratio $\frac{1}{2}$, then we can simply select the center of homothety as point T.

1974.23. Draw chord A_0B_0 perpendicular to XY and prove that angle A_0XB_0 is the one. Let Z be a point diametrically opposite to X, and AB—some other chord passing through Y. Without loss of generality, A lies on the smaller arc A_0Z, and B—on arc XB_0. Then $\angle AXB - \angle A_0XB_0 = \angle B_0XB - \angle A_0XA = \angle B_0AB - \angle A_0B_0A$. We need to prove that $\angle B_0AB > \angle A_0B_0A$. This is equivalent to $|YB_0| > |YA|$. Since angle OYA is obtuse, we have $|OA|^2 > |OY|^2 + |YA|^2$, or $|YA|^2 < |OA|^2 - |OY|^2 = |OB_0|^2 - |OY|^2 = |YB_0|^2$.

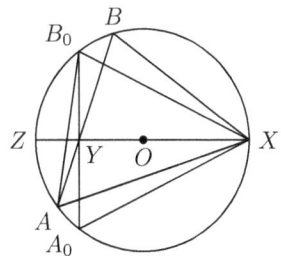

1974.24. Let A contain m letters and B—n letters; then C has $m + n$ letters. It follows from identity $CC = AABB$ that the first m letters of C coincide with A, and the last n—with B. Therefore, $C = AB$. Substituting AB instead of

C into the original identity, we get $ABAB = AABB$, and then (crossing out A on the left and B on the right), $BA = AB$.

It is sufficient now to prove the following lemma.

Lemma. *If words A and B satisfy equality $AB = BA$, then there exists word D such that both A and B can be obtained by writing D several times next to itself.*

We will do that by induction on $m + n$. Basis is trivial. Now, if $m = n$, then $A = B$, and we can simply take $D = A$. Suppose that $m > n$. Then from $AB = BA$ we obtain that the n-letter start of word A coincides with B, hence $A = BX$ for some word X. Rewriting equality $AB = BA$ as $BXB = BBX$ and crossing out B on the left, we get $XB = BX$. Induction hypothesis applied to B and X implies that these two words are "multiples" of some word D. But then A, equal to BX, also is a "multiple" of D.

1974.25. Answer: no, such a number does not exist.

Let $a = \frac{1}{3}(10^{10} - 1) = 33\ldots3$. It is easy to see that any 20-digit number that begins with eleven ones, is greater than a^2 but less than $(a+1)^2$. Indeed,

$$a^2 = \frac{1}{9}(10^{20} - 10^{11} + 10^{11} - 2 \cdot 10^{10} - 8 + 9)$$

$$= \frac{10^9 - 1}{9} \cdot 10^{11} + \frac{10^{10} - 1}{9} \cdot 8 + 1 = 11111111108888888889;$$

$$(a+1)^2 = \left(\frac{10^{10} + 2}{3}\right)^2$$

$$= \frac{10^{20} - 1 + 4 \cdot (10^{10} - 1)}{9} + 1 = 11111111115555555556.$$

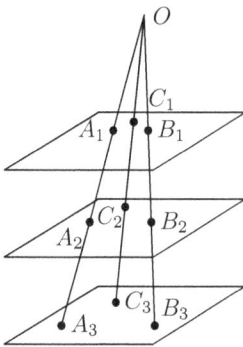

1974.26. Let $|OA_1| = a_1$, $|OB_1| = b_1$, $|OC_1| = c_1$, $|OA_2| = a_2$, and so on. We have

$$\frac{V(OA_1 B_2 C_3)}{V(OA_1 B_1 C_1)} = \frac{V(OA_1 B_2 C_3)}{V(OA_1 B_1 C_3)} \cdot \frac{V(OA_1 B_1 C_3)}{V(OA_1 B_1 C_1)}$$

$$= \frac{b_2}{b_1} \cdot \frac{c_3}{c_1} = \frac{a_2}{a_1} \cdot \frac{a_3}{a_1} = \frac{a_1 a_2 a_3}{a_1^3},$$

because pyramids $OA_1 B_2 C_3$ and $OA_1 B_1 C_3$ share common base $OA_1 C_3$ while their altitudes relate as $|OB_2| : |OB_1|$.

Doing the same for the other volumes we obtain

$$V(OA_1 B_1 C_1) : V(OA_2 B_2 C_2) : V(OA_3 B_3 C_3)$$

$$= (a_1 b_1 c_1) : (a_2 b_2 c_2) : (a_3 b_3 c_3) = a_1^3 : a_2^3 : a_3^3.$$

Therefore, all we need is to prove the inequality

$$a_1 a_2 a_3 \leqslant \frac{a_1^3 + a_2^3 + a_3^3}{3},$$

which is equivalent to the AM-GM inequality for three numbers.

1974.27. This expression can be rewritten as

$$1 - (1 - x_1)(1 - x_2)(1 - x_3)(1 - x_4).$$

Since $(1-x_1)(1-x_2)(1-x_3)(1-x_4)$ is a non-negative number that, obviously, can be equal to zero, maximum of our expression is 1.

1974.28. Assume the opposite. Project all the points onto a plane which is not perpendicular to any of the planes passing through all possible triples of the given points. Thus, the problem is reduced to its two-dimensional analog **1974.17**.

1974.29. Unfortunately, in the current formulation (the only one we have found in the archives), the claim is false.

Indeed, by the sine law, we have $|OA_{i+1}|/|OA_i| = \sin\alpha_{i-1}/\sin(\alpha_i + \alpha_{i-1})$, where α_i is the angle between rays R_i and R_{i+1} (index 13 is the same as index 1, 0—same as 12). Thus the statement of the problem is equivalent to the following inequality.

$$\prod \frac{\sin\alpha_i}{\sin(\alpha_i + \alpha_{i-1})} \leqslant \frac{1}{729}. \tag{$*$}$$

If all angles $\alpha_i = \pi/6$, then we obviously have equality here. But this point is not the point of maximum of the given expression—that can be checked using standard school calculus. Actually, it is easier to simply provide a counterexample.

Let $\alpha_1 = \alpha_2 = \cdots = \alpha_6 = \frac{\pi}{4}$, $\alpha_7 = \cdots = \alpha_{12} = \frac{\pi}{12}$. Well, the problem's condition requires that all angles are less than $\frac{\pi}{4}$, but it easily follows from continuity that if this set of angles violates the inequality, then the values very close to these will violate it as well.

Expression on the left-hand side of our inequality is then equal to

$$\frac{(\sin\frac{\pi}{4})^6(\sin\frac{\pi}{12})^6}{(\sin\frac{\pi}{2})^5(\sin\frac{\pi}{3})^2(\sin\frac{\pi}{6})^5} = 0.0016031572\ldots > 0.0013717421\ldots = \frac{1}{729}.$$

It is not entirely clear if this was simply an error made by the problem committee, or the problem was given with some additional restrictions disclosed during the contest.

Note. It is also quite easy to find a set of angles α_i (all below $\pi/4$) such that the product in ($*$) is, indeed, less than $\frac{1}{729}$. For instance, set

$$\alpha_1 = \frac{\pi}{5}, \quad \alpha_2 = \frac{\pi}{10}, \quad \alpha_3 = \alpha_4 = \cdots = \alpha_{12} = \frac{17}{100}\pi$$

produces 0.00125.

1974.30. We will prove by induction on x that this sum is always at least $x\log_2 x$, where x is the first number on the blackboard. Let us assume that the second row is the result of representation of x as the sum $a + b$. Then by induction hypothesis, the sum of the numbers written under a is at least $a\log_2 a$, and the sum under b is at least $b\log_2 b$. Thus, it would be enough to prove the inequality

$$a + b + a\log_2 a + b\log_2 b \geqslant (a + b)\log(a + b),$$

or, equivalently,

$$\frac{p\log_2 p + q\log_2 q}{2} \geqslant \left(\frac{p+q}{2}\right)\log_2\left(\frac{p+q}{2}\right)$$

where p and q are numbers $2a$ and $2b$. The last inequality immediately follows from the convexity of function $f(x) = x\log_2 x$.

1974.31. Draw a plane Π passing through satellites S_1, S_2, and the center of the planet, then find two points on the planet's surface which are the ends of the diameter perpendicular to Π. It is clear that S_1 and S_2 cannot be observed from either of those two points. Also none of the other 35 satellites can be seen from both points at the same time. Therefore, for at least one of these points, the number of satellites we can see from it is at most $[35/2] = 17$.

1974.32. Let us assume that there are some people who know each other. We will set aside everyone who knows zero people, and for the every other person in the group, we write down the number of their acquaintances to obtain sequence $s_1, \ldots, s_k$. Denote by A_i the set of all people in the group who have exactly s_i acquaintances ($i = 1, \ldots, k$). By condition, everyone is acquainted with no more than one person from A_i, and therefore cannot have more than k acquaintances. Thus, we have k different natural numbers $s_1, \ldots, s_k$, all not exceeding k, and therefore, they are numbers $1, 2, \ldots, k$ in some order. It follows then that someone has exactly one acquaintance.

1974.33. In a polygon with $2k$ sides, there are $k(2k - 3)$ diagonals. If each one of them is parallel to one of the sides, then there exists side s parallel to at least $\frac{1}{2}(2k - 3)$, and therefore to at least $k - 1$, diagonals. Then all $2k$ vertices of the polygon are the endpoints of s and those $k - 1$ diagonals. It follows that the diagonal with the largest distance from s must be a side as well, an obvious contradiction.

1974.34. First, we will make sure that every number in the first row equals 1. To do that, we find the smallest number x in the first row and double all the columns (say, k of them) that have x at their intersection with the first row. Let y be the smallest number in the first row after that; obviously, $y > x$. We can assume that $k < n$—otherwise all numbers were equal to x, and we could instead perform $x - 1$ subtraction operations on the first row to achieve our goal.

Now, we perform $y - 1$ subtraction operations on the first row. If originally the sum of the numbers in the first row was S, then now it is $S' = S + kx - n(y - 1)$, where n is the length of the table's rows. Since $x \leqslant y - 1$ and $k < n$, we have $S' < S$, while all the numbers in the row are still positive integers. Performing this procedure several times, we eventually arrive at the configuration where all the numbers in the first row are equal—otherwise the sum of the numbers in the first row decreases indefinitely while staying positive which is impossible—and thus (see above) we can achieve our goal.

Finally, when all numbers in the row are ones, we perform one more subtraction, and the entire first row is now filled with zeros.

After this, we do the same with the second row—note that the doubling operations will have no effect on the zeros in the first row, and it will not be subjected to the subtraction operations. Now both rows have only zeros in them, so we move on to the third row and so on until we have zeros in all squares.

1974.35. Answer: introduce coordinates in the square so that the origin is the common vertex of sides 1 and 2, with the opposite vertex having coordinates $(1, 1)$. Then there exists only one point A satisfying the requirements, namely, the one with coordinates $(\frac{1}{3}, \frac{1}{3})$.

First of all, note that the operations $A \mapsto A_2$, $A \mapsto A_4$ do not change the x-coordinate, and operations $A \mapsto A_1$, $A \mapsto A_3$ do not change the y-coordinate.

Therefore, we can calculate the coordinates of the point separately for x- and for y-coordinates. Namely, we can compute x-coordinate of images of A using points $A_{1313\ldots}$, then do the same for y-coordinate and points $A_{2424\ldots}$.

If x-coordinate of A is x, then x-coordinate of A_1 is $2x$, of A_3—$2x - 1$. Thus for $x = \frac{1}{3} + a$ we obtain that x-coordinate of A_1 equals $\frac{2}{3} + 2a$, and of A_{13}—$\frac{1}{3} + 4a$. Therefore, after performing these two reflections n times in a row, the x-coordinate of point $A_{1313...13}$ becomes equal to $\frac{1}{3} + 4^n a$. It is quite clear that if $a \neq 0$, then for some sufficiently large n, number $\frac{1}{3} + 4^n a$ lies outside interval $[0; 1]$. Thus $a = 0$, and x-coordinate of A is $\frac{1}{3}$. Similarly, we obtain that y-coordinate of A is $\frac{1}{3}$ as well.

1974.36. Assume the opposite. Let us remove one of the numbers and index all the others. Then the sums of form $x_1, x_1+x_2, \ldots, x_1+x_2+\cdots+x_{99}$ all have different non-zero remainders modulo 100—otherwise, when we subtract one of these sums from another, we obtain several numbers x_i, whose sum is divisible by 100.

The sum of all these remainders has remainder 50 modulo 100 ($4950 \equiv \pmod{100}$). On the other hand, this sum is congruent to $99x_1 + 98x_2 + \cdots + x_{99}$. Re-indexing these numbers with a cyclic shift we can similarly obtain that

$$99x_2 + 98x_3 + \cdots + x_1 \equiv 50 \pmod{100}.$$

Subtracting that from the previous congruence, we get

$$x_1 + x_2 + x_3 + \cdots + x_{99} - 99x_1 \equiv 0 \pmod{100},$$

that is, $x_1 \equiv -S \pmod{100}$, where S is the sum of all 99 numbers. Similarly, all other numbers x_k must be congruent to $(-S)$ modulo 100, and therefore, they are all congruent to each other, and thus equal, since they are all natural numbers less than 100. Thus, either all of them are equal to 1, and the number we removed is 101 which is not possible, or they are all equal to 2 and the sum of any fifty of them is 100.

1974.37. Answer: even numbers greater than 9, and odd numbers greater than 14.

If k is even, consider an arbitrary line that contains at least two sides AB and CD of the given polygon. Then there must exist at least four other lines containing the sides of the polygon—namely, the sides one of whose endpoints coincide with one of the points A, B, C, and D but not equal to AB or CD. That means we have at least five such lines and, therefore, k cannot be less than ten.

If k is odd, then one of the lines must contain at least three sides of the polygon. From this we similarly conclude that there are at least seven such lines, and so k is at least fourteen.

We leave to the reader the exciting task of constructing specific examples.

1974.38. If $p = 2$, then the statement is trivial—assuming the opposite, we obtain that for any two of the three numbers, one of them must be two times greater than the other one, and that is clearly impossible.

Let $p > 2$, and assume that this set of numbers does not have a common divisor greater than one; otherwise we can simply divide them all by that number.

Hence, one of these numbers is not divisible by p. If there are at least p of such numbers, then some two of them, a and b, are congruent modulo p. Denoting $\gcd(a,b)$ by d, we have that $(a - b) \vdots pd$ and $\frac{a}{d} \geqslant p + \frac{b}{d} \geqslant p + 1$, and therefore pair (a, b) satisfies the requirements.

Thus, we have no more than $p - 1$ numbers not divisible by p; consequently, there are at least two numbers which are divisible by p. Let us denote all such numbers in our set by $pa_1 < pa_2 < \cdots < pa_k$ ($2 \leqslant k \leqslant p$) and all others—by $b_1 > \cdots > b_{p-k+1}$.

If any of the numbers b_i is not divisible by one of the a_j, then we have $\frac{pa_j}{(pa_j,b_i)} > p$, and pair (pa_j, b_i) is the desired one. Thus, every b_i is divisible by each a_j, and therefore, by their least common multiple $A = [a_1, a_2, \ldots, a_k]$. Then

$$b_1 \geqslant (p - k + 1)A.$$

Also, $A \geqslant ka_1$. Indeed,

$$\frac{A}{a_1} > \frac{A}{a_2} > \cdots > \frac{A}{a_k},$$

and all these fractions are natural numbers; it follows then that the largest of them is greater than or equal to k. Furthermore,

$$b_1 \geqslant (p - k + 1)ka_1.$$

The smallest value of expression $(p - k + 1)k$ for $2 \leqslant k \leqslant p$ equals $2p - 2 \geqslant p + 1$, and therefore $b_1 \geqslant (p + 1)a_1 \geqslant (p + 1)(b_1, pa_1)$. Thus, pair (pa_1, b_1) fits the requirements.

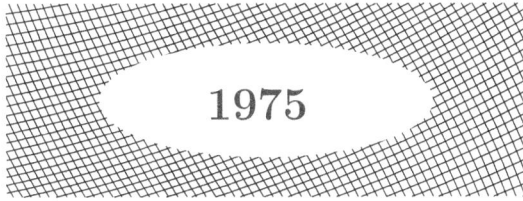

1975

1975.01. Vera can write numbers 12, 23, ..., 89, 91. Assume that none of them is the secret number. If Nick puts a minus next to $\overline{ab}$, then the secret number is $\overline{a0}$. If Nick puts minuses next to both $\overline{ab}$ and $\overline{cd}$, then the secret number is either $\overline{ad}$, or $\overline{cb}$; and by writing one of them, Vera will discover the secret number.

1975.02. If all domino halves are different (say, 0, 1, 2, 3), then the chain must contain seven zeros, seven ones, seven twos, and seven threes. Since inside the chain all numbers occur in pairs and seven is an odd number, one of the boundary numbers must be zero. Same is true for one, two, and three, which is clearly impossible since there are only two "boundary" numbers.

1975.03. Let us prove by induction on n that
$$\frac{1}{2} - \frac{3}{4} + \frac{5}{6} - \frac{7}{8} + \cdots + \frac{4n-3}{4n-2} - \frac{4n-1}{4n} = -\frac{1}{2}\left(\frac{1}{n+1} + \frac{1}{n+2} + \cdots + \frac{1}{2n-1} + \frac{1}{2n}\right).$$
Basis $n = 1$ is trivial. When substituting $n+1$ instead of n we add
$$\frac{4n+1}{4n+2} - \frac{4n+3}{4n+4} \quad \text{and} \quad \frac{1}{2}\left(\frac{1}{n+1} - \frac{1}{2n+1} - \frac{1}{2n+2}\right)$$
to the left-hand and the right-hand sides of the equality, respectively. Both these expressions are equal to $\left(-\dfrac{1}{4(2n+1)(n+1)}\right)$.

Note. Also this equality can be proved "directly":
$$\frac{1}{2} - \frac{3}{4} + \frac{5}{6} - \frac{7}{8} + \cdots + \frac{97}{98} - \frac{99}{100}$$
$$= \left(1 - \frac{1}{2}\right) - \left(1 - \frac{1}{4}\right) + \cdots + \left(1 - \frac{1}{98}\right) - \left(1 - \frac{1}{100}\right)$$
$$= -\frac{1}{2} + \frac{1}{4} - \cdots - \frac{1}{98} + \frac{1}{100}$$
$$= -\frac{1}{2}\left(1 - \frac{1}{2} + \cdots + \frac{1}{49} - \frac{1}{50}\right)$$
$$= -\frac{1}{2}\left(1 + \frac{1}{2} + \frac{1}{3} + \frac{1}{4} + \cdots + \frac{1}{49} + \frac{1}{50} - 2\left(\frac{1}{2} + \frac{1}{4} + \cdots + \frac{1}{50}\right)\right)$$
$$= -\frac{1}{2}\left(1 + \frac{1}{2} + \frac{1}{3} + \frac{1}{4} + \cdots + \frac{1}{49} + \frac{1}{50} - \left(1 + \frac{1}{2} + \cdots + \frac{1}{25}\right)\right)$$
$$= -\frac{1}{2}\left(\frac{1}{26} + \frac{1}{27} + \cdots + \frac{1}{49} + \frac{1}{50}\right).$$

1975.04. After reflection across some line ℓ intersecting the circle, point P is mapped to point P'. Then $\big||OP| - |OP'|\big| \leqslant 2$, where O is the center of the circle. Indeed, if O' is symmetric to O across ℓ, then $\big||OP| - |OP'|\big| = \big||OP| - |O'P|\big| \leqslant |OO'| \leqslant 2$.

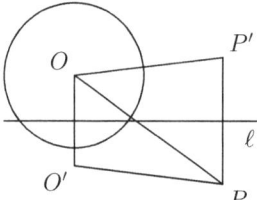

Thus, after five such reflections, the distance from a point to O can change by no more than 10, and therefore the final point cannot lie inside the circle.

1975.05. Answer: yes, he can.

Nick writes 1 as the first digit, and then with his every move, he complements each of Vera's digits (except for the last one) to 6 (that is, if Vera writes digit a, Nick writes digit $(6 - a)$). Then the sum of the digits at the end of the game will be $1 + 9 \cdot 6 + x = 55 + x$, where x is the last digit written by Nick. Since $1 \leqslant x \leqslant 5$, this sum cannot be divisible by 9.

1975.06. See solution to Question **1968.18**.

1975.07. Answer: no, he cannot.

Vera simply complements any digit written by Nick to 6, that is, if Nick writes digit a, then Vera writes digit $(6 - a)$. Then at the end, the sum of all digits will be equal to $15 \cdot 6 = 90$, and therefore the number will be divisible by 9.

1975.08. Answer: $p = 3$.

For any other odd prime number p, the second number $p^{p+1} + 2$ is divisible by 3.

1975.09. Answer: maximum value of $911 \cdot 19/9$ is attained at 911.

Suppose that the maximum value is attained for $x = \overline{abc}$. If $a \neq 9$, then replacing a by 9 changes the expression by

$$100 \cdot (9 - a)\left(\frac{1}{b} + \frac{1}{c}\right) - (10b + c)\left(\frac{1}{a} - \frac{1}{9}\right)$$

$$\geqslant 100 \cdot (9 - a) \cdot \frac{2}{9} - 99 \cdot \left(\frac{1}{a} - \frac{1}{9}\right) = (9 - a)\left(\frac{200}{9} - \frac{11}{a}\right) > 0,$$

and, therefore, increases it. Hence, $x = \overline{9bc}$. If $b \neq 1$, then replacing b by 1 changes the expression by

$$(900 + c)\left(1 - \frac{1}{b}\right) - \left(\frac{1}{9} + \frac{1}{c}\right)(10b - 10)$$

$$= (b - 1)\left(\frac{900 + c}{b} - 10\left(\frac{1}{9} + \frac{1}{c}\right)\right) > (b - 1)\left(\frac{900}{9} - 20\right) > 0.$$

Thus, $b = 1$ and $x = \overline{91c}$, and the expression turns into

$$1012\frac{1}{9} + \frac{910}{c} + \frac{10}{9}c.$$

Maximum value of this expression is attained for $c = 1$ because

$$910 + \frac{10}{9} - \frac{910}{c} - \frac{10}{9}c = (c - 1)\left(\frac{910}{c} - \frac{10}{9}\right) \geqslant (c - 1)\left(\frac{910}{9} - \frac{10}{9}\right) > 0.$$

1975.10. Place unit weights into all vertices of the hexagon. Now we can compute the center of mass H for this system in two ways.

First, splitting the weight into three pairs along sides BC, AF, DE, we have masses of 2 in points M_8, M_5, and M_6. Center of mass of $\{M_5, M_6\}$ is M_7 (with mass 4), and therefore, point H lies on segment $M_7 M_8$.

Similarly, start with the split CD, AB, FE to obtain masses of 2 in points M_1, M_2, M_4. Center of mass of $\{M_1, M_2\}$ is M_3, and therefore, H lies on segment $M_3 M_4$.

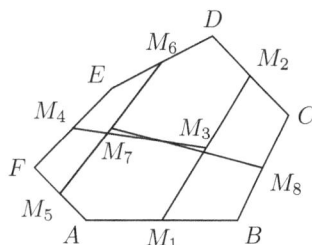

1975.11. See solution to Question **1968.18.**

1975.12. Let O be the center of the given circumference S. We can assume that point A, one of the ends of the chord, is fixed. Let $ABCD$ be the square build on chord AB. Then point D belongs to circumference S' obtained from S by 90° rotation around point A. If O' is the center of S', then lengths $|OO'|$ and $|O'D|$ are known and fixed, and therefore distance OD is largest when D lies on the extension of segment OO' beyond point O'.

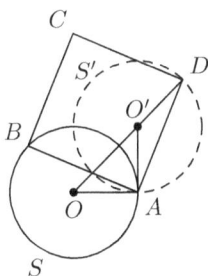

Construction: let A be an arbitrary point on S. Construct point O' such that $\angle OAO' = 90°$, $|AO'| = 1$. On ray OO', beyond point O', find point D such that $|O'D| = 1$. Using segment AD as one of the sides, construct a square which lies in the same half-plane that contains point O. (It is easy to see that this is equivalent to choosing chord AB so that it subtends a 135° angle.)

1975.13. Answer: $3 : 3 : 4$.

Let ABC be the given isosceles triangle with $|AB| = |BC|$, where altitudes BK and CL intersect at orthocenter H, and I is the incenter.

By condition, I is the midpoint of KH, K is the midpoint of AC. Further, F is the point where the incircle touches side AB; E is the projection of K onto AB. From the intercept theorem, we have $|EF| = |FL| = x$ since I is the midpoint of HK. Also, $2x = |EL| = |EA|$ since K is the midpoint of AC. From $|AK| = |AF| = 3x$, we obtain $|BA|/|AK| = |AC|/|AL| = \frac{6x}{4x} = \frac{3}{2}$. Therefore, $|AB|/|AC| = \frac{1}{2}|AB|/|AK| = \frac{3}{4}$.

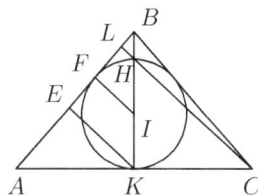

1975.14. Since there are infinitely many prime numbers, one of the progressions has to contain at least two different primes. Therefore, this progression's ratio is not an integer. It is, of course, some rational number $q = u/v$, with co-prime u and v. Consider some positive integer k such that a_1 is not divisible by v^k. Then number $a_{k+1} = u^k a_1/v^k$ cannot be an integer.

1975.15. It follows from the question's conditions that $\angle B = 60°$. Therefore, the radius of circumference $(BDFE)$ is at least $|BF|/2$, and that is the same as $|BF| \sin \angle FBE$ (since $\angle FBE = 30°$). The latter number is nothing else but the length of the perpendicular dropped from incenter F onto side AB—in other words, the inradius of the triangle.

1975.16. Add both equations to obtain $x^2 + y^2 = 81$. Coordinates of the intersection points must satisfy this equation and therefore they lie on the circumference defined by it.

1975.17. Answer: 2998.

Denote $[\sqrt{N}]$ by k. Then $k^2 \leqslant N \leqslant k^2 + 2k$ and, if $N \vdots k$, we have the following options for the value of N:

1) $N = k^2$ $(k = 1, 2, \ldots, 1000)$;
2) $N = k(k+1)$ $(k = 1, 2, \ldots, 999)$;
3) $N = k(k+2)$ $(k = 1, 2, \ldots, 999)$.

Obviously, every N belongs to just one of these sets, since N uniquely defines both k and the remainder of N/k modulo k.

1975.18. Answer: there are seven such arrangements.

Index the vertices of the polygon with numbers 1 through 100 in such a way that there are exactly 11 sides between each two vertices with indexes n and $n + 1$. This is possible because numbers 11 and 100 are co-prime. Then every move results in a token moving to the vertex with the next index (or from vertex #100 to vertex #1). Therefore, the order of the tokens along this 100-vertex cycle cannot change, and so there are exactly seven ways to arrange the seven tokens in seven fixed vertices (these arrangements differ by a cyclic shift by 1, 2, ..., or 6).

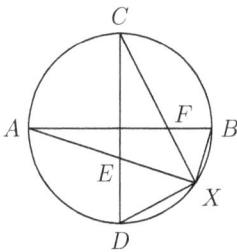

1975.19. Since XE is the bisector of angle CXD, then $|CE|/|ED| = |CX|/|DX| = \cot\alpha$, where $\alpha = \angle DCX$. Similarly, $|AF|/|FB| = |AX|/|XB| = \cot\beta = \cot(45° - \alpha)$, where $\beta = \angle XAB$. If we denote the first ratio by t, then the second one equals $(1+t)/(1-t)$, completing the proof.

1975.21. Answer: the original triangle is the one with the maximum ratio.

For any triangle, the ratio of the area to the perimeter equals half of the inradius. Thus it will suffice to show that the radius of any circle C that lies entirely inside the given triangle cannot exceed the triangle's inradius r.

Split the given triangle into three smaller triangles, connecting the center of C with the vertices. The area of each one of them is at least half of the circle's radius r' multiplied by the length of the corresponding side, because the distance from center of C to any side is at least r'. Adding these inequalities and dividing by p, we obtain $r' \leqslant S/p = r$, as desired (S and p are the original triangle's area and half-perimeter, respectively).

1975.23. Answer: 75 obtuse angles.

Let us consider three fixed rays forming angles of 120° with each other. Now, for each of them, take five rays whose directions are very close to it—altogether we have $3 \cdot 5 = 15$ rays.

To prove that one cannot have more than 75 obtuse angles, we will prove by induction that $3n$ rays on the plane cannot form more than $3n^2$ obtuse angles.

The basis $(n = 1)$ is quite obvious. For the induction step from n to $n + 1$, assume that we have some collection C of $3(n + 1)$ rays. It is clear that among any four rays you can always find two, forming a non-obtuse angle. Let us find the largest subcollection $S \in C$ of rays such that all pairwise angles between them are obtuse. Obviously, S contains two or three rays. If $|S| = 2$, add one more ray in an arbitrary manner.

Any one of the remaining $3n$ rays forms at most two obtuse angles with the rays in S, and so there are no more than $6n$ obtuse angles between rays in S and in $C \setminus S$. By induction hypothesis, there are no more than $3n^2$ obtuse angles between rays in $C \setminus S$, and no more than three obtuse angles inside S, so the total number of the obtuse angles for C does not exceed $3n^2 + 6n + 3 = 3(n + 1)^2$.

Note. In fact, we have proved the following fact. *Any graph with $3n$ vertices, which does not contain complete four-vertex subgraphs, has no more than $3n^2$ edges.*

1975.24. Answer: 198.

For every marked square, highlight the line (column or row) where this square is the only marked one. Then not all highlighted lines are columns lest the number of marked squares does not exceed 100. Similarly, not all highlighted lines are rows. Therefore, no more than 198 lines are highlighted.

An easy example of configuration with exactly 198 marked squares (or 198 highlighted lines) is the one where all the squares in some row are marked together with all the squares of some column, with the exception of the square at their intersection.

1975.25. See solution to Question **1975.14**.

1975.27. Answer: $x = 4 + 2\sqrt{3} \pm \sqrt{34 + 20\sqrt{3}}$.

Rewrite equation as

$$(x^2 - x + 1) + (x + 1) = 4\sqrt{(x + 1)(x^2 - x + 1)},$$

then divide it by $x^2 - x + 1$.

Denoting $\sqrt{\frac{x+1}{x^2-x+1}}$ by t gives $t^2 - 4t + 1 = 0$, and it follows that $t = 2 \pm \sqrt{3}$. Reversing the substitution and solving corresponding quadratic equations we get the answer presented above.

1975.28. Let O, O_1, O_2, and O_3 be the centers of the four balls B, B_1, B_2, and B_3. We will prove that B is not covered by $\mathcal{B} = B_1 \cup B_2 \cup B_3$.

Draw plane α through point O, parallel to plane passing through points O_1, O_2, and O_3. Clearly, points O_1, O_2, and O_3 all lie in plane α, or to one side of α.

Now, find point M so that vector $\overrightarrow{OM}$ has length R (R is the radius of B and B_k), is perpendicular to α, and it points to the half-space not containing points O_1, O_2, O_3. For each $k = 1$, 2, 3, we have $\angle MOO_k \geqslant 90°$, and therefore in triangle MOO_k, side MO_k is the longest. In particular, $|MO_i| > |MO| = R$. Hence, point M does not belong to any of the balls B_k. But it belongs to B, and so ball B is not covered by the union of three balls $\mathcal{B}$.

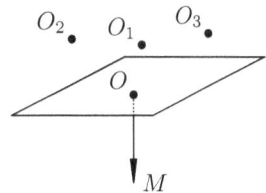

1975.30. Let α_i be the angle between the ith and the $(i+1)$th sides of the polygon, and a_i be the length of the ith side. The area of the triangle, formed by the three consecutive vertices, is equal to $a_i a_{i+1} \sin \alpha_i / 2$. Product of these areas is

$$\frac{1}{2^n}(a_1 a_2 \ldots a_n)^2 \sin \alpha_1 \sin \alpha_2 \ldots \sin \alpha_n \leqslant \frac{1}{2^n}(a_1 a_2 \ldots a_n)^2.$$

From the AM-GM inequality, we have

$$a_1 a_2 \ldots a_n \leqslant \left(\frac{1}{n}(a_1 + a_2 + \cdots + a_n)\right)^n \leqslant \left(\frac{4}{n}\right)^n,$$

since $a_1 + a_2 + \cdots + a_n$ is the perimeter of the polygon. This perimeter does not exceed the perimeter of the square, that is, 4. Thus, the product of the areas does not exceed $\left(\frac{1}{2}\right)^n \left(\frac{4}{n}\right)^{2n}$. Hence, one of the areas does not exceed $\frac{1}{2} \cdot \left(\frac{4}{n}\right)^2 = \frac{8}{n^2}$.

1975.31. Let us find angles α and β such that $a = \cos\alpha$, $b = \sin\alpha$, $c = \cos\beta$, $d = \sin\beta$. Existence of such angles immediately follows from the first two equalities. Then equality $ac + bd = 0$ implies that $\cos(\alpha - \beta) = 0$, and therefore

$$ab + cd = \frac{1}{2}(\sin 2\alpha + \sin 2\beta) = \sin(\alpha + \beta)\cos(\alpha - \beta) = 0.$$

1975.32. Answer: the second number.

We can generalize that inequality and prove it by induction. Let us replace 99 by $n > 1$ and 100 by $n + 1$. Then the basis for $n = 2$ is obvious since $2^4 = 16 < 27$. The induction step is trivial as well—it follows from the fact that if $0 < x < y$, then $2^x < 3^y$.

1975.33. Answer: two of the numbers a, b, and c are equal to 1 while the third one equals $(p - 1)$, where p is an arbitrary prime number.

Obviously, any such triple is a solution. Now, if there are two (or more) even numbers among a, b, and c, then one of the numbers $a^b + c$, $b^c + a$, and $c^a + b$ is even and greater than 2—thus it cannot be a prime. Otherwise we have two odd numbers, say, a and b. Then $b^c + a$ is even and must be equal to 2, which is possible only if $a = b = 1$. Denoting $a^b + c$ by p, we obtain $c = p - 1$.

1975.34. Answer: yes, that is possible.

Let us start with points A_1, A_2, and A_3 lying on a unit circumference with center O in the given order (clockwise) and such that angles $\angle A_1 O A_2$ and $\angle A_2 O A_3$ are both equal to $20°$. Obviously, since the measure of the arc connecting any two of the points is less than $60°$, the pairwise distances between these points are all less than 1.

Reflecting point A_1 across the perpendicular bisector of $A_2 A_3$, we will get point A_4 which lies on the same circumference beyond point A_3; obviously, angle $\angle A_3 O A_4$ is once again $20°$. Then, reflecting A_2 across the perpendicular bisector of $A_3 A_4$, we obtain next point A_5, and so on. Obviously, point A_{10} will be diametrically opposite to A_1, since $\angle A_1 O A_{10} = 9 \cdot 20° = 180°$, and therefore the distance between them equals 2.

1975.35. Answer: $k = 15$.

Indeed, consider a complete graph[106] with 30 vertices. We direct edge AB as $\overrightarrow{AB}$ (marking it with an arrow from A to B) if and only if A likes B. The overall number of

---

[106] That is, a graph where every two vertices are connected with an edge.

arrows is $30 \cdot 15 = 450$, but the number of edges equals $30 \cdot 29/2 = 435$. Therefore, some edge is marked with two opposite arrows, and that is exactly what we are looking for.

A counterexample for any $k < 15$ can be constructed like this. Arrange all the people around the circle and then make it such that each one of them likes exactly the k people following him or her in the clockwise direction.

1975.36. Let us assume the opposite, that no four vertices form such a trapezoid. Consider the circumcircle of our regular polygon. We have 35 equal arcs into which the vertices divide its circumference, and each of these arcs is assigned the length of 1. The distances between the vertices will be measured along the circumference; then, obviously, any distance between two different vertices is a natural number not exceeding 17. Since the number of pairs of different vertices of an octagon is $8 \cdot 7/2 = 28$, some two of the distances between them are equal. Denoting these two pairs by $A_1 A_2$ and $B_1 B_2$, we get only two possibilities. If all these four points are different, then $A_1 A_2 B_1 B_2$ is the required trapezoid. If some two of them coincide—say, $A_1 = B_1$—we have isosceles triangle $A_1 A_2 B_2$.

Could more than two pairs have the same distance? The answer is no. If three pairs $A_1 A_2$, $B_1 B_2$, and $C_1 C_2$ are such that $|A_1 A_2| = |B_1 B_2| = |C_1 C_2| = d$ and in every two pairs, some two points coincide (remember, no trapezoids!), then we either have $A_i = B_j = C_k = X$ or we have these pairs corresponding to the sides of a regular triangle. The former is impossible since there are only two points at distance d from point X, while in each of the three pairs distance between X and one of the points in the pair is equal to d. The latter is impossible because 3 does not divide 35—in a regular polygon with 35 sides one cannot choose three vertices forming a regular triangle.

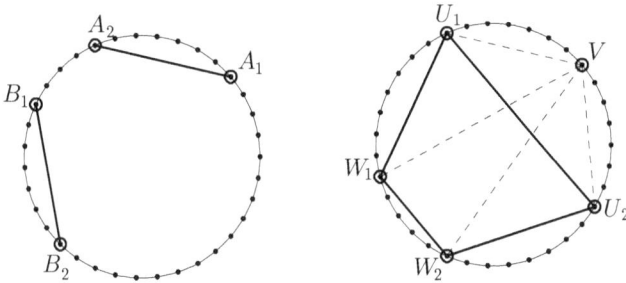

Thus, we can mentally remove the two pairs with the same distance from the list of pairs, and remove distance d from the list of distances. Therefore, we have 26 pairs and 16 distances, and the same reasoning can be applied, producing another two pairs of vertices at the same distance not equal to d. Some two of the vertices again must coincide, and we obtain another isosceles triangle. And so on, until we have 11 such isosceles triangles. In every one of these triangles, mark the "core" vertex—namely, the one with equal adjacent sides. Since $11 > 8$, some two of the "core" vertices must coincide, and so we have vertex V and two isosceles triangles $V U_1 U_2$ and $V W_1 W_2$ with different bases $U_1 U_2$ and $W_1 W_2$. Then $U_1 U_2 W_1 W_2$ is clearly the required trapezoid.

1975.37. For any two cities P and Q, denote by S_{PQ} the shortest path between them before the road closing.

Suppose that the closed road connected cities A and B. Consider any two cities X and Y, and then prove that there exists a path that (a) connects them; (b) does not contain road AB; (c) has length not exceeding 1500 km.

Let us assume that path S_{XY} passed through AB like this: $X \ldots AB \ldots Y$. Consider set U_X of all cities U such that the shortest path from U to X does not contain road AB; also define similar set U_Y. If these two sets have some city C in common, then after closing of AB, the shortest path from X to Y through C cannot exceed 1000 km.

So, it is left to examine the case when $U_X \cap U_Y = \varnothing$. Then each city belongs to either U_X or U_Y. Clearly, $A \in U_X$ and $B \in U_Y$. Assume that there exists city Z which belongs neither to U_X nor to U_Y. Then path S_{XZ} passes through road AB, and therefore it coincides with the union of path S_{XA}, road AB, and path S_{BZ}. Similarly, path S_{YZ} is the union of path S_{YB}, road BA, and path S_{AZ}. Now consider path S'_{XZ} from X to Z equal to the union of paths S_{XA} and S_{AZ}, as well as path S'_{YZ} from Y to Z equal to the union of paths S_{YB} and S_{BZ}. The sum of the lengths of paths S'_{XZ} and S'_{YZ} is less than the sum of the lengths of paths S_{XZ} and S_{YZ}. This, however, contradicts the fact that these latter paths are the shortest ones connecting the corresponding cities. This proves that each city belongs to the union of U_X and U_Y.

It follows from the problem's conditions that there exists some road CD, different from AB, such that $C \in U_X$ and $D \in U_Y$. Hence, path from X to Y, defined as the union of path S_{XC}, road CD, and path S_{DY}, does not pass through AB and has length not exceeding 1500 km.

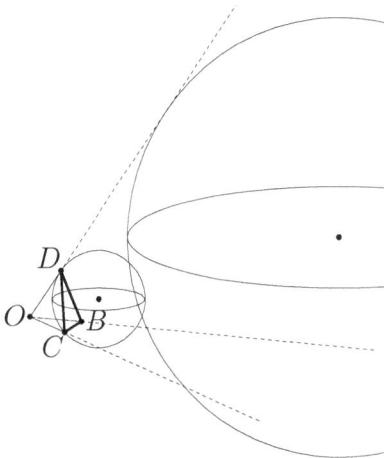

1975.38. Answer: no, that is impossible.

Mark 28 centers of all boundary squares and connect those that are the centers of adjacent squares by unit segments, forming a contour of a 7×7 square. Any cut intersects no more than two such segments, and therefore 13 cuts will "sever" no more than 26 of these connections. As a result, these 28 centers cannot be separated from each other.

1975.40. Yes, they do exist! Consider a regular tetrahedron $ABCD$ with center at O, and for each of its faces (say, BCD) draw a ball whose boundary sphere passes through three vertices of that face but at the same time does not contain point O (see the figure). After this, using homothety with different sufficiently large ratios, "push" each ball far away from point O so that all these four balls are pairwise disjoint. Then they completely "block the view" from point O—any ray issued from O must intersect one of the faces and therefore one of the balls.

1975.42. Denote $f^{-1}(x)$ by y. Then we can rewrite the condition of the question as equation $f(f(y)) + f(y) + y = 0$. Set $f(0) = 0$. Let us split the set of all real non-zero numbers into intervals of the following six types:

$$A_k : [\varphi^{3k}; \varphi^{3k+1}), \qquad \tilde{A}_k : [-\varphi^{3k+1}; -\varphi^{3k}),$$
$$B_k : [\varphi^{3k+1}; \varphi^{3k+2}), \qquad \tilde{B}_k : [-\varphi^{3k+2}; -\varphi^{3k+1}),$$
$$C_k : [\varphi^{3k+2}; \varphi^{3k+3}), \qquad \tilde{C}_k : [-\varphi^{3k+3}; -\varphi^{3k+2})$$

where k is an arbitrary integer and $\varphi = \frac{1}{2}(\sqrt{5}+1)$ is the so-called *golden ratio*, satisfying the equality $\varphi^2 = \varphi + 1$. Now we define function f via formula

$$f(x) = \begin{cases} \varphi x, & \text{if } x \in A_k \text{ or } x \in \tilde{A}_k \\ -\varphi x, & \text{if } x \in B_k \text{ or } x \in \tilde{B}_k \\ -x/\varphi^2, & \text{if } x \in C_k \text{ or } x \in \tilde{C}_k \ . \end{cases}$$

1975.44. Answer: 20.

Since $x_1^2 = 1$, $x_2^2 = x_1^2 + 2x_1 + 1$, ..., $x_{1976}^2 = x_{1975}^2 + 2x_{1975} + 1$, then adding all these equalities we obtain

$$\sum_{i=1}^{1975} x_i = \frac{1}{2}\left(x_{1976}^2 - 1976\right).$$

Since a square of an even integer (x_{1976} is even!) closest to 1976 is $44^2 = 1936$, then we have $\left|\sum_{i=1}^{1975} x_i\right| \geqslant 20$. To prove that this is the exact answer, it would suffice to construct a sequence (x_i) such that $x_{1976} = 44$, which we leave to the reader as a relatively easy exercise.

1975.45. Place weights m_a, m_b, and m_c in points A, B, and C in such a way that the center of mass coincides with point D.

Then the center of mass of $\{A, B\}$ is C_1, the center of mass of $\{B, C\}$ is A_1, and the center of mass of $\{A, C\}$ is B_1; and since $BD = 2B_1 D$, we obtain that $m_a + m_c = 2m_b$.

Now, instead of m_b, place mass $2m_b$ into point B, and consider the center of mass for the new system. Center of mass for $\{A, C\}$ is still in B_1, so we can simply place weight $m_a + m_c$ there instead. Since $m_a + m_c = 2m_b$, the center of mass of the entire system is the midpoint of BB_1.

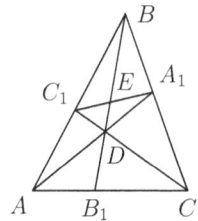

But we can compute the same point differently. Split the $2m_b$ mass in point B into two submasses of m_b. Then the center of mass of A and one-half of B is, obviously, point C_1 with mass $m_a + m_b$, and the center of mass of C with the other half of B lies at A_1 with mass $m_b + m_c$. It follows then that the center of mass of the entire system lies on $A_1 C_1$. Hence, it is at point E, which, incidentally, has to coincide with the midpoint of BB_1.

1975.46. It is enough to prove this inequality for $x_0 = 1$.

First, let us check that the inequality is true when $x_0 = x_1 = \cdots = x_n = 1$. Indeed, in this case, both sides of the inequality can be computed using formula for the geometric series, and it will suffice to show that

$$\frac{1 - \frac{1}{\sqrt{2^{n+1}}}}{1 - \frac{1}{\sqrt{2}}} \leqslant (1 + \sqrt{2}) \sqrt{\frac{1 - \frac{1}{2^{n+1}}}{1 - \frac{1}{2}}}.$$

Second, we will gradually change numbers x_k from one to zero. If we have already proved that inequality for some collection $x_0 = 1$, x_1, ..., x_n, in which $x_k = 1$ for some $k > 1$, then we can verify that it is true also for any collection in which $x_k = 0$ while all the other numbers are kept intact. For convenience, define (for the updated collection with $x_k = 0$) two sums

$$a_n = x_0 + \frac{x_1}{\sqrt{2}} + \frac{x_2}{\sqrt{2^2}} + \cdots + \frac{x_n}{\sqrt{2^n}} \quad \text{and} \quad b_n = x_0 + \frac{x_1}{2} + \cdots + \frac{x_n}{2^n}.$$

Using this notation, it remains to deduce from

$$a_n + \frac{1}{\sqrt{2^k}} \leqslant (1 + \sqrt{2})\sqrt{b_n + \frac{1}{2^k}}$$

the "updated" inequality

$$a_n \leqslant (1 + \sqrt{2})\sqrt{b_n}.$$

It is enough to show that

$$(1 + \sqrt{2})\left(\sqrt{b_n + \frac{1}{2^k}} - \sqrt{b_n}\right) \leqslant \frac{1}{\sqrt{2^k}}.$$

After multiplying both sides by $2^k\left(\sqrt{b_n + \frac{1}{2^k}} + \sqrt{b_n}\right)$, the inequality will turn into

$$1 + \sqrt{2} \leqslant \sqrt{2^k}\left(\sqrt{b_n + \frac{1}{2^k}} + \sqrt{b_n}\right),$$

which is obvious, since $b_n \geqslant x_0 = 1$ and $\sqrt{2^k} \geqslant \sqrt{2}$.

1976

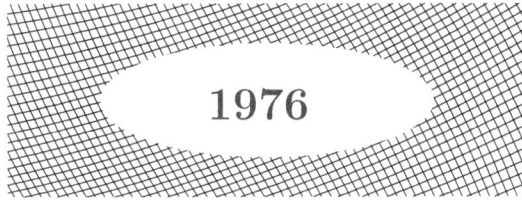

1976.01. After k jumps the flea will have moved by $1 + 2 + \cdots + k$ points. It is very easy to check that this number can never have remainder 2 modulo 3. Therefore, the overall move cannot be the move by 2 points in the counterclockwise direction. Another example: the flea will never visit the point neighboring the original one on its clockwise side, because that would represent the "shift" by $299 = 3 \cdot 97 + 2$.

1976.02. Answer: 72.
Clearly, $a^3 \geqslant 1 \cdot 2 \cdot 3 \cdot \ldots \cdot 9 > 71^3$, and therefore, $a > 71$. Finally, for the split $(1, 8, 9)$, $(3, 4, 6)$, $(2, 5, 7)$ the value of a is precisely 72.

1976.03. Answer: the school must be built in village C.
Assume that the school is built in any other point X. Moving it to point C decreases the total distance traveled by schoolchildren by $300\,|CX| - 100(|AC| - |AX|) - 200(|BC| - |BX|)$, i.e., it must decrease. Indeed, by the triangle's inequality $|AX| - |AC| \leqslant |CX|$, $|BX| - |BC| \leqslant CX$, and at least one of these inequalities is strict.

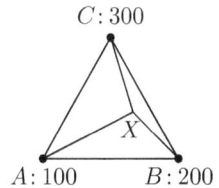

$C: 300$

$A: 100 \qquad B: 200$

1976.04. Let us measure the distance between two tokens by counting unoccupied squares between them. The winning strategy for the first player is as follows. First, he moves his token by 1 in the direction of the other token, thus making the distance equal to 27 (divisible by 3). After that, if the second player moves her token closer to the other one by x, the first player moves his token closer by $3 - x$—thus, the distance decreases by 3. If the second player moved her token away from the other token, the first player moves his token in the same direction by the same amount, thus maintaining the distance; the second player cannot do that indefinitely, sooner or later she will have to change the direction of the move, and thus the distance will decrease again, etc. Therefore, after every move by the first player, distance is always divisible by 3, and it follows he always has a move. Thus, he cannot lose.

1976.05. Answer: 21 centers.
As an example, we can take the centers of all the squares on two main diagonals.
Now, assume that we have selected less than 21 squares (centers). Then there is one selected square X that is the only one in its column (otherwise you have at least 22 centers). Assume it lies in column is C.
There is also one selected square different from X which is the only one in its row (otherwise you have at least 21). Let's call it Y, denoting its row by R.
Now take square Z at the intersection of column C and row R. It cannot be X or Y (lest we have $X = Z = Y$) or one of the selected ones (lest there is more than one selected square in the same column as X or the same row as Y). Thus, Z is not a selected square.

But then center of Z obviously cannot lie on the segment, which connects two selected centers and is parallel to the sides of the square, a contradiction.

1976.06. Let $a_1, a_2, \ldots, a_m$ be the sizes of the smaller plots, n—the size of the original plot, and S—the sum of lengths of all interior fences. Then, clearly, $4 \sum a_i = 2S + 4n$, and $\sum a_i^2 = n^2$. The second equality implies that $A = \sum a_i$ and n have the same parity, and therefore, $2A$ and $2n$ have the same remainders modulo 4. Hence, $S = 2A - 2n$ is divisible by 4, and so the sum of the lengths of all fences $S + 4n$ is divisible by 4 as well.

1976.07. Index the points with numbers 0 through 100, where point 0 is the flea's original position. Then the indexes of the points visited by the flea during the first 100 jumps are remainders of the numbers $0, 1, 1+2, 1+2+3, \ldots, 1+2+\cdots+100$ modulo 101. Since $1+2+\cdots+100 = 5050$ is divisible by 101, after the 100th jump the flea will return to the point where it started. Since there are 101 numbers on our list above, and there are two numbers equal to 0 (the first one and the last one), it follows that some number between 0 and 100 is missing, that is, the flea did not not visit the corresponding point. The length of every subsequent jump—# 101, # 102, ...—can be replaced by its remainder modulo 101. This implies that the flea will simply repeat all the previous jumps and so will visit (or not visit) the same points over and over again.

1976.09. Answer: yes, that is possible.

Introduce the coordinate system on the plane so that the original point is $(0,0)$, and the first jump is made to point $(1,0)$.

Using four jumps such that the first and the third, as well as the second and the fourth, are done using opposite directions, the grasshopper can move from point (x, y) to any point $(x \pm 2, y \pm 2)$. It means that if the difference between x and y is divisible by 4, we can use these four-jump combinations to make x and y equal; then, if these numbers are even, we can proceed to make them both zero.

Now, assume that by this moment the grasshopper made n jumps and find the smallest number k such that $n \leqslant 8k$. If necessary, let the grasshopper do a few more jumps so that the total number of jumps is $8k$.

At this point, grasshopper's x-coordinate is even, and y-coordinate is divisible by 4. With the next four jumps of odd length, it can change the x-coordinate by $\pm(8k+1) + (8k+3) + (8k+5) + (8k+7)$, while its y-coordinate will still be divisible by 4. The grasshopper will "choose" plus or minus in the expression above to ensure that the resulting x-coordinate is divisible by 4. Thus, both coordinates become divisible by 4, allowing the grasshopper to return to zero, using the four-jump combinations we described above.

1976.10. Denote the perimeter of the interior pentagon by x, the perimeter of the star by $x + y$, and the perimeter of pentagon F itself by z. Then, from the triangle's inequality we have $x < z < y$, $x + y < 2z$. So we have that numbers $p = x$, $q = z$, and $r = x + y$ are primes such that $p < q < r$, and $p + q < r < 2q$ (*).

Assuming that the sum $p + q + r$ is less than 20, we have $20 > p + q + r > 2q$. Thus, $q \leqslant 7$ and combining that with $2 \leqslant p < q$, we can see that there are only three options for q: (a) $q = 3$, (b) $q = 5$, and (c) $q = 7$.

If $q = 3$, then $p = 2$, and from inequality (*) we have $5 < r < 6$, which is impossible.

If $q = 5$, then $p = 2$ or 3; from inequality (*) we have $7 < r < 10$, which is, again, impossible.

Finally, for $q = 7$ we have $r \geqslant 11$, and $p \geqslant 2$ gives us $p + q + r \geqslant 20$.

1976.13. If this product is odd, then $x_k - y_k$ is odd for any k. But then $(x_1 - y_1) + (x_2 - y_2) + \cdots + (x_{25} - y_{25}) = 0$ must also be odd.

1976.14. This difference equals $999(a - b)$, and since 999 and 1976 are co-prime, this number can be divisible by 1976 only if $a - b$ is divisible by 1976, which is possible only for $a = b$.

1976.15. Answer: every 3 minutes.

Let us assume that the runner, the riders, and the biker move at speeds of a, b, c, and d, respectively, expressed in full circles per hour. Then the speed of the runner relative to the second rider is $-(a+c)$, meaning that they meet every $1/(a+c)$ hours. Thus, $a+c = 5$. Similarly, $b - a = 3$ and $d - c = 12$. From this, we obtain $b + d = 20$ which means that the biker and the first rider meet every three minutes.

1976.16. Answer: yes, that is possible.

Let point O be the center of this polygon. Then $\vec{X}_A = 1976 \,\overrightarrow{AO}$, and $\vec{X}_B = 1976 \,\overrightarrow{BO}$. Clearly, if point A almost coincides with a vertex of the polygon, and point B almost coincides with one of its sides' midpoints, then $AO > BO$, and subsequently, $X_A > X_B$.

1976.17. Add the equalities

$$S(BCK) = S(MKA), \quad S(CDK) = S(MDK).$$

Now, removing the areas of the parts which are present on both sides, we will get exactly the required formula.

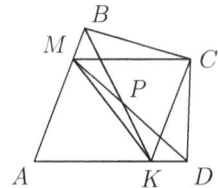

1976.18. Denote these subsets by A, B, C, and assume that regardless of how we choose two numbers from two of these subsets, their sum always belongs to the third subset.

Now let $x \in A$ and $y \in B$ be some natural numbers. Then A contains numbers $(2k+1)x + 2my$ with $k \geqslant 0$, $m \geqslant 0$, and if $m = 0$, then $k = 0$; all numbers of the form $2kx + (2m+1)y$ with $k \geqslant 0$, $m \geqslant 0$, and if $k = 0$, then $m = 0$, belong to subset B,; finally, C contains numbers $(2k+1)x + (2m+1)y$ with $k \geqslant 0$ and $m \geqslant 0$.

That is obviously true for $k = 0$ and $m = 0$. Number $x + (x + y) = 2x + y$ must be in B, and $y + (x + y) = x + 2y$ must be in A; then both $x + (2x + y) = 3x + y$ and $y + (x + 2y) = x + 3y$ must belong to C and so on.

If x is odd, then $xy + x + y = x + (x + 1)y$ belongs to A, but it is also $(y + 1)x + y$ and so it belongs to either B or C—a contradiction. Same goes for odd y. The only case left is when both x and y are even. Let us denote the maximum power of 2 that divides both numbers by 2^n, so $x = x_1 2^n$, $y = y_1 2^n$, where at least one of the numbers x_1 and y_1 is odd. Without loss of generality, we can assume that x_1 is odd. Then consider number $x_1 y_1 2^n + x_1 2^n + y_1 2^n$, equal to both $x + y(1 + x_1)$ (thus it belongs to A) and $y + x(1 + y_1)$ (so it belongs to the union of B and C). This final contradiction concludes the proof.

1976.19. Denote $a = |BC|$, $b = |AC|$, $c = |AB|$. Since $b = \frac{1}{2}(a+c)$, then

$$\frac{1}{2}(a+b+c) = \frac{3}{2}b \quad \text{and} \quad r = \frac{2S}{a+b+c} = \frac{2S}{3b} = \frac{1}{3}\left(\frac{2S}{b}\right).$$

Therefore, the inradius equals one third of the altitude dropped on AC.

1976.20. Let us prove that if $n > m$, then $\sqrt[n]{m} < \sqrt[3]{3}$. For $m = 1$, this is obvious. For $m \geqslant 2$, we can prove by induction the inequality $m^3 < 3^{m+1}$. It follows that $m^3 < 3^n$, or, equivalently, $\sqrt[n]{m} < \sqrt[3]{3}$.

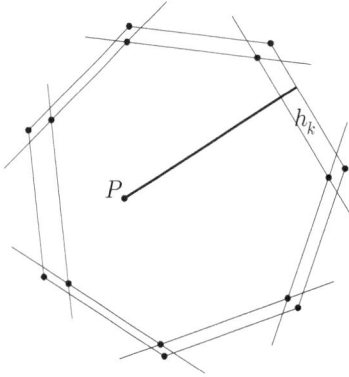

1976.22. First, consider regular polygon M with n sides. It is easy to see that the sum of the distances from point P inside M to its sides is constant. Indeed, the product of this sum by the half of the polygon's side equals the sum of the areas of the triangles obtained by connecting P with the vertices of M, that is, the area of M.

Now, move the kth side of M parallel to itself into the interior of polygon M by small distance h_k. Doing that for all sides, we will obtain the polygon M' defined by these "adjusted" sides. For any point P inside M', we have $d(P, M'_k) = d(P, M_k) - h_k$, where $d(P, M_k)$ is the distance from point P to the kth side of polygon M.

Thus, the sum of the distances from point P to the sides of M' equals the corresponding sum for M minus $\sum h_k$. That is a fixed number which is not dependent on point P.

It is left to prove that for $n \geqslant 5$ we can choose small numbers h_k in such a way that all sides of M' have different length. We will leave that to the reader.

1976.23. Let X be the set of all viceroys. There are $2^{12} = 4096$ subsets of X—from these we will discard the empty set and the X itself. After splitting the remaining 4094 subsets into 2047 pairs $(A, \bar{A})$, where $\bar{A} = X \setminus A$ is complement of A, we obtain 1000 subsets (committees). Adding to them their complements results in having 2000 "special" subsets. Since $2000 < 2047$, there has to exist a pair of subsets $(B, \bar{B})$ where both subsets are not special.

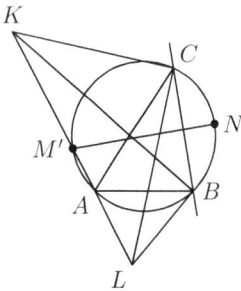

If B cannot serve as a new committee, then it does not intersect with one of the existing committees, say, C. But that means that $\bar{B}$ contains entire subset C, and therefore intersects with all the existing committees. Therefore, either B or $\bar{B}$ can be chosen as a new committee.

1976.24. Since AK is the bisector of exterior angle A, same as AL, point A lies on segment KL. Furthermore, it is easy to see that this bisector passes through the midpoint M' of arc CAB, because the bisector of angle A passes through the diametrically opposite point N, the midpoint of arc CB. Diameter $M'N$ is the perpendicular bisector for side BC, and therefore $|BM'| = |CM'|$. Now, points K, C, B, L lie on the circumference with diameter KL, and M' belongs to that diameter. Hence, M' is the center of the circumference, $|KM'| = |LM'|$ and $M = M'$.

1976.25. See solution to Question **1984.22**.

1976.26. Answer: function f is a zero function.

Let us set $g(x) = f^2(x)$, and let x and y be some arbitrary numbers; $a = y - x$. Then $g(y) = g(x) + g(a)$, and also $g(x) = g(y - a) = g(y) + g(-a)$. Since $g(a) \geqslant 0$, $g(-a) \geqslant 0$, we get $g(y) \geqslant g(x) \geqslant g(y)$. Hence, g is constant function, and that constant is, obviously, zero. Thus, $f(x) \equiv 0$ as well.

1976.27. Answer: if numbers a, b, c are non-negative, and at least one of them is different from zero, then $x = 0$; if $a = b = c = 0$, then x can be any real number (no real solutions in any other case).

Let $x > 0$ be a solution of the given equation, and assume that $a < 0$. If $x \geqslant 0$, then inequality $a - cx \geqslant 0$ implies inequality $c < 0$, which in turn implies $c + ax < 0$. And if $x < 0$, then inequality $a + bx \geqslant 0$ implies $b < 0$ and $b - ax < 0$. Thus, a has to be non-negative. Similarly, b, $c \geqslant 0$. Thus, for $x > 0$ we have $a + bx \geqslant a - cx$, $b + cx \geqslant b - ax$, and $c + ac \geqslant c - bx$; therefore, the left-hand side of our equation is greater than or equal to the right-hand side, and the equality can be attained only if $a = b = c = 0$. Similarly, for $x < 0$ the right-hand side is greater than or equal to the left-hand side. Hence, if even one of the numbers a, b, c is different from zero, then our equation has $x = 0$ as its only solution.

1976.28. Multiplying by both denominators and simplifying the result gives us
$$2abc < a^2 b + b^2 c + c^2 a + b^2 a + c^2 b + a^2 c - a^3 - b^3 - c^3 \leqslant 3abc.$$
Grouping the summands, we can rewrite this as
$$0 < (a + b - c)(b + c - a)(c + a - b) \leqslant abc.$$
The left inequality is obvious, because by the triangle's inequality, each of the three factors is positive. The right inequality can be transformed to
$$p(p - a)(p - b)(p - c) \leqslant \frac{1}{8} abcp, \qquad (*)$$
where p is the half-perimeter of the triangle.

Heron's formula shows that the left-hand side equals S^2. Using formulas $S = abc/4R$ and $S = pr$, inequality $(*)$ becomes the well-known Euler inequality $2r \leqslant R$.

1976.29. Assume that each of the sixteen possible variations occurs exactly once. Let x be the number of ones in the central 3×3 square, and a—the number of ones on the border (except for the corners).

Summing up the numbers of ones in the upper left and the lower right 4×4 squares, we have $2x + a + 2 = 16$, because each of these squares must contain 8 ones (indeed, consider, e.g., upper left square—it consists of the numbers in the upper left corners of all possible 2×2 squares). Similarly, for the lower left and upper right corners, we obtain $2x + a = 16$, a contradiction.

1976.30. See solution to Question **1976.24**.

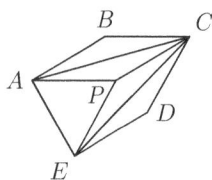

Reflect point D across line CE to obtain point P. Note that it follows from the condition of the problem that $\angle PCA = \angle BCA$, and therefore, triangles PCA and BCA are equal by **SAS** rule.

Then $|PE| = |PC| = |BC| = |AB| = |PA| = |AE|$, and point P is the circumcenter of triangle ACE. Moreover, triangle APE is equilateral and therefore, $\angle APE = 2\angle ACE = 60°$; thus, $\angle ACE = 30°$.

1976.32. Assume that each line intersects no more than two sets. Then every plane contains points from no more than three sets. Indeed, if that is not true, then we can find points A, B, C, and D such that they lie on the same plane, belong to different sets, and line AB is not parallel to line CD. Then the point where these two lines intersect must belong to the union of the first two sets and, at the same time, to the union of the third and the fourth, which is a contradiction.

Since the space cannot be covered by five planes, points of one of the sets (say, the fifth) do not all lie on the same plane. Draw plane α through some three points from the first, the second, and the third sets (α will not contain points from any other set) and select an arbitrary point P from the fourth set. Then every line passing through P and intersecting α contains points only from sets one through four. These lines cover the entire space except the plane that passes through P and is parallel to α, contradicting the choice of the fifth set.

1976.33. First, consider the case when all the variables are non-negative. If we denote by X and Y the maximum and the minimum of x_i, then, obviously, we have $X^2 \leqslant 2X$, $Y^2 \geqslant 2Y$. Hence, for non-zero values of X and Y we obtain $2 \leqslant Y \leqslant X \leqslant 2$, implying that there are only two solutions, $x_1 = x_2 = \cdots = x_5 = 2$ and $x_1 = x_2 = \cdots = x_5 = 0$.

Second, let us prove that none of numbers x_i can be negative. Assume, for convenience sake, that x_3 is the minimal of x_i and $x_3 < 0$. Then, subtracting the second equation from the first, we get $x_1 - x_3 = x_3^2 - x_4^2$. Since $x_1 - x_3 \geqslant 0$, we have $|x_3| \geqslant |x_4|$, and therefore, $x_3 + x_4 = x_5^2 \leqslant 0$, that is, $x_5 = 0$ (and only if $x_1 = x_3$). This, obviously, contradicts the fifth equation.

1976.34. Suppose that the number of losing positions $\mathcal{L}$ is greater than the number of winning positions $\mathcal{W}$. Then there exists a pair of losing positions A and B with a move from A to B. Indeed, if that is not so, then the number of positions, adjacent to the losing ones, is $n\mathcal{L}$, and each one of them is counted no more than n times (here n is that constant sum mentioned in the problem's conditions). Therefore, $\mathcal{W} \geqslant n\mathcal{L}/n = \mathcal{L}$, a contradiction.

But such pair cannot exist, because if there is a move from A to losing position B, then A is a winning position.

1976.35. Note that if two upper rows of some 3×3 square are colored with just one color, then the entire square can be repainted into that color. That is quite obvious if the third row is not entirely of the same color. Otherwise, the repainting can be done as it is shown in the table below, thus reducing this case to the previous one.

Now, let us prove the statement for any square with dimensions $n \times n$, using induction on n. The basis $n = 3$ can be checked directly with the help of the reasoning shown above.

For the induction step from n to $n + 1$, consider an arbitrary $(n + 1) \times (n + 1)$ table. By induction hypothesis, we can repaint the upper left "corner" $n \times n$ square. After this, we repaint the remaining border region in exactly the same manner we did it above for the 3×3 case.

$$
\begin{matrix}
1\ 1\ 1 \\
1\ 1\ 1 \\
2\ 2\ 2
\end{matrix}
\quad \rightarrow \quad
\begin{matrix}
1\ 1\ 1 \\
3\ 3\ 1 \\
3\ 3\ 2
\end{matrix}
\quad \rightarrow \quad
\begin{matrix}
1\ 1\ 1 \\
3\ 1\ 1 \\
3\ 3\ 2
\end{matrix}
\quad \rightarrow \quad
\begin{matrix}
1\ 1\ 1 \\
1\ 1\ 1 \\
3\ 3\ 2
\end{matrix} \ .
$$

1976.36. Begin with the following well-known fact.

Lemma. *A connected graph G has n vertices whose degrees do not exceed 3. Prove that G contains at least $\frac{1}{3}(n-1)$ non-intersecting edges.*

Let us prove this by induction on n. For $n \leqslant 4$, that is obvious. Take $n \geqslant 5$, and we have it already proved for all graphs will fewer vertices. Find a spanning tree, take one of its pendant vertices V, and find vertex X in the tree at the maximum distance from V. Vertex X must also be pendant, connected only to another vertex A.

If degree of A is 2, then remove vertices X and A from the tree to obtain the tree with $n - 2$ vertices. In that tree, by induction hypothesis, we can find at least $\frac{1}{3}(n - 3)$ non-intersecting edges. Add to them edge XA and we are done.

If degree of A is 3, then removing vertices A and X, we obtain two trees with a and b vertices, $a + b = n - 2$. Again, use induction hypothesis to find $\frac{1}{3}(a - 1)$ and $\frac{1}{3}(b - 1)$ edges in each of these trees, respectively. Adding edge XA gives us at least $\frac{1}{3}(a - 1) + \frac{1}{3}(b - 1) + 1 = \frac{1}{3}(n - 1)$ edges, proving this case as well. $\square$

Finally, apply the lemma to the graph whose vertices are the dissection triangles and edges correspond to the pairs of adjacent triangles (the ones with a common side).

1976.37. Simply examine these numbers' remainders modulo 7 and 13.

1976.38. We will need the following lemma.

Lemma. *If points $F(A)$, $F(B)$, $F(C)$ are collinear, then points A, B, C are also collinear.*

Indeed, if that is not so, then there exists point D such that $ABCD$ is a convex quadrilateral. Then quadrilateral $F(A)F(B)F(C)F(D)$ is also convex; however, three of its vertices $F(A)$, $F(B)$, $F(C)$ are collinear which is impossible, a contradiction. $\square$

Thus, we have actually proved that an inverse image of any straight line is a subset of a straight line.

Now, suppose that points A, B, C lie on line ℓ but points $F(A)$, $F(B)$, $F(C)$ are not collinear. Then the lemma implies that the inverse images of lines $(F(A)F(B))$, $(F(B)F(C))$, $(F(C)F(A))$ are all contained in ℓ. Take points $Y \notin \ell$ and $X = F(Y)$. Draw line k through point X, intersecting lines $(F(A)F(B))$ and $(F(A)F(C))$ at points M and N. As we know, $F^{-1}(M) \in \ell$, $F^{-1}(N) \in \ell$. Applying the lemma again, we obtain $Y = F^{-1}(X) \in \ell$, a contradiction.

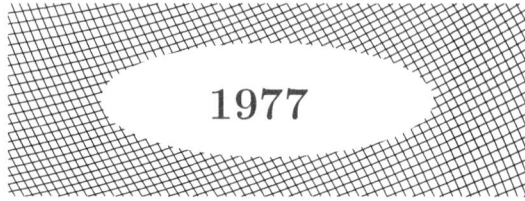

1977.01. This cannot be done. Let us denote the number of the triangles' sides on the sides of the square by a, and the number of dissection segments inside the square which are the sides of the triangles—by b. Counting the total number of all the sides of all triangles, we will have $4a + 2b = 3 \cdot 1977$. That is, however, impossible since the left-hand side is even and the right-hand side is odd.

1977.02. If there is a rook in every column, then we choose one per column and remove all rooks except these eight. If one of the columns—say, C—is empty, then each row must have a rook, namely, the one attacking the intersection of that row and column C. Then the first idea can be applied to the rows completing the proof.

1977.03. If the intersection point of these two lines is not the octagon's center, then we can move them parallel to themselves to have them both pass through the center. Then there are two different lines such that the diametrically opposite parts of the new dissection have equal areas. However, it is obvious that any such translation results in one of these two parts being expanded while the other one is diminished. This contradiction proves that the lines intersect at the center.

Now, if they are not perpendicular, rotate one of them until they are. In this new position, the two lines, obviously, satisfy the condition, which is impossible if the rotation angle is non-zero.

1977.04. See Solution **1977.08**.

1977.05. Clearly, this monotonically increasing sequence cannot "jump over" interval $[100\,000; 999\,999]$.

1977.06. The second player's strategy is as follows. He mentally splits the strip into two half-strips with six squares each. Then after every move by the first player, consisting in writing digit x into some square in one of the half-strips, he writes digit $9 - x$ into the same square in the other half-strip. At the end, the result will have form $a \cdot 10^6 + b$, where $a + b = 999\,999$. Since $10^6 - 1 = 999\,999$ is divisible by 77, we obtain

$$a \cdot 10^6 + b = a(10^6 - 1) + a + b = (10^6 - 1)(a + 1) \vdots 77.$$

1977.08. Answer: these numbers are 1, 9, 7, 7, 1, 9, 7, 7, 1, 9, 7, 7.

Consider a representation of $n = 197719771977$ such that the sum of numbers a_i is minimal. If one of numbers a_i is greater than or equal to 10, then it can be replaced by $a_i - 10$; while adding 1 to a_{i+1} (if $i < 11$), the sum of a_i will decrease. Thus, for this minimal representation, all numbers a_i, $0 \leqslant i \leqslant 10$ are, in fact, digits. Note also that in any case $a_{11} \leqslant 1$ and there exists only one representation where all numbers a_i are digits (namely, the regular decimal representation of n; the sum of the digits then is equal to 72). Thus, in all other representations, the sum of a_i is greater than 72.

1977.09. X and Y are midpoints of segments KN and LM, respectively. Then

$$2\,|AC| \leqslant 2\,|AX| + 2\,|XY| + 2\,|YC|$$
$$= |KN| + 2\,|XY| + |LM|$$
$$\leqslant |KN| + |KL| + |NM| + |LM|\,.$$

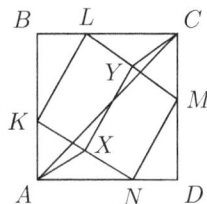

1977.12. For any point A on the boundary of the given polygon, A' is the second point of intersection of line AO with that boundary. We need to prove that, for any boundary point A, the equality $|AO| = |OA'|$ holds true.

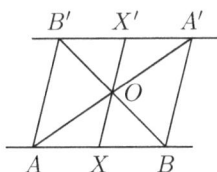

Choose some point B on the same side of P as A, located so close to A that B and B' also lie on the same side of P. X is the midpoint of AB. We know that $S(XAO) = S(X'A'O)$ and $S(XBO) = S(X'B'O)$. From $S(XAO) = S(BXO)$ we have $S(X'A'O) = S(X'B'O)$, that is, X' is the midpoint of $A'B'$. Furthermore, $AB' \parallel BA'$ because $S(ABB') = S(AA'B')$. Hence, XX' is a midline of trapezoid $ABA'B'$ parallel to AB', and therefore, $|XO| = |AB'|/2 = |OX'|$.

Thus, in trapezoid $ABA'B'$ midline XX' passes through the intersection point of the diagonals. Hence, this trapezoid is actually a parallelogram, and point O is the midpoint of the diagonal, or $|AO| = |OA'|$.

1977.13. It is easy to see that $S(BTC) > S(DTC) > S(DTE) = S(AEM) = S(APB) > S(KPB)$. Therefore, $|CT| > |KP|$ and, finally, $|CP| > |KT|$.

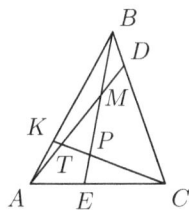

1977.14. For any set $A \subset \mathbb{N}$, denote its complement by $\bar{A}$. At least one of the sets A_1 and $\bar{A}_1$ is infinite. Let us select that set and denote it by C_1. At least one of the sets $A_2 \cap C_1$ and $\bar{A}_2 \cap C_1$ is again infinite. Select it, denote it by C_2, and so on. After n steps we obtain infinite set C_n. Any two of its elements form the required pair x, y.

1977.15. Answer: 3 or 4 vertices.

If the polygon has more than 4 vertices, then for some two of them their x-coordinates have the same parity, and the same is true for their y-coordinates. Hence, the midpoint of the segment connecting them is the grid point. Also, by convexity, it lies inside the polygon. Examples for 3 and 4 vertices are obvious, and we leave them to the reader.

1977.16. Answer: 101 broken lines.

The fact that the number of lines is at least 101 follows from the fact that if we consider all the 101 grid points on the main diagonal connecting two other opposite corners of the sheet, then a broken line of the described type cannot contain more than one of these grid points. An example with 101 broken lines is quite obvious.

1977.17. Let us assume the opposite—that segments A_k and B_k are always disjoint. We can also assume that A_1 lies to the left of B_1. Since B_2 intersects A_1, and A_2 intersects B_1, then it is clear that B_2 must lie to the left of A_2. Continuing like that to $k = 3$ and so on, we get that A_{1977} lies to the left of B_{1977}, and B_1 lies to the left of A_1, a contradiction.

1977.18. Consider a unit circumference centered at the coordinate origin O, as well as the points of intersection of this circumference with the lines passing through O and parallel to the given vectors. Then let us color the intersection point A red if vector $\overrightarrow{OA}$ is positively oriented with respect to corresponding vector $\overrightarrow{a_k}$, and blue otherwise. It follows from the problem's condition that the red and blue points on the circumference alternate while the diametrically opposite points are colored differently. Label all these points in clockwise order as A_1, A_2, ..., A_{2n}, assuming that A_1 is red. Then A_{n+1} is blue, and therefore, the number of points in sequence A_1, A_2, ..., A_{n+1} must be even. Thus, n is odd.

1977.23. Choose a natural number n such that

$$\left(x - \tfrac{1}{n}; x + \tfrac{1}{n}\right) \subset (a; b).$$

Set $t = 1 - \frac{1}{\sqrt{2}}$; then $f(x) = \{x + t\}$, where the curly brackets denote the fractional part of the number. Thus,

$$f(f(\ldots f(x)\ldots)) = \{x + nt\},$$

where function f is applied n times. Since t is irrational, the fractional parts of numbers x, $x + t$, $x + 2t$, ..., $x + nt$ are all different. Therefore, some two of them, $\{x + at\}$ and $\{x + bt\}$, differ by less than $1/n$. Hence, if $k = |a - b|$, then $|\{x\} - \{x + kt\}| < 1/n$, and so we have that $f(f(...f(x)...))$ (f is applied k times) lies in the interval $(a; b)$.

As a matter of fact, we have just proved this for any point $x \in (a; b)$.

1977.24. Let us reformulate this problem using graph theory terminology, turning points into vertices and arcs into edges. This graph G is, clearly, connected—therefore, we can find a spanning tree T.

Choose in T any vertex A and assign a non-negative integer number (level) to every vertex X of T (or G) based on the graph distance from A to X inside tree T. A is labeled with 0, vertices adjacent to it—with 1, etc. Then we mark each edge of T with two arrows, red and blue, orienting the red arrow from the lower level to the higher level and the blue arrow in the opposite direction. On all the other edges—from $G \setminus T$—we mark the arrows arbitrarily. Now consider any two vertices C and D, and connect them with path $C \to A \to D$ inside T consisting of two parts—the first one is the shortest path from C to A, where we move only along the blue arrows, and the second one, the shortest path from A to D using only the red arrows (one of these two parts could be empty). Then the color only changes once, at vertex A. Of course, if these two parts have identical edges (traveled in opposite directions), we can remove them one by one, starting from vertex A, but that is not necessary.

1977.25. Denote $f(x) = \sin x \tan x$, $g(x) = x^2$. Since $f(0) = g(0)$, it is then sufficient to prove inequality $f'(x) \geqslant g'(x)$. From $f'(0) = g'(0)$, we conclude that it is enough to prove inequality $f''(x) \geqslant g''(x)$. Indeed,

$$f''(x) = 2\frac{\sin^2 x}{\cos^3 x} + \left(\cos x + \frac{1}{\cos x}\right) > 2 = g''(x),$$

because $t + \frac{1}{t} > 2$ for any positive $t \neq 1$.

1977.26. Answer: from 4 to 8 vertices.

If the polyhedron has 9 or more vertices, then some two of them have the coordinates of the same parity for every coordinate indexed 1 through 3 (or x, y, z). Then the middle of the segment that connects them is the grid point, and by convexity, it belongs to the polyhedron as well.

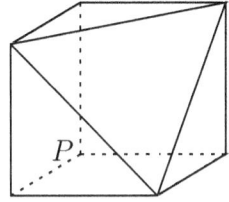

Now, we need examples of such polyhedrons with 4 to 8 vertices. They are: tetrahedron ($n = 4$), four-sided pyramid ($n = 5$), octahedron ($n = 6$), unit cube ($n = 8$) and cube without one vertex ($n = 7$) shown in the figure (also see Solution **1977.15**). The octahedron is obtained from the latter by "cutting off" one more vertex, namely P.

1977.27. Let us assume that p divides all these numbers. Then numbers

$$\binom{n+k-1}{k-1} = \binom{n+k}{k} - \binom{n+k-1}{k},$$

$$\binom{n+k-2}{k-1} = \binom{n+k-1}{k} - \binom{n+k-2}{k},$$

$$\cdots$$

$$\binom{n}{k-1} = \binom{n+1}{k} - \binom{n}{k}$$

are also divisible by p. The same can be said about all numbers $\binom{n+i}{j}$, where $i \leqslant j$ are some arbitrary non-negative integers. But one number among these is $1 = \binom{n}{0}$ ($i = j = 0$), a contradiction.

1977.28. Consider a planar section of any trihedral angle X by a plane whose normal lies within X. The section is a triangle whose angles do not exceed the corresponding dihedral angles of the pyramid. Therefore, for any trihedral angle (that is, for any vertex of the pyramid), the sum of its dihedral angles is greater than $180°$. Adding all four such inequalities gives us the required estimate.

1977.29. Let us prove by induction that for any real numbers $a_1 \geqslant a_2 \geqslant \cdots \geqslant a_{2n+1}$ the inequality

$$a_1^2 - a_2^2 + \cdots - a_{2n}^2 + a_{2n+1}^2 \geqslant (a_1 - a_2 + \cdots - a_{2n} + a_{2n+1})^2$$

holds true. The basis is obvious. To prove the induction step, replace numbers a_1 and a_2 by $a_1 - a_2 + a_3$ and a_3. Then the expression on the right will not change at all, and the expression on the left will decrease by

$$a_1^2 - a_2^2 - (a_1 - a_2 + a_3)^2 + a_3^2 = 2(a_1 - a_2)(a_2 - a_3) \geqslant 0,$$

implying that the inequality stays true. Since the second and third numbers are now equal, we can subtract them from both sides and use the induction hypothesis.

1978.01. See solution to Question **1974.03**.

1978.02. Any black triangle borders at least one white triangle. Therefore, the number of black triangles does not exceed the number of sides of white triangles, that is, three times the number of white triangles.

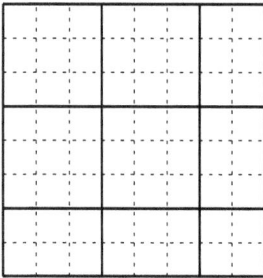

1978.03. No, it is not. $57\,599 = 240^2 - 1 = 239 \cdot 241$.

1978.04. Answer: ten kings.

Split the chessboard into nine rectangular parts as it is shown in the figure, using two horizontal and two vertical lines. If we have ten kings, then one of the parts contains two of them and clearly there is a square attacked by these two kings. On the other hand, to present the example with nine kings we simply place them into the upper left squares of these nine parts.

1978.05. Answer: no, we cannot. See solution to Question **1963.06**.

1978.06. See solution to Question **1970.09**.

1978.07. Answer: no, they cannot.

Consider these numbers' parity, that is, their remainders modulo 2. If you add up all pairwise sums of five integers a_k ($k = 1, \ldots, 5$), then, clearly, the result is equal to $4 \sum a_k$, and therefore, must be even. On the other hand, the sum of ten consecutive integers $x + 1$, $\ldots, x + 10$ equals $10x + 55$, an odd number.

1978.08. Answer: $a = 8$.

When divided by $a - 1$, $a - 2$, $\ldots$, $\left[\frac{a}{2}\right] + 1$, number a gives remainders $1, 2, \ldots, a - \left[\frac{a}{2}\right] - 1$, respectively. Compute their sum

$$1 + 2 + \cdots + \left(a - \left[\frac{a}{2}\right] - 1\right) = \frac{1}{2}\left(a - \left[\frac{a}{2}\right] - 1\right)\left(a - \left[\frac{a}{2}\right]\right)$$

$$\geqslant \frac{1}{2}\left(a - \frac{a}{2} - 1\right)\left(a - \frac{a}{2}\right) = \frac{1}{8}(a - 2)a.$$

It follows that $a \geqslant \frac{1}{8}(a - 2)a$, and so $a \leqslant 10$. Simply going through all numbers 1 to 10 we easily find the answer.

1978.10. See solution to Question **1971.09**.

1978.11. Assume the opposite and choose an arbitrary diameter on both sides of which lie exactly 50 numbers. Let us denote one of these sets of 50 numbers by A, and the other—

by B; then denote the sums of the numbers in them by $S(A)$ and $S(B)$, respectively. We can also assume that the difference $S(A) - S(B)$ is positive. Let us rotate this diameter by $2\pi/100$—then set A "loses" one number x and "acquires" number y, while set B loses y and acquires x. Obviously, the difference $S(A) - S(B)$ changes by $2(x - y)$, and that expression does not exceed $2(999 - 100) = 1798 < 1800$.

Consider interval $I = [-900; 900]$. After 50 rotations of the diameter, the difference $S(A) - S(B)$, obviously, changes the sign. Originally, it was positive and greater than 900, so it was located to the right of interval I; now it is negative and less than -900, so it lies to the left of interval I. Since every rotation changes the difference by less than 1800 (the length of I), then clearly, at some moment this difference had to lie inside this interval.

1978.12. Let us consider the largest side a of the polygon, and its vertex P, farthest from the line containing a. Draw the line parallel to a through point P. If it does not pass through any other sides of this polygon, then we can choose side a together with the two sides incident to vertex P—that will be the desired triple. And if that line contains a side of the polygon—say, PQ—then we can choose a and the two sides of the polygon adjacent to PQ. They are not parallel (otherwise the polygon would be contained between these two parallel lines), and since $n \geqslant 5$, side a would be shorter than side PQ parallel to it.

1978.13. See solution to Question **1972.15**.

1978.14. See solution to Question **1963.20**.

1978.15. Answer: the men are wrong.

Let us construct parallelograms $ABCX$, $CDEY$, and $EFAZ$ turned to the inside of the given hexagon $ABCDEF$. Then the absolute value of the difference of two sums—the sum of the lengths of the first, the third, and the fifth sides, and the sum of the lengths of the second, the fourth, and the sixth sides—equals the perimeter of triangle XYZ.

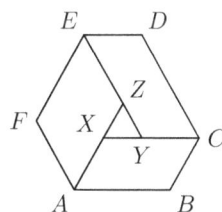

This is an equilateral triangle and its side's length is an integer. Thus, its perimeter is divisible by 3 and cannot be equal to 5.

1978.16. Yes, they can. To describe the example, we introduce the standard coordinate system. By coordinates of a square, we will mean the coordinates of its lower left corner. Now, we write 1 into every square with the sum of its coordinates divisible by 1918, and (-1) into every square with the sum of its coordinates divisible by 1978. If some square has to carry both numbers, we simply add them up and write zero. We also write zeros into every other square on the sheet.

1978.17. Let us denote these six circumferences by C_k, $k = 1, \ldots, 6$. Then draw the required seventh circumference through the following three points: point of tangency of C_1 and C_2, point of tangency of C_3 and C_4, and point of tangency of C_5 and C_6.

1978.19. First, it is obvious that the degree of the given polynomial f is even. Furthermore, if $f(p) = 2$, then $p \geqslant 2$. Let y be some large prime number such that for any $n < x = f^{-1}(y)$ we have $f(n) < f(x)$, and also f is monotonic on interval $[x; \infty)$. Then for all prime numbers from 2 to y, their inverse images lie on interval from p to x, and they are prime numbers. We have $p \geqslant 2$, and $x < y$ (because $f(x)$, obviously, cannot be equal to $x - c$), and different prime numbers from interval $[2; y]$ correspond to different prime numbers from interval $[p; x]$, then $2 = p$, $x = y$, and therefore, $f(x) = x$.

1978.20. Adding identities
$$(a_i - b_i)(a_j - b_j) = a_i a_j + b_i b_j - (a_i b_j + a_j b_i),$$
results in
$$S = \sum_{i \leqslant j}(a_i - b_i)(a_j - b_j) = \sum_{i \leqslant j} a_i a_j + \sum_{i \leqslant j} b_i b_j - \sum_{i \leqslant j}(a_i b_j + a_j b_i).$$
We also have
$$\sum_{i \leqslant j}(a_i b_j + a_j b_i) \geqslant (a_1 + a_2 + a_3)(b_1 + b_2 + b_3).$$
Assume now, without loss of generality, that $a_1 + a_2 + a_3 \geqslant b_1 + b_2 + b_3$. Then the inequality
$$S \leqslant \sum_{i \leqslant j} a_i a_j + \sum_{i \leqslant j} b_i b_j - (b_1 + b_2 + b_3)^2 \leqslant \sum_{i \leqslant j} a_i a_j \leqslant 1$$
holds true since, obviously, $(b_1 + b_2 + b_3)^2 \geqslant \sum_{i \leqslant j} b_i b_j$.

1978.22. The answer to item (b) is no. Consider a very small number ε (say, $\varepsilon = 0.01$), and circumferences of radius $\frac{1}{2} - \varepsilon$ centered at points $(0; 0)$, $(0; 1)$, $(1; 0)$, and $(1; 1)$, circumferences of radius ε centered at points $(\frac{1}{2}; 0)$, $(\frac{1}{2}; 1)$, and $(1; \frac{1}{2})$, and finally one circumference of radius $\frac{1}{\sqrt{2}} + \frac{1}{2} - \varepsilon$ centered at point $(\frac{1}{2}; \frac{1}{2})$.

1978.24. Answer: eight.

First, for every regular triangle, we must have three lines such that angles between them belong to set $A = \{0°, 60°, 120°\}$. If there is a line that does not form one these angles with any other line from our set, we can simply discard it.

Thus, there are two possibilities. First one is that we have two subsets, three lines each, such that the angles between the lines in different subsets do not belong to A. Then, we obviously have only two regular triangles.

The second case is that all the angles between our six lines belong to A. Thus, there are three groups—a_1 lines in the first one (G_1), a_2 lines in the second (G_2), and a_3 lines in the third (G_3), with $a_1 + a_2 + a_3 = 6$—such that the lines in every group are parallel to each other, and all the pairwise angles between lines belong to A.

Clearly, regular triangles formed by these lines are in one-to-one correspondence between trios of lines $\{\ell_1, \ell_2, \ell_3\}$ such that $\ell_k \in G_k$ for $k = 1, 2, 3$. Hence, the number of regular triangles equals $a_1 a_2 a_3$.

Now, from the AM–GM inequality we have
$$a_1 a_2 a_3 \leqslant \left(\frac{a_1 + a_2 + a_3}{3}\right)^3 = \left(\frac{6}{3}\right)^3 = 8.$$
An example can be constructed directly from the proof. Obviously, for the inequality to turn into equality, we need $a_1 = a_2 = a_3 = 2$. Therefore, we simply take three lines forming one regular triangle and "double" each one of them by adding a line parallel to it at a non-zero distance.

1978.27. Assume that in hexagon $ABCDEF$, diagonals AD, BE, and CF do not intersect at one point.

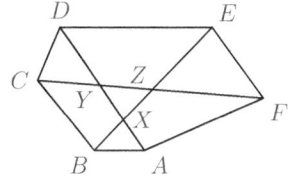

Then from the equalities

$$S(ABX) = S(XDE),$$
$$S(BCZ) = S(ZEF),$$
$$S(CDY) = S(YAF),$$

we obtain the following inequalities.

$$|AX| \cdot |BX| = |DX| \cdot |XE| > |DY| \cdot |EZ|,$$
$$|FZ| \cdot |EZ| = |CZ| \cdot |BZ| > |BX| \cdot |CY|,$$
$$|CY| \cdot |DY| = |AY| \cdot |FY| > |AX| \cdot |FZ|.$$

Multiplying them, we get

$$|AX| \cdot |BX| \cdot |FZ| \cdot |EZ| \cdot |CY \cdot |DY| > |DY| \cdot |EZ| \cdot |BX| \cdot |CY| \cdot |AX| \cdot |FZ|,$$

and that is obviously false.

1978.28. Write 10^{k-1} into every square of the kth color, for $k = 1$, 2, 3, and 4. This will guarantee that in every 2×2 subtable, the sum of the numbers equals 1111. Let S be the sum of all numbers on the boundary of the table plus double the sum of all numbers inside. Then S equals the sum of the numbers in 100×100 table plus the sum of the numbers in the central 98×98 subtable; hence, $S = (50^2 + 49^2) \cdot 1111$.

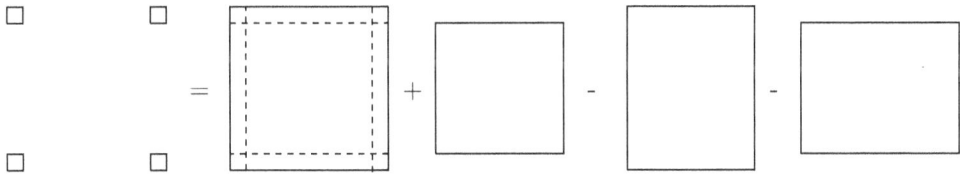

On the other hand, S equals the sum of the numbers in the corners plus the sum of the numbers in two rectangular subtables $98 \cdot 100$. It follows that the sum of the four corner numbers is $(50^2 + 49^2) \cdot 1111 - 2 \cdot 49 \cdot 50 \cdot 1111 = 1111$. That is possible only if all the four corner colors are different.

1978.29. Let P be an arbitrary point from set X, and Q—point from Y closest to P. Then by definition, $|PQ| \leqslant d_H(X, Y)$. Further, let R be the point in Z closest to Q—then $|QR| \leqslant d_H(Y, Z)$. Therefore, $|PR| \leqslant |PQ| + |QR| \leqslant d_H(X, Y) + d_H(Y, Z)$. Thus, distance from any point $P \in X$ to set Z does not exceed $d_H(X, Y) + d_H(Y, Z)$. Similarly, the distance from any point $S \in Z$ to set X does not exceed $d_H(Z, Y) + d_H(Y, X)$. Together, these inequalities prove that $d_H(X, Z) \leqslant d_H(X, Y) + d_H(Y, Z)$.

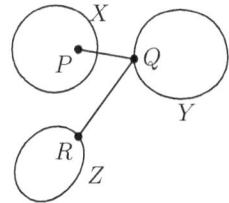

1978.30. The sum of the numbers at the vertices of the polygon becomes zero after the very first procedure and then, of course, stays equal to zero.[107]

---

[107] This is a surprisingly simple and even disappointing solution. Most likely, the jury had thought that this problem was somewhat more interesting.

1978.31. If segment AB contains x endpoints of segment CD, and segment CD contains y endpoints of segment AB, then it is easy to check that numbers x and y have the same parity. From this, it follows that adding up all numbers a_k, where a_k is the number of other segments' endpoints contained in the kth segment of our collection, results in an even number. On the other hand, the sum $1+2+3+\cdots+1978$ is odd and, therefore, the answer to the question is negative.

1978.32. Let us suppose that the sequence is periodic and also, for convenience sake, $a_1 = 1$. From the definition, we have that if binary representation of n has even number of ones, then $a_n = 0$; otherwise, we get $a_n = 1$. If the period has length p, then for sufficiently large k, equality $a_{k+np} = a_k$ is true for any $n > 0$. Set $n = 2^k q$, with q chosen in such a way that binary representation of pq has odd number of ones (we leave to the reader the proof of why that is always possible). It follows that $a_{k+np} \neq a_k$, a contradiction.

1978.33. Consider triangle K with maximum possible area, whose vertices lie on the vertices of polygon M. Then, M is entirely contained within triangle K' obtained from K by homothety with ratio (-2). To prove that, draw the lines passing through vertices of K and parallel to its sides. Clearly, image of M under any homothety with ratio $(-\frac{1}{2})$ can be translated to fit inside K because K is the image of K' under such homothety.

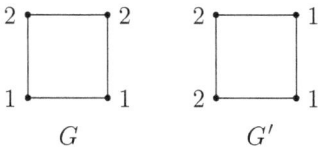

G G'

1978.34. As we have noticed before, for vertex V to change color to color #1, there must be a *clear* majority of the neighbors of V colored in #1. Otherwise, what should we do in case when there is $2k$ neighbors, k of which carry color #1 and the other k—color #2? If we were to allow re-coloring into any of the two colors, then there is an easy counterexample (see the figure).

Since every vertex here has exactly one neighbor of each color, any vertex can be re-colored however we want. Let us do that in such a way that graph G changes to G', and notice that G' is G rotated by 90° degrees counterclockwise. Thus, we can continue performing the same procedure indefinitely, and each vertex's color will change like this: $\ldots$, #1, #1, #2, #2, #1, #1, etc. This, clearly, contradicts the problem's claim.

Therefore, we must assume that the term "prevalent" means **clear** majority, i.e., in order for vertex V to change color to #1, there have to be more neighbors of color #1 than color #2 among neighbors of V.

Now, the problem can be solved using a so-called *graph duplication*. Namely, denote our graph by G, and its vertices by $V_1, V_2, \ldots, V_n$; then construct bipartite graph H as follows. Upper part of H consists of vertices $A_1, A_2, \ldots, A_n$, and lower part of vertices $B_1, B_2, \ldots, B_n$. Vertices A_i and B_j are connected by an edge if and only if graph G contains edge $V_i V_j$ (obviously, H contains edge $A_j B_i$ as well). Both parts of H have the same coloring as G's original one, meaning that vertices A_i and B_i both have the same color as V_i.

Every second, as graph G's coloring changes according to the described rules, we also similarly change the coloring of the upper part of H: each vertex A_i changes (or does not change) its color based on the prevalent color among adjacent vertices from the lower part. Thus, the new coloring of the upper part coincides with the new coloring of G.

One second later, we perform the same procedure for the lower part of H, and the lower part's coloring becomes identical to that of graph G. Continuing like that, at every moment, one of the parts of H (precisely which one can be determined based on the parity of the number of seconds) is colored identically to G, and the other part's coloring coincides with the previous second's coloring of graph G.

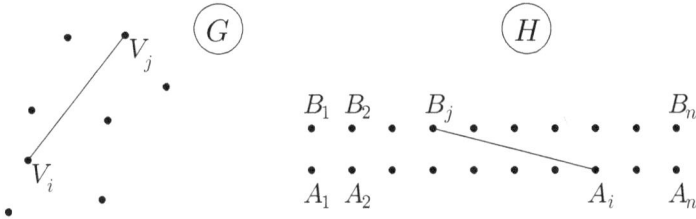

It is easy to see that with each second, the number of graph H's edges with differently-colored endpoints does not increase; moreover, it decreases if some vertex changed its color. This non-negative integer quantity cannot decrease indefinitely, and therefore, at some moment, it must reach its minimum and stop changing. After that, graph G either has the same constant coloring, or it alternates between two different colorings, proving our claim.

1978.35. Position the polygon in such a way that one of its vertices—A_1—coincides with the coordinate system origin, and side $A_1 A_2$ lies along the x-axis. Since all the lengths of the diagonals and sides are rational, then so are the cosines of all angles $\angle A_k A_1 A_m$, and therefore, from the formula

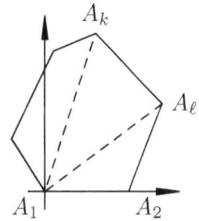

$$\cos(\alpha - \beta) = \cos\alpha \cos\beta + \sin\alpha \sin\beta \,,$$

we have that the sines of all angles $\angle A_k A_1 A_2$ are rational multiples of some number t. Thus, we can assume that all vertices A_1, A_2, ..., A_n lie on the nodes of rectangular grid with horizontal step 1 and vertical step h; we can easily move from the rational multiples of t to the integer multiples of another number h by using $h = t/N$, where N is some large integer.

Consider all the polygons obtained from M by translation via vector $\overrightarrow{A_1 P}$, where P is an arbitrary node of the grid, as well as all polygons obtained by the same translations from polygon M' symmetric to M with respect to the coordinate origin.

Now, selecting all such polygons intersecting square K gives us the collection—let us denote it by $\mathcal{C}$—that provides us with the desired covering. Indeed, consider an arbitrary polygon $N \in \mathcal{C}$ with its side XY intersecting K. Then, $\mathcal{C}$ also contains polygon N' symmetric to N with respect to midpoint of segment XY. Therefore, when we "cross over" side XY, the multiplicity of the covering does not change, and that proves that it is the same for every point inside the square.

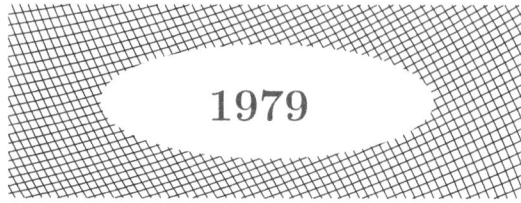

1979.01. One of the possible solutions is shown in the figure.

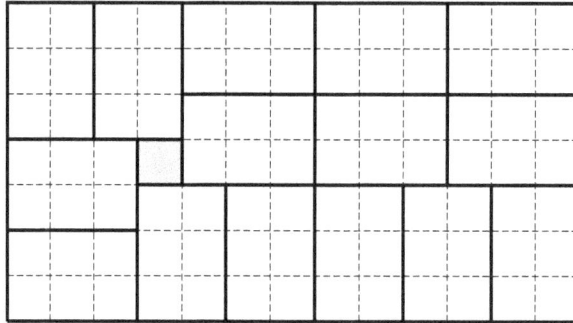

1979.02. Answer: 1957.

The sum of the digits of the birth year does not exceed 36, so this person cannot be older than 37. Hence, he was born in the year of $\overline{19xy}$, where $\overline{19xy} + 10 + x + y = 1979$. Thus, $11x + 2y = 69$, which is possible only if $x = 5$, $y = 7$ (by the way, compare this problem with Question **1962.03**).

1979.03. Assume that is not true. Split the rectangle into nine 2×2 squares and one thin strip 1×6. Then each square has no more than two black squares and the strip has at most six of them. Therefore, we have at most $2 \cdot 9 + 6 = 24 < 25$ black squares, a contradiction.

1979.04. (a) Among the three given cards we find a few with the sum divisible by 3. If one of these cards is 0, 3, 6, or 9, then our task is trivial. If we simultaneously have three cards from $A = \{1, 4, 7\}$ or three from $B = \{2, 5, 8\}$, then their sum is a multiple of 3. And if we have one from A and one from B, then the sum of those two is divisible by 3.

(b) **Answer:** five cards.

Four cards 1, 3, 4, and 7 give us an example of why four cards are not enough.

Any five cards contain either 0 or 9, or both numbers from one of the pairs $(1, 8)$, $(2, 7)$, $(3, 6)$, $(4, 5)$.

1979.05. Answer: 35 students.

Let us ask every boy to give a candy to each one of the girls he is friends with. Then, the boys give away $3x$ pieces of candy while the girls receive $2y$ of them. Thus, $3x = 2y$, then $3(x + y) = 5y$, and therefore, the number of students in the class $(x + y)$ is divisible by 5. From problem's conditions, we know that this number lies between 31 and $19 \cdot 2 = 38$, hence, it equals 35.

1979.06. Since
$$a^3 - b^3 = (a - b)(a^2 + ab + b^2) = (a - b)((a + b)^2 - ab),$$
then from $a + b$ and ab being divisible by c, we obtain that $(a + b)^2 - ab$ is divisible by c as well. Therefore, c also divides difference $a^3 - b^3$.

1979.07. Let ABC be the original triangle, $\angle A = 90°$, $\angle B = 30°$, D—the midpoint of hypotenuse BC, and E—the point of intersection of side AB with the given perpendicular. In right triangles ABC and DBE (both also possess an angle equal to $30°$), the shorter leg is half of the hypotenuse and therefore, $|BE| = 2\,|DE|$ and $|AC| = |DC|$. Hence, triangles DEC and AEC are congruent (they have equal hypotenuses and one of the legs) and consequently $|AE| = |DE| = |BE|/2$; that is, $|AE| = |AB|/3$.

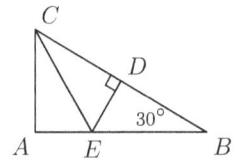

1979.08. Let us start the hourglasses simultaneously. After five minutes, when the first hourglass runs out, turn it over. Then in two minutes—when the second hourglass runs out—again turn over the first one. In two more minutes—when the first hourglass runs out—we are done. Altogether we have measured out $5 + 2 + 2 = 9$ minutes.

1979.09. See solution to Question **1974.34**.

1979.10. Answer: 16 numbers.
Consider all selected numbers except 1 as it can be added to any collection of numbers that does not already have it. For each one of them, arbitrarily choose one of its prime factors. All these numbers must be different. Since all of them are below 50, there is no more than fifteen of them (fifteen is the number of primes not exceeding 50), and the original collection has no more than 16 numbers. As an example, take fifteen primes from 2 through 47 and then add number 1.

1979.11. Answer: 27.
Let n be the given number. Sum of the digits of n^2 is 9, and each of its digits does not exceed 3 because, when we do the "long" multiplication, number n^2 is obtained as the sum of three numbers written by zeros and ones only.
Now, when we multiply n^2 by n, the product n^3 is obtained as the sum of three numbers whose digits do not exceed 3. Hence, there are no carryovers, and the sum of the digits of the result simply equals the sum of the sums of digits of these three numbers, that is, $3 \cdot 3 = 9$.

1979.12. Answer: $a = 2$, $b = 3$, $c = 4$ and $a = 0$, $b = -1$, $c = -2$.
Indeed, the equations can be rewritten as $(a - 1)(b - 1) = 2$, $(a - 1)(c - 1) = 3$, $(b - 1)(c - 1) = 6$. Multiplying the first two equations and dividing the result by the third, we obtain $1 = (a - 1)^2$, and so $a = 2$ or $a = 0$.

1979.13. Each of the segments of this broken line is a median in some right triangle (for instance, A_0B_1 is a median in triangle BB_1C). Therefore, its length equals half the length of the corresponding triangle's hypotenuse ($|A_0B_1| = |BC|/2$). Adding all these equalities gives us the desired result.

1979.14. Answer: $1000 \cdot 1979 + 1000 + 1979 - 1 = 1981978$. See solution to Question **1965.04**.

1979.15. Answer: $a = 18$. See solution to Question **1971.04**.

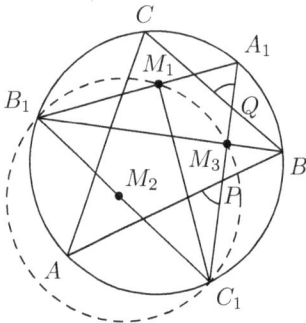

1979.16. Let $P = A_1C_1 \cap AB$, $Q = A_1C_1 \cap BC$. Then triangle BPQ is isosceles and it follows that median BM_3 coincides with bisector of angle B lying on BB_1, as well as with the corresponding altitude. Hence, $\angle C_1M_3B_1 = 90°$, and similarly, $\angle C_1M_1B_1 = 90°$, that is, M_1 and M_3 lie on the circumference with diameter B_1C_1.

1979.17. Consider numbers $(3k + 2)^2$ for all non-negative integers k; this is the desired set. Indeed

$$n^2 + p = (3k + 2)^2 \Rightarrow p = (3k + 2 - n)(3k + 2 + n).$$

Such factoring is possible only if the first factor is 1, but then $p = 3k + 2 + 3k + 1 = 6k + 3$, which is impossible for $k > 0$ because p is prime.

1979.18. When k is odd, the number is obviously divisible by 101. When k is even, after multiplying by 11, we obtain the number divisible by $A = 111\ldots11$ $(k+1$ ones). Since A is co-prime with 11, the original number is divisible by A as well.

1979.19. Multiply both sides by abc, and then factor the left-hand side to get

$$|(a - b)(b - c)(c - a)| < abc.$$

This inequality is true because $|a - b| < c$, $|b - c| < a$, $|c - a| < b$.

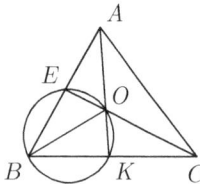

1979.20. Counting the angles shows that $\angle EOK = 120°$. Therefore, quadrilateral $BEOK$ is cyclic. Since BO is the bisector of angle B, we gave $\angle OBE = \angle OBK$. Hence, chords OE and OK have equal lengths.

1979.21. Answer: we need to take n consecutive vectors.

Let us prove that the length of the sum of n consecutive vectors is greater than the length of the sum for any other subset. Select subset M with the maximum length of the sum of the vectors in it—denote that sum as $\vec{m}$. Without loss of generality, we can assume that M contains no more than n vectors—otherwise, consider all the vectors not included in M; their sum is opposite of the sum of vectors in M, and therefore, has the same length.

Now, find all vectors that form an acute or right angle with $\vec{m}$. Obviously, they are consecutive and there are at least n of them (that is true for any non-zero vector). All of them must be included in M—otherwise, adding that vector to M would increase the length of the sum. So, we have that M has no more than n vectors but at the same time must include n consecutive vectors.

1979.22. If all these numbers are composite, then each one of them has a prime factor less than $2n - 1$. But there are no more than $n - 1$ such primes and therefore, some two of them coincide—then the two corresponding numbers from the original collection cannot be co-prime.

1979.23. Answer: it is possible only if both n and m are divisible by 4.

Assume that it is possible, and consider an arbitrary row. Any 2×2 subtable contains either 0 or 2 of its squares, and therefore, under any sign-change operation the parity of the number of ones in the row does not change. Originally, all these numbers were the same, coinciding with the parity of m; hence, these numbers must have the same parity as m for the chessboard sign arrangement as well. It is easy to see that this is possible only if m is divisible by 4. Similarly, n must be divisible by 4 as well.

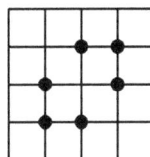

Now, if four divides both n and m, we need to show how to achieve the chessboard arrangement. It would suffice to show how that can be done in a 4×4 square. This is done by performing sign change in six 2×2 subtables whose centers are shown in the figure.

1979.24. Let us assume that all integers obtained from prime number $p = \overline{a_1 a_2 \ldots a_k}$ by circular shifts by x digits and by $x + y$ digits ($0 < x < x + y < k$) are the same.

This means that circular shift by y places produces the same number as the original number p; that is, $a_{t+y} = a_t$ for every t—we use here cyclic indexing $a_{s+k} = a_s$.

Let $d = \gcd(k, y)$; then there exist integers r and s such that $kr + ys = d$. Thus, we have $a_{t+d} = a_{t+kr+ys} = a_t$. Therefore, the digits of p can be split into several identical cycles of length d: $a = \overline{a_1 a_2 \ldots a_d}$. Obviously $p \vdots a$, implying $a = 1$.

1979.25. Answer: for $n = 10$ and $n = 11$.

First, assume that n is not divisible by 3. Denote by x the number of pairwise non-congruent isosceles triangles. We know that the number of pairwise non-congruent non-isosceles triangles also equals x. Then we have exactly nx isosceles triangles with vertices on the given points (or $n(x - 1) + \frac{1}{3}n$, if $n \vdots 3$) and exactly $2nx$ non-isosceles triangles.

Second, the total number of all triangles is $\frac{1}{6}n(n - 1)(n - 2)$, and it follows that $x = \frac{1}{18}(n - 1)(n - 2)$ (or $x = \frac{1}{18}((n - 1)(n - 2) + 4)$, if $n \vdots 3$). In addition, note that x is equal to the number of the ways n can be represented as $2a + b$; that is, $[\frac{1}{2}(n - 1)]$. It now suffices to solve the equations $(n - 1)(n - 2) = 18[\frac{1}{2}(n - 1)]$ and $(n - 1)(n - 2) + 4 = 18[\frac{1}{2}(n - 1)]$. In both cases $n = 2k$ and $n = 2k + 1$, we obtain an easily solvable quadratic equation.

1979.26. Assume that is not so. Let p_n be the nth prime number. Each of a_i is a product of at least two primes, and these primes do not repeat. Note that prime numbers from p_1 to p_{n-1} can go into prime factorization of no more than $n - 1$ of the given numbers, and therefore, there exists a_i whose prime factorization contains two prime factors greater than or equal to p_n. Then $a_i \geqslant p_n^2 \geqslant 9n^2$, because for $n \geqslant 12$ we have inequality $p_n \geqslant 3n$.

That fact is easily proved by induction on n. Basis of the induction: $p_{12} = 37 > 36$. Step: if $p_n > 3n$, then $p_{n+1} > 3n + 2$; but since p_{n+1} is not divisible by 3, we have $p_{n+1} > 3n + 3$.

1979.27. Answer: minimum area of the flap is 5.

Select the upper-right corner of the box, then lay the flap down on the bottom of the box and move it to the right until it hits the right side of the box, then up until it cannot move. Now paint on the flap the area that covers the hole.

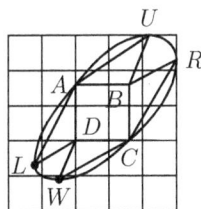

Repeat similar procedure for the other three corners. As a result, we obtain four unit squares (possibly intersecting) painted on the surface of the flap. Let $ABDC$ be the smallest rectangle covering all these squares with sides x and y (clearly, x, $y \geqslant 1$). Since the flap is convex, the rectangle lies completely inside the flap.

Now let us place the flap again into the upper-right position. Then point D coincides with the lower left corner of the hole,, and it follows that x, $y \leqslant 3$. Let U be the uppermost and R—the rightmost point of the flap (they could coincide).

Then in addition to rectangle $ABCD$ the flap must contain triangles ABU and BRC, whose areas equal $\frac{1}{2}x(3-y)$ and $\frac{1}{2}y(3-x)$, respectively.

Similarly, the flap contains triangles ADL and DCW, where L is the leftmost and W—the bottom point of the flap. Hence, the area of the flap is at least

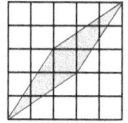

$$xy + x(3-y) + y(3-x) = 3x + 3y - xy = 9 - (3-x)(3-y) \leqslant 9 - 2 \cdot 2 = 5.$$

In the second figure (above), you can see an example of the flap that is simply a convex hull of the central unit square and two opposite corners of the box. It is easy to see that its area equals 5.

1979.28. Answer: the measures of these angles are $90°$, $\frac{\alpha}{2}$, and $90° - \frac{\alpha}{2}$.

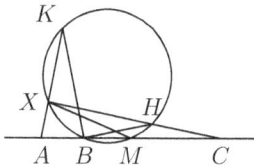

Let X be the point of intersection of lines AK and CH. Then $\angle AXC = 90°$, thus, $|MX| = |MC| = |MA|$, so $\angle HBC = \angle HCB = \angle CXM$ and points B, M, H, and X lie on the same circumference. Similarly, points B, M, K, and X lie on the same circumference, and therefore, all five points are on the same circumference.

It follows that $\angle KMH = \angle KBH = 90°$, $\angle KHM = \angle AXM = \angle XAM = 90° - \alpha/2$.

1979.29. Answer: $(x, y) = (0, 1)$ or $(2, -1)$.

In the first equation, subtract 2, then square both equations and add the results. After some quick transformations we obtain

$$x^2 + y^2 + 4 - 4x + \frac{5 + 2x^2 - 2y^2 + 8xy - 4x - 8y}{x^2 + y^2} = 0.$$

Multiplying by the denominator gives

$$x^4 + y^4 - 4x^3 - 4xy^2 + 6x^2 + 2y^2 + 8xy - 4x - 8y + 5 = 0.$$

It is easy to see that the expression on the left equals $(x^2 - 2x + 1 - y^2)^2 + (2xy - 2y + 2)^2$. Thus, both terms must be equal to zero. Equality $x^2 - 2x + 1 - y^2 = 0$ gives us $x - 1 = \pm y$. Then equality $2xy - 2y + 2 = 0$ implies that $(x - 1)y + 1 = 0$; that is, $\pm y^2 + 1 = 0$. Finally, either we have $y = 1$ and then $x = 0$, or $y = -1$ and $x = 2$.

1979.32. Answer: $y = 1$ or $y = 3$.

Assume that $y^2 + 3^y = x^2$; that is, $x^2 - y^2 = 3^y$. It follows that $x - y$ and $x + y$ are the powers of 3, namely, 3^k and 3^{y-k}. Then $y = \frac{1}{2}(3^k - 3^{y-k})$. Since $\frac{1}{2}(3^k - 3^{y-k}) \geqslant 3^{y-k-1} \geqslant 3^{(y-3)/2}$, and for $y \geqslant 7$, this number is greater than y (easily proved by induction). It remains for the reader to directly examine all values of y from 1 through 6.

1979.33. Consider all possible angles of rotation from 0 to 2π. If one of the sectors has angle measure α, and another—β, then the angles of all rotations, under which the image of the first sector intersects with the second sector, fill interval of length $\alpha + \beta \leqslant \frac{2\pi}{n^2-n+1}$. There are $n(n-1)$ such pairs, and we can see that all corresponding intervals have total length not exceeding $\frac{2\pi n(n-1)}{n^2-n+1} < 2\pi$. Therefore, there exists some angle not covered by all these intervals.

1979.34. Begin with the following lemma.

Lemma. *If the sides of a convex polygon can be paired up in such a way that the sides in each pair are equal and parallel, then this polygon is centrally symmetric.*

Indeed, it is obvious that the sides in each pair must be diametrically opposite—if they are not, then some other pair of parallel sides lies "between" them, which is clearly impossible. Now, any two neighboring major diagonals are diagonals of a parallelogram, and therefore, their intersection point divides both of them in half. It follows then that all the major diagonals intersect at the same point, which is their common midpoint. That point is the polygon's center of symmetry. □

Now, let us prove that the last face of the polyhedron is centrally symmetric. Consider an arbitrary edge a and count for each face of the polyhedron the number of its edges which are parallel and equal to a. The sum of all these numbers is even because its edge is counted exactly twice. On the other hand, each face except for the last one, has either zero or two such edges. Thus, the last face must contain another edge which is parallel and equal to a. Finally, we apply the lemma above to complete the proof.

1979.35. This is a particular case of Pascal's Theorem about six points on a conic (when one of the intersection points coincides with the center of circumference S).

The shortest but not very elementary proof uses projective transformations (*homographies*). Applying one of them, we can map the point of intersection of lines EA and BF to the infinity. This produces the configuration symmetric with respect to the center of S, and our claim follows immediately.

1979.36. See solution to Question **1970.02**.

1980

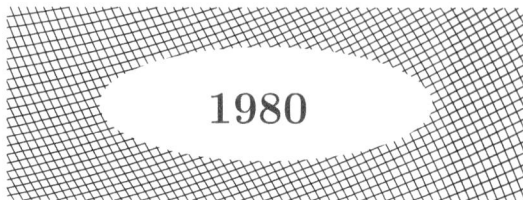

1980.01. Answer: no, that is impossible.

Otherwise the sum of the numbers in the table would be even (six times the sum in one column) but it is odd since $1 + 2 + \cdots + 30 = 31 \cdot 15 = 465$.

1980.02. Answer: there are three twelve-year-olds.

Denote the numbers of ten, eleven, twelve, and thirteen-year-olds by x, y, z, and t, respectively. Then we have $10x + 11y + 12z + 13t = 253$, $x + y + z + t = 23$, and $z = 1.5t$. These equalities give us $y + 6t = 23$, with t being even. Thus, $t = 0$ or $t = 2$. In the former case $z = 0$, $y = 23$, $x = 0$ (we will discard this answer, assuming that the camp actually exists and is not empty); in the latter case—$z = 3$, $y = 11$, $x = 7$.

1980.03. Let us split all points into pairs of symmetric points. Each one of these pairs belongs to one of the three types: red-red, blue-blue, and red-blue. It is clear that the number of red-red pairs is the same as the number of blue-blue pairs. Now, note that the sum of the distances from two points of the red-red (or blue-blue) type to point A (respectively, to point B) equals $|AB|$, and for the red-blue pair the distance from red point to A is the same as the distance from blue point to B.

1980.04. Index the coins with numbers $1, 2, \ldots, 9$. Compare group $\{1, 2, 3\}$ with group $\{4, 5, 6\}$, and then compare the heavier of these two with group $\{7, 8, 9\}$. If one of the groups is heavier than the other two, then it contains the counterfeit coins, and we can find them in one weighing.

If some two groups have equal weight, with the third one being the lightest, then both heavier groups contain one counterfeit coin each; they can be found with two more weighings.

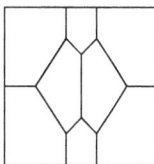

1980.05. The example is shown in the figure.

1980.06. Consider any two bus stops A and B. Then there are two bus routes XY and MN, passing through A and B, respectively. Either these two diagonals intersect at some point P and then we can get from A to B with one transfer in P, or we have a quadrilateral $XYMN$ whose diagonals XN and YM intersect at some point Q. Then one of these two diagonals, say, YM, must be a bus route, and we can get from A to B with two transfers in the following manner: $A \to Y \to M \to B$.

1980.09. Answer: no, that is not possible.

Consider some number divisible by 3. Then the numbers that stand on the positions of the same parity must be divisible by 3 as well. But there are only 660 such numbers among the given ones, which is considerably less that the half.

1980.11. Assume that there are no empty boxes. If that is not so, and there are in fact $m > 0$ empty boxes, we discard them, and then choose m pairs of candies, each one in the same box. Then if the girl chooses one of these "special" candies, the boy will choose the other one from the same pair. With all the other candies, he acts as if the "special" pairs do not exist.

The boy employs the following strategy. If the girl takes a candy from a box with only one candy, the boy will take a candy from a box which has more than one candy. If she takes a candy from a box with exactly two candies, then the boy will take the remaining candy from the same box. Finally, if the girl takes a candy from a box with more than two candies, then he will take a candy from a box with exactly one candy—such a box must exist; otherwise, there must have been non-discarded empty boxes before the girl's move.

Clearly, any two consecutive moves create one empty box which we immediately discard.

This strategy guarantees that after the boy makes his kth move, there are $2n-2k$ candies distributed among $n - k$ boxes (meaning that exactly k boxes are empty).

1980.12. Introduce standard coordinates on the sheet and then color the squares in colors 0, 1, 2, 3, 4 using the following rule. If the lower left corner of the square has coordinates (x, y), then we color it using the remainder of $x + 2y$ modulo 5.

1980.13. Answer: $(a, b, c) = (-3, -2, 2), (-3, 2, -2), (9, 4, 8), (9, 8, 4)$.

The second equation gives $a = b + c - 3$. Substituting this into the first equation, we get $9 + 2bc - 6b - 6c = 1$, or $(b - 3)(c - 3) = 5$. It suffices now to examine, one by one, all possible factorizations of 5.

1980.15. Consider any vertex P of the polygon and two triangles adjacent to the tangents from P to the circle. These triangles have indexes of different parity, and the ratio of their areas equals the ratio of distances from the given point to these two adjacent sides of the polygon.

Multiplying all these equalities, we come to the desired result.

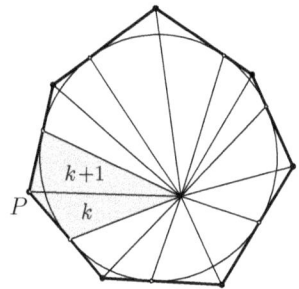

1980.16. Let O be the center of the square. If M is one of the unit squares' vertices, then $\overrightarrow{MA} = \overrightarrow{MO} + \overrightarrow{OA}$; we can also write similar equalities for other three vertices B, C, and D of this unit square. Thus, our sum equals $25\,(\overrightarrow{OA} + \overrightarrow{OB} + \overrightarrow{OC} + \overrightarrow{OD}) + \sum_M \overrightarrow{MO} = \sum_M \overrightarrow{MO}$. But this sum is $\overrightarrow{0}$, because all the summands can be split into fifty pairs of opposite vectors.

1980.17. Note that $53 + 96 = 83 + 66 = 109 + 40 = 149$. Let us introduce the following notation $a = 53$, $b = 83$, $c = 109$, $x = 149$. Then our number is equal to
$$abc + (x - a)(x - b)(x - c) = x(x^2 - (a + b + c)x + (ab + ac + bc)),$$
and therefore, it is divisible by $x = 149$ with the quotient obviously different from 1.

1980.20. Note that $\sqrt{4x + 1} < 1 + 2x$ for any $x > 0$. Thus,
$$\sqrt{4a + 1} + \sqrt{4b + 1} + \sqrt{4c + 1} + \sqrt{4d + 1} < 1 + 2a + 1 + 2b + 1 + 2c + 1 + 2d = 6.$$

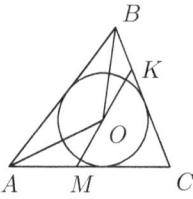

1980.21. Consider triangle BKO and BOA. Since BO is angle bisector in triangle ABC, we have $\angle OBK = \angle ABO$. Furthermore, we have $|BK|/|BO| = |BO|/|BA|$, and therefore, these triangles are similar, meaning $\angle BOK = \angle BAO$. Similarly, considering triangles OAM and BOA, we get $\angle AOM = \angle ABO$, and $\angle BOK + \angle AOB + \angle AOM = \angle BAO + \angle AOB + \angle ABO = 180°$. Hence, points M, O, and K are collinear.

1980.22. Obviously, each player takes part in at least one of any two consecutive games. Since the first player participated in 10 games, we can conclude that in total there were no more than 21 games played. It follows that there were exactly 21 games, and the third player took part in 11 of them (with the second player participating in all the games played).

1980.23. Answer: no, that is not possible.

Consider four squares adjacent to some fixed square A. If we remove any of them, the remaining three together with A form the T-shaped tetromino. It follows then that the numbers in these four squares are all congruent modulo 5. This, of course, implies that all numbers in black squares of the board—except perhaps the ones in two black corners—have the same remainders modulo 5. But it is impossible to find thirty such numbers between 1 and 64, a contradiction.

1980.24. Denote the areas of the four triangles by S_1, S_2, S_3, and S_4. Then we have $S_1^2 + S_3^2 = S_2^2 + S_4^2$; moreover, $S_1 S_3 = S_2 S_4$. Adding the doubled second equality to the first one, we obtain $S_1 + S_3 = S_2 + S_4$. From Vieta's Theorem, it follows that either $S_1 = S_2$, or $S_1 = S_4$. In any of these cases, one of the diagonals divides the other one in half.

1980.25. We can assume that $\frac{m}{n} = 0.501\ldots$. Consider expression $\frac{m}{n} - \frac{1}{2}$. On one hand,

$$\frac{m}{n} - \frac{1}{2} < 0.502 - 0.5 = 0.002 = \frac{1}{500},$$

and on the other hand,

$$\frac{m}{n} - \frac{1}{2} = \frac{2m - n}{2n} \geqslant \frac{1}{2n},$$

since $2m - n$ is a natural number. From this, we obtain $n > 250$. For $n = 251$, such a fraction exists, namely, $\frac{m}{n} = \frac{126}{251}$.

1980.26. In one weighing, we can determine the number of counterfeit coins in any given collection of coins. For convenience sake, place our nine coins into the squares of a 3×3 table. In four weighings, determine the number of counterfeit coins in the first two rows and the first two columns of the table. If one of these four weighings tells us there are two counterfeit coins in some row or column, then the results of the other weighings determine their position even without the fifth weighing. Any other case immediately gives us the pair of rows and the pair of columns that contain the two counterfeit coins; therefore, we know the four coins, two of which are the counterfeit ones. Weigh one of these coins to determine which two.

1980.27. Draw bisector AK of angle A. From the problem's conditions, it follows that angles CAK, KAB, and ABK are equal and therefore, $|AK| = |BK|$; hence, triangles ACK and ABC are similar. Then we have $|AC|/|BC| = |BK|/|AB|$, and from the bisector's property also $|AC|/(|BC| - |BK|) = |AB|/|BK|$. These two equalities immediately imply the desired one.

1980.28. Answer: the first player.
Altogether $(10+15+17)/2 = 21$ games have been played. Clearly, for any two consecutive games, the first player took part in at least one of them. Since he played 10 games out of 21, he must have played exactly all the even-numbered games. Therefore, he was the loser in the second game.

1980.29. Answer: 98 and 32.
Denote these numbers by x and y. By condition, we have digits a and b such that

$$\frac{x + y}{2} = 10a + b, \quad \text{and} \quad \sqrt{xy} = 10b + a.$$

Square these equalities and subtract the second one from the first to obtain

$$\frac{(x - y)^2}{4} = 99(a^2 - b^2).$$

Thus, $a^2 - b^2 = 11z^2$, where z is an integer. Since $a \neq b$, we have $a + b = 11$. Consequently, $a - b = z^2$; that is, either $a - b = 1$ or $a - b = 4$. The second case is not possible because $a - b$ must be odd; the first case gives us $a = 6$, $b = 5$.

1980.30. Substitute $x = 0$, $x = \frac{1}{2}$, and $x = 1$ into the given inequality to obtain three inequalities $|a + b + c| \leqslant 1$, $|a + 2b + 4c| \leqslant 4$, $|c| \leqslant 1$. It suffices now to note that

$$|a| = |(a + 2b + 4c) - 2c - 2(a + b + c)| \leqslant 4 + 2 + 2 = 8,$$
$$|b| = |(a + 2b + 4c) - (a + b + c) - 3c| \leqslant 4 + 1 + 3 = 8.$$

1980.31. See solution to Question **1963.10**.

1980.33. See solution to Question **1980.38**.

1980.34. Answer: the first player played 23 games, the second one—24, the third—25.
Denote the number of games they played by x, y, and z, respectively. Then $x+y+z = 72$. Note that all the games played by the first athlete, except maybe for the first one, happened immediately after the games that were lost by either the second or the third athlete. Also all the games lost by either the second or the third athlete, except for maybe the last one, preceded the games played by the first athlete. Thus, x differs from $(y - 12) + (z - 14)$ by no more than 1. Since $x + y + z$ is even, $x = (y - 12) + (z - 14)$. Similarly, we have $y = (x - 10) + (z - 14)$ and $z = (x - 10) + (y - 12)$. This system of equations has unique solution: $x = 23$, $y = 24$, $z = 25$. It is also quite easy to find an example.

1980.35. Answer: there are 1486 different numbers.
Consider sequence of numbers $a_k = \frac{k^2}{1980}$. Note that until index 990, the difference between two consecutive terms of this sequence is less than 1, and therefore, among the

first 990 integer parts of the sequence's terms, all the integers 0 through $495 = \frac{990^2}{1980}$ will occur. After that index, the difference between two consecutive terms of (a_k) becomes greater than one, and therefore, their integer parts will be all different. Thus, we have $496 + (1980 - 990) = 1486$ different numbers.

1980.36. Answer: no, they do not exist.

Assume the opposite. Note that the inequality is equivalent to

$$\left(f(x) - \frac{1}{2}\cos x\right)^2 < \frac{1}{4}\sin^2 x + \frac{1}{4}\cos^2 x = \frac{1}{4}.$$

Therefore, for any $x \in [0; 2\pi]$, we have inequality $|f(x) - \frac{1}{2}\cos x| < \frac{1}{2}$. Substitute values of x equal to 0, π, 2π to obtain $f(0) > 0$, $f(\pi) < 0$, $f(2\pi) > 0$, which is impossible since $2f(\pi) = f(0) + f(2\pi)$.

1980.38. Answer: n must be a power of 2.

Consider the case when one of the squares contains 1 while all the others—0. As the final result of the operations, we must obtain numbers $\frac{1}{n^2}$, but it is obvious that the denominators of all the intermediate results are powers of 2, and therefore, $n = 2^k$ for some natural k.

Now, assuming that $n = 2^k$, we need to provide an algorithm which makes all the numbers equal. Let us construct it using induction on k. Basis $k = 1$ is trivial. Assuming we already know how to make all numbers equal in any $2^{k-1} \times 2^{k-1}$ square, dissect $2^k \times 2^k$ square into four $2^{k-1} \times 2^{k-1}$ squares, and then make the numbers equal in each one of them. Now, consider quadruples of numbers with the same position in each of the four smaller squares. Obviously, these four numbers can be made equal for each quadruple, and after these operations are performed, all the numbers in the larger table will be equal.

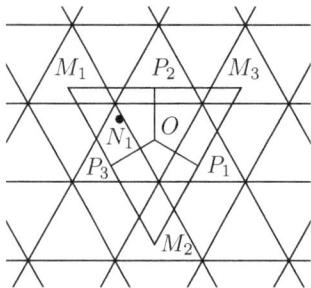

1980.39. Consider an infinite planar net[108] of a regular tetrahedron with unit edge (see the figure), where the trace of each vertex is labeled with the same letter. Let M and N be some two points on the tetrahedron's surface, and consider their images on the net. Points M_1, M_2, $\ldots$, corresponding to M, lie on the nodes of the grid of regular triangles with side 2. Consider one of the images of vertex N, namely, point N_1 inside triangle $M_1M_2M_3$. It suffices to prove that the distance from N_1 to one of the vertices of triangle $M_1M_2M_3$ does not exceed $\frac{2}{\sqrt{3}}$. Dissect this triangle into quadrilaterals $OP_1P_2M_3$, $OP_1P_3M_2$, and $OP_2P_3M_1$, where O is the center of $M_1M_2M_3$; P_1, P_2, and P_3 are the feet of perpendiculars dropped from O onto the triangle's sides.

The desired inequality holds, because point N_1 lies in one of these (congruent) quadrilaterals, and so the distance from N_1 to the corresponding point M_i does not exceed $|OM_1| = \frac{2}{\sqrt{3}}$.

1980.40. Answer: the optimal position of the cockroaches is when their positions divide the contour of the polygon in half (by length).

---

[108] What is polyhedral net? Read about it in [27].

Suppose that for this original position, the minimum of the distance is reached at points A and B, where A and B, of course, also divide the contour in half. Let us assume that for some other original position, the minimum distance is $d > |AB|$. Then obviously there exist two moments of time when the roaches define a segment parallel to AB—denote these segments $A_1 B_1$ and $A_2 B_2$—they are positioned on the different sides of AB and they are longer than AB (since their lengths are greater than or equal to d). However, portion of segment AB inside trapezoid $A_1 B_1 B_2 A_2$, on one hand, is not longer than AB, and on the other hand, cannot be shorter than both bases of the trapezoid, a contradiction.

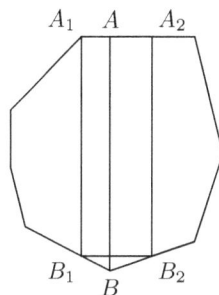

1980.41. Answer: $3k$.

It is obvious that the team has no more than k members who went to the All-Union Olympiad two years ago, no more than k additional members who went to the All-Union Olympiad one year ago, and finally, no more than k members who joined the team just this year.

1980.42. Consider the left-hand side expression as a function of x. This is a quadratic trinomial with positive leading coefficient—therefore, its maximum value is attained at one of the endpoints of interval $[0; 1]$. Thus, it is enough to prove the inequality for $x = 0$ and $x = 1$. For $x = 0$, the left-hand side does not exceed $3y^2 z^2 \leqslant 3$. For $x = 1$, we get $3y^2 + 3z^2 + 3y^2 z^2 - 2yz(1 + y + z) \leqslant 3$. Similar reasoning shows that it is enough to prove this inequality for $y = 0$ and $y = 1$. For $y = 0$, it is obvious, and for $y = 1$, we have a very simple inequality $3 + 4z^2 - 4z \leqslant 3$ that is clearly true for any $z \in [0; 1]$.

1980.43. Answer: this point is the triangle's center of mass.

Let us denote the sides' lengths by a, b, and c, and the distances from some point to the sides by d_a, d_b, and d_c, respectively. Then

$$\sqrt[3]{abcd_a d_b d_c} \leqslant \frac{ad_a + bd_b + cd_c}{3} = \frac{2S}{3},$$

where S is the triangle's area. Equality is attained only when $ad_a = bd_b = cd_c$, that is, the point is the center of the mass.

1980.44. Let us start with $3k - 2$ empty groups and start adding people to these group one by one. We will prove that anyone who is not already assigned to a group can always be added to one of the groups. Indeed, the number of people who like the same composer or the same writer or the same artist cannot exceed $3(k - 1) = 3k - 3$; therefore, there is at least one of the $3k - 2$ groups containing none of these people. Thus, we can always perform the assignment. Then we take the next person, assign him or her to a group, and so on.

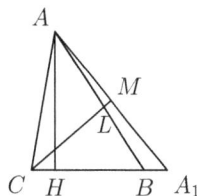

1980.45. Let $|BC| \leqslant |AC| \leqslant |AB|$ be the sides of the given triangle, AH—its altitude, and CL—one of the angle bisectors. Find point A_1 on ray CB such that $|CA_1| = |CA|$. Since AB is the largest side, angle $\angle ABC$ is acute, and therefore, angle $\angle A_1BA$ is obtuse and $|A_1A| \geqslant |AB| \geqslant |AC| = |A_1C|$.

Hence, in triangle A_1AC, altitude AH dropped onto A_1C is greater than or equal to altitude CM dropped onto A_1A. But line CM is an angle bisector in isosceles triangle A_1AC, and so we have $|CM| \geqslant |CL|$, and $|AH| \geqslant |CM| \geqslant |CL|$.

1980.46. Let us prove by induction on the number n of polygon's vertices a somewhat more general fact; namely, that such a dissection is possible for any n if the coloring uses all three colors.

Basis $n = 3$ is trivial. For the step of induction, consider polygon with $n + 1$ sides and vertex A with its neighbors B and C colored differently (such a vertex must exist!). Cutting off triangle ABC, we will get a polygon with n sides whose vertices are either colored in all three colors (then we can use the induction hypothesis) or only in two colors. In the latter case, we will put back triangle ABC and dissect the polygon into triangles AV_kV_{k+1}, where $A = V_0$, V_1, V_2, ..., V_n are the vertices of our polygon listed in the clockwise order.

1980.47. Answer: the maximum number of parallelepipeds is 52.

Dissect the cube into twenty seven smaller $2 \times 2 \times 2$ cubes, and color them black and white in the chessboard pattern so that the corner cubes are all black. Then any parallelepiped contains exactly two black and two white unit cubes. Since there are eight more black unit cubes than the white ones, then regardless of how the non-overlapping parallelepipeds are placed, at least eight black cubes will not be covered. Hence, we cannot have more than $(6^3 - 8)/4 = 52$ parallelepipeds. We leave to the reader the exciting task of constructing an example.

1980.48. Without loss of generality, we can assume that the coordinate system on the plane is such that $O = (0;0)$, $A_1 = (1;0)$, $B_1 = (0;1)$. From the given conditions, we have that the four points with both coordinates ± 1 do not belong to polygon F. Since F is convex, it follows that all points with absolute value of both coordinates greater than or equal to 1 do not belong to F as well. For example, if point C with both coordinates greater than or equal to 1 lies in F, then an entire triangle A_1B_1C, including point $(1, 1)$, must lie inside F, and that is not true.

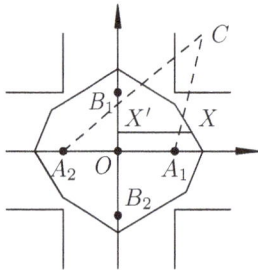

Now, we know that square K with vertices $(\pm 1, \pm 1)$ is covered by the images of square $A_1B_1A_2B_2$ when translated by vectors $\overrightarrow{OA_1}$, $\overrightarrow{OA_2}$, $\overrightarrow{OB_1}$, and $\overrightarrow{OB_2}$; thus, K is covered by the images of F under the same four translations.

Let us consider point $X(x, y) \in F$ possessing one coordinate with absolute value greater than 1 and the other coordinate with absolute value less than 1. For convenience sake, assume that $0 \leqslant y < 1 < x$, and consider point $X'(0, y)$. Obviously, it lies inside F, and therefore, segment XX' is contained in F. Since the length of XX' is at least 1, then point X is covered by F translated by vector $\overrightarrow{OA_1}$.

1981

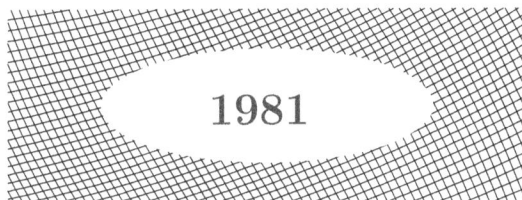

1981.01. Say, for example, Nick had x A's, y B's, z C's, and t D's. Then from the point average condition, it follows[109] that $5x + 4y + 3z + 2t = 5t + 4x + 3y + 2z$ and therefore, $x + y + z + t = 4t$. So the total number of grades is divisible by 4 and cannot be equal to 54.

1981.02. No, that is not possible. If it were, all the digits would be standing on the places of the same parity.

1981.03. The answer is shown in the figure.
Note. That is the graph of the vertices and edges of one of the regular polyhedrons, namely, an icosahedron.

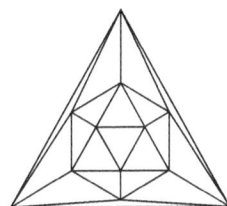

1981.04. Answer: 72 squares.
Dissect the large square into thirty-six smaller 2×2 squares. In each of them, at least two unit squares must be painted over, and so the number of painted squares has to be at least 72. The chess black-and-white coloring provides an example with exactly 72 painted unit squares.

1981.05. Yes; for example, $n = 111\,111\,111$.

1981.06. It suffices to prove that one ruble can be paid out with the given coins. But one ruble can simply be paid with a few coins of the same value—otherwise, the total sum would be less than four rubles (we only have four types of coins).

1981.07. Number $a = 11\ldots1$ (n ones) has the same remainder modulo 9 as n. Therefore, $a - n$ is divisible by 9. It suffices to note that the number in the statement equals

$$\frac{9a}{81} - \frac{n}{9} = \frac{a - n}{9}.$$

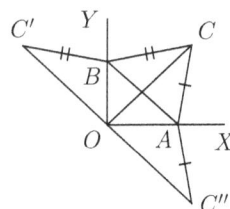

1981.08. Reflect point C across both rays to obtain points C' and C''. Then the perimeter of triangle ABC is equal to the length of broken line $C'ABC''$, that is, at least the length of segment $C'C''$, or $2\,|OC|$.

1981.11. If all the digits are different, then their sum is $0 + 1 + 2 + \cdots + 9 = 45$, and therefore, the number is divisible by 3. However, for any natural n, number $n^2 + 1$ cannot be divisible by 3.

1981.12. Vertices of a cube can be colored black and white in such a way that for every edge its endpoints have different color. Consider now the difference d between the sum of

---

[109] Reminder: in the Soviet school grading system $A = 5$, $B = 4$, $C = 3$, and $D = 2$.

the numbers in white vertices and the sum of the numbers in the black vertices. After every procedure described in the question, this difference does not change—i.e., d represents a so-called *invariant*. Initially $d = \pm 1$, but if all the numbers become divisible by 3, then d must be divisible by 3 as well. Therefore, such a configuration cannot be achieved.

1981.13. Answer: no, such a number does not exist.

Indeed, removal of the first digit a in $(k+1)$-digit number is equivalent to subtracting $10^k a$. We know that our number decreases by $1980x$, where x is the resulting number. But $10^k a$ cannot be divisible by $1980 = 11 \cdot 180$, since prime number 11 divides neither a nor 10^k.

1981.16. It is easy to see that together with any number x, this collection must contain number $m + 1 - x$ as well. If $x = \frac{1}{2}(m+1)$ is an integer, then it is easy to see that $m = 3$. Therefore, m is even, and it follows that all numbers in the collection can be split into pairs of numbers adding up to $m + 1$.

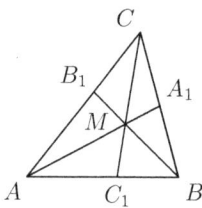

1981.17. We will use the following lemma (a so-called Gergonne's Theorem).

Lemma. *Points A_1, B_1, and C_1 are chosen on sides BC, CA, and AB of triangle ABC, respectively, so that lines AA_1, BB_1, and CC_1 intersect at point M. Then*

$$\frac{|A_1 M|}{|A_1 A|} + \frac{|B_1 M|}{|B_1 B|} + \frac{|C_1 M|}{|C_1 C|} = 1.$$

To prove it, place weights a, b, and c in the vertices A, B, C, respectively, in such a way that their center of mass is M. Then

$$\frac{|A_1 M|}{|A_1 A|} = \frac{a}{a+b+c}, \quad \frac{|B_1 M|}{|B_1 B|} = \frac{b}{a+b+c}, \quad \frac{|C_1 M|}{|C_1 C|} = \frac{c}{a+b+c}.$$

Adding these equalities, we obtain the desired result. □

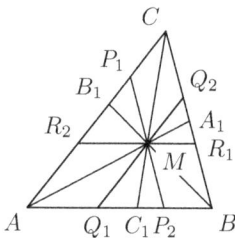

To solve our problem, we define points A_1, B_1, and C_1 on the sides BC, CA, and AB as intersections of those sides with lines AM, BM, and CM, respectively. Then

$$\frac{|P_1 P_2|}{|BC|} = \frac{|AM|}{|AA_1|} = 1 - \frac{|A_1 M|}{|AA_1|}.$$

Doing the same for the other fractions and adding them together, we use the lemma to obtain the desired result.

1981.19. We begin with the simple fact that the sum of 1981 consecutive integers starting with n equals $1981n + 990 \cdot 1981 = 1981(n + 990)$. Now, choose $n = 1981^2 - 990$.

1981.20. We can assume that the square's side equals 1. Since the lengths of the rectangles' sides do not exceed 1, each one of the ratios is greater than or equal to the area of the corresponding rectangle. Now, note that the sum of the areas is 1.

1981.21. Consider quadratic polynomial $f(t) = t^2 + yt + xz$. We know that $f(x) < 0$, and therefore, this polynomial has two real roots. Thus, its discriminant $D = y^2 - 4xz$ is positive.

1981.22. Let M and N be the midpoints of sides BC and AB, M_2—the midpoint of BA_1, K—intersection of segments C_1B_1 and C_2B_2. It will suffice to prove the congruence of two triangles with common vertex K.

Since $A_2B_2 \parallel AM \parallel A_1C_1 \parallel C_2M_2$, then from intercept (aka Thales') theorem, we have lines AM and A_1C_1 dividing segment B_2C_2 into three equal parts. Similarly, lines B_1C_1 and CN divide segment B_2C_2 into three equal parts. Therefore, point K lies on segment AM, and our triangles are congruent (use **SAS** rule).

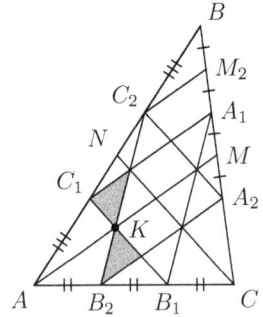

1981.23. Using the fact that $|AB|^2$ and $|AC|^2$ are unequal integers, we have

$$|AB| - |AC| = \frac{|AB|^2 - |AC|^2}{|AB| + |AC|} \geqslant \frac{1}{|AB| + |AC|} > \frac{1}{|AB| + |AC| + |BC|} = \frac{1}{p}.$$

1981.24. Answer: 34 moves. Color all black squares below the diagonal L, connecting the upper-left and the lower-right corners, into red and blue colors so that any two adjacent squares have different color. The coloring of the other half of the board is obtained by the reflection of the above coloring across line L. Now, with every move, the checker changes the color of its square except when it crosses diagonal L.

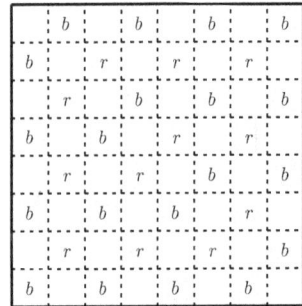

Because there are eight more blue squares than the red ones, the checker has to make at least three moves onto the squares it has already visited. Hence, the minimum number of moves is at least $31 + 3 = 34$. We leave to the reader finding an example of such a route.

1981.25. Begin with the following lemma, easily verifiable via the area computation.

Lemma. *Inside an angle the locus of points, whose sum of the distances to the sides of the angle is fixed, is a segment perpendicular to the angle's bisector.*

Denote the given quadrilateral by $ABCD$, and point of intersection of lines AB and CD—by K. Then consider points X and Y inside the quadrilateral such that XY is perpendicular to the bisector of angle AKD. It follows from the lemma that the sums of distances from X and Y to the lines AB and CD are equal; hence, the sums of the distances to lines BC and AD are equal as well. The same lemma shows that this is possible only if $BC \parallel AD$. But then for any point Z inside the quadrilateral, not lying on line XY, the sum of the distances to BC and AD—and consequently the sum of the distances to AB and CD—is the same as for point X. From the lemma, we obtain that point K does not exist, and therefore, the quadrilateral is a parallelogram.

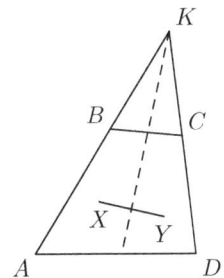

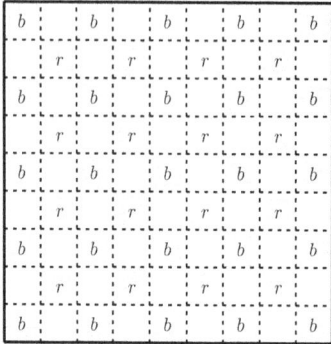

1981.27. Answer: 48 moves.

Let us color all black squares in red and blue in alternating order. We will have 25 blue and 16 red squares. With every move, the checker changes the color of its square, and in order for its route to contain 25 blue squares, it has to be at least 49 squares long, that is, to have at least 48 moves. Example for 48 moves is quite straightforward.

1981.28. Sequence $b_n = a_n - n$ is non-decreasing since $a_{n+1} \geqslant a_n + 1$. Then the problem's condition can be rewritten as equality $b_n = b_{2n+b_n}$. It follows that b_1 occurs infinitely many times in (b_n) and therefore, this sequence is constant. Hence, there exists number C such that $a_n - n = C$ for all n.

1981.29. Consider the following chain of identities.

$$2(a+b)(c+d)(ac+bd-A) = (a+b)(c+d)(2ac+2bd-2A)$$
$$= (a+b)(c+d)(2ac+2bd-(b^2-a^2)-(d^2-c^2))$$
$$= (a+b)(c+d)((a+c)^2-(b-d)^2)$$
$$= (a+b)(c+d)(a+b+c-d)(a-b+c+d)$$
$$= (a+b)(c+d)\left(a+b-\frac{A}{c+d}\right)\left(c+d-\frac{A}{a+b}\right)$$
$$= [(a+b)(c+d)-A]^2.$$

1981.31. Arbitrarily place one hexadecagon on top of another, and then consider sixteen rotations of the upper polygon by the angles which are integer multiples of $360°/16$. If after every one of these rotations we have no more than three coincidences of the marked vertices, then altogether we will have no more than 48 coincidences. At the same time each vertex of the upper polygon passes over each of the vertices of the lower polygon, and therefore, the number of coincidences must be $7 \cdot 7 = 49$.

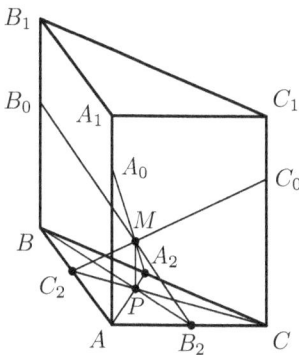

1981.34. Let us select on sides BC, CA, AB of triangle ABC points A_2, B_2, C_2, respectively, in such a way that lines AA_2, BB_2, CC_2 intersect at point P. Then point A_2 lies on A_0M and $d/a = A_2P/A_2A$. Adding these equalities for all vertices of the triangle and using Gergonne's Theorem (see Solution **1981.17**), we get the desired result.

1981.35. Answer: 53 moves.

Solution is virtually identical to Solution **1981.24**.

1981.36. First, find a power of five 5^k such that $5^k - 1$ is divisible by 2^{100} (existence of such exponent k follows from Euler's Theorem). We can assume that the number of digits in the decimal representation of 5^k is greater than 100; otherwise, consider exponent $2k$, or $3k$ and so on.

Denote $a = 5^{k+100}$, and then we have that the difference $a - 5^{100} = 5^{100}(5^k - 1)$ is divisible by $5^{100}2^{100} = 10^{100}$; that is, the first hundred places of decimal representations of numbers a and 5^{100} coincide. Note that $5^{100} < 10^{70}$, because $2^{100} = 1024^{10} > 10^{30} = 2^{30}5^{30}$, and therefore, $2^{70} > 5^{30}$.

From this, we obtain that no more than 70 of the first one hundred places of decimal representation of a are filled with non-zero digits.

1981.37. Answer: $p = 1, 2$.

Consider the remainders modulo 5. For $p = 1$ and $p = 2$, we get 11 and 13. For $p \geqslant 3$, the number equals $2^{4k} + 9 = 16^k + 9$, which is always divisible by 5.

1981.38. Answer: $\sqrt{3}$.

Note that in a triangle with angle $60°$ the distance from that angle's vertex to the point, where the incircle touches one of the angle's sides, equals $\sqrt{3}r$ (here r denotes the inradius of the triangle).

Therefore, to maximize the sum of the radii of these circles, it is necessary to maximize the sum of these distances. It is the same as minimizing the sum of the distances between tangency points of these circles with each of the sides of the given triangle; that sum equals the perimeter of triangle $A_1B_1C_1$. It is well known that this perimeter is minimal when points A_1, B_1, C_1 coincide with the midpoints of the sides of triangle ABC. In that case, the sum of the radii is equal to $\sqrt{3}$.

1981.39. Answer: the first player wins if $n^2 \leqslant m < n^2 + n$ for some natural n, and loses if $n^2 + n \leqslant m < (n+1)^2$.

If $m = n^2$, the winning strategy for the first player is as follows. With his first move he takes one stone, with his second move he complements the other player's first move to 3, with his third move he complements the other player's second move to 5, and so on. Since $n^2 = 1 + 3 + \cdots + (2n-1)$, the last stone will be taken by the first player. If $m = n^2 + n - 1$, the first player takes one stone, then complements the opponent's move to 4, then to 6, and so on. In other words, at move $\# k$ he complements the second player's move to $2k$ instead of $2k - 1$. The first player wins because $n^2 + n - 1 = 1 + 4 + 6 + 8 + \cdots + 2n$. For all intermediate values, the first player also wins. Indeed, if $m = n^2 + i$ $(0 \leqslant i \leqslant n-1)$, then, after taking one stone at first, for the next i moves (with indexes $k = 2, 3, \ldots, i+1$), he complements the second player's move to $2k$ and not to $2k - 1$.

Let us prove that in all other cases it is the second player who wins. Complementing the first player's moves to 2, 4, 6, ..., $2n$, he will take the last stone if $m = n^2 + n$, and complementing them to 3, 5, 7, ..., $2n+1$, he wins for $m = n^2 + 2n = (n+1)^2 - 1$. If $m = n^2 + n + i$ $(0 \leqslant i \leqslant n)$, he complements to $2k + 1$ exactly i times out of n.

1981.40. Suppose that such numbers a, b, c, d exist. Then the equality

$$(a - b\sqrt{3})^4 + (c - d\sqrt{3})^4 = 1 - \sqrt{3} \qquad (*)$$

also must hold true. Indeed, after expanding the original equality, the left-hand side can be represented as $x + y\sqrt{3}$, where x and y are rational numbers, and then the left-hand side in $(*)$, when expanded, must coincide with $x - y\sqrt{3}$. Furthermore, if $x + y\sqrt{3} = 1 + \sqrt{3}$, then $x = 1$, $y = 1$—this follows from rationality of numbers x and y.

But equality $(*)$ is impossible since number $1 - \sqrt{3}$ is negative.

1981.41. Without loss of generality, we can assume that each row and each column were repainted no more than once, and that in total there were x columns and y rows repainted. Furthermore, we can also assume that they were the first x columns and the first y rows. Then the number of squares that changed their color equals $A = x(8 - y) + y(8 - x)$.

Let us prove that this number is greater than or equal to the doubled number of black squares among the repainted—it would follow then that the number of new black squares will not be less than the number of black squares which became white.

Case 1. $x, y \leqslant 4$. The number of black squares among the repainted does not exceed $B = 2x + 2y$, and $A - 2B = 2(2x + 2y - xy) \geqslant 0$, since $xy - 2x - 2y = (x - 2)(y - 2) - 4 \leqslant 0$.

Case 2. $x, y \geqslant 4$. We can assume that the repainted lines are the last $8 - x$ columns and the last $8 - y$ rows because repainting them gives the same result. Then $8 - x$, $8 - y \leqslant 4$, and this is reduced to Case 1.

Case 3. $x \leqslant 4$, $y \geqslant 4$. Then $A - 32 = 8x + 8y - 2xy - 32 = 2(4 - x)(y - 4) \geqslant 0$, that is, A is greater than or equal to twice the overall number of the black squares, immediately proving the desired inequality.

1981.42. Answer: no.

Let us assume that these six numbers exist. Then every three-digit number can be obtained from them in exactly one way. Indeed, each six-digit number generates $\binom{6}{3} = 20$ three-digit numbers, and so from the six of them we get 120 numbers, that is, exactly the total amount of three-digit numbers of the required form.

Without loss of generality, we can also assume that one of the given numbers is $a = 123456$. From a we obtain, in particular, all numbers of form $12X$.

Now consider all six-digit numbers that produce three-digit numbers of form $1X2$. Clearly, 2 in them must be the very last digit, otherwise we would be able to produce number of form $12X$, and there would be a repetition. Moreover, if 1 is the first digit, then it generates four numbers $1X2$, and if it is not, then it generates less than four such numbers. There are exactly four numbers of form $1X2$, and that means that our collection either has one more number (in addition to 123456) that starts with 1 or has two numbers that end with 2.

Consider the first case (the second case can be reduced to it by reversing the order of digits and then permuting them).

From two six-digit numbers that begin with 1, we get all twenty three-digit numbers that begin with 1, and therefore, in all four remaining six-digit numbers, 1 occupies one of the last two places. Now let us count how many three-digit numbers that end with 1 we can obtain.

Any number that has 1 in the next to last place produces six of those; any number that ends with 1—even more. Thus, we obtain at least 24 three-digit numbers ending with 1, leading to the contradiction as there are only 20 of them.

1981.43. Answer: $\lfloor (p + 1)/2 \rfloor$ operations.

Write $(+1)$ to the left of the given sequence, and then compute for the resulting sequence the number S of the "sign changes", that is, the number of pairs of neighboring numbers of different sign. Each operation decreases S by no more than 2. Obviously, S cannot exceed p and for the sequence with alternating signs, the required number of operations is at least $[(p+1)/2]$. At the same time, this number of operations is always enough because if you have more than one sign change, then there is an operation which decreases the number of sign changes by 2 (in case of just one sign change, one operation is also enough).

1981.44. Yes, they do exist. Take four balls that all touch each other externally, and find the fifth one which is inscribed in the space between them, touching all the four original balls. Then the centers of these five balls will serve as the desired five points.

1981.45. Let us make substitutions $a = \cos x$, $b = \cos y$, $c = \cos z$. Then we need to prove the inequality

$$\cos x \sin y \sin z + \cos y \sin x \sin z + \cos z \sin x \sin y - \cos x \cos y \cos z \leqslant 1,$$

which holds true because the left-hand side is the formula for $-\cos(x+y+z)$.

1981.46. Consider polynomial

$$F(x_1, x_2, \ldots, x_7) = \left[1 - \left(\sum_{i=1}^{7} a_i x_i^2\right)^2\right]\left[1 - \left(\sum_{i=1}^{7} b_i x_i^2\right)^2\right]\left[1 - \left(\sum_{i=1}^{7} c_i x_i^2\right)^2\right]$$

of degree 12 in seven variables $x_i \in \mathbb{Z}_3 = \{0, 1, 2\}$—that is, residues modulo 3. The values of this polynomial are also computed modulo 3. Then, if we add up the values of F for all 3^7 possible septuplets of variables, the result equals zero—that follows from the fact that after expanding, every monomial in F contains at least one variable that comes into it with the exponent not greater than one.

On the other hand, $F(0, 0, \ldots, 0) = 1$. Therefore, there exists one other septuplet $\{x_i\}$ for which $F(x_1, x_2, \ldots, x_7) \neq 0$, that is, $F(x_1, x_2, \ldots, x_7) = 1$, because the three factors of F can only take values of 0 and 1 modulo 3. This means that

$$\sum_{1}^{7} a_i x_i^2 = \sum_{1}^{7} b_i x_i^2 = \sum_{1}^{7} c_i x_i^2 = 0.$$

Now remove the triples (a_i, b_i, c_i) for which $x_i = 0$. To prove that this produces the desired outcome, once again use the fact that $x^2 \equiv 1 \pmod 3$ for any $x \not\equiv 0 \pmod 3$.

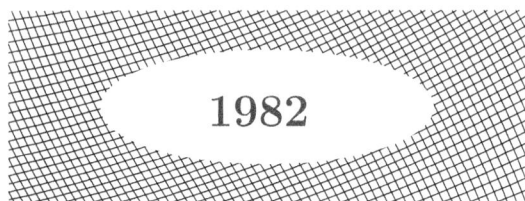

1982

1982.01. Let us write out all seven-digit numbers, from which the given number X can be obtained, in the following order. First, we write the ten numbers we get by adding one of the ten digits at the beginning of X. Then, we write the ten numbers obtained by writing one of the digits into the space between the first and the second digits of X. Then, the ten numbers obtained by inserting one digit between the second and the third digits, and so on. Finally, the last ten are obtained by adding a digit after the very end of X.

We have written 70 numbers but some of them, clearly, repeat. Let us remove the numbers which coincide with the ones written before them—then we will have the required list where each number occurs exactly once.

Now, we just need to determine how many numbers we have to remove. In fact, there will be exactly six of them—because we remove one per each ten numbers except for the first ten. More precisely, from the kth ten, we remove the numbers where the inserted digit coincides with the one immediately preceding it (that is, the $(k-1)$th digit of X). Indeed, let us denote the $(k-1)$th digit of X by a. So if the number was obtained by inserting a after $(k-1)$th digit of X, then clearly we already had it in the previous ten as it was obtained by inserting the same digit a into the space before $(k-1)$th digit of X. If we inserted some other digit b in that space, then the number obtained did not occur before as all the numbers which were produced by inserting digits anywhere before the $(k-1)$th digit of X have digit a in the kth position, not b.

Therefore, our list contains $70 - 6 = 64$ numbers.

1982.02. No, it could not. After the 25th jump, the grasshopper will have shifted by an odd number of centimeters relative to its original position.

1982.03. Assume the opposite. Then, in each of the five rows, we have some prevailing color, red or blue. Say, we have three rows with prevailing red color, so each one has three red squares. We can assume that one of these rows is the first row, and in it, the first three squares are red. Then in the two remaining rows, the fourth and the fifth squares are red, and they form the desired quartet of squares.

1982.04. Since $770 = 2 \cdot 5 \cdot 7 \cdot 11$, in order for $x(770 - x)$ to be divisible by 770, one of the factors must be divisible by 2, one—by 5, and so on. Note also that if one of the factors is divisible by a divisor of 770, then the other factor is also divisible by the same number.

Thus, if the product is divisible by 770, then each of the factors is divisible by 770, which is impossible for two natural numbers under 770.

1982.05. If we add up all sums of the numbers on the faces, the result equals double the sum of all numbers. This shows that the sum of the numbers on each face must be equal to 26.

An example is shown in the figure below.

1982.06. The first player wins. With his first move, he splits the pile of 100 stones into two piles with 63 and 37 stones respectively. After this, his strategy is to maintain the following property: the number of stones in the largest pile has to be a power of 2 decreased by 1.

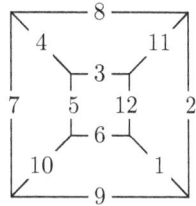

1982.07. Answer: $(15, 55)$, $(25, 65)$, $(35, 75)$.

Let x and y be the remainders of a and b modulo 9. Then $a = 9y + x$, $b = 9x + y$, $0 \leqslant x < 9$, $0 \leqslant y < 9$. Since $a > 9$, $b > 9$, and $a \neq b$, we have $y > 0$, $x > 0$, and $x \neq y$. The last digits of a and b are the same, and therefore, $8(y - x) = a - b \vdots 10$; hence, $y - x \vdots 5$, and $|y - x| = 5$. From this we obtain all possible pairs of x and y: $(1, 6)$, $(2, 7)$, $(3, 8)$.

1982.09. Factoring 30030, we obtain $30030 = 2 \cdot 3 \cdot 5 \cdot 7 \cdot 11 \cdot 13$. Then $a + b = 30030$ and $ab \vdots 30030$ for some natural a and b. Since $ab \vdots 2$, one of the numbers a and b is divisible by 2, and from $a + b \vdots 2$, it follows that both a and b are divisible by 2. Similarly, a and b are both divisible by 3, 5, 7, 11, and 13. Thus, both natural numbers a and b are divisible by 30030, and so their sum must be at least 60060, a contradiction.

1982.10. Let X_1, X_2, ..., X_{1982}—our points, A and B—two diametrically opposite points on the circumference such that line AB contains none of X_i. Then either A or B satisfies the conditions. Indeed, from the triangle's inequality, we have $AX_i + BX_i > AB$ for any index i. Adding these inequalities, we obtain

$$(AX_1 + AX_2 + \cdots + AX_{1982}) + (BX_1 + BX_2 + \cdots + BX_{1982}) > 2 \cdot 1982,$$

and therefore, at least one of the two summands in parentheses must be greater than 1982.

1982.11. Answer: no, it could not.

Introduce the coordinate system so that the x-axis is directed along the grasshopper's first jump, with the length of the first jump equal to 1. Then the parity of the x-coordinate changes only after every odd-numbered jump. Thus, after the last jump, it has changed 991 times, and therefore, became odd.

1982.12. Index the boys and then denote by A_n the set of the girls who danced with the nth boy. If we assume that the statement of the problem is false, then for any m and n, either $A_m \subset A_n$, or $A_n \subset A_m$. Therefore, we obtain the chain

$$A_{i_1} \subset A_{i_2} \subset A_{i_3} \subset \cdots \subset A_{i_k},$$

where A_{i_k} is the largest of A_i. But then, since A_{i_k} does not coincide with the set G of all girls, there exists a girl who did not dance with any boy—that would be any girl from $G \setminus A_{i_k}$.

1982.13. Squaring the inequality gives us

$$1 + a^2 b^2 < a^2 + b^2 \implies (1 - a^2)(1 - b^2) < 0.$$

1982.14. It would suffice to prove that the sum of any three angles at vertices A, B, and C is greater than the fourth angle at vertex D. Indeed, the sum of the angles A, B, and C is at least the sum of the angles in triangle ABC that is equal to $180°$; measure of angle D is, of course, less than $180°$.

1982.15. Number written by 1982 twos has remainder 4 modulo 9. Suppose that for some integers x and y, number $xy(x + y)$ has remainder 4 when divided by 9. Then this number has remainder 1 modulo 3, and it is easy to see that both numbers x and y must have remainders 2 modulo 3, and therefore, one of the remainders 2, 5, or 8 modulo 9. In all these cases, number $xy(x + y)$ has remainder 7 modulo 9, a contradiction.

1982.16. If x is the number of the seventh graders, then the number N_7 of points they gained in total does not exceed $x \cdot 2x + \frac{1}{2}(x^2 - x)$. Similarly, we have $N_6 \geqslant \frac{1}{2}2x(2x - 1)$. So we obtain

$$2x^2 + \frac{1}{2}(x^2 - x) \geqslant \frac{7}{5}x(2x - 1),$$

and it follows that $x \leqslant 3$.

Now, it is easy to see that $x = 3$, $N_7 = 21$, $N_6 = 15$. Therefore, the total number of points is $21 + 15 = 36$, implying there were nine participants since $36 = 9 \cdot \frac{8}{2}$.

1982.17. Here is one example:

12	9	2
1	6	36
18	4	3

1982.19. There are two cases. Either all three polynomials have a common root (then the proof is trivial), or there are three numbers $a < b < c$ such that each two of them are roots of one of these polynomials. Say, b is a root of the first and the second quadratic polynomials. Then the value of the third polynomial at point b is negative, and thus, the value of the sum of all three polynomials at b is negative. This, clearly, proves the existence of the root.

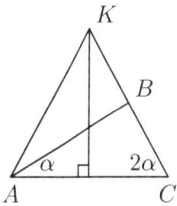

1982.20. Denote angle CAB by α. Then

$$\angle ACB = 2\alpha, \quad \angle ABC = \pi - 3\alpha.$$

Since $|AC| > |CB|$, we have $\angle ABC > \angle CAB$; that is, $\pi - 3\alpha > \alpha$, and $2\alpha < \frac{\pi}{2}$. Hence, the perpendicular bisector of segment AC intersects ray CB at point K such that $|KC| = |KA|$.

Segment AB is the bisector of angle $\angle CAK$, therefore

$$|CB| : |BK| = |CA| : |AK| = 2|CB| : |AK|,$$

or, equivalently, $|BK| = |AK|/2 = |KC|/2$. Hence, bisector AB in triangle CAK is also a median, and thus, an altitude as well, and so $\angle ABC = 90°$.

1982.21. Answer: yes, we will.

Indeed, sooner or later some four consecutive terms in the sequence will repeat. Moving from these repeating quadruples backward, we can easily see that first quadruple to repeat must be 1, 9, 8, and 2. Let us find the four numbers preceding the second of these two quadruples—those are 3, 0, 4, and 4.

1982.22. Let A, B, C, D, E, and F be consecutive vertices of the polygon. Angle ACE is divided by diagonals from vertex C into 176 parts and therefore, $\angle ACE \geqslant 176°$. For the same reasons, $\angle CAE \geqslant 2°$ and $\angle CEA \geqslant 2°$. It follows that $\angle ACE = 176°$, and thus, the angle between any two neighboring diagonals from any vertex is equal to $1°$.

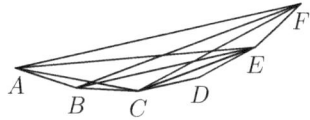

Furthermore, since angles $\angle BEC$ and $\angle BFC$ are the same and equal to $1°$, then points B, C, E, and F lie on one circumference. Similarly, all other vertices of the polygon, except possibly A and D, lie on the same circumference. But since $\angle FAE = \angle FBE = 1°$, point A also lies on the same circumference. And since $\angle CAD = \angle CFD = 1°$, point D also lies on that circumference. It is also clear that the polygon's vertices divide that circumference into equal arcs with angle measure $2°$.

1982.23. In any column, at least three out of five squares have the same color. Let us mark that triple as well as the corresponding three rows. Since there are 41 columns, and the number of possible combinations of three rows is $\binom{5}{3} = 20$, then by the pigeonhole principle, some triple has to be marked at least three times, and we are done.

1982.24. We will call the parts, into which the plane is dissected by these lines, *tiles*.

Draw circle ω containing all the points of intersection of the given lines. On each line L, mark two oppositely oriented rays so that both of them contain all the intersection points on L. Altogether we have $4n$ rays.

Now, let us index these rays in the counterclockwise direction, in ascending order of the angles they form with x-axis: a_1, ..., a_{2n}, a_{2n+1}, ..., a_{4n}. It is clear that rays a_k and a_{2n+k} belong to the same line $L_k = L_{2n+k}$. It is also easy to see that the points where these rays intersect ω go in the same order as the rays. Thus, each angle tile is formed by a pair of rays with consecutive indexes.

If the angle, defined by rays a_k and a_{k+1}, is a tile, then the point of intersection of lines L_{k+1} and L_k lies on ray a_k further away from its vertex than the point of intersection of lines L_{k-1} and L_k. Hence, the angle formed by rays a_{k-1} and a_k is not a tile—it is intersected by line L_{k+1}. Neither is the angle formed by rays a_{k+1} and a_{k+2}.

Therefore, no more than $2n$ parts of dissection are angle tiles. Also, if we have exactly $2n$ such parts, then arcs cut by these angles on ω must alternate. Say, the angle formed by rays a_k and a_{k+1} is a tile. Then angle defined by rays a_{2n+k} and a_{2n+k+1} is also a tile, that is, two vertical angles formed by lines L_k and L_{k+1} are not intersected by any other line. Obviously, that is impossible for $n > 1$, and therefore, no more than $2n - 1$ dissection parts are angle tiles.

1982.28. After transforming the fractions so they have the same denominator, we obtain equality equivalent to $(a + b - c)(b + c - a)(a + c - b) = 0$. If, for instance, $c = a + b$, then the first fraction equals (-1), and two others are equal to 1.

1982.29. Compute the kth power of $1 + \sqrt{1982}$. After combining the like terms, we have $a + b\sqrt{1982}$, where a and b are some integers. Doing the same with number $1 - \sqrt{1982}$ we obtain $a - b\sqrt{1982}$. Multiplying these two powers produces equality $(-1981)^k = a^2 - 1982b^2$;

that is, numbers a^2 and $1982b^2$ differ by 1981^k. All that is left is to denote the smaller of these two numbers by n.

1982.30. Take any three chips A, B, and X. Reflecting X with respect to A and then B, we end up translating X by vector $2\overrightarrow{AB}$. Now let us select chips A, B, and C, not lying on one straight line. Clearly, applying translations by vectors $2\overrightarrow{AB}$, $2\overrightarrow{AC}$, and $2\overrightarrow{BC}$, it is possible to move any chip into some fixed circle of sufficiently large radius R. Consider convex polygon P with $n-2$ sides, one of whose vertices lies on point A and its angle at that vertex lies inside angle BAC. Now, choose a very small radius r and construct circles of radius r centered at each vertex of P. Radius r must be so small that replacing every vertex of P with an arbitrary point lying in the corresponding circle always results in a polygon that is still convex.

Finally, consider the homothety with center A and such that the images of all small circles have radii greater than R, and the image of P with added vertices B and C is a convex polygon with n sides. Clearly, we can move all chips except B and C inside the images of these small circles—after that all chips are positioned in the vertices of a convex polygon.

1982.34. Answer: $\angle BAK = 90°$.

On ray CA select some point F located beyond point A. Then $\angle FAB = \angle DAB$, and therefore, projection onto plane ADC maps line AB into the bisector of angle FAD. Since that bisector is perpendicular to AK, then the plane that projects onto it must be perpendicular to AK. Thus, $AB \perp AK$.

1982.35. Let us index the vertices clockwise from 0 to $n-1$ and label the kth vertex with $\cos(k\pi/n + \alpha)$, $k = 0, 1, 2, \ldots, n-1$. Choose α as some more or less arbitrary angle so that none of the numbers above are equal to zero. This is the required labeling.

Hint: These numbers are, in fact, projections of the vectors leading from the origin to the vertices of this regular polygon, onto some axis passing through its center.

1982.36. Assume that some red segments AB and CD intersect. Replace them with two non-intersecting red segments AC and BD. It is easy to see that the number of intersections between red and blue segments will not increase, and the sum of the lengths of red segments will decrease. Since there are only finite number of ways to connect $2n$ red points with n segments, after a few such replacements, we will achieve the configuration where no two red segments intersect.

Now consider any red segment AB. Chord AB divides the circumference into two arcs, each of them containing odd number of selected points. Hence, segment AB must intersect at least one segment which must be blue.

1982.37. Yes, it does. Example: $k = 120$.

1982.38. Each one of these vectors $\overrightarrow{A_i A_j}$ can be represented as difference $\overrightarrow{OA_j} - \overrightarrow{OA_i}$, where O is the center of the square. Hence, if we consider the parity of the x-coordinate of the sum of all vectors, it will be the same as for the sum of all vectors from O to all marked centers. It is left to check (which is quite easy) that the above-mentioned coordinate is odd.

1982.39. Answer: 25 segments.

Split the points into two groups of five and connect by segments every two points belonging to different groups. We now have 25 segments with vertices on the given ten points that do not form triangles.

Now let us prove that we cannot have more than 25 such segments. Indeed, take a point with the maximum number n of segments issuing from it. Group the n endpoints of these segments. There are no segments connecting any two points in this group—otherwise there would be a triangle formed by the segments. Thus, each segment has one of its endpoints in one of the remaining $10 - n$ points, and therefore, the total number of segments cannot exceed $n(10 - n) \leqslant 25$.

1982.40. Answer: rice dumpling costs 13 kopecks and cabbage dumpling costs 9 or 11 kopecks.

Denote the cabbage dumpling's price in kopecks by a, and rice dumpling's price by b (with $b > a > 1$). We have $8b \geqslant 100$ and $9a \leqslant 101$, that is, $b \geqslant 13$ and $a \leqslant 11$. If Pete bought x rice dumplings, and Vicky bought y rice dumplings ($x \leqslant 8$, $y \leqslant 9$), then

$$xb + (8 - x)a = 100, \quad yb + (9 - y)a = 101.$$

It follows that $8a - 100 \vdots (b - a)$, $9a - 101 \vdots (b - a)$. Hence,

$$a - 1 = (9a - 101) - (8a - 100) \vdots (b - a),$$
$$92 = 9(8a - 100) - 8(9a - 101) \vdots (b - a).$$

The first divisibility implies that $b - a \leqslant a - 1 \leqslant 10$, and the second gives us that $b - a$ could only be equal to 1, 2, or 4. From inequalities $b \geqslant 13$, $a \leqslant 11$, we see that case $b - a = 1$ is impossible, and if $b - a = 2$, then $a = 11$, $b = 13$ (and $x = 6$, $y = 1$). Finally, if $b - a = 4$, then $a = 9$, 10, or 11, and condition $a - 1 \vdots (b - a)$ is satisfied only if $a = 9$; in that case, $b = 13$, $x = 7$, $y = 5$.

1982.41. Let I be the incenter of the triangle. Since angles BDI and BKI are right angles, points I, B, D, and K lie on the same circumference. If K and I belong to different half-planes defined by BC, then $IBKD$ is an inscribed quadrilateral and

$$\angle BDK = \angle BIK = \angle BAI + \angle ABI = \frac{\alpha}{2} + \frac{\beta}{2} = 90° - \frac{\gamma}{2},$$

where α, β, and γ are angles of triangle ABC. Note that $\angle CDE = 90° - \frac{\gamma}{2}$ since CDE is isosceles. Therefore, $\angle BDK = \angle CDE$, and lines DK and DE must coincide.

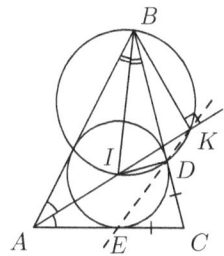

The case when K and I belong to the same half-plane can be proved in absolutely identical manner.

1982.42. We will use induction on $n+k$. Its basis requires us to prove that b_2 is divisible by b_1^2, or $a_1 a_2 \vdots a_1^2$. It follows from $a_2 - a_1 \vdots a_1$ that $a_2 \vdots a_1$.

Now for the induction step. We need to prove that $a_{n+1}a_{n+2}\ldots a_{n+k}$ is divisible by $a_1 a_2 \ldots a_k$. Obviously,

$$a_{n+1}a_{n+2}\ldots a_{n+k} = a_{n+1}a_{n+2}\ldots a_{n+k-1}(a_{n+k} - a_n) + a_n a_{n+1}\ldots a_{n+k-1},$$

where the second summand is divisible by $a_1 a_2 \ldots a_k$ (by induction hypothesis). Thus, it is left to prove the same for the first summand. But $a_{n+1}a_{n+2}\ldots a_{n+k-1}$ is divisible by $a_1 a_2 \ldots a_{k-1}$, and difference $a_{n+k} - a_n$—by a_k.

1982.43. Let A_1, A_2, ..., A_n be the given points on the plane, and convex polygon $M = A_1 A_2 \ldots A_m$—their convex hull. It is easy to see that, for $i > m$, the circumcircle of the triangle with vertex A_i cannot contain all other points.

For instance, let us prove that for triangles of form $A_x A_y A_i$, where x, $y \leqslant m$. It is clear that point A_i lies in one of the triangles $A_x A_y A_z$ ($z \leqslant m$). Then $\angle A_x A_y A_z < \angle A_x A_y A_i$, and therefore, point A_z does not lie inside the circumcircle of triangle $A_x A_y A_i$.

Thus, it is enough to consider only the triples of points which are the vertices of the convex hull M. It is obvious, that if some circumference contains all vertices of M, then it also contains all other points, and so we do not have to take points A_{m+1}, ..., A_n into account.

Next, we prove that there exists such a triangulation of polygon M that its $m - 2$ triangles represent all triangles whose circumcircles contain any points of our set (we will call such triangles *proper*).

Let diagonal $A_k A_\ell$ be included in proper triangle $A_k A_\ell A_i$. This diagonal divides the plane into two half-planes $P \supset A_i$ and P'. Obviously, for any vertex $A_x \in P$ we have $\angle A_k A_x A_\ell > \angle A_k A_i A_\ell$, and therefore, diagonal $A_k A_\ell$ is included in exactly one proper triangle in half-plane P. Select in P' vertex A_j such that angle $\angle A_k A_j A_\ell$ is the smallest for all vertices from P'. Since point A_j lies inside the circumcircle of triangle $A_k A_i A_\ell$, we have $\angle A_k A_i A_\ell + \angle A_k A_j A_\ell > 180°$. It is easy to see now that $A_k A_j A_\ell$ is the only proper triangle with side $A_k A_\ell$ that lies in half-plane P'. Thus, if a diagonal is included in a proper triangle, then there are actually two such triangles, and they lie in different half-planes separated by that diagonal.

For any side $A_k A_{k+1}$ of convex polygon M, there exists exactly one vertex A_i such that triangle $A_k A_{k+1} A_i$ is proper; it is the vertex A_i for which angle $\angle A_k A_i A_{k+1}$ is minimal. If $A_k A_i$ is a diagonal, and not a side, then we can construct the second proper triangle with side $A_k A_i$ and so on. As the result of these operations, polygon M is triangulated in such a way that all the triangles of this dissection T are proper. The number of the triangles in T is $m - 2$.

Let us prove by induction on m (number of vertices of polygon M), that there exist no other proper triangles. Basis $m = 3$ is obvious. To prove the step, note that triangulation T must contain a triangle of form $A_{t-1}A_t A_{t+1}$. Then it follows from the reasoning above that there are no other proper triangles with vertex A_t. We now "cut" that triangle off the polygon. By induction hypothesis, the "remaining" polygon has no proper triangles which are not members of triangulation T.

1982.44. Assume that such a split into two subsets A and B is possible, and that $2 \in A$. Let us denote the smallest number in B by n, then find the smallest prime number p which does not divide n. Then $p < n$. Indeed, if $n > 2$ is divisible by every prime number less than n, then n itself must be prime. But in that case, $n + 1$ is not prime, with $ab = n + 1$, $a, b < n$. It follows that $a, b \in A$, but then $n = ab - 1$ must belong to A.

Lemma. *If $x < p$, then for any number $b \in B$ number xb also belongs to B.*

Indeed, it is enough to prove that for prime x. If $xb \in A$, then number $\frac{n}{x}$ also lies in A. But $(xb)(\frac{n}{x}) = x \cdot x$, obviously contradicting the definition of the split. $\square$

Further, since n is not divisible by p, there exists $k < p$ such that p divides $kn + 1$. Then $\frac{kn+1}{p}$ belongs to A, and $(\frac{kn+1}{p}) \cdot p - 1 = kn$ lies in A as well, a contradiction.

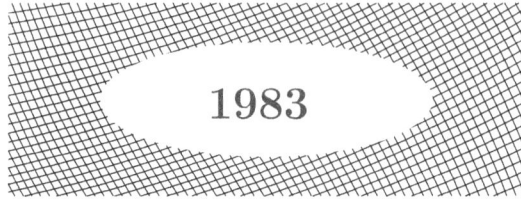

1983.01. Answer: at most 24 players can qualify.

In each game, the participants share exactly 1 point between themselves. Thus, altogether they gained $\frac{1}{2}(30 \cdot 29) = 435$ points. Each qualified player gained at least $0.6 \cdot 29 = 17.4$ points. Actually, it must be at least 17.5 points since the point total is always an integer or a half-integer. Therefore, the number of qualified players cannot be greater than $\frac{435}{17.5} < 25$. To show that 24 is possible, consider a tournament in which some 24 players drew all matches among themselves and defeated everyone else—then each one of them gained exactly 17.5 points.

Note. A different (and also interesting) question would arise if we added a condition that the tournament had no draws. The answer then is different. We leave it to the reader as an exercise.

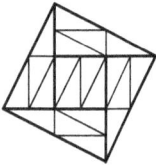

1983.02. From four tiles, we can make a 20×20 square as well as a right triangle with legs equal to 20 and 40. Arranging four such triangles around that square, we get the required tiling.

1983.03. Answer: Flick and Glick tell the truth, while Blick and Plick are liars.

From Blick's statement, we know that either Blick or Flick tell the truth. Thus, Plick is a liar, and from his second statement, it follows that Glick tells truth. Hence, Blick is a liar, and Flick tells the truth.

1983.04. This is a simple case of **1983.10**.

1983.05. Using the time machine shifts the month's number by 4 modulo 12. Since $12 \vdots 4$ and at the same time 26 is not divisible by 4, a trip claimed by the baron is not possible.

1983.06. Answer: 7, 4, 3, 0.

Denote these digits by $a > b > c > d$. If $d > 0$, then the largest and the smallest numbers are $\overline{abcd}$ and $\overline{dcba}$, respectively. Further,

$$10477 = \overline{abcd} + \overline{dcba} = 11(91a + 91c + 10b + 10d),$$

which is impossible since 11 does not divide 10477.

If $d = 0$, then

$$10477 = \overline{abc0} + \overline{c0ba} = 1001a + 110b + 1010c = 10(100a + 11b + 101c) + a,$$

and we have $a = 7$ and $11b + 101c = 347$. This equality holds true only if $c = 3$, $b = 4$.

1983.07. If a is divisible by an odd prime number $p = 2k + 1$, then $a - 1$ is divisible by $p - 1 = 2k$, and therefore, a must be odd!

1983.08. Let $a < b < c < d < e$ be the lengths of the smaller sticks. Assuming the opposite, we have

$$c \geqslant a + b, \; d \geqslant b + c \geqslant a + 2b, \; e \geqslant c + d \geqslant 2a + 3b.$$

Then the length of the large stick is at least $5a + 7b \geqslant 12 \cdot 17 = 204$ centimeters, a contradiction.

1983.09. Let K be the point of intersection of segments AE and BC, and L the point of intersection of DE and BC. Triangles ABK and DCL are congruent (**ASA** rule) and therefore, $|AK| = |DL|$. Angles EKL and ELK are equal as exterior angles for these triangles. Hence, triangle KEL is isosceles with $|KE| = |LE|$. Consequently, $|AE| = |DE|$.

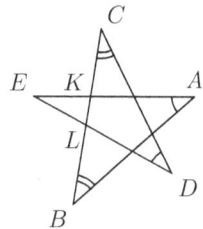

1983.10. Prove that we can change the sign of one of the numbers without changing any of the other signs. To do that, index the numbers clockwise with $1, 2, \ldots, 8$. Then changing the signs for the triples $(1, 2, 3)$, $(2, 3, 4)$, $(4, 5, 6)$, $(5, 6, 7)$, $(7, 8, 1)$ results in changing the sign only for the eighth number.

1983.12. Answer: no, that is not possible.
Let us start with an empty triangle, and then, one by one, we add the blue points. Each new blue point lies inside some existing triangle with blue vertices, so we can connect it with them by segments. This operation increases the number of "blue" triangles by 2. After adding 10 blue points, the original triangle becomes dissected into 21 blue triangles. Since there are only 20 red points, one of these blue triangles must contain at least two of the red points.

1983.14. Select two corner squares of the same color (say, white), one of which is on the lower side of the sheet, and the other one—on the upper side. We now construct a sequence of polygons P_n. The first polygon P_1 consists solely of the lower side white corner square. Every next polygon P_{k+1} is obtained from P_k by adding one more square—first we add, one by one, all the squares of the first (lowermost) row, then we do the same with the squares of the second row, and so on. The last polygon contains all squares of the sheet except for the upper side's white corner square.

Let us monitor the difference d_k between the numbers of black and white squares inside P_k. In the beginning, $d_1 = 1$, at the end it equals (-1), and with every "move" from index k to index $k + 1$, this difference changes by ± 1. Therefore, it is obvious that at some moment we have $d_k = 0$. Then the corresponding polygon P_k contains equal number of black and white squares, and the same is true for its complement.

1983.15. Let us arrange the sticks in the descending order of their lengths. If it is impossible to form a quadrilateral, then the second stick is longer than the sum of the three shortest ones, and therefore, longer than 57 cm. Similarly, the first stick is longer than the sum of the second, the third and the fourth, and thus, longer than $19 + 19 + 57 = 95$ cm. Then the total length of the original stick is greater than $19 \cdot 3 + 57 + 95 = 209$ cm, a contradiction.

1983.17. Transform the original expression as follows.

$$2^{58} + 1 = 2^{58} + 1 + 2^{30} - 2^{30} = (2^{29} + 1)^2 - 2^{30}$$
$$= (2^{29} - 2^{15} + 1)(2^{29} + 2^{15} + 1)$$
$$= 5 \left(\frac{2^{29} - 2^{15} + 1}{5} \right) (2^{29} + 2^{15} + 1).$$

1983.18. Perform these operations for all squares from the union of some column and some row. It is easy to see that the result will be exactly the change of the sign for only one number—the one at the intersection of the selected column and the selected row. Thus, we can change the sign for any number in the table without affecting any other number.

1983.19. Let $S(n)$ denote the sum of the digits of number n. Represent number a as $a = x_1 + y_1$, where all digits of x_1 are even, and y_1 is written only with zeros and ones. Similarly, we represent $b = x_2 + y_2$.

Obviously, each of the numbers $5y_1$ and $5y_2$ is written only by zeros and fives, and $S(y_1) = S(y_2)$. Numbers $5x_1 = \frac{1}{2}(10x_1)$ and $5x_2 = \frac{1}{2}(10x_2)$ differ only by some permutation of their digits, and therefore, $S(5x_1) = S(5x_2)$. In addition, all the digits in these numbers do not exceed 4. Thus, there are no carryovers in both additions $5x_1 + 5y_1$ and $5x_2 + 5y_2$, and subsequently,

$$S(5a) = S(5x_1) + S(5y_1) = S(5x_2) + S(5y_2) = S(5b).$$

1983.20. Answer: 18.5 points.

Let us split all 24 games into six parts each consisting of four consecutive games. Take any one of these parts. The winner could not have gained more than 3.5 points, and there is only one way to gain 3.5 points—his point gains in these four games must be 1, 0.5, 1, 1. If that happened, and this part is not the last one, then in the next four-game part, its first game (obviously, it has an odd number) must end with the winner's loss, and in the last three games of this part, he could not have gained more than 2.5 points. So, in any non-final four-game part, the winner either gains no more than 3 points, or he gains no more than 6 points in the eight games of this part together with the one that follows it. Hence, in all six parts together, the winner gains at most $3 \cdot 5 + 3.5 = 18.5$ points.

An example with 18.5 points gained—the winner wins all odd-numbered games, as well as the very last game of the match; all the other games are drawn.

1983.21. Let S be the area of regular hexagon $A_1 A_2 A_3 A_4 A_5 A_6$. We assume that

$$S(A_1 O A_2) + S(A_3 O A_4) + S(A_5 O A_6) \geqslant \frac{S}{2}.$$

Now consider equilateral triangle ABC with sides equal to the sides of this hexagon. Let triangles $AC'B$, $BA'C$, and $AB'C$ be congruent to triangles $A_1 O A_2$, $A_3 O A_4$, and $A_5 O A_6$, respectively.

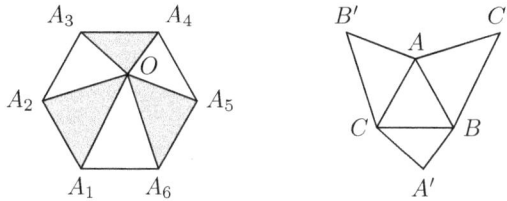

Then the area of hexagon $AB'CA'BC'$ is greater than or equal to $S/2+S(ABC)=2S/3$.

1983.22. From Thales' (intercept) Theorem, we have

$$\frac{|A_1C_1|}{|BC_1|}=\frac{|CM|}{|MB|}=\frac{|CM|}{|MA|}=\frac{|B_1C_1|}{|AC_1|}.$$

Hence, $|A_1C_1|\cdot|AC_1|=|B_1C_1|\cdot|BC_1|$. Then

$$\frac{S(AA_1C_1)}{S(ACM)}=\frac{|A_1C_1|\cdot|AC_1|}{|AM|\cdot|MC|}=\frac{|B_1C_1|\cdot|BC_1|}{|MB|\cdot|MC|}=\frac{S(BB_1C_1)}{S(BMC)}.$$

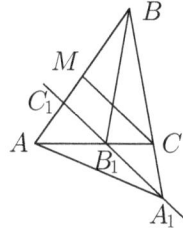

Here we have used the fact that the ratio of the areas of two triangles with pairs of respectively parallel sides equals the ratio of the products of those sides.

Since $S(ACM)=S(BMC)$, we obtain $S(AA_1C_1)=S(BB_1C_1)$.

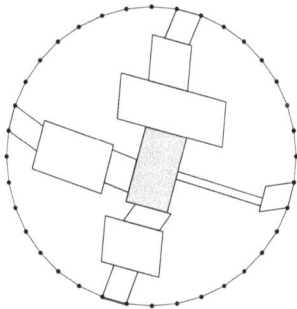

1983.23. Consider two opposite sides of the polygon. They must be connected to each by other by a chain of adjacent parallelograms. Now, consider the other pair of opposite sides perpendicular to the sides of the previous pair. In the same manner, they also determine a chain of parallelograms. These two chains must intersect—it is clear that their intersection is a parallelogram that has to be a rectangle with the sides parallel to the selected quadruple of the polygon's sides. Now, altogether we can select one hundred such quadruples of sides, and all the corresponding rectangles, of course, must be different.

1983.24. Let us index these points by numbers 1 through n. Then, we can assume that when points collide, they do not bounce off of each other but go through each other, keeping their velocities while "swapping" their indexes. If T is the time that one point needs to go around the circle exactly once, then after this amount of time passes, all the points return to their original places, although their indexes will be permuted in some order. Let us denote that permutation by η.

After one more time period of length T passes, the points again return to their original positions with the same index permutation performed in superposition with the first one—thus, the overall permutation is equal to $\eta^2=\eta\circ\eta$.

Therefore, after k periods of length T pass, we will have the permutation of indexes equal to η^k. If $k=n!$ (n factorial), this permutation returns all indexes to their original values, and therefore, all points return to their original positions.

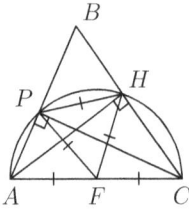

1983.25. Answer: $60°$.

If F is the midpoint of segment AC, then

$$|FH| = |FP| = |FA| = |FC| = |AC|/2 = |PH|.$$

Thus, triangle FPH is equilateral. Since the inscribed angle is half of the central angle, therefore, we obtain $\angle HAB = \angle HFP/2 = 30°$, and then $\angle B = 60°$.

1983.26. Answer: there exists only one "prime" square—the one with vertices $(2; 2)$, $(2; 3)$, $(3; 2)$, $(3; 3)$.

It is obvious that all the coordinates are greater than or equal to 2. If for some side, both x- or y-coordinates of its endpoints are greater than 2, then all coordinates of integer points on this side are greater than 2, and therefore, must be odd. Since the midpoint M of that side is an integer point, its coordinate is again odd. Then the midpoint of the half-segment between M and one of the side's endpoints X is again an integer point, and therefore, its coordinates are odd. Take the midpoint of the segment between this new point and X, and so on. Thus, we obtain an infinite sequence of different integer points on that side, which is impossible.

Hence, at least one end of each side of the square lies on one of the lines $x = 2$ or $y = 2$. It is easy to see that the sides must be parallel to the coordinate axes. It follows immediately that any side's length must be 1 and only one square fits these restrictions.

1983.27. Obviously the overall number of candies must be finite; it is also always bounded from above.

Indeed, if in the beginning each kid had at most $2n$ candies, then each of them would leave for himself or herself no more than n of his or her own candies (including, if there is one, a candy from the teacher), and then would receive from the neighbor on their left no more than n of their candies. Thus, the number always stays under $2n$.

From this, we can conclude that at some point the teacher will stop handing out the extra candies—otherwise, the overall number of candies would increase indefinitely, contradicting the $2n$ upper bound we just proved. Therefore, from some moment on, the numbers of candies that the kids have must become even and stay even.

If at some moment the kids had $2a_1$, $2a_2$, $\ldots$, $2a_n$ candies, then at the next moment they have $a_n + a_1$, $a_1 + a_2$, $\ldots$, $a_{n-1} + a_n$ candies. If we evaluate the sum of the squares of these numbers, then it is easy to see that this expression decreases unless all numbers a_k are equal. Since this expression is a non-negative integer, it cannot decrease indefinitely, and therefore, at some moment all the numbers must become equal.

1983.28. Answer: the inequality is true only for $n = 2$, 3, 4, and 5.

The difference between the left-hand and the right-hand sides of the inequality is

$$\sum_{i=1}^{n-1}\left(x_i - \frac{x_n}{2}\right)^2 + \frac{1}{4}(5-n)x_n^2.$$

This expression is non-negative for $n = 2$, 3, 4, 5; for any other n, set $x_n = 2$, $x_1 = x_2 = \cdots = x_{n-1} = 1$ to provide the counterexample.

1983.29. Assume, without the loss of generality, that radius of K_1 does not exceed radius of K_2, and denote the points of tangency between circles K_1 and K_2 with lines ℓ_1 and ℓ_2 by A, B, C, and D. It is clear that $AC \parallel BD$ and lines A_1C_1 and B_1D_1 are parallel to AC and tangent to circles K_1 and K_2 at points E and F, respectively.

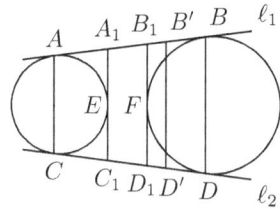

Then $|AA_1| = |A_1E|$, $|CC_1| = |C_1E|$, $|BB_1| = |B_1F|$, $|FD_1| = |DD_1|$. It is clear that $|AA_1| = |CC_1| < |BB_1| = |DD_1|$, unless points E and F coincide. Construct points B' and D' such that A_1 is the midpoint of segment AB', and C_1 is the midpoint of segment CD'. Then line $B'D'$ is parallel to line BD while not coinciding with it. Trapezoid $AB'D'C$ is circumscribed, and therefore, trapezoid $ABDC$ is not, a contradiction.

1983.31. Consider remainders modulo 101.

1983.34. Define unit vectors $\vec{a}$, $\vec{b}$, $\vec{c}$ issuing from the vertex of the angle along its edges. Now, the two directions of the angle bisectors, perpendicular to each other, are $\vec{a} + \vec{b}$ and $\vec{b} + \vec{c}$.

Then $(\vec{a} + \vec{b})(\vec{b} + \vec{c}) = \vec{0}$, that is, $\vec{a}\,\vec{b} + \vec{b}\,\vec{c} + \vec{c}\,\vec{a} = -1$. This implies

$$(\vec{a} + \vec{c})(\vec{b} + \vec{c}) = \vec{0}, \quad (\vec{a} + \vec{b})(\vec{a} + \vec{c}) = \vec{0}.$$

Hence, all three angle bisectors of the face angles are pairwise perpendicular, and the proof follows.

1983.35. Answer: $9°$, $85°30'$, $85°30'$.

Let A be the vertex of the pyramid, H—the foot of the altitude, O—common center of its inscribed and circumscribed spheres, ABC—some side face, and F—projection of point O onto plane ABC. Since O is the pyramid's circumcenter, point F is the circumcenter of triangle ABC, and since O is the pyramid's incenter, triangles BHC and BFC are congruent (**SSS** rule). Thus, we obtain that $\angle BAC = \frac{1}{2}\angle BFC = \frac{1}{2}\angle BHC = \frac{360°}{20\cdot2} = 9°$. Hence, $\angle ACB = \angle ABC = 85°30'$.

1983.37. The set of vertices of regular polygon with twenty sides can be represented as the union of the sets of vertices of four regular pentagons.

One of these sets must contain at least three marked vertices; finally, any three vertices of a regular pentagon form an isosceles triangle.

1983.38. Since the problem is clearly invariant with respect to affine transformations,[110] we can assume, without loss of generality, that the triangle is equilateral and its area equals 1.

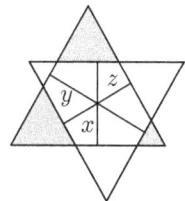

If the center of symmetry lies outside of the triangle, then the area of intersection is zero. So the center is inside the triangle; let us denote the distances from this point to the sides of the triangle by x, y, and

---

[110] Affine transformation is the bijection that maps lines to lines; a general affine transformation of the coordinate plane is defined by formulas $x_1 = ax + by$, $y_1 = cx + dy$, where $ad - bc \neq 0$.

z, respectively. Then $x + y + z = h$, where h is the length of the triangle's altitude. If all three distances are less than $\frac{h}{2}$, then the intersection is the hexagon whose area equals $1 - (\frac{h-2x}{h})^2 - (\frac{h-2y}{h})^2 - (\frac{h-2z}{h})^2$. Applying AM-GM inequality for three numbers, and using $x + y + z = h$, we obtain the desired inequality. And if the intersection is a parallelogram, then its area does not exceed even one half of the area.

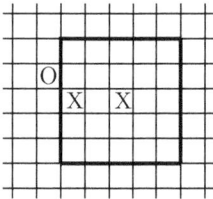

1983.39. Answer: yes, the first player wins in the errorless game.

We should emphasize here that the second player does not have any objectives in the game except for preventing the first player from reaching his.

After each player made one move, we can assume that the first player's move was made to square $a3$ of 5×5 table, and the second player's move was made outside of that table.

Then the first player makes the second move to $c3$ (see the figure). Define set S as the collection of seven squares $\{a2, b2, c2, b3, a4, b4, c4\}$. If the second player's next move is outside of S, then the first player puts an X into $b3$, and after the next move of the second player, one of the triples $a2$, $b2$, $c2$ and $a4$, $b4$, $c4$ is completely empty. The first player then puts X into $b2$ (or $b4$) and wins with his next move.

If the second player writes O into one of the seven squares of S, then the first player writes X into $e3$ and after the second player's next move, at least one of the triples $a1$, $c1$, $e1$ and $a5$, $c5$, $e5$ is empty. The first player then writes X into the center square of the empty triple, creating the fork that wins the game.

1983.40. Answer: $10°$.

Construct segment BN on ray AM such that $|BN| = |BC|$. Then it follows from $|BM|/|BN| = |CB|/|CN|$ that in isosceles triangle BNC, line CM is the bisector of angle C. Since $\angle BCN = 20°$, we obtain $\angle BCM = 10°$.

1983.41. Answer: such sequence exists if and only if n is not a power of two.

Set $s_k = a_1 + a_2 + \cdots + a_{k-1}$. Then equalities from the condition can be rewritten as $s_k = s_{2k}$ for any k; we also, obviously, have $s_1 = 0$. Since there are n independent variables s_k and n equations for them, then $a_1 = a_2 = \cdots = a_n = 0$ is equivalent to $s_1 = s_2 = \cdots = s_n$, which is in turn equivalent to the fact that for every two indexes k and m, there exist powers of two 2^a and 2^b such that $2^a k \equiv 2^b m \pmod{n}$. In other words, for any k, there exists power of two 2^c such that $2^c k$ is divisible by n. Our claim follows now from using this fact for $k = 1$.

1983.42. The second layer is constructed as follows. Tile the plane with 2×2 squares, with each of them covered by either two horizontal or two vertical tiles. Then we lay down the third layer. Namely, we connect centers of the neighboring squares, that are not yet covered by a 1×2 tile, with a segment. Obviously, each center has two segments coming out of it. Curves, formed by these segments, can be divided into two types, (a) finite closed curves, and (b) infinite non-self-intersecting curves. It is easy to see that the curves of type (a) must have even length. Therefore, each curves of any type can be split into non-intersecting unit segments, and these segments define the tiles of the third layer. Finally, the fourth layer is uniquely determined by the first three—for each square X there is exactly one of its neighbors Y such that two squares X and Y do not form a tile already included in one of the three previous layers.

1983.43. Let us assume that a_k is the first term of the sequence not satisfying the given inequality. Then it is clear that $a_k - 1$ does not belong to the sequence, and thus, there exists index s such that $a_s + 2s + 1 = a_k$. Denote

$$n = \left[\frac{a_s + 1}{2 + \sqrt{2}}\right] = a_s - \left[\frac{a_s + 1}{\sqrt{2}}\right], \text{ since } \frac{1}{2 + \sqrt{2}} + \frac{1}{\sqrt{2}} = 1,$$

and we have $a_n + 2n \leqslant (2 + \sqrt{2})n < a_s + 1$.

To every natural number x between $a_s + 1$ to a_k, we assign the term of sequence (a_i) that "generated" number x, i.e., either $a_i = x$ or $a_i + 2i = x$. It follows from the inequalities above that to each x, we assigned at most one number from the interval between $n + 1$ and k; therefore, $a_k - a_s \leqslant k - n$, or $a_k - (a_s - n) \leqslant k$ (∗).

Let $w = [(a_s + 1)/\sqrt{2}]$. Then $a_s > \sqrt{2}w - 1$, and therefore, $s \geqslant w$. It follows that

$$\frac{a_s + 2s + 1}{2 + \sqrt{2}} \geqslant \frac{\sqrt{2}w - 1 + 2w + 1}{2 + \sqrt{2}} = w.$$

Hence, we have proved that $a_k/(2 + \sqrt{2}) \geqslant w$, or, equivalently, $a_s - n \leqslant a_k/(2 + \sqrt{2})$. Adding this to inequality (∗), we obtain $a_k \leqslant \sqrt{2}k$.

1983.44. Answer: $10(n^2 - 1)$.

Consider an arbitrary segment S in that road system (let us assume that it is horizontal), starting and ending outside of a town. Suppose there are a and b (with $a \geqslant b$) roads perpendicular to S that have one of their endpoints inside S while lying entirely below and above S, respectively.

If we start gradually translating S in the downward direction, we would not increase the total length of the system until one of the two things happen. First case is that S overlaps some other existing road-segment, and the second case is that S merges with another segment to become a new, longer road. In the former case, the total length of the system decreases, and in the latter case, we can continue on doing the same but with the new, extended segment, unless, of course, one of its endpoints is a town.

Thus, we can transform our road system into another one, whose total length does not exceed the original total length, with all roads beginning and ending in the towns. It suffices now to remember that any connected graph with k vertices has at least $k - 1$ edges.

1984

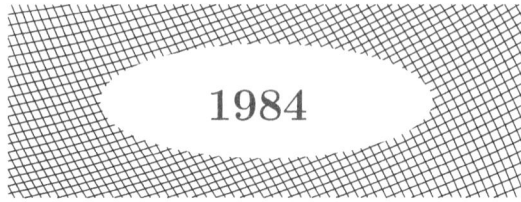

1984.01. Cross out several digits so that the remaining number is

$$419841984\ldots1984 \quad \text{(quartet `1984' repeats 90 times.)}$$

Its sum of the digits equals $4 + 90 \cdot (1 + 9 + 8 + 4) = 1984$.

1984.02. Here is one of the examples.

1	−1	−1	1
−1	1	1	−1
1	−1	−1	1
−1	1	1	−1

1984.03. The difference between the distance from any such point to A and the distance from that point to B equals the length of segment AB. Clearly, the sum of 45 numbers of form $\pm|AB|$ cannot be equal to zero.

1984.04. Let us assume the opposite. Any two of the nine squares of 3×3 table can be covered by the given figure; therefore, they must all have different colors, which is impossible.

1984.05. Let us color the sectors into black and white, alternating the colors so that any two adjacent sectors are colored differently.

Then the difference between the sum of the numbers in black sectors and the sum of the numbers in white sectors is an *invariant*—that is, it does not change under the described operations. Originally, that difference was 2, and therefore, it cannot vanish.

1984.06. The first player will place the multiplication sign between 1 and 2, and then he will mentally split all yet unoccupied spaces, starting with the space between 3 and 4, into pairs of consecutive spaces. Then he will use the following strategy. When the second player places a sign into one of the spaces of some pair, the first player responds with the multiplication sign in the second space of that pair (if it is not already occupied; as you will see, it can be only occupied by a multiplication sign). If the second player places a sign between 2 and 3, the first player writes the multiplication sign into any unoccupied space.

As a result, in the end the players always end up with a sum of several even numbers.

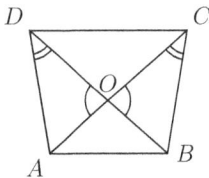

1984.08. These equalities imply that $|AC| = |BD|$, $|AD| = |BC|$. Then triangles ABC and BAD are congruent and $\angle OCB = \angle ODA$. Therefore, triangles OAD and OBC are also congruent and we have $|OB| = |OA|$.

1984.09. (b) The sum of the eight numbers in any two neighboring columns is zero, because they are the union of the sets of vertices of two 2×2 squares. Furthermore, the sum of the eight numbers in the first and the third columns is zero, because they are the union of the sets of vertices of two 3×3 squares.

If we denote the sum of the numbers in the ith column by b_i, then we have $b_1 + b_2 = b_2 + b_3 = b_1 + b_3 = 0$, and solving this system we get $b_1 = b_2 = b_3 = 0$. Similarly, $b_4 = 0$.

1984.10. Rewrite the inequalities as follows: $7H > 5D$, $25H < 18D$, where H and D are the prices of a Humpty and of a Dumpty, respectively. Thus, we have $7H \geqslant 5D + 1$ and $25H \leqslant 18D - 1$.

Multiplying these inequalities by 25 and 7, respectively, we obtain $175H \geqslant 125D + 25$ and $175H \leqslant 126D - 7$. Subtracting, we get $D \geqslant 32$. From the first inequality, we now obtain $H \geqslant 23$, and now it follows that $3H + D \geqslant 101$.

1984.12. Color in black all the unit squares on the sheet for which both the number of the column and the number of the row have a non-zero remainder modulo 3. Obviously, these black squares form one hundred 2×2 tiles. It is also clear that any tile on the sheet cannot intersect more than one black tile. Since we have 100 black tiles, one of them does not intersect any of the 99 tiles we cut out. This black tile is the one we can now cut.

1984.13. Answer: no, they cannot.
Assume the opposite. Note that each of the segments (side or diagonal) serves as a side of three different triangles formed by other segments. Consider segment with length 2. For each of the three triangles that use it as a side, the difference of the two other sides has to be less than 2 (triangle's inequality). But among other numbers, there are only two non-intersecting pairs of numbers with difference 1, a contradiction.

1984.14. Answer: the sum equals 2.
After transforming the original equation

$$\frac{1}{x^2 + 1} + \frac{1}{y^2 + 1} = \frac{2}{xy + 1}$$

to the equal denominators and then simplifying, we get $(x - y)(xy - 1) = 0$. Since $x \neq y$, we have $xy = 1$, and that implies that the third summand is 1, and the sum total is 2.

1984.15. From triangle ABC we can construct parallelogram $ABXC$. Then $2|AM| = |AX|$. Also $|AX| = |DE|$, because triangles ABX and DAE are congruent by **SAS** rule ($|AB| = |DA|$, $|BX| = |AC| = |AE|$, and $\angle ABX = \angle ABC + \angle ACB = \angle DAE$). Therefore, $2|AM| = |DE|$.

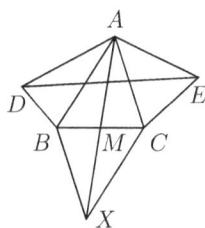

1984.16. Let $d = |AB|$. Distances from any red point to points A and B differ by d, and the same is true for any blue point. Thus, the two sums in the problem's statement differ from each other by the sum of 45 summands of form $\pm d$, which is different from zero because 45 is odd.

1984.17. See the solution to **1984.12**, and let us color black the same one hundred 2×2 tiles. Again, one of them does not intersect any of the 99 cut 2×2 squares. This black tile is the one we can cut.

1984.18. Answer: eight colors.

Assume there exists a coloring that uses only seven colors. Consider an arbitrary 2×3 rectangle. All of its six squares are of different color. Also, the colors of the squares, adjacent to the midpoints of the longer sides of this rectangle, must be different from all these six colors, and therefore, they must be the same color. It follows then that any two squares in the same row (or column), which are separated by exactly two squares, must have the same color.

4	8	2	6
3	7	1	5
2	6	4	8
1	5	3	7

The entire infinite sheet then "falls apart" into nine separate one-color square grids with square size 3. It is easy to see that any two of these grids have neighboring (by side or by corner) squares. This implies that any two grids must be colored differently, requiring nine colors.

For an example with eight colors, simply tile the plane with 4×4 squares colored as shown in the figure.

1984.19. Let us assume the opposite. Index the rows with letters a, b, ..., and the columns—with numbers 1, 2, ... Now, consider corner square $a1$. Say, it is covered by the horizontal 1×2 tile. Then square $a2$ must be covered by the vertical tile, and therefore, square $b2$ is covered by the horizontal tile, and so on, until we come to the opposite corner of the table, where we immediately run into a contradiction.

1984.20. We have $99\,999(a - f) + 990(\overline{bc} - \overline{de}) \; \vdots \; 271$. Since prime number 271 divides $99\,999$ but does not divide 990, it follows that $\overline{bc} - \overline{de} \; \vdots \; 271$, which is possible only if $\overline{bc} = \overline{de}$.

1984.21. Assume that $x_2 > x_1$. Then the first equation implies $x_3 < x_2$, the second equation implies $x_3 > x_4$ and so on, and finally, $x_2 < x_1$, a contradiction. Thus, $x_1 = x_2$, and it follows that other variables x_i are equal to each other as well.

1984.22. Let A, B, C, and D be the vertices of the given trapezoid, $BC \parallel AD$. We can prove that $|OA| + |OB| + |OC| > |OD|$ for any point O. Indeed, $|OA| + |OB| \geqslant |AB|$, $|AB| = |CD|$, and $|OC| + |CD| \geqslant |CD|$; furthermore, at least one of these inequalities is sharp. All other cases are almost identical.

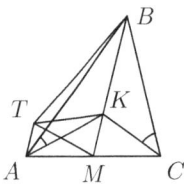

1984.23. It would suffice to prove that $BM \perp AC$. Assume the opposite, for instance, that $\angle BMC < \pi/2$. Reflect point C across BM to obtain point T. Then MB is the exterior bisector of angle $\angle TMA$ at the vertex of isosceles triangle ATM, hence, $MB \parallel AT$. Furthermore, $\angle KTB = \angle KCB = \angle KAB$, and therefore, trapezoid $BKAT$ is inscribed, and as such, isosceles.

It means that the midpoint of base KB is equidistant from the ends of the other base, that is, from points A and T. But the perpendicular bisector of segment AT intersects line BM (parallel to AT) at the only point M that is not the midpoint of BK. This contradicts the original assumption.

1984.24. Split $\mathbb{N}$ into consecutive hundreds, and in each one of them find an element of set A. Consider the differences between two consecutive elements of A—they cannot exceed 200, and therefore, they must repeat. Hence, there are numbers a, b, c, d belonging to A such that $a - c = b - d$ or, equivalently, $a + d = b + c$.

1984.27. Answer: 1984.

Let the answer be $n = 13x_1 + 73y_1 = 13x_2 + 73y_2 = 13x_3 + 73y_3$, where $x_1 > x_2 > x_3$. Then $13(x_1 - x_2) \vdots 73$, and $x_1 - x_2 \geqslant 73$. Similarly, $x_2 - x_3 \geqslant 73$. Therefore, $x_1 \geqslant x_3 + 146 \geqslant 147$ and $n \geqslant 13 \cdot 147 + 73 = 1984$. If $n = 1984$, then we have three desired representations with $x_1 = 147$, 74, and 1.

1984.28. Answer: ABC is equilateral triangle.

Let $\alpha \geqslant \beta \geqslant \gamma$ ($\angle BAC = \alpha$) be $\triangle ABC$'s angles, with altitudes intersecting at point H. Then HA_1BC_1 is inscribed, thus, $\angle HA_1C_1 = \angle HBC_1 = 90° - \alpha$; similarly, $\angle HA_1B_1 = 90° - \alpha$. Then $\angle B_1A_1C_1 = 180° - 2\alpha$, and the measures of the angles of ortho-triangle $A_1B_1C_1$ are equal to $180 - 2\alpha$, $180 - 2\beta$, $180 - 2\gamma$. Obviously, that is the ascending order and therefore, from similarity, we have $\alpha = 180° - 2\gamma$, $\gamma = 180° - 2\alpha$, hence, $\alpha = \gamma = 60°$.

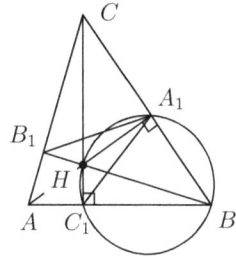

1984.30. Indeed,

$$a^4 + b^4 + c^4 = a^2(b+c)^2 + b^2(a+c)^2 + c^2(a+b)^2$$
$$= 2(a^2b^2 + b^2c^2 + c^2a^2) + 2abc(a+b+c)$$
$$= (a^2 + b^2 + c^2)^2 - a^4 - b^4 - c^4.$$

Hence, $2(a^4 + b^4 + c^4) = (a^2 + b^2 + c^2)^2$.

1984.31. See solution to **1984.22**

1984.33. Denoting our parallelogram by $ABCD$, we have

$$|AC| \cdot |BD| > |\overrightarrow{AC} \cdot \overrightarrow{DB}| = |(\overrightarrow{AB} + \overrightarrow{BC})(\overrightarrow{AB} - \overrightarrow{BC})| \geqslant |AB|^2 - |BC|^2.$$

1984.34. Using substitution $x = y - 1$, change the given polynomials to $y^8 - 7y^7 + \cdots$ and $ay - (a - b)$. Accordingly, change the operation of multiplication by $x + 1$ to multiplication by y, and differentiation by x—to differentiation by y. Clearly, the free term $a - b$ is then the result of these operations applied to polynomial $(-7y^7)$. At some moment, the degree of that polynomial changed from seven to six when the leading coefficient was multiplied by 7, and therefore, became divisible by 49. The further operations cannot change this property, thus, from that moment on, the leading coefficient must be divisible by 49. Hence, $a - b$ is divisible by 49.

1984.35. Answer: 1984.

This sum is equal to

$$\pm x_1 \pm 1 \pm x_2 \pm 2 \pm \cdots \pm x_{63} \pm 63 = \pm 1 \pm 1 \pm 2 \pm 2 \pm \cdots \pm 63 \pm 63,$$

where sixty-three signs among "$\pm$" are pluses, and other sixty-three are minuses. The maximum possible value of this expression equals $-1 - 1 - 2 - 2 - \cdots - 31 + 31 + 32 + 32 + \cdots + 62 + 62 + 63 + 63 = 1984$, and it is attained when $x_1 = 63$, $x_2 = 62$, ..., $x_{63} = 1$.

1984.36. Assume that such an edge does not exist. Only three numbers 1, 99, and 100 can be next to 50; also, numbers 99 and 1 cannot have any other common neighbor. Thus, the second end of the vertical edge across from 50 is labeled with 100. Suppose that 1 is placed to the right of 50, 99—on the left. Then the position of numbers 2 and 51 is uniquely defined. Furthermore, moving around the prism, we can determine the position of all the other numbers. After the full circle, coming back from the other side of the prism, we obtain that 99 cannot be at the position we already determined in the beginning.

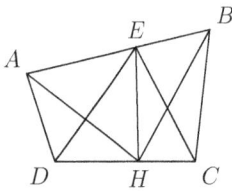

1984.37. Since

$$S_{AEH} : S_{BEH} = |AE| : |BE| = |DH| : |CH| = S_{DHE} : S_{CHE},$$

segment EH divides the areas of triangles ABH and DEC in equal ratios. Therefore, $S_{BEH} = S_{CHE}$ and $S_{AEH} = S_{DHE}$. It follows that points B and C are equidistant from line EH, as well as points A and D. Hence, $BC \parallel EH \parallel AD$.

1984.38. With the first move, player #1 writes 7, thus preventing player #2 from immediately writing a perfect square; from that moment on, he only writes either 7 or 8 to the right of the previous number. If player #2 adds some digit on the left, then the result ends with 7 or 8, and therefore, cannot be a perfect square.

Now let us prove that player #1 can always choose between 7 and 8 so that regardless of what digit player #2 adds on the right, the new number will not be a perfect square. Indeed, if for both choices player #2 can create a perfect square, that would mean that there are two perfect squares different only in the last two digits (and the difference in the tens would be at most 1). Then, clearly, the difference between these two perfect squares cannot exceed 20, but for any two consecutive squares, each of them written by more than two digits, their difference is at least $121 - 100 = 21$.

1984.39. Answer: no, that is impossible.

Assume that it can be done. The final quartet of points is such that for every point, the difference of its coordinates is divisible by 3. This property is invariant under the described operations (as well as the inverse operations), and therefore, the initial quartet would have to possess it as well.

1984.40. Consider the smallest non-zero number a among the given ten numbers. After the permitted operation, this number does not decrease unless the arithmetic mean operation involves a zero; in that case, it can decrease, but (a) there will be one less zero; (b) the new value of a will be greater than or equal to $\frac{a}{2}$. It follows that only nine operations can decrease the value of a. Furthermore, nine operations cannot decrease a by the factor greater than $2^9 = 512$. Thus, the answer is $\frac{1}{512}$.

1984.41. Consider a segment of length

$$S = a_1 + a_2 + \cdots + a_k = b_1 + b_2 + \cdots + b_n.$$

Then let us divide this segment using $k-1$ red points into smaller red segments with lengths $a_1, a_2, \ldots, a_k$; also, using $n-1$ blue points, divide it into smaller blue segments with lengths

$b_1, b_2, \ldots, b_n$. Now, into the table's square, at the intersection of the ith row and the jth column, write the number equal to the length of the intersection of the ith red segment and the jth blue segment. It is easy to see that this is, indeed, the desired number placement.

1984.42. Assume that such a route is possible. Then let us consider the "border" of the square formed by the points one of whose coordinates equals 0, 1, 11, or 12. It is easy to see that the two border points cannot serve as consecutive stops on the chip's route. Thus, if the route goes through all 169 points, then no more than 85 of them can belong to the border. However, there are $169 - 81 = 88$ border points—this contradiction proves that such a route is impossible.

1984.43. Let us consider the case when this circumference—we will denote it by S—lies outside triangle OCD. If we define C' and D' as intersections of the sides of the angle with the line tangent to S at point A, then S is inscribed in triangle $OC'D'$. It is obvious that this triangle is the image of triangle OCD under a homothety with center O.

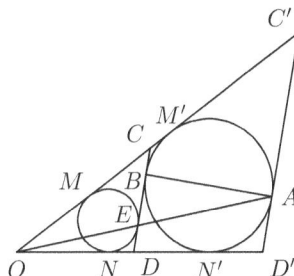

Therefore, E is the point where the incircle of triangle OCD touches side CD. Let M, M', N, N' be the points of tangency of these two circumferences with lines OC and OD.

From the property of tangents we have

$$2|BC| + |BE| = |BC| + |CE| = |M'C| + |CM| = |M'M| = |OM'| - |OM|$$
$$= |ON'| - |ON| = |N'N| = |N'D| + |DN|$$
$$= |DB| + |DE| = 2|DE| + |BE|.$$

Thus, $|BC| = |DE|$.

1984.44. Consider the set A of all natural numbers whose ternary representation (that is, base 3) contains only digits 0 and 1. Obviously, any natural number x can be uniquely represented as a sum of two numbers from A, and these two numbers are different if and only if x has at least one ternary digit 1. Also, clearly, $n \in A$ if and only if ternary representation of $2n$ has only digits 0 and 2, that is, exactly when $2n$ is not the sum of two different numbers from A.

1984.46. Let $SA_1A_2\ldots A_n$ be the given pyramid. Cut its side surface along the longest edge (say, SA_1) and unfold it onto the plane. It follows from the problem's conditions that point S lies inside polygon $A_1A_2\ldots A_nA_1'$. Let B be the second point of intersection of line SA_1' with broken line $A_1A_2\ldots A_nA_1'$. This point breaks $A_1A_2\ldots A_nA_1'$ into broken line $A_1\ldots B$ of length a and broken line $A_1'\ldots B$ of length b. Inequalities $SA_1 < a + SB$, $A_1'S + SB = A_1'B < b$ give us that $2SA_1 = SA_1 + SA_1' < a + SB + A_1'S < a + b$. It is left to note that the last number is the perimeter of the pyramid's base.

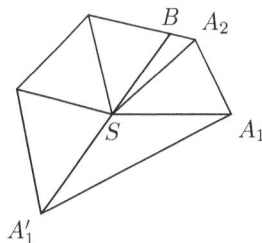

1984.47. Consider identity

$$a = \frac{3a}{3} = \frac{x^2 + 2y^2}{3} = \left(\frac{x \pm 2y}{3}\right)^2 + 2\left(\frac{x \mp 2y}{3}\right)^2.$$

It is obvious that numbers x and y are divisible or not divisible by 3 simultaneously. If their remainders modulo 3 are the same, choose the upper signs, and if different—the lower signs. Then both fractions are integers.

1984.51. Consider polynomial

$$P(x) = (x-a)(x-b)(x-c)(x-d)(x-e).$$

Expanding, we get $P(x) = x^5 + px^4 + qx^3 + rx^2 + sx + t$. From Vieta's formula we get

$$p = -(a+b+c+d+e),$$

$$q = \frac{1}{2}\left((a+b+c+d+e)^2 - (a^2+b^2+c^2+d^2+e^2)\right),$$

with both of these numbers divisible by n. Adding equalities $P(a) = 0$, $P(b) = 0$, ..., $P(e) = 0$, we obtain that

$$a^5 + b^5 + c^5 + d^5 + e^5 + 5t + r(a^2+b^2+c^2+d^2+e^2) + s(a+b+c+d+e)$$

is divisible by n. Now we use equality $t = -abcde$ to complete the proof.

1984.52. Define function

$$S(a, b, c, d, e, f) = 2a + 4b + 6c + 8d + 10e + 12f.$$

Now, if x_1, ..., x_7 are some seven consecutive terms of our sequence, then the last digits of numbers $S(x_1, \ldots, x_6)$ and $S(x_2, \ldots, x_7)$ are the same—that can be easily verified. It follows that for any k the last digit of $S(x_{k+1}, \ldots, x_{k+6})$ is always the same. For the original six numbers, we have $S(1, 0, 1, 0, 1, 0) = 18$, and $S(0, 1, 0, 1, 0, 1) = 24$. These values have different last digits, and we are done.

1985

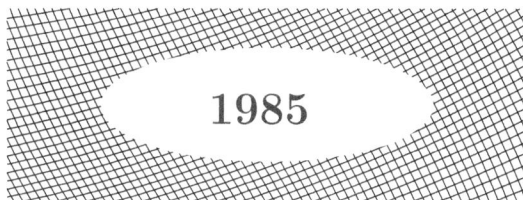

1985.01. Split these 68 coins into 34 pairs and compare the coins in each pair—that requires 34 weighings. Among the 34 heavier coins, we can find the heaviest one using $34-1 = 33$ weighings, and then similarly among the 34 lighter coins we find the lightest one with 33 more weighings. Total number of weighings in this procedure is $34 + 33 + 33 = 100$.

1985.02. The sum of the digits of this number is equal to 285. Thus, it is divisible by 3 but is not divisible by 9. Hence, it is not a perfect square.

1985.03. It is clear that the distance between B and C is greater than the distance between B and A—otherwise the traveler would return to A. Similarly, the next town's distance from C is greater than or equal to the distance from B to C, and so on. Thus, the distance that is traveled at each leg of the trip becomes larger and larger. Thus, if the traveler ever comes back to A, then on the next day he will travel to B again, and that distance will once again be AB, which is not possible.

1985.04. For example, take the following set of 998 ones, 2, and 1000:

$$\{1, 1, 1, \ldots, 1, 1, 2, 1000\}.$$

1985.05. In each shoe size, one type—right or left—comprises at least one half of the boots. Let us write those three types down. One of them, say, the left, repeats at least twice, for instance, in sizes 9 and 10.

Since the number of left boots in these two sizes is at least 50 (among two hundred boots of sizes 9 and 10, there are no more than 150 right ones), we get at least 50 good pairs.

1985.06. See solution to Question **1985.18.**

1985.07. Answer: no, it will not. **Hint:** Consider the remainder of the number on the blackboard modulo 3, and how it changes each second.

1985.08. Since OM is the median of isosceles triangle ABO, line MO is a perpendicular bisector of segment AB. Therefore, sides BC and AC are equal, and CM is the altitude. Angle bisector BL passes through the point of intersection of two altitudes, and so BL is also an altitude. Hence, sides AB and BC are equal as well.

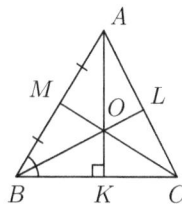

1985.09. Answer: $n = 1$, $p = 2$, $q = 3$ or $n = 1$, $p = 3$, $q = 2$.

Finding the common denominator, we get $n(p + q + 1) = pq$. Number n is not divisible by p, otherwise $n(p + q + 1) \geqslant p(p + q + 1) > pq$. Similarly, n is not divisible by q. Therefore, $p + q + 1$ is divisible by both p and q, hence, $(p + 1) \vdots q$ and $(q + 1) \vdots p$. This is only possible if $p = 2$, $q = 3$ (or vice versa), and then $n = 1$.

1985.11. Denote the grasshoppers by letters A, B, and C. There are six ways of how they can be arranged along the line (from left to right): ABC, BCA, CAB, ACB, BAC, and CBA. Let us call the first three "regular" and the others—"irregular". It is easy to see that with each jump, the type changes. Thus, after 1985 jumps, the type of the configuration will differ from the original one.

1985.13. First, find all the angles shown in the figure. Since triangles ABP and ADQ are congruent (equal legs and hypotenuses), it follows that

$$\angle QAD = \angle PAB = \frac{1}{2}(90° - \angle QAP) = 15°.$$

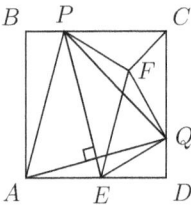

Since E lies on the perpendicular bisector of segment AQ, triangle AQE is isosceles, and therefore, $\angle AQE = \angle QAE = 15°$. Triangle PQF is obviously symmetric to triangle AQE with respect to the angle bisector of AQP, and therefore, $EF \parallel AP$, implying $\angle PEF = \angle EPA = 30°$.

Triangles PQF and PQC are isosceles (with base PQ), and therefore, F lies on the bisector of angle C, hence, $\angle QCF = \angle PCF = 45°$. Furthermore,

$$\angle EQF = \angle EQA + \angle AQP + \angle PQF = 15° + 60° + 15° = 90°.$$

Finally,

$$\angle EPF = \angle EPQ + \angle QPF = 30° + 15° = 45°,$$
$$\angle CPF = \angle CPQ - \angle QPF = 45° - 15° = 30°.$$

Now note that triangle FEP is similar to triangle FPC (two equal angles) with the ratio of similarity equal to $\sqrt{2}$, since $|FP| = |FQ|$ and $\frac{|FE|}{|FQ|} = \sqrt{2}$ (triangle EQF is right and isosceles). Therefore, $|FE| : |FP| = |FP| : |FC| = \sqrt{2}$, and $|FE| = 2|FC|$.

1985.14. Answer: the original number is equal to 25.

For $n \geqslant 26$, the sequence is strictly monotonically increasing, because

$$n^2 - 600 - n = (n - 25)(n + 24) > 0.$$

For $n = 24$, all terms, starting with the second one, are equal to (-24). If $n \leqslant 23$, then $n^2 - 600 < -71$, and at the next step, we will obtain number greater than $71^2 - 600 = 4441$, and after that, the sequence will be monotonically increasing. Therefore, the last case left—$n = 25$—is the only possible solution. Indeed, $25 = 25^2 - 600$.

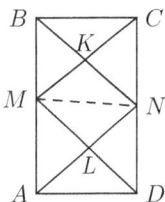

1985.15. Since the areas of triangles AMD and AMN are equal, and triangle ALM is their common part, then we have

$$S_{ALD} = S_{NLM} < S_{NLMK}.$$

1985.16. Answer: 497 numbers.

If number x is selected, then from the following seven numbers we can choose at most one among $x + 1$, $x + 4$ or $x + 6$. Therefore, no more than two can be selected from any eight consecutive numbers. All numbers 1 through 1984 can be split into 248 such octuplets, and so no more than 496 can be selected. Adding 1985, the total count cannot be greater than 497. That is exactly how many numbers would be selected if we took all the numbers of form $4k + 1$—difference between any two of them is divisible by 4 ,and therefore, is not a prime.

1985.17. Assume the opposite, and consider all red points (if any). Each one of them is connected either only with blue points, or only with green points. In the former case, repaint that red point green, and in the latter case repaint it blue. Then we obtain a configuration without red points.

Denote by k the number of segments coming out of every point, and by B and G— the numbers of blue and green points, respectively. But then the number of all segments equals kB, and at the same time it equals kG, that is, $B = G$, which is impossible because $B + G = 1985$.

1985.18. Let us assume that in one (left) stack, the first k containers are already arranged in the desired order—#1 at the bottom, then #2, and so on, through #k. Then we will do the following. Take all the containers above #k off that stack, move them to the right stack, then take all the containers above (and including) #$(k+1)$ in the right stack and move them to the left stack. Thus, after these two operations, we now have the first $(k+1)$ containers arranged in the proper order in the left stack. Finally, we do not need more than one operation for the last container #n; therefore, we can achieve the objective with $2n - 1$ operations.

1985.19. Suppose that arithmetic progression (a_n), $a_n = a + nd$ contains perfect square $x = y^2$. Then all numbers of form $(y + kd)^2$ are the terms of the same progression.

1985.20. Answer: $(0, 0, 0)$; $\left(\frac{1}{\sqrt{8}}, \frac{1}{\sqrt{8}}, \frac{1}{\sqrt{8}} \right)$; $\left(-\frac{1}{\sqrt{8}}, -\frac{1}{\sqrt{8}}, -\frac{1}{\sqrt{8}} \right)$.

Subtract the second equation from the first one and then use the well known formula for the difference of two cubes.

$$(x - z)\big((x + y)^2 + (x + y)(y + z) + (y + z)^2\big) = z - x.$$

Since the second factor on the left is always non-negative, we obtain $x = z$. Similarly, $x = y$, and therefore, $8x^3 = x$, and the answer follows.

1985.21. Answer: $\angle DAB = 57°30'$, $\angle ABC = 100°$, $\angle BCD = 72°30'$.

Find on ray BD point D' such that $BD' = BA = BC$. Then $2\angle BD'A = 180° - 65° = 115°$, $2\angle BD'C = 180° - 35° = 145°$, and therefore, $2\angle AD'C = 115° + 145° = 260°$. Now we know that $2\angle ADC = \angle ADB + \angle CDB = 260°$. That means $D = D'$ (since each of the angles $\angle AXB$, $\angle CXB$ decreases if we slide X along ray BD away from B). From this we have $\angle BAD = 115°/2$, and $\angle BCD = 145°/2$.

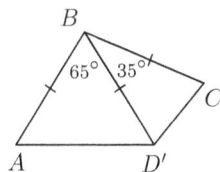

1985.22. The inequalities in the problem's statement mean that quadratic polynomials $a_1x^2 + 2b_1x + c_1$ and $a_2x^2 + 2b_2x + c_2$ have non-positive discriminants, while their leading coefficients are positive. Therefore, these polynomials are non-negative for any real x.

In addition, $5x^2 + 6x + 2 > 0$ for all real x. Hence, the sum of these three polynomials is always positive and therefore, its discriminant (equal to $4(b_1 + b_2 + 3)^2 - 4(a_1 + a_2 + 5)(c_1 + c_2 + 2)$) is negative.

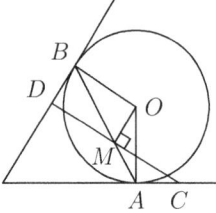

1985.23. Diameters of the circumcircles of triangles AOM and BOM are equal, because $|AO|/\sin \angle AMO = |BO|/\sin \angle BMO$. Furthermore, $\angle CMO = \angle CAO = \angle DBO = \angle DMO = \pi/2$, and therefore, OC and OD are the diameters of the circumcircles of quadrilaterals $ACOM$ and $DMOB$ (the order of these points can be different). Thus, $|OC| = |OD|$ and then $|CM| = |MD|$, since altitude OM of isosceles triangle COD is also a median.

1985.24. Answer: the first player wins.

At any moment of the game, all the tokens can be mentally split into contiguous groups separated by stretches of empty squares. With his first move, the first player puts the token on the rightmost square, and then with every move, he makes sure that any "empty" stretch has an even length.

If he can do that, then he can always make a move. Indeed, if the second player puts token X into one of the gaps between two groups, then it is always possible to place another token next to X so that both smaller gaps now have even length (perhaps, zero). If the second player moved last token in a group to the neighboring square, then the first player moves that token again in the same direction. Finally, if the second player moved some other token of the group to the same square, then the first player moves the next token of the group by the same distance.

Therefore, the first player can always make a move. Since the game cannot last longer than $625 = 25 \cdot 25$ moves, the second player must lose.

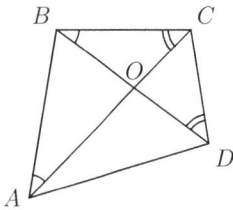

1985.25. Triangles CBO and CAB are similar. Therefore, $|CO| \cdot |CA| = |CB|^2$. Similarly, triangles CBO and DBC are similar, and $|BO| \cdot |BD| = |CB|^2$. Thus, $|CO| \cdot |CA| = |BO| \cdot |BD|$. Hence, the powers of points C and B with respect to the circumcircle of triangle AOD are equal; therefore, the tangents from C and B to this circle have equal length.

1985.26. From the last two equations, we get $y - cz = z - by$, or $y(b+1) = z(c+1)$; similarly, $y(b+1) = x(a+1)$. Hence, triple $\{x, y, z\}$ is proportional to triple $\{(b+1)(c+1), (c+1)(a+1), (a+1)(b+1)\}$ with some non-zero ratio. Then the last triple also satisfies this system; in particular, the first equation, that is,

$$(b+1)(c+1) = b(a+1)(c+1) + c(a+1)(b+1),$$

implying $2abc + ab + bc + ac = 1$.

Note. Existence of a non-trivial solution of homogeneous system of k linear equations in k variables is equivalent to the system's determinant being zero; that is, $1 - ab - bc - ac - 2abc = 0$.

1985.27. Let O be some arbitrary point inside quadrilateral $ABCD$. It is clear that one of the angles AOB, BOC, COD, and DOA (without loss of generality, let it be angle AOB) is greater than or equal to $90°$. Then $|AO|^2 + |BO|^2 \leqslant |AB|^2 \leqslant 7^2 = 49$, and therefore, either $|AO|^2 < 25$, or $|BO|^2 < 25$, implying that point O lies inside one of the circles with radius 5 centered at vertices A and B.

1985.28. Let $a + b + c = p$, $ab + bc + ac = q$, and $abc = r$. Then numbers $-a$, $-b$, and $-c$ are the roots of equation
$$0 = (x + a)(x + b)(x + c) = x^3 + px^2 + qx + r.$$
Since p, q, and r are positive, all the roots of this equation must be negative, and therefore, a, b, $c > 0$.

1985.29. For each $i \geqslant 1$, we have
$$\frac{1}{x_{i+1}} = \frac{1}{x_i - x_i^2} = \frac{1}{x_i(1 - x_i)} > \frac{1 + x_i}{x_i} = 1 + \frac{1}{x_i}.$$
From these inequalities, it follows that $1/x_{1001} > 1000 + 1/x_1 = 2000$, that is, $x_{1001} < 0.0005$.

1985.31. Let us assume that $a \neq 1$. Then
$$a^{xyz} = b^{yz}c^{yz} = (ca)^z(ab)^y = aba^za^yca = a^{z+y+2}bc = a^{x+y+z+2}.$$
Therefore, $x + y + z - xyz = -2$.

1985.33. Point O is the intersection of the given lines, and points K, L, M, and N are the feet of perpendiculars dropped from O onto edges AB, AC, CD, and BD, respectively; F is the incenter of triangle ABC. Using theorem of three perpendiculars, we obtain that $FK \perp AB$, $FL \perp AC$, and therefore, points K and L are the points where incircle of ABC touches sides AB and AC. Thus, $|AK| = |AL|$. Similarly, $|BK| = |BN|$, $|CM| = |CL|$, and $|DM| = |DN|$. Adding these equalities we get $|AB| + |CD| = |AC| + |BD|$. The other equality is proved analogously.

1985.34. Consider graph of function $f(x) = \sqrt[4]{x^4 + 1}$. It divides 3×3 square in two parts with areas S_1 and S_2. Denote the remaining part of the undergraph of f by S_3 (see the figure), and we have
$$\int_0^3 \sqrt[4]{x^4 + 1}\, dx = S_1 + S_3.$$
Since function $g(x) = \sqrt[4]{x^4 - 1}$ is the inverse function for $f(x)$, we obtain
$$\int_1^3 \sqrt[4]{x^4 - 1}\, dx = S_2.$$
Therefore, the sum of our two integrals equals $S_1 + S_2 + S_3$.

Since S_3 is less than the area of rectangle $ABCD$, which in turn does not exceed 0.0001, we have the desired inequality.

1985.37. Take, for instance, number $15\,841\,584\,158\,415\,841\,584$.

1985.38. Denote the triangle formed by these three lines by T, and the center of the incircle of the original triangle S by I.

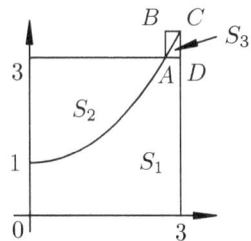

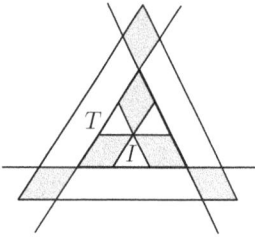

From formula $S = pr$, we obtain that the inradius of S is $r = 2$ cm. Thus, the distance from I to the sides of S is 2 cm, and to the sides of T—1 cm. Thus, triangle T is homothetic to S with ratio $\frac{1}{2}$ (with center of the homothety I), and its area equals 25 cm². Draw lines parallel to sides of T passing through point I.

They cut three parallelograms off triangle T with a total area less than 25 cm². It is left to note that these parallelograms are congruent to the parallelograms described in the problem's statement.

1985.39. All trios can be split in two types: those where each team had exactly one win, and those where the teams' results are 2, 1, and 0. Denote the number of triples of the first type by a, and of the second type—by b. Then $a + b = \binom{15}{3} = 455$. On the other hand, there are 49 trios where the given teams have exactly one win; in every Type I trio, we have three such teams. Therefore, each Type I trio is counted three times, and each Type II trio is counted once. Thus, $3a + b = 49 \cdot 15 = 735$. Solving these two equations, we obtain $a = 140$.

1985.40. See solution to Question **1978.28**.

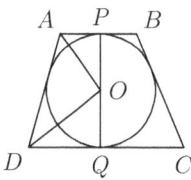

1985.41. Let $ABCD$ be the given trapezoid with bases $a = |AB|$ and $b = |CD|$, O—the center of inscribed circumference, P and Q—points where that circumference touches bases AB and CD. Also denote $x = |AP|$, $y = |BP|$, then $a = |AB| = x + y$. Note that AO and DO are the bisectors of angles A and D, and therefore, $\angle OAD + \angle ODA = \frac{1}{2}(\angle A + \angle D) = 90°$; hence, $\angle AOD = 90°$.

It follows that

$$\triangle APO \sim \triangle AOD \sim \triangle OQD, \text{ and } |DQ| = |OQ| \cdot |OP|/|AP| = R^2/x.$$

Similarly, $|CQ| = R^2/y$, and then

$$b = |DQ| + |CQ| = R^2/x + R^2/y = \left(\frac{1}{x} + \frac{1}{y}\right)R^2.$$

Multiplying, we obtain

$$ab = (x + y)\left(\frac{1}{x} + \frac{1}{y}\right)R^2 = \left(2 + \frac{x}{y} + \frac{y}{x}\right)R^2 \geqslant 4R^2.$$

For the last step, we use well known inequality $\frac{x}{y} + \frac{y}{x} \geqslant 2$.

1985.42. Consider difference $d_i = 500a_i - i$. Clearly, $d_1 = 499$, and $d_m = -500a_m$. It is obvious, that with transition from i to $i + 1$, this difference either increases by 499 or decreases by 1. Let n be the first index, for which d_n is non-positive. Obviously, it has to be equal to zero, that is, $a_n = n/500$.

1985.43. Note that $(f_k f_{k+2})^{-1} = (f_k f_{k+1})^{-1} - (f_{k+1} f_{k+2})^{-1}$. Then we have

$$\frac{1}{f_1 f_3} + \frac{1}{f_2 f_4} + \cdots + \frac{1}{f_{98} f_{100}}$$
$$= \left[\frac{1}{f_1 f_2} - \frac{1}{f_2 f_3} \right] + \left[\frac{1}{f_2 f_3} - \frac{1}{f_3 f_4} \right] + \cdots + \left[\frac{1}{f_{98} f_{99}} - \frac{1}{f_{99} f_{100}} \right]$$
$$= \frac{1}{f_1 f_2} - \frac{1}{f_{99} f_{100}} = 1 - \frac{1}{f_{99} f_{100}} < 1.$$

1985.44. Since the terms of the sequence do no take more than 100 different values, one can find two equal numbers among the first 101 terms. Thus, $a_{m+q} = a_m$ for some natural m and q, where $m + q \leqslant 101$. Since every term is uniquely determined by the previous one, the sequence must have a cycle of length q, starting from the mth term, and so $a_{n+q} = a_n$ for any $n \geqslant m$. By induction, it follows that $a_{n+kq} = a_n$ for any $n \geqslant m$ and any natural k. Find the smallest k for which $kq \geqslant m$, and substitute $n = kq$ to obtain $a_{2n} = a_{n+kq} = a_n$. It is left to show that $n \leqslant 100$. Assuming that $n \geqslant 101$, we have $kq = n \geqslant 101 \geqslant m + q$, and it follows that $(k-1)q \geqslant m$, contradicting the minimality property of index k.

1985.47. Let us call a trio of teams *good* if in the matches among them each team has exactly one win, and *bad* otherwise.

(a) Each team is included in $9 \cdot 8 + 10 \cdot 9/2 = 81$ bad trios as a winner of the trio (team that defeated both other teams of the trio) or as a loser of the trio (team that lost both matches). Of course, each bad trio has exactly one winner and one loser. Therefore, the number of bad trios is $81 \cdot 20/2 = 810$. Number of all trios is $20 \cdot 19 \cdot 18/6 = 1140$, and therefore, the number of good trios equals $1140 - 810 = 330$.

(b) It is sufficient to note that for each team A there are no more than 81 trios, where A is a winner or a loser of the trio.

1985.51. Consider the remainders of the terms of our sequence modulo $a_{n-6} > 1$. From the recurrence equation we obtain that $a_{n-5} \equiv a_{n-4} \equiv 1 \pmod{a_{n-6}}$. Therefore, $a_{n-3} \equiv 2$, $a_{n-2} \equiv 3$, $a_{n-1} \equiv 7$, $a_n \equiv 22 \pmod{a_{n-6}}$. It follows that $a_n - 22 \;\vdots\; a_{n-6}$, and since $1 < a_{n-6} < a_n - 22$ for $n > 10$, this is a composite number.

1985.52. For any group G of several boys, we can define group G' of all girls who are friends with an odd number of boys from G.

We know that every such group G' has an even number of girls. If the overall number of boys (and girls) is n, then we have $2^n - 1$ non-empty groups of boys and $2^n - 1$ non-empty groups of girls. But the number of non-empty groups containing an even number of girls (it equals $2^{n-1} - 1$) is less than that, and therefore, there exist two different groups of boys G and H such that $G' = H'$. It follows that for every girl in the class, the number of her friends in G has the same parity as the number of her friends in H. Consider the symmetric difference of G and H: that is, the group of boys who belong to exactly one of groups G and H but not to both of them. They form the group we need.

1986

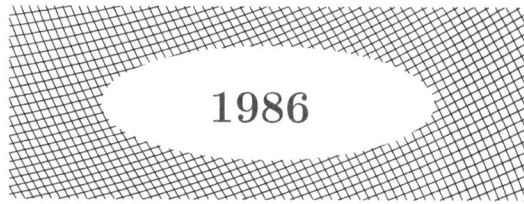

1986.01. Let us mark the cards that we flip during each operation by underlining them. Then the sequence of required operations can be presented as follows.

$$(7, 8, 9, \underline{4, 5, 6, 1, 2, 3}) \to (\underline{7, 8, 9, 3, 2, 1}, 6, 5, 4) \to$$

$$\to (1, 2, 3, \underline{9, 8, 7, 6, 5, 4}) \to (1, 2, 3, 4, 5, 6, 7, 8, 9).$$

1986.02. Each queen attacks no fewer than 22 squares: its own, seven squares in the same column, seven in the same row, and at least seven in the diagonals. There remain at most 42 squares which are not under attack, and therefore, each queen attacks at least one of the remaining 43 queens.

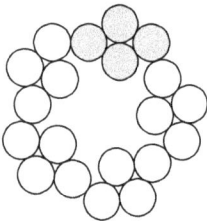

1986.03. From $77a = 34a + 43a = 43b + 43a = 43(a + b)$, it follows that $43(a + b)$ is divisible by 77. Therefore, $a + b$ is also divisible by 77, proving it is a composite number.

1986.04. Four nickels can be arranged into a "rhombus", where two of the nickels touch three of them, and the other two touch only two other nickels. The required arrangement now can be obtained if we arrange several rhombuses in a "ring" so that different rhombuses do not overlap and touch only at the "vertices" (see the figure).

1986.05. Denote these numbers by a_1, a_2, ..., a_{55}. Then $a_1 = -a_4$, because $a_1 + a_3 + a_2 + a_4 = a_2 + a_3$. Similarly, $a_1 = -a_4 = a_7 = \cdots = -a_{52} = a_{55}$, that is, $a_1 = a_{55}$. Hence, any two neighboring numbers are equal, and so all the numbers around the circle are the same, so it immediately follows that they are all zeros.

1986.06. (a) **Answer:** 7 639 128. (b) **Answer:** no, it does not.

First, the decimal representation cannot contain a zero. Second, the number must be even. Third, it cannot be divisible by 5—otherwise it would end with a zero. We have eight digits left; however, their sum is not divisible by 3, despite the fact that one of these digits is 3. This proves that such a number does not exist.

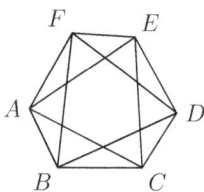

1986.09. Consider triangles ABC and ABF. Since they have common side AB with their third vertices lying on the same side of line AB, correspondence between their vertices in their congruence must be as follows: $\triangle ABC \cong \triangle BAF$ (other cases are possible but then the triangles will be isosceles and congruence $\triangle ABC \cong \triangle BAF$ still holds). In the same way, we establish congruence in the following equalities

$$\triangle ABC \cong \triangle BAF \cong \triangle EFA \cong \triangle FED \cong \triangle CDE \cong \triangle DCB.$$

That means the angles of the hexagon are equal, while the sides' lengths alternate.

Now consider triangles ABC, CDE, and EFA. Their sides AC, CE, and EA are equal, and therefore, triangle ACE is equilateral. Triangles ABC, CDE, and EFA have common side with it. They also are exterior with respect to this triangle—therefore, they map into each other under rotation around this triangle's center by $120°$. This rotation maps segments AD, EB, and CF into each other; hence, they have equal length.

1986.11. The snail crawls along the edges of the triangular grid shown in the figure. Also, if it started at point O, then after every hour, it arrives at a grid point. Assuming that the snail's trip lasted for $n + x$ hours, where $0 < x < 1$, it is obvious that in x hours the snail is not able crawl from one grid point to another.

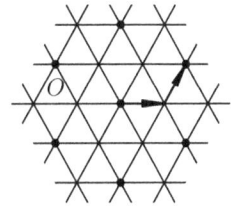

1986.13. Let us index these clubs by digits 1 through 5. There are 32 subsets of the set of these five clubs (digits). We will split them into 10 collections:

$$[1]: \; \emptyset \subset \{1\} \subset \{1,2\} \subset \{1,2,3\} \subset \{1,2,3,4\} \subset \{1,2,3,4,5\};$$
$$[2]: \; \{2\} \subset \{2,5\} \subset \{1,2,5\} \subset \{1,2,3,5\};$$
$$[3]: \; \{3\} \subset \{1,3\} \subset \{1,3,4\} \subset \{1,3,4,5\};$$
$$[4]: \; \{4\} \subset \{1,4\} \subset \{1,2,4\} \subset \{1,2,4,5\};$$
$$[5]: \; \{5\} \subset \{3,5\} \subset \{3,4,5\} \subset \{2,3,4,5\};$$
$$[6]: \; \{1,5\} \subset \{1,3,5\};$$
$$[7]: \; \{2,3\} \subset \{2,3,5\};$$
$$[8]: \; \{2,4\} \subset \{2,4,5\};$$
$$[9]: \; \{3,4\} \subset \{2,3,4\};$$
$$[10]: \; \{4,5\} \subset \{1,4,5\}.$$

Notice that in every collection, for any two subsets, one of them contains the other. More exactly, any subset contains the one written to the left of it, and so each collection represents a chain of subsets where for any subsets A and B in that chain, if subset A is written somewhere to the right of B, then $B \subset A$.

Now, for every student s, define subset C_s of all clubs s belongs to. Since there are eleven students, some two of subsets C_s belong to the same subset collection listed above, and therefore, one of them contains the other.

1986.15. Suppose that $c \geqslant b \geqslant a$, $S = \frac{1}{a} + \frac{1}{b} + \frac{1}{c}$. Obviously, $a \geqslant 1$. If $a \geqslant 3$, then $S \leqslant \frac{1}{3} + \frac{1}{3} + \frac{1}{4} = \frac{11}{12} < \frac{41}{42}$, since case $a = b = c = 3$ is obviously excluded.

Now, if $a = 2$, then $b \geqslant 3$. If $b = 4$, then $S \leqslant \frac{1}{2} + \frac{1}{4} + \frac{1}{5} = \frac{19}{20} < \frac{41}{42}$, since case $a = 2$, $b = c = 4$ is also excluded. And if $b = 3$, then $\frac{1}{c} < 1 - \frac{1}{a} - \frac{1}{b} = \frac{1}{6}$, and therefore, $c \geqslant 7$. Hence, $S \leqslant \frac{1}{2} + \frac{1}{3} + \frac{1}{7} = \frac{41}{42}$.

1986.16. Reflect point B across angle bisector CD to obtain point M on segment CE. Counting the angles, we see that triangles DEM and AED are isosceles, and therefore, $|BD| = |DM| = |DE| = |AE|$.

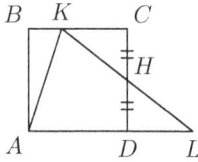

1986.18. Extend segment KH to the intersection with AD at point L. It is easy to see that $\angle BKA = \angle KAD$, and therefore, it is sufficient to show that angles $\angle KAD$ and $\angle AKH$ are equal; that is, to prove that triangle ALK is isosceles. Let $|AB| = a$; then $|KC| = 2a/3$, $|CH| = a/2$. From Pythagoras' Theorem, we have $|KH| = 5a/6$. Obviously, right triangles KCH and LDH are congruent, and $|AL| = |AD| + |DL| = |AD| + |KC| = 5a/3$, $|KL| = 2|KH| = 5a/3$.

1986.19. Let us connect every pair of stones by a thin thread. Then, every time we split some pile of stones in two, we snap all the threads between the stones going to different piles. The number we write on the board is exactly the number of the snapped threads. The original number of threads is $25 \cdot 24/2 = 300$. Since eventually all of them will be snapped, with each one snapped exactly once, this proves the required equality.

1986.20. Answer: 198.

Denote our number by $\overline{abc}$. Then we have $100a + 10b + c = 11(a+b+c)$, or $89a = 10c + b$. The right-hand side is less than 100, implying $a = 1$, then $c = 8$ and $b = 9$.

1986.21. Let P be the intersection of AE and MB, and Q— the intersection of AK and BH. It is easy to see that $\angle CBM = \angle ABH$ and $\angle BAE = \angle KAD$. Since $\angle ABP = 90° - \angle CBM = 90° - \angle ABH = \angle AHB$, triangles ABP and AHQ have equal corresponding angles. In particular, $\angle APB = \angle AQH$, and it follows that the interior quadrilateral is inscribed.

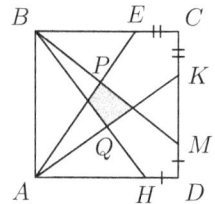

1986.22. Rewrite the equality as $1001a + 999b = 1$. It is clear now that we can take $a = 500$, $b = -501$.

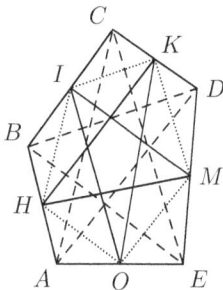

1986.23. The length of broken line $ACEBDA$ is equal to twice the perimeter of pentagon $HIKMO$ (e.g., $|AC| = 2|HI|$, $|CE| = 2|KM|$, and so forth). It suffices to prove now that double perimeter of this pentagon is greater than the sum of its diagonals. To do that, we add up five triangle's inequalities $|HI| + |IK| > |HK|$, $|IK| + |KM| > |IM|$, and so on.

1986.24. See solution to Question **1963.26**.

1986.25. Let us prove by induction on n that for any $n = 1$, 2, ..., 33 we can find two disjoint groups of teams—n teams in the first group and $5n + 1$ teams in the second—such that the second group contains all the teams that played with the teams from the first group (in particular, that would imply that the teams of the first group did not play each other). For the basis,

$n = 1$, consider group #1 consisting of one arbitrary team A, and group #2 of all six teams that A played against.

Now, assume that for some $n < 33$, these groups exist. Since the two groups together contain an odd number of teams, there exists team A which (a) does not belong to any of these groups, and (b) on the first day, played a match with one of the teams from the second group. Further, there can be no more than five teams outside these groups that played team A. We add team A to the first group, and these five (or less) teams—to the second group. Also, if necessary, add a few more teams (from those not yet included in our two groups) to the second group so that it contains exactly $5n + 6 = 5(n + 1) + 1$ teams. Since $n < 33$, that is possible, and so we have updated the required groups for $n + 1$.

Now, consider our two groups for $n = 33$. Since $6n + 1 = 199 < 200$, there exists a team belonging to neither of these two groups. Adding it to the 33 teams of the first group, we obtain 34 teams that did not play each other.

1986.28. Answer: $a = \frac{1}{42}$, $b = \frac{1}{7}$, $c = \frac{1}{3}$, $d = \frac{1}{2}$.

Subtracting the third equation from the fourth, the second from the third, and the first from the second, we obtain: $d = 1 - d$, $c = d - cd$, $b = cd - bcd$. From this we find d, then c and b.

1986.29. Answer: $a = 2.5$.

Let us translate triangles $AA'A''$ and $BB'B''$ mapping vertices A and B to point C (see the figure). Consider triangle $C'C''D$ whose sides have lengths 3, 4, and 5, and therefore, it is a right triangle. By construction, point C is the center of this triangle's circumcircle with radius a. Hence, C is the midpoint of the hypotenuse, and so $a = |C'C''|/2 = 2.5$.

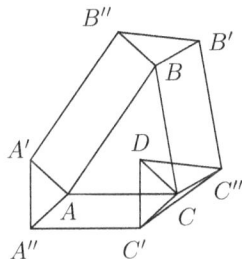

1986.31. Answer: $a = 500$, $b = -1$, $c = -500$.

Rewrite the equality as

$$1000(1001a + 999c) = 1 - 999 \cdot 1001b.$$

In order for the right-hand side to be divisible by 1000, we can take, for example, $b = -1$. Then we have $1001a + 999c = 1000$, that is, $2a + 999(a + c) = 1000$. This equality is true, for instance, when $a = 500$, $a + c = 0$, or $c = -500$.

1986.34. Answer: 4.

From $(ab)^2 = (a + b)^2 \geqslant 4ab$, we have $ab \geqslant 4$. Equality is attained only if $a = b = 2$.

1986.35. Rewrite the equation as

$$x^{1985}(x - 1) = 1986^{1985}(1986 - 1).$$

Since $x = 1986$ is, obviously, a solution, and the left-hand side is a monotonically increasing function for $x > 1$, it follows that the equation has exactly one root.

1986.37. Answer: $30°$.

Let H, X, Y be the feet of the perpendiculars dropped from point B onto plane ACD and onto lines AC and AD, respectively. Denote the length of edge AB by s. Then $|AX| = |AB|\cos 45° = s/\sqrt{2}$, $|AY| = |AB|\cos 60° = s/2$.

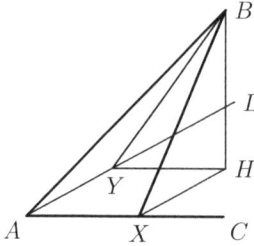

From the three perpendiculars theorem we have $HX \perp$ AC and $HY \perp AD$, that is, $AXHY$ is a rectangle. Thus, $|AH| = \sqrt{|AX|^2 + |AY|^2} = s\sqrt{3}/2$. It follows that $\cos \angle BAH =$ $|AH|/|AB| = \sqrt{3}/2$, $\angle BAH = 30°$.

1986.38. It is easy to see that

$$\frac{1}{a_n} = \frac{1}{a_1 a_2 \ldots a_{n-1}} - \frac{1}{a_1 a_2 \ldots a_n}.$$

Transforming each summand in the same manner, we obtain

$$\frac{1}{a_1} + \frac{1}{a_2} + \cdots + \frac{1}{a_{100}} = \frac{2}{a_1} - \frac{1}{a_{100}} = 1 - \frac{1}{a_{100}} < 1.$$

1986.39. Let BC be the longest side of the polygon (or one of them). Consider strip T bounded by perpendiculars to BC erected at points B and C. Clearly, T contains points of some other side a of the polygon. It is not possible that both endpoints of a lie outside T—that would contradict our choice of BC. Thus, one of the vertices belongs to T—it will serve as our vertex A.

1986.40. Without loss of generality, we can assume that the length of Martian year is 100 days. Then, in every Martian's life, there was exactly one New Year's Day. Since the number of all Martians is odd, that means that on the first day of some Martian year, the population of Mars was odd.

Similarly, on the second day of some (possible, same) Martian year, the population of Mars was odd, and so on. Thus, in the history of Mars there were at least hundred "odd" days.

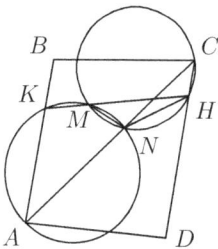

1986.41. The statement is true even if $ABCD$ is simply a trapezoid with $AB \parallel CD$. Let N be the second point of intersection of the two given circumferences. Note that $\angle NCH = \angle NMH = 180° - \angle NMK = \angle NAK$. Therefore, lines CN and NA coincide.

1986.42. Assume that route AB was closed and that the shortest route from X to Y now has the length of $2n + 1$ landings. Let us index the airports on this route from $X_0 = X$ to $X_{2n+1} = Y$. We used to have the shortest route from X to X_{n+1} with no more than n landings, as well as the route from X_n to Y with no more than n landings. It is clear that both of them used connection AB; moreover, we can assume that both times the airplane flew from A to B. Suppose that the number of landings on the route X—X_{n+1} from X to A is equal to s_1, and from B to X_{n+1}—t_1; on the route X_n—X_{2n+1} from B to X_{2n+1} is equal to s_2, and from X_n to A—t_2.

Obviously, $s_1 + 1 + t_1 \leqslant n$, $s_2 + 1 + t_2 \leqslant n$. Adding these inequalities, we get $(s_1 + t_2) + (s_2 + t_1) \leqslant 2n - 2$, and therefore, one of the summands on the left-hand side does not exceed $n - 1$. Therefore, there exists route from X to X_n or from X_{n+1} to X_{2n+1} not including connection AB and shorter than n, a contradiction.

1986.43. Answer: 10000. See solution to Question **1969.18**.

1986.44. Our task can be restated as follows: we are asked to project the skeleton of a four-dimensional hypercube $3 \times 3 \times 3 \times 3$ onto the plane. Let us describe it in more details.

On the plane, fix point O (as the origin of coordinate system) and four vectors $\vec{p}, \vec{q}, \vec{r}, \vec{s}$. Now, compute all vectors of the form $n\vec{p} + m\vec{q} + k\vec{r} + l\vec{s}$, where n, m, k, $l \in \{0, 1, 2, 3\}$ and have their endpoints comprise set S. Now, through every point in S, draw four lines parallel to vectors $\vec{p}, \vec{q}, \vec{r}$, and $\vec{s}$. Each of these lines will, obviously, contain three more points from S. The only thing that is left now is to make sure that no other points from S lie on these lines. For this, we select vectors $\vec{p}, \vec{q}, \vec{r}, \vec{s}$ one by one to satisfy the following requirements: $\vec{q} \nparallel \vec{p}$; $\vec{r} \nparallel m\vec{p} + n\vec{q}$ for m, $n \in \{-3, -2, \ldots, 2, 3\}$, $m^2 + n^2 \neq 0$; $\vec{s} \nparallel m\vec{p} + n\vec{q} + k\vec{r}$ for m, n, $k \in \{-3, -2, \ldots, 2, 3\}$, $m^2 + n^2 + k^2 \neq 0$. The figure below shows such an arrangement of lines and points.

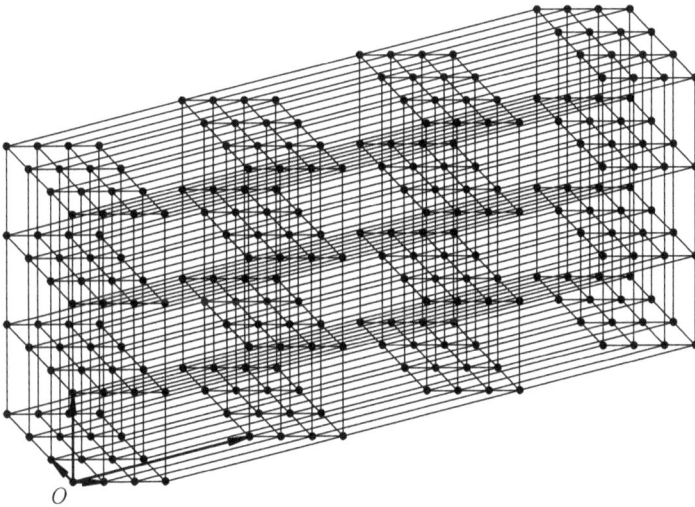

1986.45. Answer: the first player wins.

Let us prove that if the dimensions of the sheet are two unequal integers greater than 3, then the first player has a winning strategy.

The main idea is the strategy that we will call "return to the diagonal". If the current end of the cut is in node u, then we can draw the down-and-left diagonal $\mathcal{D}$ from node u. If the player who has to make the current move (we will call him A) extends the cut in the leftward or downward direction, then the other player (we will name her B) can return to $\mathcal{D}$ with her move in the downward or leftward direction, respectively, returning the current end of the cut to diagonal $\mathcal{D}$.

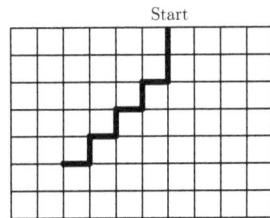

Player A now can cut only leftward or downward and so on. Thus, player B can use this method to move along diagonal $\mathcal{D}$ as far as she wants, provided the end of the cut does not touch the first vertical from the left (first after the edge) or first horizontal from the bottom. Also, the end of the cut returns on diagonal $\mathcal{D}$ only after player B's moves, and player A is always able to move only leftward or downward (of course, he can make some other move, but then he immediately loses).

An absolutely identical strategy exists for any diagonal parallel to one of the four possible directions.

Now, assume that the horizontal side of the sheet is longer than the vertical whose length is n. Then the first player makes her first cut downward from the upper edge of the sheet at the point which is $n - 1$ units away from the upper left corner.

The second player has to continue the cut downward lest he loses immediately. After that the first player makes the leftward cut, and then employs the diagonal strategy for the down-and-left diagonal $\mathcal{D}$ that passes through the lower left corner C of the sheet. This strategy will guarantee that eventually after one of the first player's moves, the endpoint of the cut is the node located two units up and two units right from point C. Then it is the second player's turn to move, and it is obvious that regardless of his choice the first player wins immediately after that.

1986.50. Consider any number $c > 0$, and prove that $c \in A$. It is clear that there exist numbers a and b such that $a < b < c$, and $[a; b] \subset A$. Similarly, interval $[c - b; c - a]$ contains interval $[x; y] \subset A$. In particular, $x \in A$. Since $c - x \in A$, then $c = (c - x) + x$ belongs to A.

1986.51. Answer: 610.

Let B_n be the set of the natural numbers $m \neq 1$ for which Step 1 is executed exactly n times, and b_n be the cardinality (size) of this set.

For instance, $b_1 = b_2 = 1$ (because $B_1 = \{2\}$, $B_2 = \{4\}$, $B_3 = \{3, 8\}$). It is clear that the number of even elements in B_n equals b_{n-1}, and the number of odd elements in B_n equals b_{n-2}. Therefore, $b_n = b_{n-1} + b_{n-2}$, meaning that sequence (b_n) coincides with Fibonacci series. In particular, $b_{15} = 610$.

1986.53. Answer: $16 + 64\sqrt{2}$.

Let us color the odd-numbered verticals (columns) in black. Then we have 45 black squares, and at least 44 of them cannot be the last square on the king's route. Any non-vertical move made from any of these squares moves the king to a white square.

Since there are only 36 white squares, the number of the vertical moves must be at least eight. Similarly, we can show that there must be at least eight horizontal moves. Therefore, the length of the route cannot exceed $8 + 8 + (80 - 16)\sqrt{2} = 16 + 64\sqrt{2}$. An example of a route with that exact length is shown in the figure.

1986.54. Let us choose points A, $B \in M_1 \cup M_2 \cup \cdots \cup M_n$ in such a way that the distance between them is the largest possible. Consider projections of polygons M_1, ..., M_n onto line AB. The length of the projection of any polygon is not greater than its diameter, and therefore, the sum of the lengths of the projections does not exceed the length of segment AB. Therefore, the projections cannot cover the entire segment, and we can take a perpendicular to AB erected at any point not covered by the projections.

1986.55. The condition guarantees us that equation $F(x) = x$ has a solution only if $x = 1$. Assume, however, that $F(1) > 1$ (the other case is similar). Then it follows from continuity of $F(x)$ that for any x we have $F(x) > x$, and so $F(F(...F(1)...)) > 1$, regardless of the number of times we apply F. This contradiction concludes the proof.

1986.56. Set

$$A = \frac{1}{1 + \sqrt{3}} + \frac{1}{\sqrt{5} + \sqrt{7}} + \cdots + \frac{1}{\sqrt{9997} + \sqrt{9999}}$$

$$= \frac{\sqrt{3} - 1}{2} + \frac{\sqrt{7} - \sqrt{5}}{2} + \cdots + \frac{\sqrt{9999} - \sqrt{9997}}{2}$$

$$B = \frac{1}{\sqrt{3} + \sqrt{5}} + \frac{1}{\sqrt{7} + \sqrt{9}} + \cdots + \frac{1}{\sqrt{9999} + \sqrt{10001}}$$

$$= \frac{\sqrt{5} - \sqrt{3}}{2} + \frac{\sqrt{9} - \sqrt{7}}{2} + \cdots + \frac{\sqrt{10001} - \sqrt{9999}}{2}.$$

Then $A + B = (\sqrt{10001} - 1)/2 > 48$. Since each term in A is greater than the corresponding term in B, we obtain $A > 24$.

1986.57. Denote $\alpha = \angle BAD$. Then BAX and XCY are isosceles triangles with angles $\alpha/2$ at the base. If O is the in-center of triangle XCY, then center angles $\angle XOC$ and $\angle COY$ equal α. Thus, after rotation by angle α around O, segment YC maps to segment CX. Since points D and B lie on the extensions of these segments and $|CD| = |AB| = |BX|$, we have point D mapping to point B. Therefore, $\angle DOB = \alpha$, and, moreover, $\angle BCD = \alpha$. It follows that points D, C, O, and A all lie on the same circumference.

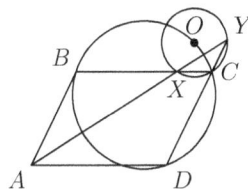

1986.58. Use the fact that function $1/(1 + x^3 + \sqrt{1 + x^6})$ is odd.

$$\int_{-1}^{1} \frac{dx}{1 + x^3 + \sqrt{1 + x^6}} = \int_{-1}^{0} \frac{dx}{1 + x^3 + \sqrt{1 + x^6}} + \int_{0}^{1} \frac{dx}{1 + x^3 + \sqrt{1 + x^6}}$$

$$= \int_{0}^{1} \frac{dx}{1 - x^3 + \sqrt{1 + x^6}} + \int_{0}^{1} \frac{dx}{1 + x^3 + \sqrt{1 + x^6}}$$

$$= \int_{0}^{1} \left(\frac{1}{1 - x^3 + \sqrt{1 + x^6}} + \frac{1}{1 + x^3 + \sqrt{1 + x^6}} \right) dx$$

$$= \int_{0}^{1} \frac{(2 + 2\sqrt{1 + x^6}) \, dx}{(1 + \sqrt{1 + x^6})^2 - x^6} = \int_{0}^{1} 1 \, dx = 1.$$

1986.59. Consider a sphere, inscribed into the contour of regular dodecahedron D, and the circles which are the intersections of this sphere and the faces of D. Under stereographic projection onto a plane, these circles will be mapped into circles, for which the required property will follow from the fact that each face of D has five edges.

1986.60. It is easy to see that $u_1 + u_2 + u_3 = v_1 + v_2 + v_3$, and $u_1 u_2 + u_2 u_3 + u_3 u_1 = v_1 v_2 + v_2 v_3 + v_3 v_1$. Since $u_1 u_2 u_3 = v_1 v_2 v_3$, polynomials $(x - u_1)(x - u_2)(x - u_3)$ and $(x - v_1)(x - v_2)(x - v_3)$ coincide. Therefore, the sets of their roots coincide as well.

1986.61. Let us prove that if the playing field is the square sheet with dimensions $n \times n$ with $n > 6$, the second player always wins.

We introduce coordinate axes parallel to the grid lines with the lower left corner of the sheet being the origin with coordinates $(0,0)$. Without loss of generality, we can assume that the first player begins with a cut from the lower edge of the sheet at the coordinate $k \leqslant n/2$ (if not, we can rotate or flip the sheet). Obviously, if $k = 1$, then the second player immediately wins with his next move. Otherwise, the second player response is to continue the cut upward.

Let $k = 2$. In this case, after the second player's response, the end of the cut is on the main diagonal $x = y$, and it is clear that the first player can only continue the cut upward or rightward. Then the second player "returns" the end of the cut to the main diagonal and so on. Eventually the first player will have to make a move from the node with coordinates $(n - 2, n - 2)$. Obviously, any move from that node loses.

In all other cases, the second player uses the same "return to the diagonal" method, choosing a diagonal that does not end in the corner of the sheet.

Consider, for example, case $k = 3$, and suppose that the first player's second move was the rightward cut, to node $(4, 2)$. Then the second player continues with the rightward cut, moving to $(5, 2)$. The first player has to respond with the upward cut.

Draw diagonal $\mathcal{D}$ in the up-and-right direction starting from node $(5, 2)$. Using the "return to the diagonal" method, the second player leads the cut to node $(2n - 2, 2n - 5)$. From that node, the first player can only move upward, and the second player then responds with one more upward cut, moving to node $(2n-2, 2n-3)$.

Now, the first player can only move leftward or upward, and then, using the symmetric cut the second player moves to node $(2n - 3, 2n - 2)$. Now, the first player can only move leftward, and the second player continues to node $(2n - 5, 2n - 2)$. After this, the first player can only respond with downward and leftward cuts, and the second player responds with symmetric leftward and downward cuts, respectively, moving along left-and-down diagonal that ends up in node $(5, 2)$. From this position, the first player has to cut downward, then the second cuts downward to $(3, 2)$ and wins with his next move.

Now, consider the case when with his second move the first player cuts either leftward or upward. Then the second player moves to node $(2, 3)$, the first player moves upward, the second player responds with upward cut to $(2, 5)$. Here is when we use inequality $n \geqslant 7$.

After this, the second player uses "return to the diagonal" method for the up-and-right direction continuing to the node in the second topmost horizontal line, acting just like in the previous case but in the opposite direction. Finally, he descends down-and-left to node $(5, 2)$. From that position, the first player wins with the next move.

And finally, the last case, when $3 < k \leqslant n/2$. The strategy here is quite similar to the one for $k = 3$.

Before the second move of the first player, the end of the cut is in node $(k, 2)$. Consider diagonal going up-and-right through node $(k, 2)$. If the first player moves rightward or upward, the second player returns back to this diagonal and eventually ends up in node $(n - 2, n - k)$ on the second rightmost vertical line.

Note that $n - k \leqslant n - 4$. Both players make upward moves, after which the second player will follow the "return to the diagonal" strategy along the up-and-left diagonal until the cut hits the second topmost horizontal line. Then the first player has to make the leftward cut, the second player does the same, and follows the down-and-left diagonal ending up in node $(2, k)$. There the first player has to make the downward cut, and the second player repeats that by moving to $(2, k - 2)$. From there, the first player can only cut in the rightward or downward direction, so the second player follows the right-and-down diagonal strategy, coming to node $(k - 2, 2)$. After that, regardless of the first player's next move, the second player can finish the game.

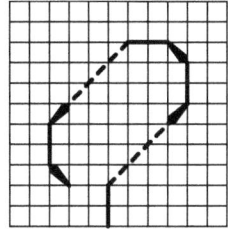

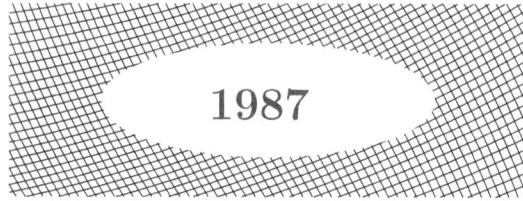

1987.01. One of the many possible ways to achieve that is as follows. Start with adding 1 nine times to the first row, then six times—to the second row, and three times—to the third row. Then subtract 1 nine times from the first column, six times—from the second column, and three times—from the third column.

1987.02. Answer: no, that is not possible.

The number of bills of the given value, and the sum that can be paid by them, always have equal remainders modulo 9. One million and half-million have different remainders modulo 9, giving us the answer.

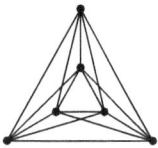

1987.03. The diagram is shown in the figure.

1987.04. Answer: the number of boys exceeds the number of girls by one.

If we denote the numbers of boys and girls by x and y, respectively, and the prices of a doughnut and a muffin by a and b, respectively, then we obtain $xa + yb = xb + ya - 1$; that is, $(x - y)(b - a) = 1$. It follows that $x - y = 1$.

1987.05. Answer: 1001 tickets.

To show that buying 1000 tickets is not always enough, consider the "stretch" of tickets from $000,001$ to $001,000$. This is exactly one thousand of "unlucky" tickets.

To prove that buying 1001 tickets is always enough, assume that the first ticket we buy is $\overline{abcdef}$. That means that our 1001 tickets are all inside the two thousand tickets interval from $\overline{abc000}$ to $\overline{xyz999}$, where $\overline{xyz} = \overline{abc}+1$. If we can always find two lucky ticket numbers such that (a) the first number, S, is in the first thousand of that interval; (b) the second one, T, is in the second thousand; and (c) difference $T - S$ does not exceed 1001, then the proof is over. Indeed, in that case, it is clear that any 1001 consecutive tickets inside the above-mentioned interval contain either S or T.

These two lucky numbers are $S = \overline{abcabc}$ and $T = \overline{xyzxyz}$. Properties (a) and (b) are obvious. The following computation proves (c).

$$T - S = \overline{xyz000} + \overline{xyz} - \overline{abc000} + \overline{abc}$$

$$= 1001 \cdot \overline{xyz} - 1001 \cdot \overline{abc} = 1001 \cdot (\overline{xyz} - \overline{abc}) = 1001.$$

1987.06. With her first move, Ann places X into the central square. After that, to every move by Bob, she responds with the move symmetrical to his with respect to the center of the board. Then, after the game is over, for each column (or row) with more O's than X's, the column (row) symmetric to it, obviously, has more X's than O's (and vice versa).

Furthermore, it is clear that the middle column and the middle row have more X's than O's. Therefore, in total we have Ann and Bob receive 10 and 8 points, respectively.

1987.08. Let O be the point of intersection of BK and CH. It follows from the angle equality that $|CO| = |OB|$. Then $|HO| = |OK|$, and triangles COK and BOH are congruent. Thus, BK is an altitude in our triangle. And since BK is also a median, we have $|AB| = |BC|$. On the other hand, altitudes BK and CH are equal, and therefore, $|AB| = |AC|$.

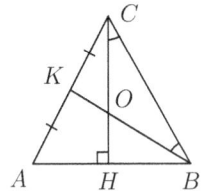

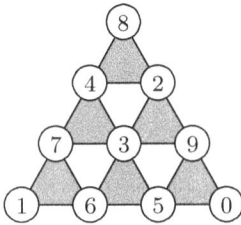

1987.10. It is obvious that the difference between the number of dillers and the number of dallers always has the same remainder modulo 11. At the beginning, this difference is 1, so it can never become zero.

1987.12. Answer: yes, that is possible.

For example (from top to bottom and from left to right): 8; 2, 4; 9, 3, 7; 0, 5, 6, 1.

1987.13. Denote the midpoints of the sides of the given quadrilateral $ABCD$ by K, L, M, and N. Segments KL and MN are equal and parallel because they are the midlines in triangles ABC and ADC with common side AC.

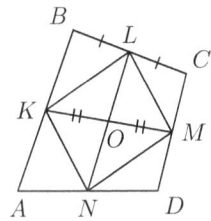

Thus, quadrilateral $KLMN$ is a parallelogram, and its diagonals KM and LN intersect at point O, the midpoint of both of these diagonals. Since the perimeters of quadrilaterals $BKOL$ and $LCMO$ are equal, it follows that $|BK| = |CM|$, or $|AB| = |CD|$. Similarly, $|AD| = |BC|$.

1987.16. Assume that is not so, and consider coloring of the cube's vertices in black and white such that any two vertices connected by an edge are colored differently. Then any edge of our broken line either connects two vertices of the same color or coincides with one of the major diagonals of the cube. Obviously, in our closed broken line, there must be at least two edges connecting the vertices of different colors. Then both of them are major diagonals, in which case they would intersect at the center of the cube, a contradiction.

1987.17. If we assume that the length of the road is always less than 100 kilometers, then on every month, the company builds at least $1/100^{10}$ more kilometers of the road. Therefore, within 100^{11} months, the highway will certainly be finished.[111]

1987.18. Given point A and line L, draw two arbitrary lines through A which intersect L at points B and C.

Erecting perpendiculars to them in points B and C, we obtain lines intersecting at point D. It is easy to see that projections of points A and D onto L are symmetric with respect to the midpoint of segment BC.

---

[111] So this will take a "little" longer than the remaining lifespan of our Sun.

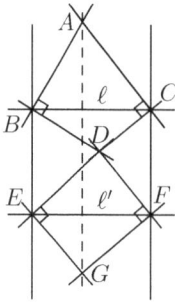

Now, construct rectangle $BCFE$ (where side FE is chosen at random) and repeat the same procedure: from point D and segment EF construct new point G. Obviously, line AG is the required perpendicular.

1987.19. Answer: no, that is not possible.

Color all the squares on the board in black and white, chessboard style, so that the lower left square is black. Then 26 checkers stand on black squares and 24—on white. When a checker makes a move, the color of the square it stands on does not change.

But the left half of the board has only 25 black squares—therefore, it is not possible to move all the checkers there.

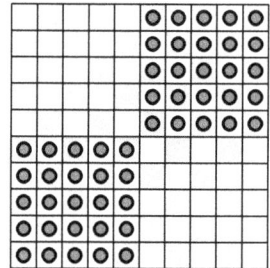

1987.20. Assume that $A = x + 2y + 5z + 10t + 20u + 50v + 100w$, and $x + y + z + t + u + v + w = B$. Then, multiplying the second equality by 100, we get $100B = 100x + 50(2y) + 20(5z) + 10(10t) + 5(20u) + 2(50v) + 1(100w)$, showing us how B rubles can be paid out with A coins.

1987.21. Indeed, $(1 + ab)^2 + (1 + cd)^2 + a^2c^2 + b^2d^2 \geqslant (1 + ab)^2 + (1 + cd)^2 + 2abcd = 1 + (ab + cd + 1)^2 \geqslant 1$.

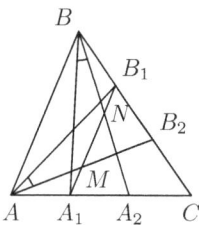

1987.22. Let N and M be the midpoints of segments A_2B and B_2A, respectively. It is easy to see that points A_1, B_1, N, and M lie on one straight line, parallel to AB. Moreover, $|A_1N| = |AB|/2 = |B_1M|$ from midline theorem. In triangles AMB_1 and BNA_1, we have equal sides A_1N and B_1M, equal angles A and B, as well as equal altitudes, dropped from vertices A and B. Therefore, these triangles are congruent and $|AB_1| = |BA_1|$. (We leave the proof of congruence to the reader.)

Therefore, the diagonals in trapezoid AA_1B_1B are equal; hence, it is isosceles. It follows that the original triangle is isosceles as well.

1987.23. Let us prove by induction on the number of branches that the length of the process as well as the total number of the crows that fly away does not depend on the order of the *evictions*.

Suppose that the tree has n branches, and the top one is occupied by p crows. Let the process on the lower $n - 1$ branches take exactly t minutes, during which exactly q crows fly off the top, $(n-1)$th branch. Then the number of crows that fly from the nth branch equals exactly $p + q - 1$. That takes $p + q - 1$ additional minutes; therefore, the entire process for n branches takes $t + p + q - 1$ minutes.

1987.24. Answer: no, such an arrangement does not exist.

In fact, these cubes cannot be arranged (while complying with the "neighboring" condition) into any spacial shape whose width along any of the three dimensions is less than 8.

Assume the opposite. Note that if some four cubes originally formed 2×2 square, then they would have to form the same square (more precisely, a rectangular parallelepiped $2 \times 2 \times 1$) in the spatial arrangement. As a corollary, any three cubes that used to form an L-shape must form an L-shape in the spatial arrangement as well.

Consider the unit cubes, adjacent to the lower side of the 8×8 square. They cannot form the same strip of length 8 in space, and therefore, there exist three of them which formed 1×3 rectangle before, but form an L-shape after the rearrangement.

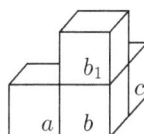

Let us denote these three unit cubes a, b, and c (b is adjacent to both a and c), and for convenience sake, assume that the L-shape formed in the spatial arrangement is horizontal. Consider also unit cubes a_1, b_1, and c_1 that in the original square were positioned above the cubes a, b, and c, respectively.

Cube b_1 in space must form an L-shape with each pair of cubes (a, b) and (b, c). Clearly, that is possible only if it shared a common horizontal face with b; in other words, b_1 is the neighbor of b from above or from below—without loss of generality, we can assume it is positioned above b. Then the position of cubes a_1 and c_1 is uniquely determined—they must be positioned directly above a and c. Similarly, cubes a_2, b_2, and c_2, positioned in the original square above a_1, b_1, and c_1, form spatial L-shape directly above L-shape a_1, b_1, and c_1 (because positions under a_1, b_1, and c_1 are already occupied by cubes a, b, and c). Continuing in the same manner, we show that the cubes forming the column from cube b up to the upper edge of the original square have to form the identical vertical column of height 8. This is an obvious contradiction.

1987.25. Answer: no, that is not possible.
Obviously, each column must always contain exactly four black squares.

1987.26. Answer: $\frac{100}{101}$.
Let us denote by a_n the fraction which uses n twos. From obvious recurrent equation $a_n = (2 - a_{n-1})^{-1}$, it is easy to show (by induction) that $a_n = \frac{n}{n+1}$.

1987.27. Denote the centers of these circumferences by O and O'. From property of inscribed angles, we have

$$\angle MAB + \angle MBA = \angle AO'B/2.$$

Since $OA \perp O'A$, $OB \perp O'B$, we have $\angle AO'B = 180° - \angle AOB$; that is, $\angle XAB + \angle YBA + \angle AOB/2 = 90°$. Therefore, the measure of arc $XBAY$ is $180°$, that is, XY is a diameter.

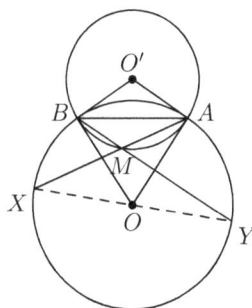

1987.29. Answer: 96433469.
Let $n = \overline{a_1 a_2 \ldots a_n}$ be that largest number. Let us prove that if $a_{k-1} < a_k$, then there are no more than two digits to the right of digit a_k in the decimal representation of n. Indeed, in this case $a_k \geqslant a_{k-1} + 1$ and also $a_k < \frac{a_{k-1} + a_{k+1}}{2}$, that is, $a_{k+1} > 2a_k - a_{k-1}$. Thus

$$a_{k+1} \geqslant 2a_k - a_{k-1} + 1 \geqslant a_k + 2 \geqslant a_{k-1} + 3.$$

Similarly, we get

$$a_{k+2} \geqslant 2a_{k+1} - a_k + 1 \geqslant a_{k+1} + 3 \geqslant a_{k-1} + 6\,,$$

$$a_{k+3} \geqslant 2a_{k+2} - a_{k+1} + 1 \geqslant a_{k+2} + 4 \geqslant a_{k-1} + 10\,.$$

In the same way, we can show that if $a_{k+1} > a_k$, then there are no more than two digits to the left of digit a_k. Therefore, number n cannot have more than eight digits, and if the number of digits is exactly eight, then $a_3 > a_4 = a_5 < a_6$. It follows from all this that $9 \geqslant a_8 \geqslant a_7 + 3 \geqslant a_6 + 5$; hence, $a_6 \leqslant 4$ and therefore, $a_5 \leqslant 3$. Furthermore, $2a_7 < a_6 + a_8 \leqslant 13$, and $a_7 \leqslant 6$. Similarly, $a_1 \leqslant 9$, $a_2 \leqslant 6$, $a_3 \leqslant 4$, and $a_4 \leqslant 3$. Thus, $n \leqslant 96433469$.

1987.30. Denote the stars obscured by the cloud by A_1, A_2, ..., A_{25}, and the visible stars by B_1, B_2, ..., B_{25}. Add triangle's inequalities for all triangles $B_i B_j A_k$, that is,

$$|B_i B_j| \leqslant |B_i A_k| + |B_j A_k|\,, \ 1 \leqslant i,j,k \leqslant 25\,, \ i \neq j\,.$$

The left-hand side of the result is equal to $25T$, where T is the sum of all pairwise distances between visible stars. The right-hand side contains each segment $B_i A_k$ exactly 24 times, and so it does not exceed $24(S - T)$. It follows that $25\,T \leqslant 24(S - T)$, or $49\,T \leqslant 24\,S$, or, equivalently, $T \leqslant 24\,S/49 < S/2$.

1987.35. Answer: yes, such a number exists.

Note that for odd n, number $a^n + b^n$ is divisible by $a + b$. Thus, it suffices to find an odd integer n such that $n + (n+1)$ is divisible by 1987. Take, for example, $n = 993$.

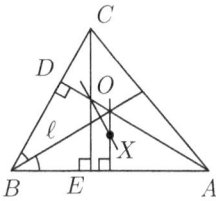

1987.37. Denote the bisector of angle ABC by L, and the circumcenter of triangle ABC by X. In right triangle ABD angle B equals $60°$, and therefore, $|AB| = 2|BD|$. Thus, the midpoint of AB and point D are symmetric with respect to L. Hence, the perpendicular bisector of segment AB is symmetric to altitude AD.

Similarly, perpendicular bisector to CB is symmetric to altitude CE. It follows that points X and O are symmetric with respect to L, and $XO \perp L$. It is easy to see now that the common bisector of angles AOE and COD is also perpendicular to L, and therefore, point X lies on this bisector.

1987.38. The proof immediately follows from inequality

$$\sqrt{(a+c)(b+d)} \geqslant \sqrt{ab} + 1\,.$$

To prove it, square it and then divide by $ab + cd = ab + 1$. We obtain $ad + bc \geqslant 2\sqrt{ab} = 2\sqrt{abcd}$, which is true by virtue of the AM-GM inequality.

1987.39. Answer: the first player wins.

Using his five moves, he can exhaust all odd digits; moreover, after his last (fifth) move he can guarantee that the intermediate result is odd—if he writes an odd digit, he should use $+$ or $\times$ sign depending on the parity of the previous intermediate result, and if he writes an even digit, then he should use $\times$ sign. Then before the last move, his opponent is faced with an even intermediate result and an even digit to use, meaning that the final result has to be even.

1987.40. Represent any complex apartment exchange as some permutation of apartments. It is well known that any permutation can be represented as composition of several disjoint cyclic exchanges. Now, each cyclic permutation can be thought of as a rotation of a regular polygon, with apartments placed in its vertices. Any permutation-rotation of regular polygon with n vertices (by angle $\alpha = 2\pi k/n$) can be represented as a superposition of two line symmetries (reflections) such that the angle between the two axes of symmetry equals $\alpha/2$. Finally, note that any such symmetry is but a collection of two-vertex swaps.

1987.41. Consider angle with measure $60°$, and then on its sides select points B_i at distance OA_i from the vertex, where even-numbered points are selected on one side of the angle and odd-numbered points—on the other. Then we need to prove that

$$|B_1B_2| + |B_3B_4| + |B_5B_6| < |B_2B_3| + |B_4B_5| + |B_6B_1|.$$

This inequality is obtained by adding up three obvious inequalities (see the figure).

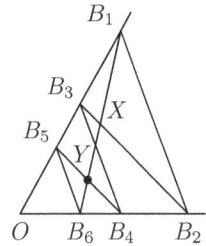

$$|B_1B_2| < |B_1X| + |XB_2|;$$
$$|B_3B_4| < |B_3X| + |XY| + |YB_4|;$$
$$|B_5B_6| < |B_5Y| + |YB_6|.$$

1987.42. Denote 991 by x and rewrite our expression as

$$(x - 2)(x + 10)(x + 16) + 320 = x^3 + 24x^2 - 108x$$
$$= x(x + 6)(x + 18) = 991 \cdot 997 \cdot 1009.$$

These three numbers are prime, so we have the required factorization.

1987.43. The boundary of any figure, formed by the cut parts of the plane, is comprised of several arcs of the circumferences (convex or concave with respect to the interior of the figure). For each such figure F and for any $R > 0$, we will define $L(R)$ as the difference between the sum of the lengths of convex arcs and the sum of the lengths of concave arcs of radius R on the boundary of F. It is clear that for any fixed radius R, value of $L(R)$ does not change when the parts are rearranged, because a convex arc can be "canceled out" only by a concave arc of the same radius and vice versa, and therefore, this value is the same for all figures that can be obtained from the fixed collection of parts.

Denote by S the total area of all pieces. Assume the opposite, that they were somehow rearranged to form several circles. Then

$$S = \frac{1}{2} \sum R \cdot L(R),$$

where the summation is done over all radii R for which $L(R) \neq 0$. If we can show that, in fact, $S < \frac{1}{2} \sum R \cdot L(R)$, we would be done.

Indeed, consider the original figure H formed by all the pieces. For each circumference, draw the radii

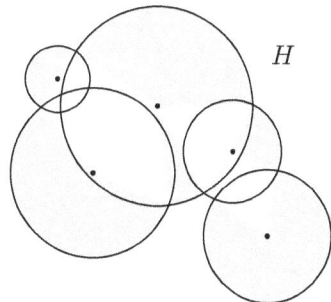

connecting the center of this circumference C with all the points on the common boundary of C and H, in effect, *shading* some part of H within C. Note that the sum of the areas of shaded portions (all of them being open circular sectors) is equal to $\frac{1}{2} \sum R \cdot L(R)$. Let us prove that these sectors do not intersect. Assuming the opposite, some two sectors contain common point P. Then it belongs to some radii $O_1 A_1$ and $O_2 A_2$ of the corresponding circumferences, where $R_1 = |O_1 A_1|$ and $R_2 = |O_2 A_2|$.

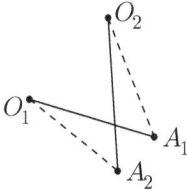

Adding two triangle inequalities $|O_1 A_2| \leqslant |O_1 P| + |P A_2|$ and $|O_2 A_1| \leqslant |O_2 P| + |P A_1|$, we obtain $|O_1 A_2| + |O_2 A_1| \leqslant R_1 + R_2$. Hence, $|O_1 A_2| \leqslant R_1$ or $|O_2 A_1| \leqslant R_2$. If, for instance, $|O_2 A_1| \leqslant R_2$, then point A_1 lies inside the second circumference or on it, but that means that radius $O_1 A_1$ could not have been drawn during the shading.

This proves that the sectors do not intersect, and the sum of their areas equals the area of their union. It suffices now to note that these sectors (even with their boundaries included) do not cover the entire figure—simply examine a small neighborhood of any of its corners. Thus, the area of the figure is strictly less than the sum of the sectors' areas; that is, $S < \frac{1}{2} \sum R \cdot L(R)$.

1987.44. (a) It is easy to see that the palace always has a room (which we will call the *keyroom*), such that after closing it, the palace is split into several isolated parts each of them containing no more than half of all the rooms.

Place one guard into the keyroom; his job is to stay there for the duration of the entire search, guaranteeing that the thief will not be able to sneak from one part of the palace to another. In the meanwhile, the other guards search the isolated parts we mentioned above one by one. Thus, by obvious induction, we proved that n guards can always catch the thief in the palace with no more than 2^n rooms. The basis $n = 1$ is self-evident.

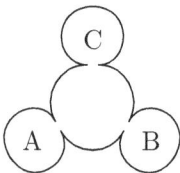

(b) Consider sequence (a_n) defined by equations $a_1 = 4$, $a_{n+1} = 3a_n + 1$. Now, let us construct (by induction) the palace with a_n rooms, in which n guards cannot catch the thief. For $n = 1$, such palace with four rooms is shown in the figure. Further, if we already have a floor plan for a palace with a_n rooms, in which n guards cannot catch the thief, then build three copies of such palace A, B, C as well as one more room with three doors connecting that room with each of the three palaces.

Let us prove that in such palace, with $a_{n+1} = 3a_n + 1$ rooms, $n+1$ guards cannot catch the thief. We can assume that originally the guards are in the center room, and that they have a strategy to catch the thief. In order to prove that the thief can escape, we will assume that this strategy is known to the thief. By definition, the guards can catch the thief only if they are in the same room with him in one of the three smaller palaces, say, A. Assume that originally, the thief is not in palace A (e.g., he is currently in palace B). By induction hypothesis, we know that n guards cannot catch the thief inside the smaller palace—that means that all $n + 1$ guards must be inside one of the three palaces in order to catch the thief there. Hence, during the stage of the strategy lasting until all the guards are gathered inside palace A, the thief avoids the guards by staying inside palace B. Then the thief reviews the strategy to find out which one of the small palaces is the one where all $n + 1$ guards plan to gather next. If that palace is A or C, the thief does not need to do anything but continue avoiding the guards inside B. If that palace is B, then at the moment when all guards are inside A the thief moves to palace C, where he follows the level n strategy to avoid the guards (no more than n of them). After the guards gather inside B, the thief reviews the plan again to find out what is the next smaller palace which he should avoid at the next stage and so on.

It follows that there exists a palace with $a_5 = 364$ rooms with no winning strategy for five guards.

(c) We will call a chain of consecutive adjacent rooms of the palace an *enfilade*. Let us prove that there exists an enfilade (we will call it the *key enfilade*) that splits the palace into components, each containing less than one third of all rooms.

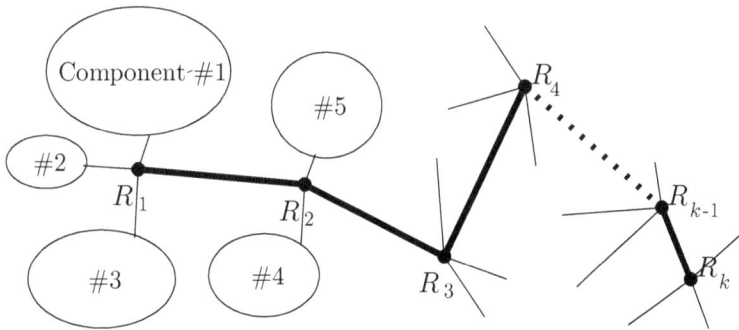

If the palace possesses room E such that among the components defined by it there are two containing at least one third of all rooms each, then declare room X as the initial enfilade. Otherwise, take an arbitrary room E and make the same declaration.

Now, if one of the components that E defines contains at least one third of all rooms—obviously, there are no more than two such components—then we extend E by adding the room that lies in that component and is immediately adjacent to E. The components defined by this expanded enfilade E, clearly, have the smaller sizes. Continuing to extend E, we eventually must end up with a *key enfilade*.

Finally, the winning strategy. Place the first guard into the first room R_1 of the key enfilade $E = (R_1 R_2 \ldots R_k)$; then the other guards search all the components defined by E adjacent to room R_1 (in the example, shown in the figure, these are components #1, #2, and #3). Then the first guard moves to the next room R_2, the other guards search the components adjacent to it, and so on.

Therefore, if n guards can always catch the thief in any palace with no more than p_n rooms, then $n + 1$ guards can always catch the thief in any palace with no more than $p_{n+1} = 3(p_n + 1)$ rooms. One guard can always catch the thief in a palace with three rooms, and thus, we can set $p_1 = 3$. It is left to note that $p_6 = 1092 > 1000$.

1987.47. Color the vertices of the cube in black and white so that no two vertices of the same color are connected by an edge. Then our sum of products does not exceed xy, where x is the sum of the numbers on the black vertices, and $y = 1 - x$ is the sum of the numbers on the white vertices. Now, simply use the inequality $x(1-x) \leqslant 1/4$.

1987.48. Consider remainders modulo 1987. We have $a_n \equiv a_1$, $a_{n-1} \equiv -a_2$. From the formula $a_k + a_{k+1} = a_{k+2}$ it follows by induction that $a_{n-i} \equiv (-1)^i a_{i+1}$ for $i = 0, 1, 2, \ldots,$ $n-1$. For $i = n-1$, we get $a_1 \equiv (-1)^{n-1} a_n \equiv (-1)^{n-1} a_1$. Since a_1 is not divisible by 1987, then $n-1$ must be even.

1987.51. Begin with indexing the cards by numbers 0 through $2n$. Then the following property is invariant under the described operations: for any three consecutive cards with indexes a, b, and c, expression $a - 2b + c$ is divisible by $2n+1$. Then it follows that the indexes of the first and the second cards in the deck determine the entire card arrangement. Since the number of such index pairs is $2n(2n+1)$, the number of possible deck arrangements one can obtain cannot be greater than that.

1987.54. Consider point x_0 such that $g(x_0) = x_0$. Its existence follows from a simple lemma.

Lemma. *For any continuous function $h : [0; 1] \to [0; 1]$ there exists point x such that $h(x) = x$.*[112]

Indeed, consider function $\tilde{h}(x) = h(x) - x$. This function is also continuous, its value at $x = 0$ is, obviously, non-negative, and its value at $x = 1$ is non-positive. It follows from the intermediate value theorem that at some point $z \in [0; 1]$ we have $\tilde{h}(z) = 0$, or, equivalently, $h(z) = z$. $\square$

Now, let $x_1 = f(x_0)$, $x_2 = f(x_1)$, and so on. Then, first, $g(x_n) = x_n$ for each n, and, second, sequence (x_n) is monotonic—since function $f(x)$ is increasing. Monotonically increasing (or decreasing) sequence bounded from both above and below must have limit a, and it is easy to show that a is the common fixed point for both f and g.

1987.56. Let us use induction on T. Basis $T = 1$ is obvious. Now, for the induction step from $T - 1$ to T. Since every $(T-1)$-tuple of consecutive terms of the sequence is complemented by one number (by adding it to the right of $(T-1)$-tuple) to a T-tuple, and there are no more than T such collections (T-tuples), there must be no more than T different $(T-1)$-tuples. If the number of these collections actually does not exceed $T - 1$, then by induction hypothesis, the sequence is periodic. So, we can assume that the number of different $(T-1)$-tuples is exactly T. It follows that if we have two identical $(T-1)$-tuples, then they are complemented to two identical T-tuples. In other words, the $(n+T)$th term x_{n+T} is uniquely determined by the sequence of previous $T-1$ terms x_{n+1}, x_{n+2}, $\ldots$, x_{n+T-1}. From this, the periodicity follows immediately because the number of $(T-1)$-tuples is finite, and therefore, eventually one of them repeats.

1987.57. Note that

$$|OA|/|OC| = (|AB| \cdot |AD|)/(|BC| \cdot |CD|),$$

since both ratios are equal to $S(ABD)/S(CBD)$. Similarly,

$$|OB|/|OD| = (|AB| \cdot |BC|)/(|AD| \cdot |CD|).$$

---

[112] Such point x is called a *fixed point* of function f.

Adding these inequalities, we obtain

$$\frac{|OA|}{|OC|} + \frac{|OB|}{|OD|} = \frac{|AB|}{|CD|}\left(\frac{|AD|}{|BC|} + \frac{|BC|}{|AD|}\right) \geqslant 2\frac{|AB|}{|CD|}.$$

Take three more similar inequalities with

$$\frac{OB}{OD} + \frac{OC}{OA}, \quad \frac{OC}{OA} + \frac{OD}{OB}, \quad \text{and} \quad \frac{OD}{OB} + \frac{OA}{OC}$$

as their left-hand sides, add them all up and then divide by 2.

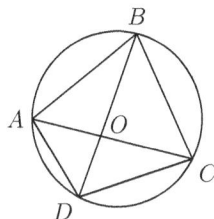

1987.60. Denote the subsets by A_1, A_2, ..., A_s. For each k from 1 to m consider the set C_k of all chains of subsets

$$B_1 \subset B_2 \subset \cdots \subset B_m = \{1, 2, \ldots, m\},$$

such that each B_i contains exactly i elements, and subset B_{a_k} coincides with A_k. It is obvious that the number of these chains is equal to the product of the number of the chains of form

$$B_1 \subset B_2 \subset \cdots \subset B_{a_k} = A_k,$$

and the number of the chains of form

$$A_k = B_{a_k} \subset B_{a_k+1} \subset \cdots \subset B_m.$$

These two numbers are equal to $a_k!$ and $(m - a_k)!$, respectively. Since by condition, no two subsets A_k contain each other, then sets C_k are disjoint, and therefore, the number of chains in their union (which is equal to $\sum a_k! (m - a_k)!$) does not exceed the number of all possible chains, that is, $m!$. Hence,

$$\sum_{k=1}^{s} a_k! (m - a_k)! \leqslant m! \implies \sum_{k=1}^{s} \frac{1}{\binom{m}{a_k}} \leqslant 1.$$

1988

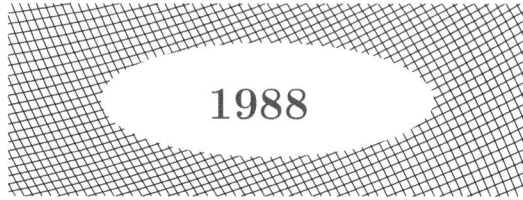

1988.01. However many operations we perform, the number in the central square must always be equal to the sum of the numbers in the corners. Indeed, every permitted operation increases exactly one of the corner numbers by one, and it always increases the central number by one as well. In the beginning, the condition is certainly true as zero in the center equals the sum of four zeros in the corners. Thus, it remains true after any number of permitted operations. It is left to note that in the given table, the sum of the corner numbers is 22, while the number in the center is 18.

1988.02. The number of points gained by a player can be equal to 0, 1, 2, ..., 27, 28, or 30 (it is not possible to get exactly 29 points—if 29 numbers in two sequences coincide, then the last thirtieth numbers must coincide as well). Hence, if for all thirty players these numbers are different, then each of these thirty values must be present. It follows that someone had to gain exactly 30 points.

1988.03. Answer: yes, that is possible.
Example: 51, 1, 52, 2, 53, ..., 49, 100, 50.

1988.04. Answer: no, such integers do not exist.
For any two non-zero numbers, either their sum or their difference has absolute value greater than that of both of these numbers.

1988.05. Answer: no, that is not possible.
Suppose that after n operations we have $n + 1$ piles with three stones each. Since n stones were discarded, then $3(n + 1) = 1001 - n$; that is, $4n = 998$, and that is impossible.

1988.06. Answer: yes, it is possible.
If the painter walks from the original room to any other room W with white floor and then walks back using the same route, then the floor in room W changes its color from white to black with no other changes elsewhere. The painter can use this procedure to change the color in 32 rooms out of 64 and to obtain the chessboard coloring.

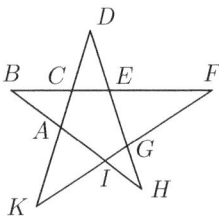

1988.09. If $ab - cd = p$, then a, b, c, d are divisible by p, and therefore, $p = ab - cd$ is divisible by p^2. It follows that $p = 1$.

1988.10. Assume that all these inequalities hold true. Consider triangles ABC, CDE, EFG, and so on, forming the ends of the star's "rays". In any triangle, against the larger side lies the larger angle, and therefore, $\angle BAC > \angle BCA = \angle DCE > \angle DEC = \angle FEG > \cdots > \angle KAI = \angle BAC$, a contradiction.

1988.11. Let us prove that if this process goes on for sufficiently long time, then the cards with numbers 25, 24, ..., 14, as well as one of the cards with 13 will eventually stop moving (we will say that they have *stabilized*). Indeed, both cards with 25 cannot be passed from the very beginning. Next, if cards with 25, 24, ..., n ($n \geq 14$) have already stabilized, then no more than 24 people are holding them. Therefore, after several minutes, a card with $n-1$ must be held by a student for whom it is the larger number. From this moment on that card stops moving. For $n-1 = 13$, the second card with 13 cannot stabilize, since each student cannot hold more than one stabilized card. Therefore, that card must be constantly passed around until it comes to the person who already has the other card with 13.

1988.12. Answer: the first player wins.

The winning strategy consists of always taking 1 or 2 matches in such a way that the number of the remaining matches is divisible by 3.

1988.13. $\frac{x}{(1+y)} + \frac{y}{(1+x)} \leq \frac{x}{(x+y)} + \frac{y}{(y+x)} = 1.$

1988.14. Assume the opposite. Then $\angle CNH = \angle LNP = 60°$, and it follows that $\angle HCN = 30°$ (by counting the angles in triangle CNH) and $\angle ACB = 60°$. In triangle CPM, angle P equals $60°$, angle C equals $30°$, and therefore, angle M equals $90°$. Thus, median BM is also an altitude in triangle ABC, implying that it is isosceles; since its angle C equals $60°$, this triangle is actually equilateral. Hence, AH, BM, and CL pass through the same point (triangle's center), contradicting the conditions.

1988.18. Consider all possible ways to color the vertices of one square. Now, let us count the number of segments with both endpoints of red color, by counting them separately for every square on the sheet. Obviously, each red segment inside the sheet is counted exactly twice. Then there are 41 such segments on the boundary—hence, the overall sum is odd. On the other hand, squares of types 1, 2, 4, 5, 6 (see the figure) contribute an even number into this sum, and only squares of type 3 contribute an odd number.

Therefore, there exists at least one (actually an odd number of) square of type 3.

1988.19. From the second equation, we get $a+b+c = ab+bc+ac$, and it follows (together with equality $abc = 1$) that $(a-1)(b-1)(c-1) = 0$. Thus, one of the factors in this product must be equal to zero.

1988.20. In right triangle ABB_1, angle at vertex A equals $30°$, and therefore, $|BB_1| = |AB|/2 = |BC_2|$. Moreover, $\angle B_1BC_2 = 60°$. Hence, triangle B_1BC_2 is equilateral, and it follows that $\angle AC_2B_1 = 120°$. Similarly, $\angle AB_2C_1 = 120°$. Let X be the intersection of segments B_1C_2 and C_1B_2.

Summing up all angles of AB_2XC_2, we obtain $\angle B_2XC_2 = 90°$.

1988.21. Answer: number $111 \ldots 11995125$ (it begins with 94 ones).

1988.22. Arrange n differently colored points around the circle—these points correspond to the bricks. Then the order of the bricks' colors is always the same as the order of the points around the circle, with possible change of direction. Thus, the arrangement of the bricks can be obtained from the point arrangement by choosing the first point (n ways to do that), and the direction (two ways; clockwise or counterclockwise). That gives us $2n$ possible arrangements.

1988.23. Answer: yes, that is true.

Let us assume that every morning, every two people who live in the same apartment shake each other's hand, and denote the total number of these handshakes by H.

We will show that after every move, the value of H decreases. If on some day, a crowded apartment with $n \geqslant 15$ people is "disbanded", with these people moving to the apartments with $a_1, a_2, \ldots, a_n$ inhabitants, then the value of H decreases by

$$\frac{n(n-1)}{2} - (a_1 + a_2 + \cdots + a_n) \geqslant 105 - 104 = 1 > 0,$$

since $a_1 + a_2 + \cdots + a_n \leqslant 119 - n \leqslant 104$.

Finally, since H is a non-negative integer, its value cannot decrease indefinitely.

1988.24. Simply use the inequality

$$\frac{1-x}{1+x} \cdot \frac{1-y}{1+y} \geqslant \frac{1-(x+y)}{1+(x+y)}.$$

1988.25. Answer: $(2, -1, -1, -1)$ or $(-2, 1, 1, 1)$, and their permutations.

Adding the first and the third equations, we obtain $(a+c)(b+d) = -2$, implying $|a+b+c+d| = 1$. At the same time,

$$a^2 + b^2 + c^2 + d^2 = (a+b+c+d)^2 - 2(ab+cd+ac+bd+ad+bc) = 7,$$

which shows that one of the numbers has absolute value of 2 and all others—1. The rest of the solution is quite obvious.

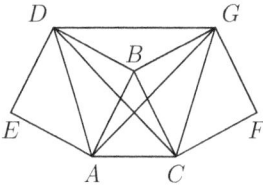

1988.26. Triangles ABG and DBC are equal since $|AB| = |DB|$, $|BG| = |BC|$, and $\angle ABG = 90° + \angle ABC = \angle CBD$. Therefore, diagonals AG and DC in trapezoid $ADGC$ are equal as well, so the trapezoid is isosceles; that is, $\angle CAD = \angle ACG$. Hence, $\angle CAB = \angle CAD - 45° = \angle ACG - 45° = \angle ACB$, and triangle ABC is isosceles.

1988.27. Rewrite the equation as

$$f(x) = (x-b)(x-c) + (x-a)(x-c) + (x-a)(x-b) = 0.$$

By condition $f(a) = (a-b)(a-c) > 0$, $f(b) = (b-a)(b-c) < 0$, and $f(c) = (c-a)(c-b) > 0$. Therefore, each of the intervals $]a; b[$ and $]b; c[$ contains at least one root of the equation. On the other hand, quadratic equation has no more than two roots, and so they must lie inside these intervals.

1988.29. Expression $(a + b + c)^{13}$ equals the sum of monomials of type $a^i b^j c^k$, where $i + j + k = 13$, and i, j, $k \geqslant 0$. Let us prove that each of the monomials is divisible by abc. That is obvious if i, j, $k > 0$. Assume the opposite, that one of the exponents, say, k equals zero. If $j \geqslant 4$, then $a^i b^j \vdots a^i b^{j-3} c \vdots abc$. And if $j < 4$, then $i \geqslant 10$ and $a^i b^j \vdots a^{i-9}(a^3)^3 b^j \vdots a b b^3 \vdots abc$. Finally, if only one exponent (say, i) is different from zero, then $i = 13$ and $a^{13} = a \cdot a^3 \cdot (a^3)^3 \vdots ab \cdot b^3 \vdots abc$.

1988.30. Let $\vec{x}$, $\vec{y}$, and $\vec{z}$ be the vectors defined by the edges of P coming out of one of its vertices. Then the vectors of the diagonals are $\pm\vec{x} \pm \vec{y} \pm \vec{z}$.

Now, let us rewrite the equality of the diagonals' lengths $|\vec{x}+\vec{y}+\vec{z}|$ and $|\vec{x}+\vec{y}-\vec{z}|$ using scalar product

$$0 = (\vec{x} + \vec{y} + \vec{z})^2 - (\vec{x} + \vec{y} - \vec{z})^2$$
$$= (\vec{x}, \vec{x}) + (\vec{y}, \vec{y}) + (\vec{z}, \vec{z}) + 2(\vec{x}, \vec{y}) + 2(\vec{x}, \vec{z}) + 2(\vec{y}, \vec{z})$$
$$- \big((\vec{x}, \vec{x}) + (\vec{y}, \vec{y}) + (\vec{z}, \vec{z}) + 2(\vec{x}, \vec{y}) - 2(\vec{x}, \vec{z}) - 2(\vec{y}, \vec{z})\big)$$
$$= 4(\vec{x} + \vec{y}) \cdot \vec{z},$$

and it follows that $\vec{x} \cdot \vec{z} + \vec{y} \cdot \vec{z} = 0$. Similarly, $\vec{x} \cdot \vec{y} + \vec{x} \cdot \vec{z} = 0$ and $\vec{x} \cdot \vec{y} + \vec{y} \cdot \vec{z} = 0$. Adding the last two equalities and subtracting the first, we get $2\,\vec{x} \cdot \vec{y} = 0$, that is, $\vec{x} \perp \vec{y}$. Similarly, $\vec{x} \perp \vec{z}$ and $\vec{y} \perp \vec{z}$, proving that the parallelepiped is rectangular.

1988.34. Answer: $g(x + f(y)) = x/2 + y + 5/2$.

Substitute $y = 0$ in $f(x + g(y)) = 2x + y + 5$ to obtain that $f(x)$ can be expressed as $f(x) = 2x + c$, where c is some constant. Then $2x + y + 5 = f(x + g(y)) = 2x + 2g(y) + c$, and so we have $g(y) = y/2 + 5/2 - c/2$. Finally,

$$g(x + f(y)) = \frac{x + f(y)}{2} + \frac{5}{2} - \frac{c}{2} = \frac{x}{2} + y + \frac{5}{2}.$$

1988.35. Answer: no, that is not possible.

Indeed, if the product of these two numbers is a perfect square, then 2 goes into the prime factorization of these numbers with exponents of equal parity. At the same time, any one hundred of consecutive natural numbers contains a number of form $2k + 1$ and a number of form $4m + 2$.

1988.36. Answer: $1 + \sqrt{7/3}$.

Denote the vertex of the pyramid by A, the foot of the altitude by H, the center and the radius of the circumscribed sphere by O and R, respectively, and the center and the radius of the inscribed sphere by I and r, respectively. Then select one of the side faces ABC and denote the projection of point I onto plane ABC by F, and the midpoint of BC by L.

From Pythagoras' Theorem for triangle AFI, we have

$$|AF|^2 = (R + r)^2 - r^2 = R^2 + 2Rr.$$

Triangles AFI and AHL are similar, and therefore

$$\frac{r^2}{R^2 + 2Rr} = \frac{|IF|^2}{|AF|^2} = \frac{|HL|^2}{|AH|^2} = \frac{|HL|^2}{(R + 2r)^2}.$$

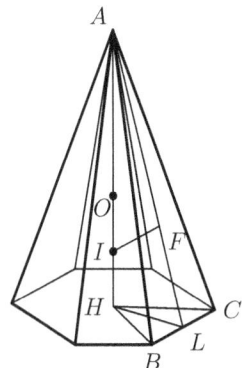

Hence, $|HL|^2 = r^2(R+2r)/R$. From Pythagoras' Theorem for triangle OHB we have

$$|HB|^2 = |OB|^2 - |OH|^2 = R^2 - 4r^2.$$

Now, from HBC being an equilateral triangle, we obtain

$$\frac{r^2(R+2r)}{R(R^2-4r^2)} = \frac{|HL|^2}{|HB|^2} = \frac{3}{4},$$

which implies $4r^2 = 3(R^2 - 2rR)$. This is quadratic equation in R/r, with its positive root being $1 + \sqrt{7/3}$.

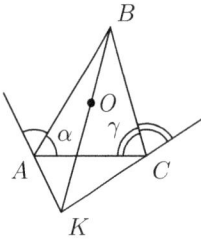

1988.37. Let $\alpha = \angle BAC$, $\gamma = \angle BCA$. Then $\angle KAC = \pi - 2\alpha$, $\angle KCA = \pi - 2\gamma$. Therefore, $\angle AKC = 2\alpha + 2\gamma - \pi$. Point B is the intersection of the bisectors of exterior angles A and C of triangle AKC, and so it lies on the bisector of angle AKC. Hence, $\angle AKB = \alpha + \gamma - \frac{\pi}{2}$, and from counting the angles in triangle AKB we get $\angle ABK = \frac{\pi}{2} - \gamma$. Also $\angle ABO = \frac{\pi}{2} - \gamma$, because ABO is an isosceles triangle with vertex angle $\angle AOB = 2\gamma$. It follows that point O lies on BK.

1988.38. Consider numbers $y_1 = x_1 - x_2$, ..., $y_6 = x_6 - x_1$. If an odd number of them are negative, then the inequality is obvious. It is not possible for all numbers y_i to be negative, since $y_1 + \cdots + y_6 = 0$. It is left to investigate the cases when there are two or four negative y_i. In the latter case, we can replace all x_i by $1 - x_i$, and, if necessary, using the cyclic permutation, we can reduce this case to the former (with two negative y_i) with $y_1 < 0$. If at the same time $y_2 < 0$ or $y_3 < 0$, then

$$y_1 y_2 \cdots y_6 \leqslant y_4 y_5 y_6 \leqslant \left(\frac{y_4 + y_5 + y_6}{3}\right)^3 \leqslant \left(\frac{x_4 - x_6}{3}\right)^3 \leqslant \left(\frac{1}{3}\right)^3 < \frac{1}{16}.$$

Cases $y_5 < 0$ and $y_6 < 0$ are absolutely identical. And if $y_4 < 0$, then

$$y_1 y_2 \cdots y_6 \leqslant (y_2 y_3)(y_5 y_6) \leqslant \left(\frac{y_2 + y_3}{2}\right)^2 \left(\frac{y_5 + y_6}{2}\right)^2 \leqslant \left(\frac{1}{2}\right)^4 = \frac{1}{16}.$$

1988.39. Take, for example, numbers $a = 4001$ and $b = 8001$. Difference of any powers of these two numbers is not zero, and at the same time it is divisible by 4000.

1988.40. Let us prove that by induction on the number of the cities N. The basis is obvious. For $N > 2$, let us start driving from city to city, beginning at some arbitrarily chosen location. Clearly, sooner or later we must return to a city we have already visited, thus creating a cycle of $k \geqslant 2$ cities. Declare these k cities, together with the corresponding roads, as one large "city" L to obtain the system of $N + 1 - k$ cities connected by $2N - k - 1$ roads (note that if inside L there exists a road which is not included in the cycle, then this immediately gives us the road we can close). Since $2N - k - 1 \geqslant 2(N + 1 - k) - 1$, using the induction hypothesis completes the proof.

1988.41. It follows from the problem's conditions that quadrilateral $AKLD$ is inscribed. Therefore, $\angle ADL + \angle AKL = 180°$. Hence, $\angle BKL + \angle BCL = 180°$, meaning that $BCLK$ is also inscribed. So we have $\angle ABL = \angle LCK$, and consequently, $\angle BLA = \angle CKD$.

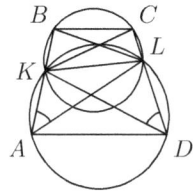

1988.42. Answer: the second player wins.

Denote by $v(n)$ the largest exponent of a power of 2 that divides n (also set $v(0) = \infty$). In the beginning, if we compute this function for the piles' sizes, we have $v(100) = v(252) = 2$.

Let us prove that the second player can play to ensure that if a and b are the sizes of the piles after her move, then $v(a) = v(b)$.

It is clear that in this case the first player will not be able to take the last match.

If $v(a) = v(b)$ before the first player's move, then his move always breaks that property. Indeed, if he takes d matches from the first pile, then $b \,\vdots\, d$ and $v(d) \leqslant v(b) = v(a)$. If $v(d) < v(a)$, then $v(a - d) = v(d) < v(b)$. And if $v(d) = v(a) = k$, then $a = 2^k x$, $d = 2^k y$, where x and y are odd; then $a - d = 2^k(x - y)$ is divisible by 2^{k+1} and $v(a - d) > v(b)$.

Now, the second player is presented with a and b such that $v(a) \neq v(b)$, say, $v(a) > v(b) = k$. Then she takes 2^k matches from the first pile (if the first pile is already empty, then she takes the entire second pile and wins right away).

1988.43. Answer: no.

To any word $A = a_1 a_2 \ldots a_n$ we assign number $T(A) = a_1 + 2a_2 + \cdots + na_n$. It is easy to check that after any permitted operation the remainder of $T(A)$ modulo 3 does not change. Since $T(01) = 2$, and $T(10) = 1$, it is impossible to obtain one from the other.

1988.44. Answer: surprisingly enough, this time the baron could be telling the truth. Such a forest is, actually, possible.

Let us consecutively build forests Γ_1, Γ_2, Γ_3, ..., so that in each of them there are exactly 10 birches at the distance of 1 km from every fir (denote this property by $(*)$).

To build Γ_1, we take one fir O and ten birches planted on the circumference centered at O with radius 1 km.

Now, assume that we already have forest Γ_k. Consider some circumference of radius 1 km with center O and ten points M_0, M_1, ..., M_9 on this circumference. Then construct Γ_{k+1} as the union of images of Γ_k under ten translations by vectors $\overrightarrow{M_0 M_i}$, $(i = 0, 1, \ldots, 9)$, plus the set of firs at the points obtained from all birches in Γ_k using translation by $\overrightarrow{M_0 O}$.

It is easy to see that for "general position" choice of points M_i, the resulting forest Γ_{k+1} satisfies property $(*)$. Note that Γ_1 has 1 fir and 10 birches, Γ_2—20 firs and 100 birches, Γ_3—300 firs and 1000 birches, and so on, and finally Γ_{11} consists of $11 \cdot 10^{10}$ firs and 10^{11} birches. Since $11 \cdot 10^{10} > 10^{11}$, forest Γ_{11} satisfies all the required conditions.

Note. Alas, this still sounds quite suspicious; the outrageously huge number of trees (two hundred and ten billion!) would cause anyone to distrust the baron yet again. However, there exists a significantly smaller forest possessing property $(*)$, with more firs than birches.

This time, build Γ_1 from two firs and two birches planted in the opposite corners of the square with side 1 km. Forest Γ_{k+1} is obtained by adding to Γ_k its image $\tilde{\Gamma}_k$ under translation by some "general position" vector of length 1 km. We also change all firs to birches and vice versa in $\tilde{\Gamma}_k$. It is obvious that Γ_9 satisfies property (∗), but the numbers of firs and birches in that forest are equal (512 firs and 512 birches). However, if we apply to Γ_9 the operation we have described in the first proof, it will provide us with a quite realistic forest consisting of merely $10 \cdot 1024 + 512 = 10\,752$ trees.

1988.46. Find the checker A that made it back to its original square before any other. Then before the last move of A, none of the other checkers were occupying their original squares. Indeed, each one of them must have been moved at least once—otherwise, A would not be able to traverse the entire board. On the other hand, it follows from the definition of A that none of them have yet returned to the original position. Same is true for A as well.

1988.47. Simply use the inequality

$$\frac{1}{x} + \frac{1}{y} \geqslant \frac{4}{x+y}$$

several times.

1988.48. Let us choose any two submissions and call the stations in one of them "red" and in the other—"blue". By definition, it is possible to drive from any red gas station A to some blue gas station B, and from B—to some red gas station C. Then $C = A$, otherwise it would be possible to reach C from A, contradicting the conditions of the tender.

If you can reach two different blue gas stations B and C from A, then it follows that from both of them one can get to A, but then there exists a way to drive from B to C—again, a contradiction. Therefore, there exists a one-to-one correspondence between the red and the blue gas stations, and so the sizes of both submissions must be the same.

1988.49. The solution begins with the following obvious fact.

Lemma. *If segment X that connects two sides of trapezoid is parallel to the bases of trapezoid and passes through the point of intersection of its diagonals, then that point divides X in half.*

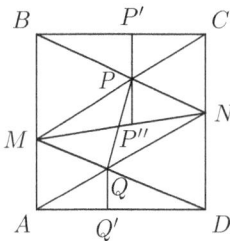

Suppose some line passing through P is parallel to AB and intersects segments BC and MN in points P' and P''.

If $BM \geqslant CN$, then segment $P'P''$ does not exceed (by length) the midline of trapezoid $BMNC$, that is, $|P'P''| \leqslant (|BM| + |CN|)/2$. By the lemma, $|P'P| \leqslant (|BM| + |CN|)/4$. For similarly defined point Q' we have $|Q'Q| \leqslant (|AM| + |DN|)/4$. Add these two inequalities to obtain $|P'P| + |Q'Q| \leqslant |AB|/2$. Finally, using obvious inequality $|P'P| + |PQ| + |QQ'| \geqslant |AB|$, we get $|PQ| \geqslant |AB|/2$.

1988.50. Let k be the largest number occurring infinitely many times among terms of the sequence, and N—the natural number such that $a_i \leqslant k$ for $i \geqslant N$. Choose any $m > N$ such that $a_m = k$. We intend to prove that m is the period for the given sequence, namely, that $a_{i+m} = a_i$ for any $i \geqslant N$.

First, assume that $a_{i+m} = k$. Since $a_i + a_m$ is divisible by $a_{i+m} = k$, $a_m = k$ and $0 < a_i \leqslant k$, we have $a_i = k = a_{i+m}$.

If $a_{i+m} < k$, then choose some $j \geqslant N$ such that $a_{i+m+j} = k$. We have already proved that $a_{i+j} = k$. Since $a_{i+m} + a_j$ is divisible by $a_{i+m+j} = k$, $a_{i+m} < k$, $a_j \leqslant k$, we have that $a_{i+m} + a_j = k$; in particular, $a_j < k$. Finally, $a_i + a_j$ is divisible by $a_{i+j} = k$, $a_j < k$, $a_i \leqslant k$, and therefore, $a_i + a_j = k$. Hence, $a_{i+m} = k - a_j = a_i$.

1988.53. Answer: $100\sqrt{2}$ meters.

Since the snail made 99 left turns, we can split its path into 100 sections of integer (possibly, zero) length (in meters) and such that inside each section the snail made only right turns. It is easy to see that the distance between the start and the finish of each section cannot exceed $\sqrt{2}$. Therefore, the distance traveled by the snail is less than or equal to $100\sqrt{2}$. We leave the example construction to the reader.

1988.54. Answer: $f(500) = 1/500$.

Since $f(999) = f(f(1000)) = 1/f(1000) = 1/999$, our function attains both values 999 and $1/999$, and therefore, there exists number a such that $f(a) = 500$. Then

$$f(500) = f(f(a)) = \frac{1}{f(a)} = \frac{1}{500}.$$

1988.57. Let BB_1 and CC_1 be the altitudes of triangle ABC. It is clear that point B_1 lies on the circumference with diameter BN, and therefore, the power of point H with respect to this circumference equals $BH \cdot HB_1$.

Similarly, the power of point H with respect to the second circumference equals $CH \cdot HC_1$. Since points B, C, B_1, C_1 lie on the same circumference (with BC being its diameter), then $BH \cdot HB_1 = CH \cdot HC_1$. Hence, the powers of point H with respect to these two circumferences are equal, and therefore, it lies on their radical axis PQ.

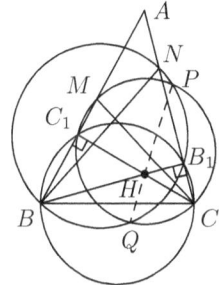

1988.58. First, a very helpful lemma.

Lemma. *If $P(x) + kP'(x) \geqslant 0$ for any x, then $P(x) \geqslant 0$ for any x.*

Indeed, if $P(x_0) < 0$ for some x_0, then since $P(x)$ cannot have a local minimum a with $P(a) < 0$, we can conclude that $\lim\limits_{x \to \pm\infty} P(x) = -\infty$. This, however, is impossible because both the leading coefficient and the degree of polynomial $P(x)$ are the same as the ones for $P(x) + kP'(x)$. $\square$

Now, consider polynomials $Q(x) = P(x) - 2P'(x) + P''(x)$ and $R(x) = P(x) - P'(x)$. By condition $Q(x) + Q'(x) = P(x) - P'(x) - P''(x) + P'''(x) \geqslant 0$ for any x, and it follows from the lemma that $Q(x) \geqslant 0$. Further, $Q(x) = R(x) - R'(x)$, and using the lemma again gives us $R(x) \geqslant 0$ for any x. Finally, applying the lemma for the third time, we obtain that $P(x) \geqslant 0$ for any x.

1988.60. Let $M = A_1 A_2 \ldots A_n$ be the given polygon, with its ith side $A_i A_{i+1}$ ($A_{n+1} \equiv A_1$). Consider vectors $\vec{v}_i = \overrightarrow{A_i A_{i+1}}$, and also add to this collection vectors $\vec{w}_i = -\vec{v}_i =$

$\overrightarrow{A_{i+1}A_i}$. Now, place all these vectors at the origin and index them in the clockwise direction beginning with $\vec{v}_1$: $\vec{u}_1$, $\vec{u}_2$, ..., $\vec{u}_{2n}$. After this consider the terminal points of vectors $\vec{u}_1$, $\vec{u}_1 + \vec{u}_2$, $\vec{u}_1 + \vec{u}_2 + \vec{u}_3$, ..., $\vec{u}_1 + \vec{u}_2 + \cdots + \vec{u}_{2n} = \vec{0}$. It is easy to see that they form a convex centrally symmetric polygon N with $2n$ vertices. Further, for each side of M we have two equal and parallel to it sides of N; moreover, projection of N onto any line is twice as long as the projection of M onto the same line.

It follows then that expression $\sum a_i/d_i$ has the same value for polygon M as for polygon N. Thus, it is sufficient to prove our inequalities for any convex and centrally symmetric polygon N. Set $N = B_1 B_2 \ldots B_{2n}$, and denote by h_i the projection of N onto the direction perpendicular to the ith side of N. Then $d_i h_i \geqslant S(N)$ (S here denotes the area function), and we have

$$\sum_{i=1}^{2n} \frac{a_i}{d_i} \leqslant \sum_{i=1}^{2n} \frac{a_i h_i}{S(N)} = \sum_{i=1}^{2n} \frac{4S(OB_i B_{i+1})}{S(N)} = 4\,,$$

where O is the center of symmetry for N.

The left inequality follows from $\sum a_i/d_i \geqslant \sum a_i/D$, where D is the diameter of N, that is, the largest distance between its vertices. Since $\sum a_i$ is the polygon's perimeter, we also have $\sum a_i > 2D$.

1988.61. Clearly, finding a solution for item (b) is enough. Let us assume that the painter can start and end his walk in any room of the castle; also when the painter starts the walk in some room, he does not begin with repainting it. We can also assume that repainting happens when the painter enters the room, not when he leaves the room. It would suffice to prove that in $2n^2$ steps, the painter can repaint any set of rooms. Also we assume, for convenience sake, that the rooms he wants to repaint are currently painted black (so his goal is to paint them white).

Obviously, he can repaint all the rooms if needed. Thus, we will assume that currently there are both white and black rooms.

First, consider the case when n is even. Let R be some arbitrary route visiting all rooms of the castle exactly once and ending at the start. Now, index all the white (only the white!) rooms of the route in the order they are visited.

The painter starts moving in a *zig-zag* manner as follows. He starts in the white room #1, follows route R for x steps, then y steps ($y < x$) in the opposite direction, then z steps ahead ($z > y$), and so on. These numbers are chosen in such a way that the rooms where the painter makes a turn (changes the direction) are the white rooms. Thus, the painter turns to move back in the white rooms with odd indexes, and turns to move forward in the white rooms with even indexes.

Clearly, the route ends either in the initial room, if the number of white rooms on this route is even, or in the last room of the route if that number is odd. As a result, all white rooms are visited exactly once, and all black rooms—either once or thrice. Note that the black portions, which are traveled backward and forward, alternate.

Now, we only need to make sure that the total length of the portions traveled thrice does not exceed the total length of the portions traveled once. If that is not so, then the route must be started in the next (on the route) white room.

For odd n, consider a very similar route visiting all the rooms except one; say, except corner room A. If A is white, then the method used above fits here as well. If room A is black, then select room B—one of the rooms adjacent to A. Temporarily change the color of room B and construct the painter's route R which results in the white coloring. Since we have several ways to choose rooms A and B, we can assume that B is neither the start nor the finish of R. Now, let us adjust the route—when the painter enters room B, he leaves the route, steps into room A, immediately returns back, and continues moving along R. After this, all rooms except B become white. But since the real color of B is opposite to the one we used when constructing route R, then room B also ends up colored white.

1988.62. Answer: yes, it is possible.

Note that $(x+1)+(x+6)+(x+8) = 3(x+5)$ and $(x+2)+(x+3)+(x+7) = 3(x+4)$. Therefore, any eight consecutive numbers $x+1, x+2, \ldots, x+8$ can be split into two such groups. Since numbers 1 through 1000 can be divided into 125 octuplets of consecutive integers, the rest of the construction is obvious.

1988.63. Split the board into $2n-1$ diagonals parallel to the "North-East" direction (that is, they are parallel to the main diagonal connecting the lower-left and the upper-right corners). We will call a diagonal *empty*, if there are no beetles sitting on its squares. Let A be the non-empty diagonal closest to the upper left corner. First, A cannot lie below main North-East diagonal—otherwise the leftmost of the beetles on A can crawl to the square on its left. Second, A must be completely occupied by the beetles. Indeed, if it contains empty squares, then there are two neighboring squares one of which is empty and the other is occupied by a beetle. Depending on which of these two squares is closer to the upper-right corner, the beetle can crawl either in the leftward or the upward direction while keeping the required property intact.

Therefore, entire diagonal A is occupied by the beetles, and since their number is less than n, it must be located above the main diagonal. Similarly, the non-empty diagonal B closest to the lower-right corner must (a) lie below the main diagonal, and (b) be completely occupied by the beetles.

Let A contain k squares and B—m squares. Then there are $2n-1-k-m$ diagonals between A and B, and they must accommodate the remaining $n-1-k-m$ beetles. Note that $2n-1-k-m > 2(n-1-k-m)+1$, that is, the number of diagonals between A and B exceeds twice the number of beetles sitting on them by more than 1. It follows that between A and B, there exist two neighboring empty diagonals. Let us denote them by C and D, and without loss of generality, assume that C is shorter than D and lies above D.

Now, consider diagonal X closest to C among non-empty diagonals lying above C. Such diagonals exist because A is non-empty and lies above C. Since two diagonals under X are longer than X and empty, the rightmost of the beetles on X can simply crawl rightward.

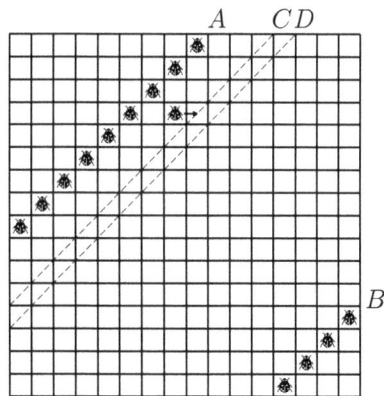

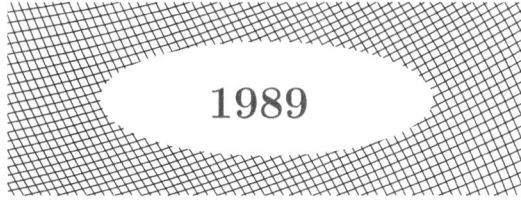

1989.01. Answer: 33 questions. The number of questions, obviously, attains maximum when every question is included in no more than two problem sets. The overall number of questions then equals $4 \cdot 6 + 3 \cdot 6/2 = 33$.

1989.02. Indeed, if we replace the first three digits of any lucky ticket by their complements to 9, then we obtain the ticket with the sum of its digits equal to 27, and vice versa.

1989.03. When the toy train moves along the rail of Type 1 it turns 90° clockwise, when it moves along the rail of Type 2—90° counterclockwise. Therefore, in any closed track, the difference between the number of Type 1 rails and the number of Type 2 rails must be divisible by 4. Obviously, if one of the rails of Type 1 is replaced by a rail of Type 2, this divisibility property will no longer hold.

1989.04. See solution to **1989.09**

1989.05. Example: 166667 and 333334.

1989.06. Answer: the second player wins.
      To do that, he simply reflects the moves of the first player with respect to the center of the board, changing X to O and vice versa unless he can immediately win. The fact that such an opportunity must eventually arise can be easily checked by simple direct examination of what happens within the central 2×2 square.

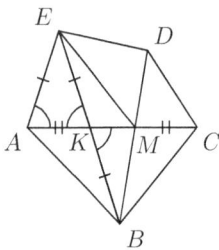

1989.08. Segments EM and BC are equal as corresponding sides in congruent triangles AME and KCB.

1989.09. Split all the stones into thirty-two pairs, then with the first 32 weighings compare the weights in all pairs.
      The lighter stones we put aside, focusing on the 32 heavier stones. Split them again into sixteen pairs, compare weight in pairs using 16 weighings and so on. Thus, we need $63 = 32 + 16 + 8 + 4 + 2 + 1$ weighings to determine the heaviest stone S. The second heaviest stone must be among the six stones that have been compared with S during these operations, because every other stone is lighter than some stone other than S. Now, five more weighings to find the heaviest one among these six stones, and therefore, $68 = 63 + 5$ weighings are enough.

1989.10. Answer: $a = b = c = \pm 10$.
      Subtracting the second equality from the first, then using the formula for the square of

the sum, we obtain

$$0 = 100 - 100 = a^2 + 2b^2 - 2bc - 2ab + c^2 = (a-b)^2 + (b-c)^2,$$

hence, $a = b = c$. Now, from the second equality we get $a^2 = 100$.

1989.11. Answer: no, that is impossible.

Consider the edge of the stack that is not covered by $[1\text{–}101]$ side of any of the polygons—there are at least two such edges. Then the sums of the numbers on the vertical edges of the corresponding face of the stack must have different parity and therefore, cannot be equal.

1989.12. Answer: $1944 = 2^3 \cdot 3^5$.

If this number has a prime factor greater than 3, then it is greater than or equal to $5 \cdot 600 = 3000$. Examine all numbers of form $2^k 3^n$ directly, and we are done.

1989.13. Let us introduce quantity S equal to the sum of the squares of numbers in the collection. Then S increases if at least one of the numbers a and b is not equal to zero, and does not change if both numbers vanish.

The very first operation deals with non-zero numbers, so S increases during that step and does not decrease afterwards. Thus, the original collection cannot occur again.

1989.14. Since everyone claims that their right neighbor is a mathematician, it follows that the left neighbor of every physicist is a liar, and the right neighbor of every liar is a physicist.

Therefore, the number of liars is equal to the number of physicists. The number of physicists is n, and the liars are evenly split between physicists and mathematicians. Thus, n is even.

1989.15. Note that $\angle COD = \angle OAD + \angle ODA = \angle OAD + \angle OAB = \angle BAD$.

Therefore, triangles COD and BAD are congruent (**SAS** rule); hence, we have $\angle ABD = \angle BDC$. It follows that lines AB and CD are parallel. We leave to the reader as an exercise proving that lines BC and AD are not.

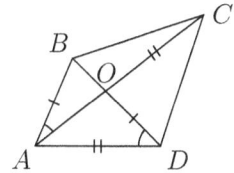

1989.16. If absolute values of numbers x, y, and z are greater than or equal to 1, then $x^2 + y^2 + z^2 \geqslant x + y + z$, and our inequality is trivial. If one of them, say, x, has absolute value less than 1, then

$$x^2 + y^2 + z^2 \geqslant y^2 + z^2 \geqslant |yz| \geqslant |xyz|.$$

1989.18. Answer: yes, the librarian can do that (moreover, that can be done regardless of the number of volumes in the collection).

Let us prove this by induction on the number of volumes n. We can assume that in the beginning the volumes are arranged in ascending order. If there are only two volumes, then the statement is trivial. Now, denote the required sequence of swaps for n volumes by S_n. Then the librarian can achieve the desired goal for $n + 1$ volumes in the following manner. She starts with the first swap in S_n (the $(n + 1)$th volume is the last one, and therefore, it does not prevent her from doing any swap for the first n volumes; that book can be considered as temporarily separate from the first n volumes); then she proceeds to "move" the $(n+1)$th volume from the right end of the shelf to the left, transposing it with all the volumes on the way (that requires n swaps). Then this volume once again can be thought of as a separate from volumes 1 through n, and so the librarian can perform the second swap from sequence S_n. After that she again performs n swaps, "moving" the $(n+1)$th volume from the left end to the right. Then follows the third swap from S_n, and so on. Finally, the same n "moving" swaps are performed after the last swap of S_n, and the construction of sequence S_{n+1} is over.

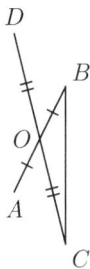

1989.19. Answer: the second player wins.

His strategy is as follows. With every move he shifts the pawn to the node symmetric to the one it is currently on, with respect to the center of the square. Let us prove that this strategy satisfies the rules of the game. Let O be the center of the square, and suppose that the players' last moves were A, B, C, and D. Then $|AB| < |BC|$, $|OB| + |OC| > |BC| > |AB| = 2|OB|$. Therefore, $|OC| > |OB|$, and consequently, $|CD| > |BC|$.

1989.20. Answer: yes, such a set exists.

Take arbitrary hundred different natural numbers and multiply them all by the same number P equal to the product of the sums of all quintuplets from the original collection.

1989.21. Solution 1. Let us show that all these numbers have the same sign. For example, that $xy > 0$. Indeed,

$$\frac{1}{z} = -\frac{1}{x} - \frac{1}{y} = -\frac{x+y}{xy} = \frac{z}{xy},$$

and therefore, $xy = z^2 > 0$.

Solution 2. Consider cubic polynomial $f(t)$ with roots x, y, and z. By Vieta's Theorem, the coefficients of this polynomial at monomials t^2 and t vanish, and so we have $f(t) = t^3 + a$. Therefore, f has only one real root, that is, $x = y = z$ which is, clearly, impossible.

1989.22. Let $d = b - a$. Then $a^2 + 1 = (a+d)(a-d) + d^2 + 1$, and therefore, $d^2 + 1$ is divisible by $a + d = b$. But if $d \leqslant \sqrt{a}$, then $d^2 \leqslant a$ and $d^2 + 1 \leqslant a + 1 < b$ ($b \neq a + 1$ because $a^2 + 1$ is not divisible by $a + 1$), a contradiction.

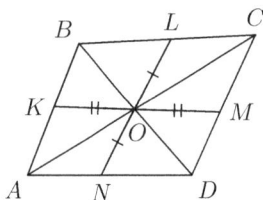

1989.23. Suppose that $|OC| \geqslant |OA|$ and $|OD| \geqslant |OB|$. Then, under symmetry with respect to point O, triangle OAB is mapped inside triangle OCD, but at the same time point K is mapped to M. That is only possible if $|OC| = |OA|$ and $|OB| = |OD|$; in other words, $ABCD$ must be a parallelogram.

1989.24. Let us denote the number of columns containing the pawns by k; these columns contain a_1, a_2, ..., a_k pawns, respectively. Compute the product for every pawn of the first column—obviously, the sum of these numbers does not exceed ma_1. Similar inequalities are true for all other columns. Therefore, the sum of all products does not exceed $ma_1 + ma_2 + \cdots + ma_k = m^2$. But if the number of pawns, for which the product is at least $10m$, is greater than $\frac{m}{10}$, then this sum would exceed m^2.

1989.25. Answer: if $k = 1989$, then the tournament had two participants, and if $k = 1988$, then there could only be two or three participants.

Suppose that the number of players was n, participant in the last place gained a points, the next one gained ap points, ..., and the winner gained ap^{n-1} points. The winner could not gain more than $2k(n-1)$ points, and all other participants took at least $k(n-1)(n-2)$ points. Comparing these numbers, we see that if $n \geqslant 4$, then $2k(n-1) \leqslant k(n-1)(n-2)$. On the other hand, $ap^{n-1} > a + ap + \cdots + ap^{n-2}$. Hence, $n < 4$. Moreover, if $p \geqslant 3$, then the points of the winner would be more than twice the total of all other players, which is impossible for $n = 3$. Therefore, $n = 3$ is possible only for $p = 2$; that is, players have a, $2a$, and $4a$ points, and the overall total $6k$ must be divisible by 7. Hence, k must be divisible by 7—and since 7 does not divide 1989, we can conclude that $n = 2$ is the only option in that case. Finally, finding an example for $k = 1988$ and $n = 3$ is quite easy.

1989.26. Answer: this is possible only for even n. Consider all points labeled with 1. There are two lines passing through each one of them, and therefore, n must be divisible by 2.

Now, let n be some even number. Index the lines by numbers 1 through n. For each pair of indexes i and j less than n, let us label the intersection of the ith and the jth lines with a positive remainder of $i + j$ modulo $n - 1$ (meaning that instead of 0 we write $n - 1$).

Finally, label the intersection of the nth line and the ith line with the remainder of $2i$ modulo $n - 1$.

1989.28. Let K, L, M, and N be the points where the incircles of triangles ABX and BCX touch the sides of triangle ABC. Then half-perimeter of triangle ABC equals $|AK| + |KX| + |BM| + |CM|$; also that same half-perimeter minus the length of BC is equal to AX. Therefore, X is the point where the incircle of ABC touches side AC.

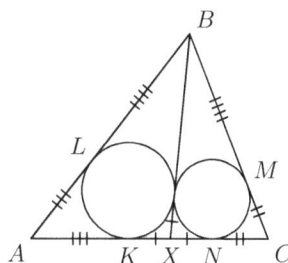

1989.30. Answer: $1 - \frac{1}{(1+\cos 72°)}$.

Let p be the exterior perimeter of the star, and x—the unknown perimeter of the interior pentagon $ABCDE$. We know that $p + x = 1$.

Since all the vertex angles of the star are the same, all angles in pentagon $ABCDE$ are the same, and therefore, all the triangles ("star rays") are isosceles with 36° angle at the vertex and 72° angles at the base. It follows that $x = p \cos 72°$, and finally, $x + \frac{x}{\cos 72°} = 1$.

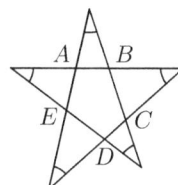

1989.31. Answer: yes, it is possible.

Example: place numbers 1, 0, 1, 0, 1, 0, 1, 0, 1, 0 on the diagonal that runs from the lower left corner to the upper right corner of the table. Then write 1 into all the squares above the diagonal, and (-1)—into all the squares below the diagonal.

1989.34. Let us, for the sake of simplicity, set $AB = 6$. If XK and XM intersect diameter AB at point C and D, respectively, then $|AC| = |CD| = |DB| = 2$.

Clearly, chord KM gets longer as angle $\angle KXM = \angle CXD$ increases; that angle reaches its maximum when X is the midpoint of semicircle AB. It is left to prove that $5\,|KM| = 3\,|AB|$. Indeed, we have

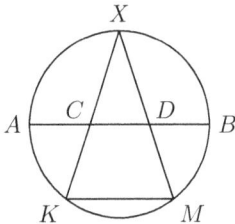

$$|XC| = \sqrt{10}, \;\; \sin \angle KXM = \frac{3}{5}, \;\; |KM| = \frac{18}{5}.$$

1989.37. Answer: yes, it can.

For instance, take the operation which maps any pair of natural numbers to their greatest common **odd** divisor.

1989.38. It is quite easy to check that the sum of these three numbers equals zero.

1989.40. It is easy to see that $x^2 \geqslant b^2 \geqslant y$ and $x \geqslant a^2 \geqslant y^2$. Thus, every point (x, y) that possesses that property belongs to the shaded area shown in the figure. Conversely, if $x^2 \geqslant y$ and $x \geqslant y^2$, then we can set $a = y$ and $b = x$.

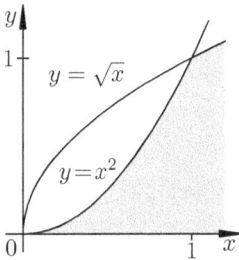

1989.41. Denote the pyramid's vertex by O and let OB be the shortest of the side edges of the pyramid. Let us prove that faces ABO and CBO are congruent. Using the law of sines for triangles ABO and BCO we have

$$\sin \angle BAO = \frac{|BO| \cdot \sin \angle BOA}{|BA|} = \frac{|BO| \cdot \sin \angle BOC}{|BC|} = \sin \angle BCO.$$

Hence, either angles $\angle BAO$ and $\angle BCO$ are equal (and then triangles ABO and CBO are congruent), or they are supplementary angles adding up to $180°$. In the latter case, we obtain that one of the angles BCO and BAO is not acute, and therefore, side BO is the longest one in triangle ABO or in triangle CBO. This contradicts the choice of BO as the shortest side edge of the pyramid.

1989.44. Using properties (1) and (2) one after the other we obtain

$$(a * b) * c = -(c * (a * b)) = -((c * a) * b)$$
$$= b * (c * a) = -(a * (b * c)) = -((a * b) * c),$$

and so it follows that $(a * b) * c = 0$. But we also know that $a * b$ can take any integer value x, and the same is true for $x * c$, a contradiction.

1989.47. Let $t > 1$ be the square root of one of the roots of the given quadratic equation. Then $At^4 + (C - B)t^2 + (E - D) = 0$; that is, $At^4 + Ct^2 + E = Bt^2 + D$. Denote the left-hand and right-hand sides of this equality by F and G ($F = G$), respectively. If we define $f(x) = Ax^4 + Bx^3 + Cx^2 + Dx + E$, then $f(\pm t) = F \pm tG$. Since $t > 1$, these two numbers have opposite signs and therefore, polynomial f must have a root on interval $[-t; t]$.

1989.48. Let O be the circumcenter of triangle BMC. Then

$$\angle BOC = \angle BOM + \angle MOC$$
$$= 180° - 2\angle BMO + 180° - 2\angle CMO$$
$$= 360° - 2\angle BMC = 180° - \angle BAC,$$

and therefore, quadrilateral $ABOC$ is inscribed.

Equality $|BO| = |OC|$ implies that AM is the bisector of angle $\angle BAC$. It is left to note that CM is the bisector of angle $\angle ACB$, because $\angle BCA = \angle BOM = 2\angle BCM$ (the last equality follows from angle $\angle BOM$ being the central angle, and angle $\angle BCM$ being the inscribed angle subtending the same arc).

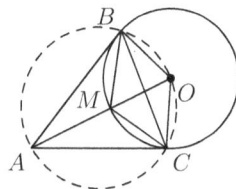

1989.49. Let us prove the inequality by induction on n. The basis for $n = 2$ is obvious.

Now, assume that $n > 2$, and all numbers from the given collection do not exceed some number N. We can even assume that the collection contains all natural numbers not exceeding N whose prime divisors are not greater than n. Then split the collection into two subcollections: the numbers with prime divisors not exceeding $n - 1$, and the others (this second subcollection is empty if n is composite).

The sum s of reciprocals of all numbers of the first subcollection does not exceed $n - 1$ by the induction hypothesis. The sum of reciprocals of all numbers from the second subcollection does not exceed

$$s\left(\frac{1}{n} + \frac{1}{n^2} + \cdots + \frac{1}{n^k}\right) = s \cdot \frac{1}{n} \cdot \frac{1 - 1/n^{k+1}}{1 - 1/n} < \frac{s}{n-1} \leqslant \frac{n-1}{n-1} = 1,$$

where k is the maximum exponent for prime number n in factorizations of all numbers of the original collection. Now, add the inequalities for the two subcollections to prove the induction step.

1989.50. Assume that such a placement exists. Let 2^a be the minimum sum of numbers in a row of the table. Then $2^a \geqslant 1 + 2 + \cdots + k = \frac{1}{2}k(k+1)$. On the other hand, the sum of all numbers in the table must be divisible by 2^a, and so we obtain that $\frac{1}{2}k^2(k^2 + 1)$ is divisible by 2^a. If k is odd, then $\frac{1}{2}k^2(k^2 + 1)$ is also odd. If k is even, then 2^a must divide $\frac{1}{2}k^2$, but $\frac{1}{2}k^2 < \frac{1}{2}k(k+1) \leqslant 2^a$, a contradiction.

1989.51. Assume the opposite. Then at any moment there exist a row and a column entirely filled by the pawns of the same color. Originally the color of that "cross" was white, and in the end it has to be black. Now, note the obvious fact that repainting and moving one pawn cannot create a cross of the color different from the one we just had on the board.

1989.52. Answer: 18.

If necessary, we can cut the quadrilateral along its diagonal and flip one of the resulting triangles so that the sides with lengths 1 and 8 are neighbors. Now, sides with lengths 4 and 7 are also neighbors, and therefore, the area of our quadrilateral does not exceed $1 \cdot \frac{8}{2} + 4 \cdot \frac{7}{2} = 18$.

Further, since $1^2 + 8^2 = 4^2 + 7^2$, we can "glue" together two right triangles with legs 1, 8 and 4, 7, respectively, to obtain a quadrilateral with area 18. Thus, the value of 18 can be attained.

1989.53. The first player wins by adding 1 with his every move. Then, if the number was even, it becomes odd and then, obviously, the second player makes it even again by adding some, necessarily odd, divisor of this new number. Therefore, this strategy guarantees that the number before the first player's move is always even. Since the first player always adds 1, the only situation where he can lose is if the number he had before his move was 19891989—but that, as we have shown, is not possible.

1989.54. Answer: 56 words.

It is easy to show that, using the described operations, all digits 1, except for possibly one, can be moved to the beginning of the word. Thus, every word has a synonym of form $00\ldots00$ or $11\ldots100\ldots010\ldots0$ (begins with a block of a ones, then follows block of b zeros, where either of numbers a and b could be equal to zero; then a block that consists of one or zero digits 1, and finally a few—possibly, none—zeros).

Words of the second type correspond to the solutions of inequality $a+b \leqslant 9$ with $a, b \geqslant 0$; there are $55 = 1 + 2 + \cdots + 10$ of them. It is left to prove that for different values of a and b, the words shown above cannot be synonyms of each other. That fact is obvious for two words with different values of a. Now, consider T—number of zeros with an odd number of ones written to the right of them. It is quite easy to see that the described operations do not affect the value of T. For the words shown above, $T = b$, and therefore, if two of them have different value of b, then they cannot be synonyms.

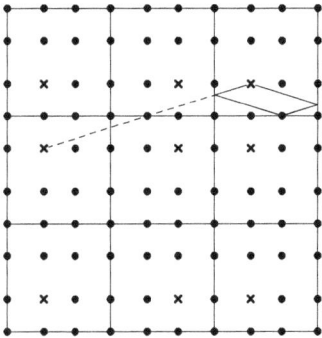

1989.55. Answer: yes, Professor Jones can do it.

Fifteen students are enough. Introduce the coordinate system with its axes along the midlines of the room, assuming that the side of the room has length 2. Let Smith stand in the point with coordinates (x_0, y_0), and assume, for convenience sake, that x_0, $y_0 \geqslant 0$. Then Jones places the students in points with coordinates (a, b), where $a \in \{\pm x_0, \pm 1\}$ and $b \in \{\pm y_0, \pm 1\}$, and, of course, $(a, b) \neq (x_0, y_0)$.

Let us prove that Smith cannot see any of his reflections. To do this, tile the entire plane with the mirror images of the room (see the figure). Then any ray of light that connects Smith with "Smith", reflecting from the walls of the room, corresponds to a straight line segment that connects Smith and one of his mirror-image phantoms. It is easy to see that the midpoint of such segment coincides with the phantom of one of the students, that is, the ray passes through one of the students placed by Jones.

1989.57. Since 10^7 has remainder 1 modulo 239, our number has the same remainder modulo 239 as the sum of the seven-digit numbers forming it. That sum is equal to $1 + 2 + \cdots + 9\,999\,999 = \frac{1}{2}10^7(10^7 - 1)$, and therefore, is divisible by 239.

1989.58. Difference of the right- and left-hand sides of the inequality is at least
$$(1 - x^2)(1 - y) + (1 - y^2)(1 - z) + (1 - z^2)(1 - x) \geqslant 0.$$

1989.60. Use induction on the row index. It is easy to check that the numbers in the first four rows are integers. To prove that some number x in the kth row (with $k > 4$) is an integer, consider the numbers shown in the figure. They are all (except possibly for x) non-zero integers.

Now, note that
$$ex + 1 = gh = (ae + 1)gh - aegh = bcgh - aegh$$
$$= (de - 1)(ef - 1) - aegh = e(def - f - d - agh) + 1,$$

Therefore, x equals $def - f - d - agh$, and so it is also an integer.

$$\begin{matrix} & a & \\ b & & c \\ d & e & \\ g & h & f \\ & x & \end{matrix}$$

1989.61. Consider sequence (b_k) such that $b_1 = \arctan a_1$, and $b_{k+1} = b_k + \arctan(1/k)$. Then, as we can easily check, using the sum formula for tangent, we have $a_k = \tan b_k$. It suffices now to use the fact that increasing sequence (b_k) is tending to infinity, while the difference between its consecutive terms tends to zero.

1989.62. Both equalities are equivalent to the quadrilateral $OABC$ being exscriptible[113]—i.e., there exists a circle which touches segments AO and CO as well as rays AX and CY (see the figure on the left). Indeed, if such a circle exists, then $|AM| + |AN| = |PM| + |QN| = |RM| + |SN| = |CM| + |CN|$ and $|AO| + |AB| = |QO| + |PB| = |RO| + |SB| = |CO| + |CB|$.

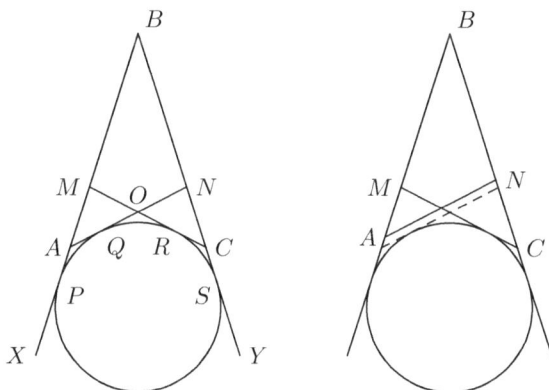

Inversely, suppose that $|AM| + |AN| = |CM| + |CN|$. It is obvious that segment AN can be translated in the direction parallel to itself so that the resulting "adjusted" quadrilateral is exscriptible. If, for example, we end up shifting AN "down" (see figure on the right), then expression $|AM| + |AN|$ increases and $|CN| + |CM|$ decreases, and therefore, for this quadrilateral, the equality $|AM| + |AN| = |CM| + |CN|$ does not hold true, a contradiction. Thus, the original quadrilateral must be exscriptible. If $|AO| + |AB| = |CO| + |CB|$, then the proof is absolutely identical.

---

[113] Meaning that there exists a circle that touches all sides of the quadrilateral (or their extensions) from outside.

1989.63. Answer: For all $k < 100$ such that $k + 1$ is not divisible by 8.

First of all, note that if one of the arcs is contained in another, the larger arc can be shrunk to coincide with the smaller one. Further, mark on the circumference all the endpoints of the arcs, then orient all the arcs in the clockwise direction. Again, by shrinking the arcs, we can obtain the configuration where for each marked point P, there is at least one arc that begins at P.

Now, let us measure the length of an arc by the number of gaps between the marked points contained in this arc. Simply put, distance between any two neighboring marked points is equal to 1. Let the length of arc α, beginning at point M_1, be a, and the length of arc β, beginning at the next marked point M_2—b. Then $b \geqslant a$, otherwise arc β would contain arc α without coinciding with it. Traveling along the circumference in the clockwise direction, we will obtain closed sequence of such inequalities, from which we can conclude that the lengths of all arcs are the same and equal to some number $m \leqslant n$, where n is the number of all marked points.

Denote by a_i $(i = 1, \ldots, n)$ the number of arcs beginning at point M_i. Then the number of arcs intersecting with arc beginning at M_i equals $a_{i-m} + a_{i-m+1} + \cdots + a_{i+m} - 1$, which, by condition, must be equal to k. Therefore, for any i we have $a_{i-m} + a_{i-m+1} + \cdots + a_{i+m} = k+1$. Adding these equalities for $i = 1, 2, \ldots, n$, we get

$$100(2m + 1) = (k + 1)n.$$

Hence, number $k + 1$ divides $100(2m + 1)$, and therefore cannot be divisible by 8. All other values of $k < 100$ are acceptable; examples are easily constructed, and we will leave them to the reader.

1989.64. Denote by m number $[\frac{1}{2}(n+1)] \geqslant \frac{1}{2}n$, and consider the square with dimensions $m \times m$ inscribed into our triangle, whose sides form angles $45°$ with the triangle's rows (see the left figure). Let us prove that this square contains at least $\frac{1}{4}n$ different numbers.

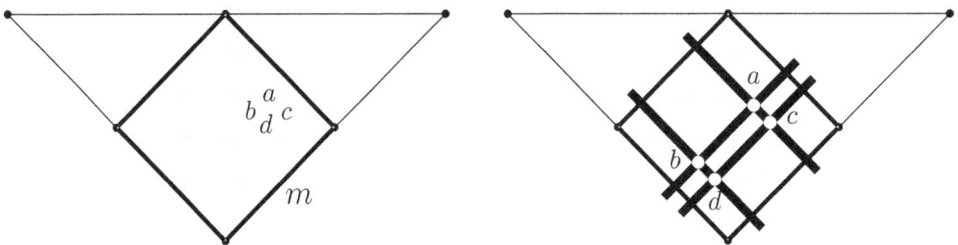

First, it is relatively easy to show that for any four numbers a, b, c, and d from this square lying on the intersections of two "rows" and two "columns" of this square (a is above b and c, d is below b and c) the inequality $ad < bc$ holds true. For the simplest case, when the two "rows" are next to each and the same is true for the "columns", we have $ad = bc - 1$, so the inequality holds (see the left figure). To prove the inequality for the general case, where the rows' indexes are $r_1 < r_2$ and the columns' indexes are $c_1 < c_2$ (see the right figure), we multiply the "simplest" case inequalities for all pairs of "rows" with indexes $r_1 + k$ and $r_1 + k + 1$ (with $r_1 + k + 1 \leqslant r_2$) and pairs of "columns" with indexes $c_1 + \ell$ and $c_1 + \ell + 1$ (with $c_1 + \ell + 1 \leqslant c_2$).

Now, assume that the square contains less than $\frac{1}{4}n$ different numbers. Then in any "row" (remember, its length is $m \geqslant \frac{1}{2}n$) we have quite a few pairs of equal numbers—namely, at least $\frac{1}{2}m$ such pairs. For every one of them record the pair of their "column" indexes. Altogether, at least $m \cdot \frac{1}{2}m = \frac{1}{2}m^2$ such index pairs must be recorded. On the other hand, there are at most $x = \frac{1}{2}m(m-1)$ such pairs of indexes. Since $x < \frac{1}{2}m^2$, there exists a pair of "rows" and a pair of "columns" such that among the four numbers on their intersections, the ones in the same "row" coincide—using the notation above, $a = c$ and $b = d$. Obviously, that contradicts inequality $ad < bc$.

1989.69. Answer: three solutions.

Obviously, $x = 0$ is one of the solutions. All other solutions lie in the intervals $[-3; 0)$ and $(0; 3]$. Because both functions $f(x) = \sin(\sin(\sin(\sin(\sin(x)))))$ and $g(x) = \frac{1}{3}x$ are odd, numbers of solutions in these two intervals are equal. Consider interval $[0; 3]$. Computing second derivative, we can see that function $f(x)$ is convex, and therefore the equation has no more than two solutions, one of which—$x = 0$—is already found. The existence of the second solution follows from the fact that for small positive values of x we have $f(x) > \frac{1}{3}x$, and $f(x_0) < \frac{1}{3}x_0$ at $x_0 = 3$. Therefore, open interval $(0; 3]$ contains exactly one solution; hence, the overall number of solutions is equal to 3.

1989.70. Let k be the original number. It is easy to check that with n operations, it is possible to transform k to any natural number from $2^n(k-1) + 1$ through $2^n k$. If this interval does not contain any fifth powers, then it lies between two consecutive fifth powers N^5 and $(N+1)^5$ for some natural number N. Therefore,

$$\frac{(N+1)^5}{N^5} > \frac{2^n k}{2^n(k-1)} = \frac{k}{k-1}.$$

On the other hand, ratio $(N+1)^5/N^5$ tends to 1 as $N \to \infty$, and therefore, for all N greater than some N_0, this ratio is less than $\frac{k}{k-1}$. Choose n such that $2^n k > N_0$; then interval $[2^n(k-1) + 1; 2^n k]$ contains a perfect fifth power that can be obtained by n operations of our microcalculator.

1989.71. Let us fix some natural number k and consider an arbitrary natural number n. Then

$$|a_{n+1} + a_k - a_{n+k+1}| < \frac{1}{n+k+1},$$

and

$$|a_n + a_{k+1} - a_{n+k+1}| < \frac{1}{n+k+1}.$$

Therefore,

$$|a_{n+1} + a_k - a_n - a_{k+1}| \leqslant \frac{2}{n} \Rightarrow |(a_{n+1} - a_n) - (a_{k+1} - a_k)| \leqslant \frac{2}{n}$$

for any natural n.

Letting n go to infinity, we obtain that the limit of sequence $(a_{n+1} - a_n)$ equals $a_{k+1} - a_k$. But then for any other natural m, this limit must also be equal to $a_{m+1} - a_m$. Since the limit exists, it is unique, and so we have $a_{m+1} - a_m = a_{k+1} - a_k$ for any k and m, and therefore, sequence (a_n) is an arithmetic progression.

1989.72. Answer: the second player wins.

The second player mentally splits all numbers from 0 through 999 into 125 octuplets of consecutive numbers, and then acts as follows. After the first player writes on the blackboard his first number in some octuplet, the second player writes out the smallest possible number from the same octuplet that can be obtained by taking three, four, or five matches. Indeed, since this is the first time a number from this octuplet is written, it follows that it must be greater than or equal to $x + 4$, where $x + 1$ is the smallest number of the octuplet.

After that, the second player splits the remaining six numbers of that octuplet into three pairs in ascending order. If the first player ever writes out one of these numbers, the second player writes the other number from the same pair.

It is easy to see that the matches taken by the second player during the first "visit" to an octuplet are enough to make all the other three moves within that octuplet (we can assume that the second player puts those matches in a separate small pile in order to use them later only for the moves within that octuplet). The only "suspect" case is when the second player has only three matches, but that can only happen if during the first "visit" to the octuplet $\{x + 1, \ldots, x + 8\}$, the first player wrote $x + 4$ and the second player followed that with $x + 1$. Then the remaining six numbers are split into pairs $(x + 2, x + 3)$, $(x + 5, x + 6)$, and $(x + 7, x + 8)$, and therefore, the three matches are enough for the second player to make all her future moves inside this octuplet.

Hence, the second player always has an answering move, guaranteeing her the win.

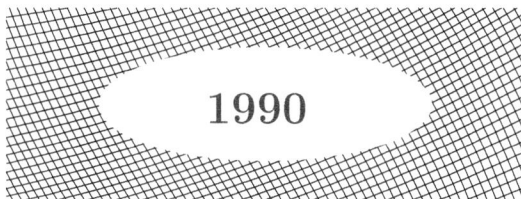

1990.01. The two page numbers written on one sheet are always consecutive, and there-fore, their sum is odd. Adding up 25 odd numbers, Nick will get an odd number as well.

1990.02. Put fifty coins on each pan of the scales. If the scales are in equilibrium, then the remaining coin is counterfeit, and we can use the second weighing to determine whether it is heavier or lighter than a genuine coin.

If one of the pans—say, the left one—is heavier, then let us take those "left" coins off the scales, divide the coins from the right pan into two piles of twenty-five coins, and compare them using the second weighing. If the scales are balanced, then it would mean that the right pan did not have the counterfeit coin; hence, it was in the left pan, and we can claim that the counterfeit coin was heavier. Otherwise, the counterfeit coin was in the right pan, and therefore, it is lighter.

1990.03. No, that is not possible, because 39 cannot be represented as a sum of several numbers, each equal either to 5 or 11.

1990.04. Tom wins. To do that, with every move, Tom subtracts the last digit, making the number divisible by 10. Obviously, after that, Jerry cannot obtain zero with his next move. Also, when Jerry subtracts a non-zero digit from such number, he necessarily makes the last digit different from zero, thus enabling Tom to continue with his strategy.

1990.05. Let x, y, and z be the quantities of the problems solved by exactly one, two, and three kids, respectively. Then $x + y + z = 100$ and $x + 2y + 3z = 180$. Now, we simply subtract twice the first equation from the second one.

1990.06. See solution to Question **1990.19** with boys as knights and girls as knaves.

1990.07. Answer: John lives in apartment 217.

If John's apartment's number is $10a + b$ where $1 \leqslant b \leqslant 10$, then $a + 1$ is the floor number and $10a + b + a + 1 = 239$. It follows that $11a + b = 238$, and therefore, $b = 7$ and $a = 21$.

1990.08. Answer: 29 chairs.

It is obvious that it is not possible for all 30 chairs to be occupied simultaneously. Now, to construct an example, let k be an arbitrary integer between 0 and 28. Suppose that k leftmost chairs are occupied, and the other chairs are empty. Then we can bring in two more guests. The first of them sits on the $(k+2)$th chair, and the second comes after that to occupy the $(k+1)$th chair, followed by the first newcomer walking away. After that, we will have $k + 1$ leftmost chairs occupied with other chairs empty. Now, perform this procedure for $k = 0$, then for $k = 1$, for $k = 2$, and so on, until $k = 28$.

1990.09. Answer: yes, he can.

Indeed, if we denote computer actions by arrow $\rightarrow$, and Theo's actions by double arrow $\Rightarrow$, then Theo's strategy is as follows

$$123 \rightarrow 225 \rightarrow \underline{327} \rightarrow 429 \rightarrow 531 \Rightarrow 135 \rightarrow 237 \Rightarrow \underline{327} \rightarrow \cdots,$$

and then Theo repeats his actions from five moves prior.

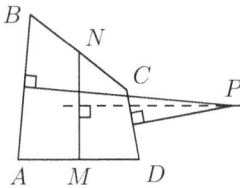

1990.10. Since P lies on the perpendicular bisectors to segments AB and CD, we have $|AP| = |BP|$ and $|DP| = |CP|$. Triangles ADP and BCP are congruent (**SSS** rule), and therefore, their medians PM and PN are equal. Thus, P is equidistant from points M and N; that is, it lies on the perpendicular bisector of segment MN.

1990.11. Consider the following three cases.

1) heights of all rectangles adjacent to the lower (upper) side of the square do not exceed 1;
2) one of the rectangles adjacent to the lower (upper) side of the square has height exceeding 1 and width not exceeding 1;
3) both height and width of one of the rectangles adjacent to the lower (upper) side of the square exceed 1.

In cases (1) and (2) the statement is obvious. And case (3) cannot be true for both upper and lower sides.

1990.13. Split the square into four quarters. If the sums of the numbers in them have the same signs, then it is obvious that one of these numbers cannot have absolute value exceeding 25. If there are two quarters with sums a and b of opposite sign, let us connect these two quarters by a chain of 25×25 squares so that each two squares in the chain overlap by a 24×25 rectangle. Then the sums of the numbers in any two consecutive squares of the chain differ by no more than 50. On the other hand, since numbers a and b have opposite signs, then this chain must have two consecutive squares such that the sum of the numbers in one of them is positive and in the other one—negative. It is obvious that in one of these squares, the absolute value of the sum of the numbers cannot exceed 25.

1990.14. Answer: no, that is not possible.

The sum of the page numbers on one sheet is always congruent to 3 modulo 4. Therefore, the sum of twenty-four such numbers always has remainder 0 modulo 4, while number 1990 does not.

1990.16. Answer: $a = 12$, $b = 13$, $c = 57$.

The solution immediately follows from equality $(b - a)(a + b - 1) = 24$.

1990.17. Consider two arbitrary towns A and B, and assume that there is no road $A \rightarrow B$, as well as no town C such that both roads $A \rightarrow C$ and $C \rightarrow B$ exist. The roads from A lead to 40 towns $A_1, A_2, \ldots, A_{40}$, and there are 40 towns $B_1, B_2, \ldots, B_{40}$ with roads that go from them to B; all these 82 towns must be different. There are 1600 roads coming out of towns A_i; at the same time, the overall number of roads connecting towns A_i does not exceed $20 \cdot 39 = 780$, and the number of roads that go from A_i to the remaining 19 towns cannot be greater than $20 \cdot 38 = 760$. Since $1600 > 1540 = 780 + 760$, there exists a road $A_i \to B_j$, and so we have the path $A \to A_i \to B_j \to B$.

1990.18. Put one coin aside and split the remaining coins into three piles, 34 coins in each one. Use two weighings to compare pile # 1 and pile # 2, and then pile # 2 and pile # 3. Obviously, they cannot all be equal. Assume that the first weighing showed that pile # 1 is heavier than pile # 2; also assume that the second weighing had the pans equal. Then either all coins in piles # 2 and # 3 are genuine, and therefore, pile 1 has some counterfeit coin(s) that must be heavier, or both piles # 2 and # 3 have exactly one counterfeit coin, lighter than any genuine one. To distinguish between these two cases, determine whether pile # 3 contains exactly one counterfeit coin or not. That can be done in one weighing by splitting it in two halves with seventeen coins each and comparing their weights.

All other cases are either absolutely similar or can be analyzed in the same manner.

1990.19. Select several knaves (call them *super-knaves*) in such a way that no two super-knaves are acquainted with each other, and every knave is acquainted with at least one super-knave. This can be done as follows. Take one arbitrary knave L_1 and declare him a super-knave. If there exists at least one knave who does not know L_1, declare this knave L_2 the second super-knave and so on. If at the $(k + 1)$th step, there exists a knave who is acquainted with none of $L_1, \ldots, L_k$, add him to the group of super-knaves and denote him by L_{k+1}. Eventually this process will have to stop when every knave is acquainted with at least one of L_i.

Further, the number of knaves, obviously, does not exceed the sum $a_1 + a_2 + \cdots + a_n$, where n is the number of super-knaves, and a_i is the number of knaves acquainted with L_i, including L_i himself.

Now, denote by b_i the number of knights acquainted with L_i. We know that $b_i \geqslant a_i$. Also it is easy to see that no knight can be acquainted with two super-knaves—otherwise these two super-knaves would be acquainted with each other. Hence, the number of all knaves cannot exceed $a_1 + a_2 + \cdots + a_n \leqslant b_1 + b_2 + \cdots + b_n$, which, in turn, does not exceed the number of all knights.

1990.20. Answer: 1706 pairs.

Consider standard grid on the coordinate plane and draw line ℓ defined by equation $y = \sqrt{2}x$. Pair (m, n) satisfies the inequalities if and only if point $(n + 1, m)$ lies below ℓ, and point $(n, m + 1)$—above ℓ; that is, if ℓ intersects the grid square whose left lower corner has coordinates (n, m). Now, we simply need to count how many squares inside region $K = \{1 \leqslant x, y \leqslant 1000\}$ are intersected by line ℓ. This line intersects 1000 horizontal grid lines ($y = 1$, $y = 2$, $\ldots$, $y = 1000$) and 707 vertical ones ($x = 1$, $\ldots$, $x = [\frac{1000}{\sqrt{2}}] = 707$). Therefore, altogether it intersects 1707 grid lines. Since ℓ does not pass through the grid nodes with the only exception of $(0, 0)$, it intersects 1706 unit squares whose lower left corner belongs to K.

1990.21. Answer: no, it cannot.

One of these numbers (say, y) is greater than 5, and therefore, $y! \vdots 10$. Then, obviously, $x! \vdots 10$; hence, $x \geqslant 5$. But then the sum $x! + y!$ is divisible by 4 and cannot end with 1990.

1990.22. Answer: there is no such triangle.

If the length of median BM is 1, then from the triangle's inequality we obtain $|AB| + |BC| > |AC|$, and therefore, $|AB| + |BC| \geqslant |AC| + 1$. On the other hand, $\frac{|AC|}{2} + 1 > |AB|$ and $\frac{|AC|}{2} + 1 > |BC|$; hence, $\frac{|AC|}{2} + \frac{1}{2} \geqslant |AB|$ and $\frac{|AC|}{2} + \frac{1}{2} \geqslant |BC|$. So we have $|AC| + 1 \geqslant |AB| + |BC|$ and consequently, $|AB| + |BC| = |AC| + 1$, $|AB| = |BC|$. Thus, BM is also the altitude, and from Pythagoras' Theorem, we get $|AM|^2 = |AB|^2 - 1$. It follows that $2|AB| - 1 = |AC| = 2\sqrt{|AB|^2 - 1}$. Solving this equation for $|AB|$, we obtain $|AB| = \frac{5}{4}$, a contradiction.

1990.23. Let a be one of the terms of the arithmetic progression with difference d. Then all the terms of this progression of the form $a + 10^k d$, where $10^k > a$, have the same sum of the digits.

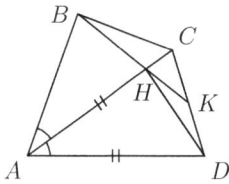

1990.24. Let K be the intersection of line BH and side CD. Congruence of triangles AHD and ABC gives us equality $|AB| = |AH|$. Therefore, the isosceles triangles ABH and ACD are similar, and $\angle HCD = \angle BHA = \angle CHK$; that is, $|CK| = |HK|$. The perpendicular bisector to segment CH is parallel to HD and passes through point K; hence, K is the midpoint of CD.

1990.25. Increase every fraction on the left-hand side of this inequality by replacing its denominator by $1 + abc$. Then, clearly, it is sufficient to show that
$$a + b + c \leqslant 2 + 2abc.$$
This inequality is obvious if each of the numbers a, b, and c is less than or equal to $\frac{2}{3}$. Thus, we only need to deal with the case when one of the numbers—say, a—is greater than $\frac{2}{3}$. Then
$$a + b + c \leqslant 2 + bc \leqslant 2 + 2 \cdot \frac{2}{3}bc \leqslant 2 + 2abc.$$
The first inequality is obtained by adding inequalities $a \leqslant 1$ and $b + c \leqslant 1 + bc$ (this one follows from $(1 - b)(1 - c) \geqslant 0$).

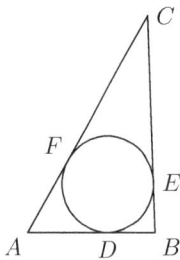

1990.27. Answer: $x = y = z = 0$ or $x = y = z = 3$.

Obviously, x, y, and $z \geqslant 0$. Subtracting the second equation from the first, we obtain $(x - z)(x + z + 6) = 0$, and therefore, $x = z$. Similarly, $x = y$. Hence, $2x^2 = 6x$, and so $x = 0$ or $x = 3$.

1990.28. Answer: $|BC| = 13$.

Let E and F be the points where the circumference touches sides BC and CA. Then $|AD| = |AF| = 5$, $|BD| = |BE| = 3$, and $|CE| = |CF| = x$; hence, $|AB| = 8$, $|BC| = x + 3$, and $|CA| = x + 5$. From the cosine law for $\triangle ABC$ we have
$$8^2 + (x + 5)^2 - 8(x + 5) = (x + 3)^2.$$
Solving this equation, we obtain $x = 10$ and, therefore, $|BC| = 13$.

1990.31. Let $a < b < c < d$ be these natural numbers. Then $a \geqslant 1$, $b \geqslant 2$, $c \geqslant 3$, $d \geqslant 4$, and

$$\frac{1}{ab} + \frac{1}{ac} + \frac{1}{ad} + \frac{1}{bc} + \frac{1}{bd} + \frac{1}{cd} \leqslant \frac{1}{2} + \frac{1}{3} + \frac{1}{4} + \frac{1}{6} + \frac{1}{8} + \frac{1}{12} = \frac{35}{24} < 2.$$

Hence, $cd + bd + bc + ad + ac + ab < 2abcd$.

1990.32. Answer: (a) yes; (b) no.

(a) It can be done without using any square more than four times. Start with arranging four 1×1 squares to form one 2×2 square S_1. Then take three more 2×2 squares to form a 4×4 square S_2, in which S_1 occupies the lower-left (LL) quarter. Then add three more 4×4 squares to form square S_3, in which S_2 occupies upper-right (UR) quarter. And so on, consecutively choosing quarter "locations" from the sequence LL, UR, LL, UR, ...

(b) Let us start with the following simple "helper" statement.

Lemma. *Every vertex of any square lies on the side of some other, larger, square.*

Consider any side s of square P with length 2^k. Then it follows from inequality $1 + 2 + 4 + \cdots + 2^{k-1} < 2^k$ that side s cannot be fully covered by the smaller squares used in the tessellation, and therefore, some portion of s must be covered by a larger square P_s. Of course, P_s covers one of the endpoints of s, protruding beyond that vertex—let us denote it by A_s.

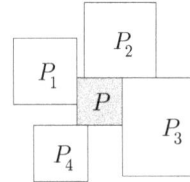

Hence, we have a mapping from the set of four sides of P to the set of four vertices of P: each side s is mapped to vertex A_s. Clearly, two sides cannot map to the same vertex, and therefore, each vertex of P is covered by exactly one such larger square P_s, and each square P_s "owns" exactly one vertex. As a consequence, the tessellation around P must look as it is shown in the figure. $\square$

Now, consider the smallest square M participating in our tessellation. Then there must be no gaps between squares M_s lest we get an obvious contradiction with M being the smallest square. Now, select the largest square among M_s—denote it by M'—and find vertex A of M' lying on the contour of M. Then A (as a vertex of M') violates the lemma.

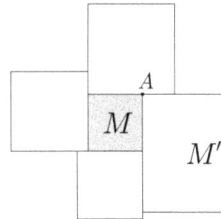

1990.33. No, that is not possible. Indeed, from the well-known divisibility criteria, in order for a number to be divisible by 11, the alternating sum of its digits must be divisible by 11. In our case, this sum is odd, and its absolute value cannot exceed $4+5+6-1-2-3 = 9 < 11$; thus, it cannot be divisible by 11.

1990.35. Since polynomial $F(x)$ has integer coefficients, $F(a) - F(b)$ is divisible by $a - b$ for any different integers a and b. Thus, 5 divides $F(7) - F(2)$, and so $F(7)$ is divisible by 5.

Similarly, 2 divides $F(7) - F(5)$, and it follows that $F(7)$ is divisible by 2. Thus, $F(7)$ is divisible by 10.

1990.37. For the convenience of terminology, let us call the 2×2 squares covering the board *tiles*.

(a) There are nine horizontal 2×10 rectangles. Thus, one of them contains at least seven tiles, and it is obvious that one of them is completely covered by some two others.

(b) Mark twelve "white" and four "black" squares as it is shown in the figure on the left. First, among the given tiles, consider twelve that cover the white squares, and then there are exactly eight uncovered squares on the boundary of the board. Take eight more tiles that cover those squares, and altogether we have twenty tiles covering the entire two-squares thick border of the board. Consider now four tiles that cover the four central squares. In the central 6×6 square, they cover 16 squares, and for each of the remaining 20 squares, we can find a tile covering it. Therefore, we have 24 tiles covering that central 6×6 square. In total, we have selected 44 tiles covering the entire board; hence, one of the given 45 tiles can be removed.

(c) **Answer**: this smallest value is 39.

The figure on the right shows an example of 38 centers of 2×2 squares covering the entire board while no 37 of them satisfy the same condition.

Full solution for 39 is tedious, long, and not very interesting. We leave it to the readers— perhaps they can find a better one.

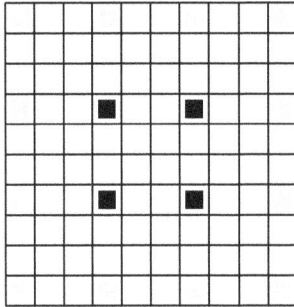

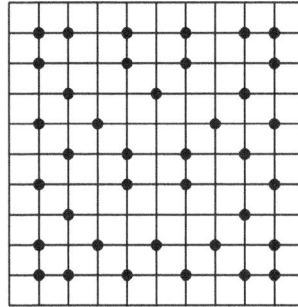

1990.41. Note that $x^n > 1$, and therefore, $x > 1$. From that we obtain $x^n = x^{n-1} + \cdots + 1 > n$. Multiply the given equation by $x - 1$ to get $(x - 1)x^n = x^n - 1$. Hence, $x - 1 = \frac{x^n - 1}{x^n} = 1 - \frac{1}{x^n}$, which, obviously, is less than 1 and greater than $1 - \frac{1}{n}$.

1990.42. Consider tetrahedron $T = A_1 A_2 A_3 A_4$ with edge length a. For each index $1 \leqslant k \leqslant 4$, take vertex A_k and construct homothetic image of T with ratio $\frac{1}{2}$ using point A_k as the center of the homothety. Cutting off all these four "half-tetrahedrons" (with edge length $\frac{a}{2}$) produces octahedron with edge length $\frac{a}{2}$.

Using this method, start with octahedron with edge 1 and four tetrahedrons with edge 1, and put them together to form a tetrahedron with edge 2. Then add three more tetrahedrons and one octahedron with edge lengths 2 to form the next tetrahedron with edge 4. Again, add three tetrahedrons and one octahedron with edge lengths 4, and so on.

Obviously, this dissection does not have more than four identical polyhedrons.

1990.43. Identity $b(a^2 + ab + 1) - a(b^2 + ba + 1) = b - a$ implies that $a - b$ is divisible by $b^2 + ba + 1$, which, obviously, is possible only if $a - b = 0$.

1990.44. One by one, exclude the segments which are completely covered by other segments. Then, starting from the leftmost segment, move the right ends of the segments to the left so that the segments' intersections would be zero-length segments (that is, points). The left halves of the segments can only diminish during these operations. The rest of the proof is obvious.

1990.45. Denote the point where segments AP and BD meet by R'. Then it would suffice to show that quadrilateral $R'PCQ$ is inscribed. That follows from the angle equality $\angle BCR' = \angle BAR' = \angle BQP$ (the first step—from symmetry, the second—from quadrilateral $ABPQ$ being inscribed).

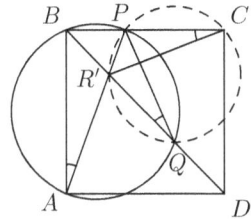

1990.46. Let us use induction on n. The basis $n = 1$ is obvious. Now, let us split the collection of all given subsets into two subcollections: the first one consisting of all subsets that contain number n, and the second one comprising all other subsets. Using the induction hypothesis for $n - 1$, we can easily compute the sum of the squares of products for both subcollections. The first one gives us $n^2((n-1)! - 1) + n^2$, and the second one—$(n! - 1)$. Add these two numbers to obtain $(n+1)! - 1$.

1990.47. Let us use the following well-known fact—the area of any triangle with its vertices on the nodes of the grid is a half-integer; in other words, it is equal to $\frac{n}{2}$, where n is an integer. Then $|AC| \cdot |AD| \sin \angle DAC$, as well as $|BC| \cdot |BD| \sin \angle DBC$, is an integer. It follows that $\big| |AC| \cdot |AD| - |BC| \cdot |BD| \big| = \frac{m}{\sin \alpha}$, where $\alpha = \angle DAC = \angle DBC$, and m is a positive integer; the desired inequality immediately follows.

1990.48. Let us prove this by induction on the number of roads.

The basis is obvious. Now, assume that all the roads except one, which connects cities X and Y, are already colored as required. Let 1 and 2 be the two colors out of the given ten such that for city X there is a road of color 1 but no road of color 2 coming out of it; and vice versa, for city Y there is a road of color 2 but no road of color 1 coming out of it.

Consider route R which (a) starts at city X, (b) uses only roads of colors 1 and 2, and (c) passes through the maximum possible number of roads. This route cannot pass through some city twice, and it cannot end in city Y. Now, repaint all roads in R, swapping colors 1 and 2, after which we can paint road XY in color 1.

1990.49. Answer: $p+q-1$. The example should be pretty clear from the figure below, which illustrates the solution for $p = 3$, $q = 5$.

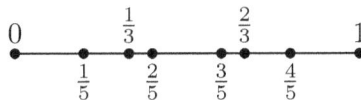

Suppose that the pie can be cut into $m < p+q-1$ parts $K_1, \ldots, K_m$, so that there is a way to distribute these pieces equally between p guests, and then there is a way to do the same with q guests.

Now, construct graph G with vertices $A_1, \ldots, A_p$ (corresponding to p guests from the first scenario) and $B_1, \ldots, B_q$ (corresponding to q guests from the second scenario). Connect vertices A_i and B_j by an edge if and only if these two guests, in the respective scenarios, received pieces of pie that share some common portion. Graph G has $p + q$ vertices and m edges (each piece corresponds to one edge). Since $m \leqslant p+q-2$, this graph is not connected. Consider any component of connectedness of G, and all pieces of the pie that correspond to its edges. These pieces together add up to several $\frac{1}{p}$ th portions of the pie and several $\frac{1}{q}$ th portions of the pie, but not the entire pie, which is impossible since p and q are co-prime.

1990.50. Answer: yes, that is possible.

Arrange twenty integers $x_1, x_2, \ldots, x_{20}$ around the circle so that the differences $x_2 - x_1$, $x_3 - x_2, \ldots, x_{20} - x_{19}, x_1 - x_{20}$ produce the original collection $1, 2, \ldots, 9, 10, -1, -2, \ldots$, $-9, -10$. That is possible, because the sum of the numbers in that collection is zero. It is easy to see that the transposition of two neighbors in this new collection is equivalent to the operation described in the problem being applied to the original sequence. So, using these transpositions, one can attain sequence $x_1, x_{20}, x_{19}, \ldots, x_2$, meaning that on the original circle, we get exactly what is desired.

1990.54. Answer: no, Alex cannot do that.

The main idea of Serge's strategy (which we will denote $(*)$) is as follows. He tries to surround some region $\mathcal{R}$ on the board by black squares so that the king would not be able to move from a square inside $\mathcal{R}$ to a square outside $\mathcal{R}$. Serge also needs region $\mathcal{R}$ to contain either empty or white squares—that will, obviously, guarantee that Alex cannot achieve his goal.

Suppose that Alex makes his first move somewhere within the upper-right quarter (other cases are identical). Then Serge paints square $(2, 2)$[114] black. Now, there are two main cases determined by Alex's second move.

Case 1. Second move by Alex is made outside of 2×2 square in the lower-left corner, or to square $(1, 1)$.

If this move is not $(1, 1)$, then we can assume, without loss of generality, that the square that Alex painted with his second move is located above the main diagonal connecting lower-left and upper-right corners of the board.

Serge answers with painting black square $(1, 2)$. Alex must paint $(2, 1)$—otherwise Serge takes that field and achieves his goal (see strategy $(*)$ above). Then Serge answers to $(3, 2)$. Alex must take square $(3, 1)$, Serge takes $(4, 2)$ and so on, and the process, obviously, ends with Serge surrounding some portion of or the entire first row of the board.

Case 2. Second move by Alex is $(1, 2)$ (painting $(2, 1)$ is countered with the strategy symmetric to the one outlined below).

Serge answers with $(2, 3)$, threatening to take squares $v = (1, 3)$ and $h = (2, 1)$. Consider now region $\mathcal{R}$ comprising two first rows and two first columns of the board. This region can be naturally split into the horizontal portion H (two first rows) and the remaining vertical portion V. If Alex moves to H but not to h, Serge takes h; same goes for V and v. If Alex takes v, then Serge takes the next square in the second column, thus "moving" the threatened square v one unit up; if he takes h, then Serge takes the next square in the second row, thus "moving" the threatened square h one unit right.

---

[114] That is, the square at the intersection of column 2 and row 2.

Thus, after every two moves, we either have one of the ends of the threatened region closed by Serge, or extended by one square, maintaining the same position on the end where Alex chose to make his move. Eventually, the players run out of board and Serge has his "surrounded" region $\mathcal{R}$.

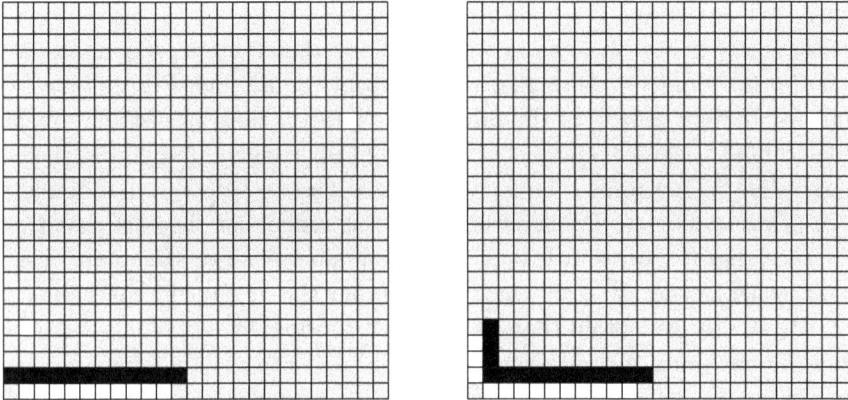

1990.55. Reflect vertex C across diagonal BD, then see **1990.47**.

1990.56. Answer: 124 swaps.

Consider any volume indexed a_1. It stands on the a_2th place, volume a_2—on the a_3th place and so on, until some volume a_n happens to stand on the a_1th place. Then volumes $a_1, \ldots, a_n$ form a cycle. (Cycle can consist of just one volume if it stands in its proper place.) Obviously, this collected works is thus split into several cycles.

If we swap two volumes from the same cycle, that cycle is then transformed (cut) into two smaller cycles. If we swap two volumes from different cycles, then these cycles merge together into one larger cycle. In the end, when all volumes stand on their proper places, there are 100 cycles of length 1. Hence, all the cycles have to be cut, and each such operation (i.e., a cut) can be done only if the cycle has two volumes of different parity.

Thus, if the original arrangement had one cycle of 50 volumes with odd numbers and 25 cycles of length two, then at least 25 merge operations are necessary, because we cannot cut a cycle which has no volumes of different parity. Furthermore, since we start with 26 cycles and we must end up with 100 cycles, the number of cuts (each cut increases the number of cycles by one) must exceed the number of merges (each merge decreases the number of cycles by one) by 74. Therefore, there must be no less than $25+74 = 99$ cuts and 25 merges, resulting in at least 124 swaps.

Assume that originally there were a non-trivial cycles (of length greater than 1) consisting only of even-numbered volumes, and b non-trivial cycles consisting only of odd-numbered volumes. Then it is possible to make no more than $\max(a,b) \leqslant 25$ merges to ensure that all non-trivial cycles have volumes of both parities. Each cycle with x volumes, some two of which have different parity, can be cut into x trivial cycles in $x - 1$ cuts. Indeed, swapping volumes a_k and a_{k+1} in cycle $(a_1, a_2, \ldots, a_n)$ cuts out one trivial cycle with volume a_{k+1}. Obviously, this cut can be done in such a way that the other cycle still has volumes of different parity (as long as this cycle is still non-trivial).

Therefore, 99 cuts are always enough to cut all non-trivial "odd-and-even" cycles into trivial ones; hence, we can perform no more than 25 merges and then no more than 99 cuts to reach the desired arrangement with 100 trivial cycles; this proves that 124 swaps are always enough.

1990.57. Let us assume the opposite—for any $k = 1, 2, \ldots, m$ there exists number x_k such that $F(x_k)$ is not divisible by a_k. It follows that there exist numbers $d_k = p_k^{\alpha_k}$, where p_k are prime numbers, such that a_k is divisible by d_k, but $F(x_k)$ is not. If among these numbers we have powers of the same prime integer, then all of them except for the smallest one can be discarded—indeed, if $F(x)$ is not divisible by that power, then of course it is not divisible by any other. Thus, we obtain collection $\{d_1, d_2, \ldots, d_s\}$, consisting of pairwise co-prime numbers. It follows from Chinese remainder theorem that there exists number N such that for any $k = 1, 2, \ldots, s$ this number has the same remainder modulo d_k as x_k (in other words, $N \equiv x_k \pmod{d_k}$). From this, using the divisibility property $F(m) - F(n) \vdots m - n$, we obtain that $F(N)$ is not divisible by any of d_k, and therefore, is not divisible by any of a_k, a contradiction.

Note. By the way, the just mentioned divisibility property is valid for any polynomial with integer coefficients, but not necessarily for a polynomial with rational coefficients taking only integer values for integer arguments (e.g., for $P(n) = \frac{1}{2}n(n-1)$).

1990.58. Denote the given 22 points by $0 \leqslant a_1 < a_2 < \cdots < a_{21} < a_{22} \leqslant 1$. Split them into pairs $(a_1, a_2), (a_3, a_4), \ldots, (a_{21}, a_{22})$, and from each pair select a point at random. Let us denote them by $x_1, x_2, \ldots, x_{11}$, while denoting the other points of the respective pairs by $y_1, y_2, \ldots, y_{11}$. Also index these pairs in the descending order of their differences, that is,

$$|x_1 - y_1| \geqslant |x_2 - y_2| \geqslant \cdots \geqslant |x_{11} - y_{11}|.$$

Now, perform in the given order the "merge" operations, starting from merging x_1 with x_2, then merging the result with x_3, the result of that operation—with x_4 and so on. Do the same for points $y_1, y_2, \ldots, y_{11}$. This procedure produces points

$$a = 2^{-10}(x_1 + x_2 + 2x_3 + \cdots + 2^9 x_{11}),$$
$$b = 2^{-10}(y_1 + y_2 + 2y_3 + \cdots + 2^9 y_{11}).$$

Let us prove that the selection of (x_k) can be done in such a way that $|a - b| \leqslant 2^{-10} < 0.001$. Note that the selection of points x_k in pair (a_{2k-1}, a_{2k}) is equivalent to selecting the corresponding sign in the right-hand side of the expression

$$2^{10}(a - b) = \pm 2^9 |x_{11} - y_{11}| \pm 2^8 |x_{10} - y_{10}| \pm \cdots \pm |x_2 - y_2| \pm |x_1 - y_1|.$$

We intend to show that for some sequence of signs, the sum in the right-hand side of this equation belongs to $[-1; 1]$. Choose the first sign in that expression arbitrarily and the second sign opposite to the first one. It is easy to see that the absolute value of the sum of the first two terms does not exceed $2^8 |x_{10} - y_{10}|$. Now, choose the next sign opposite to the sign of that sum to guarantee that the sum of the first three terms has the absolute value not exceeding $2^7 |x_9 - y_9|$, and so on. Finally, the last sign choice guarantees that the absolute value of the entire sum does not exceed $|x_1 - y_1| \leqslant 1$.

1990.59. Let us call two natural numbers x and y, both less than number $b = a^n - 1$ and co-prime with it, *equivalent* if there exists natural number k such that $(x - a^k y) \vdots b$. That relation will be denoted as $x \sim y$.

It is easy to verify that this relation is indeed the equivalency, that is,

$$x \sim x$$
$$x \sim y \Rightarrow y \sim x$$
$$x \sim y \ \& \ y \sim z \Rightarrow x \sim z$$

for any numbers x, y, and z (as long as all relations above are defined). Also note that the set of numbers equivalent to any given number x coincides with

$$x, \ ax, \ a^2 x, \ \ldots, \ a^{n-1} x \pmod b.$$

Since the set S of all natural numbers less than b and co-prime with it can be split into such subsets (equivalency classes), it follows that size of S is divisible by n.

1990.60. Color the remaining halves red. Let us apply to each of these red segments homothety with ratio 3 centered in its midpoint. It would suffice to show that segment S is covered by these "stretched" segments. Assume the opposite, that there exists a point $x \in S$ not covered by those segments. Then the distance from x to every red (original) segment is greater than that segment's length. But in that case, point x is, clearly, not covered by the original segments as well.

1990.61. Answer: no.

Let us assume that such hexagon $ABCDEF$ exists. If all diagonals from vertex F are equal in length, then $|AD| = |AE| = |BD| = |BE|$. Therefore, isosceles triangles ABD and ABE are congruent. But then sides AB and DE must intersect, which is not possible.

1990.62. Answer: no, that is not possible.

Denote the number of red, blue, and green rooks by a, b, and c, respectively. Count the numbers of pairs of rooks attacking each other: x such pairs of red and blue colors, y—of blue and green colors, and z—of green and red colors, respectively.

Then we have the inequalities $2a \leqslant x \leqslant 2b \leqslant y \leqslant 2c \leqslant z \leqslant 2a$. Therefore, $a = b = c$, and the total number of rooks must be divisible by 3.

1990.63. Note that H is the center of the escribed circle of triangle ACF touching side AF, because that point lies on both the bisector of external angle A and the bisector of angle $\angle ACF$. Therefore, FH is the bisector of angle $\angle AFB$, and we have $\angle AFH = \frac{1}{2}\angle AFB$.

Similarly, $\angle AFG = \frac{1}{2}\angle AFC$, and it follows that $\angle GFH = \frac{1}{2}(\angle AFB + \angle AFC) = 90°$.

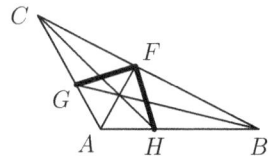

1990.64. Let us call a directed graph (graph whose edges are oriented, or directed) *strongly connected* if it is possible to reach any of its vertices from any other by moving along the edges of the graph complying with the edge orientations.

For any subset W of the set V_G of all vertices of graph G, denote by $G(W)$ the graph with vertex set W and edge set obtained by restricting E_G (the edge set of G) on W. A component of strong connectedness for directed graph G is the maximum (by inclusion) subset $W \subset V_G$ such that graph $G(W)$ is strongly connected. Maximality property simply means that you cannot add another vertex to W so that $G(W)$ is still strongly connected.

Thus, V_G is obviously split into the components of strong connectedness; of course, each vertex of G is included into exactly one such component.

It is easy to see that for every two components of strong connectedness U_1 and U_2, all edges of graph G connecting these components are oriented in one direction (either all of them go from U_1 to U_2, or vice versa).

Now for the solution. We have a directed graph G with 100 vertices such that every two vertices are connected with one edge.

It is given to us that graph G is not strongly connected. Let $W_1, \ldots, W_m$ be its components of strong connectedness. Since every two vertices of G are connected by an edge, these components can be indexed in such a way that if $i < j$, then all edges of G are oriented from component W_i to component W_j (e.g., see Question **1965.17** and use the fact that graph of these components cannot contain oriented cycles). There are two cases.

Case 1. $m > 2$. Take any vertex $v \in W_2$ and change the direction of all edges incident to it. Then graph $G' = G(W_1 \cup W_m \cup \{v\})$ is, clearly, strongly connected. From any vertex $u \in G$ outside of G', we can reach W_m, and we can get to u from W_1—thus, G is now strongly connected.

Case 2. $m = 2$, meaning that G has exactly two components of strong connectedness W_1 and W_2. Suppose W_1 contains at least four vertices. Since graph $G(W_1)$ is strongly connected, it follows that for each vertex in W_1 there is an edge that starts in it, leads to some other vertex in W_1, and at least one other edge which leads to W_2. From $|W_1| \geqslant 4$, we obtain that there exists vertex $v \in W_1$ serving as the end of at least two edges starting in vertices from W_1. Thus, v has at least two edges that end in it and at least two edges that start in it. Let $W_1' = W_1 \setminus \{v\}$, and $G' = G \setminus \{v\}$.

If graph $G(W_1')$ is strongly connected, then it is easy to see that the change of direction for all edges incident to v makes graph G strongly connected.

And if graph $G(W_1')$ is not connected, then graph G' has more than two components of strong connectedness. It follows from Case 1 that we have vertex $u \in G'$ such that changing directions for all edges incident to u makes graph G' strongly connected. Let us go back to vertex v. Since it has at least two incoming and at least two outgoing edges, then even after changing the direction of edge uv, we have an edge coming out of v and an edge going into v. Therefore, G becomes a strongly connected graph.

This approach proves the problem for any complete directed, but not strongly connected, graph G with seven or more vertices. Such graph either has three or more components of strong connectedness or there are exactly two components with one of them containing at least four vertices.

Note. If we consider the "smaller" graphs, then clearly, a component of strong connectedness cannot contain exactly two vertices. Thus, for graph G with three or five vertices, the statement of the problem is also true. It is not true, however, for some directed graphs with four or six vertices. A counterexample must have exactly two components of strong connectedness, each one with either one or three vertices. There are only two such graphs with four vertices, and only one with six vertices—they are shown in the figures below.

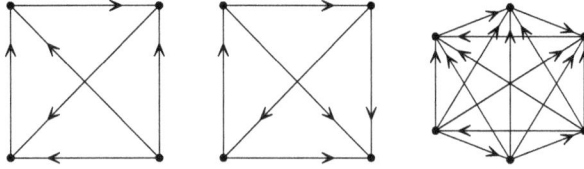

1990.65. Let us show that any straight line $x + ny = cn$ with $n \in \mathbb{N}$, $c \in \mathbb{R}$, intersects the graph of function $f(x)$ in exactly one point. If we assume that on the graph there are two such points (x_1, y_1) and (x_2, y_2), then

$$c - \frac{x_1}{n} = y_1 = f(x_1) = f(x_1 + nf(x_1)) = f(cn)$$

$$= f(x_2 + nf(x_2)) = f(x_2) = y_2 = c - \frac{x_2}{n}.$$

Hence, $x_1 = x_2$ and the points coincide. Furthermore, if function $f(x)$ is not constant, then there exist two numbers a and b such that $b \in (a, a + f(a))$ and $f(b) \neq f(a) = f(a + nf(a))$. From intermediate value (or Cauchy-Bolzano) theorem, we know that for some sufficiently large n line defined by equation

$$y = \frac{x}{n} + \frac{f(a) + f(b)}{2}$$

must intersect the graph of function $f(x)$ in at least two points, a contradiction.

1990.66. Let us denote these numbers (in clockwise order) by a_0, a_1, ..., a_{n-1}, and their sum by $s > 0$. Let $\{x_k\}_{k \in \mathbb{Z}}$ be a sequence of numbers such that $x_{k+1} - x_k = a_{k \bmod n}$. Then the transformation of the original sequence described in the problem is equivalent to the following transformation of $\{x_k\}$: two neighboring numbers x_k and x_{k+1} swap places with each other, and at the same time this swap is also performed for the numbers in all pairs (x_{k+mn}, x_{k+mn+1}) for $m \in \mathbb{Z}$. Since there exists exactly one permutation of sequence $\{x_k\}$ which arranges its terms in ascending order, we can obtain at most one sequence of non-negative numbers a_0, a_1, ..., a_{n-1}.

Let us prove that this unique arrangement can be realized. Consider possible changes in the finite subsequence x_1, x_2, ..., x_n. Any permutation of its terms is possible, as well as the following transformation: when $k \vdots n$, numbers x_1 and x_n swap places, then one of them is increased by s while the other is decreased by s. Thus, when we do it in the reversed order, we can subtract s from one of these numbers while adding s to the other. Therefore, as we gradually decrease the maximum term of the sequence and increase the minimum term of the sequence, we end up obtaining the sequence where the difference between the maximum and the minimum is less than s. After that, all that remains is to permute the numbers so they go in the ascending order.

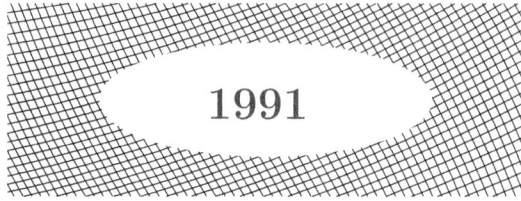

1991

1991.01. Consider the students with equal amounts of nails and bolts but with the amount of screws different from the amounts of nails. There must be at least $40-15-10 = 15$ of them. On the other hand, for any such student, the amount of screws she has must be different from the amount of bolts.

1991.02. Answer: no, that is not possible.

Since the number of Ivan's food stamps did not change, he exchanged two stamps for three the same number of times he exchanged three stamps for two. Therefore, the overall number of stamps he gave away in the process must be divisible by 5.

1991.03. Let a, b, c, and d be the velocities of the cars and t—the time from the start to the meeting of A with C (or B with D). Since the sums $at + ct$ and $bt + dt$ are equal to the length of the track (denote it by s), then $a + c = b + d$ and $a - b = d - c$. Assuming that $a > b$ and that cars A and B caught up one to another for the first time at moment T, we obtain $aT = bT + s$ and $(a - b)T = s$. Therefore, $(d - c)T = s$; hence, the first time when cars C and D catch up with each other is also T.

1991.04. Answer: 6 days.

Assume that the baron was able to make that claim on August 7. Let a_k be the number of ducks the baron shot on the kth of August. For convenience sake, let us treat July 30 and July 31 as (-1)th and 0th of August. Then $a_k < a_{k+2}$ and $a_k > a_{k+7}$. Therefore,

$$a_0 < a_2 < a_4 < a_6 < a_{-1} < a_1 < a_3 < a_5 < a_7 < a_0,$$

a contradiction. Now, we present an example of the baron being able to say that sentence for six days in a row. For instance,

$$a_0 = 0, \ a_2 = 1, \ a_4 = 2, \ a_6 = 3, \ a_{-1} = 4, \ a_1 = 5, \ a_3 = 6, \ a_5 = 7,$$

and $a_k = 100$ for $k < -1$.

1991.05. Answer: yes, he can.

Nick should break both the first and the third sticks the same way: into parts of lengths 50 cm, 25 cm, and 25 cm. If Vera broke the second stick into parts of lengths $x \geqslant y \geqslant z$, then $x \leqslant 100$, and y, $z \leqslant 50$, and it is possible to form triangles out of triples $(x, 50, 50)$, $(y, 25, 25)$, and $(z, 25, 25)$.

1991.06. No, not necessarily. Consider the following tournament. Teams A_1, A_2, A_3 gained exactly one point in matches between them, and the same happened in the triples B_1, B_2, B_3 and C_1, C_2, C_3. Also all teams A_i defeated all teams B_j, all teams B_j defeated all teams C_k, and finally all teams C_k defeated all teams A_i.

1991.07. An example is shown in the figure.

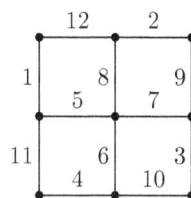

1991.08. Answer: 12 divers.

Let us assume that when divers took their shares, k of them took the half and m took the third, and that at the end n pearls were sacrificed to Poseidon. Then the overall number of pearls is $2^k \cdot (3^m/2^m) \cdot n = 2^{k-m} \cdot 3^m \cdot n$. But

$$10^6 \geqslant (2^{k-m} \cdot 3^m \cdot n)^2 \geqslant 4^{k-m} \cdot 3^{2m} \geqslant 3^{k+m},$$

and it follows that $k + m \leqslant 12$. It is easy to construct an example with twelve divers. Suppose they have brought 3^6 pearls. The first diver took one third, and $2 \cdot 3^5$ pearls remained; then the second diver took one half which left 3^5 pearls in the pile; then the third took one third again and so on; finally, the twelfth took one half of the remaining two pearls.

1991.09. Assume that is not so. Ask all the gnomes and elves to leave the table. Then obviously, between any two neighbors, there will be exactly one empty chair. Thus, the number of seats at the table is an even number.

1991.10. Note that $\angle BAD > \angle DBA$, and therefore, $|BD| > |AD| = 3$. From the triangle's inequality we have $|BC| + |CD| > |BD|$, and then $1 + |CD| > 3$, that is, $|CD| > 2$.

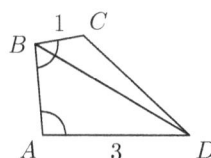

1991.11. Answer: $n = 19$.

If $n > 20$, then the collection consisting of N numbers -1 gives us the counterexample. If $10 < n < 19$, consider the collection of $n - 1$ numbers 2 and one number -1. But if $n = 19$ and $a_1 \leqslant a_2 \leqslant \cdots \leqslant a_{19}$, then we have

$$a_1 + a_2 + \cdots + a_{10} > a_{11} + \cdots + a_{19},$$

and therefore, $a_1 > 0$. It follows that all other numbers are positive as well.

1991.12. Suppose that it is impossible to reach B from A by railroad while visiting no more than two other towns of the way, and we cannot travel from C to D by highway satisfying the same restriction. By the way, some two of these four towns could be the same. Consider only roads connecting these four towns. It is easy to see that one of the modes of transportation connects all of them, and therefore, it is possible to travel from every town of these four to any other, visiting, of course, at most all the other towns of the four (no more than two!) on the way.

1991.13. See solution to Question **1991.19**.

1991.15. Answer: 24, 42.

Clearly, number x can only have two digits. Let $x = 10a + b$, then $(10a + b)(10b + a) = 1000 + ab$, and therefore, $10ab + a^2 + b^2 = 100$. Hence, $ab < 10$ and $a^2 + b^2 \vdots 10$. It is easy to see that we only have the following cases to examine: $\{1, 3\}$, $\{1, 7\}$, and $\{2, 4\}$. Obviously, only the last one fits.

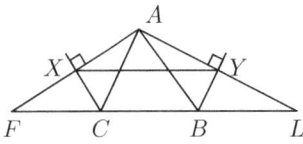

1991.17. Lines AX and AY intersect BC at points F and L, respectively. Then, CX is both an angle bisector and an altitude of triangle ACF. Therefore, triangle ACF is isosceles, and X is the midpoint of AF. Similarly, Y is the midpoint of AL, and XY is the midline of triangle FLA. Then, we have

$$|XY| = \frac{1}{2}|FL| = \frac{1}{2}|FC| + \frac{1}{2}|CB| + \frac{1}{2}|BL| = \frac{1}{2}|AC| + \frac{1}{2}|CB| + \frac{1}{2}|BA|.$$

1991.18. Note that if $f(x) = x^2 + x + 1$ and $g(x) = x^2 - x + 1$, then $g(x) = f(x-1)$. Since $a^3 \pm 1 = (a \pm 1)(a^2 \mp a + 1)$, the given fraction equals

$$\frac{1 \cdot 2 \cdot 3 \cdot \ldots \cdot 99 \cdot (2^2 + 2 + 1) \ldots (99^2 + 99 + 1)(100^2 + 100 + 1)}{3 \cdot 4 \cdot \ldots \cdot 101 \cdot (2^2 - 2 + 1) \ldots (99^2 - 99 + 1)(100^2 - 100 + 1)}$$

$$= \frac{2(100^2 + 100 + 1)}{(2^2 - 2 + 1) \cdot 100 \cdot 101} = \frac{3367}{5050}.$$

1991.19. Mark with stars the squares for which both the row number and the column number are odd.

Now, if the dissection employs x triminoes ⊟, and y tetrominoes ⊞ and ⊕, then $3x + 4y = (2n-1)^2$. Since, obviously, no polyomino can cover more than one star, we have $x + y \geqslant n^2$. Hence, $4x + 4y \geqslant 4n^2$, and therefore, $x \geqslant 4n^2 - (2n-1)^2 = 4n - 1$.

1991.20. Answer: yes, such collections do exist. For example, consider the following two collections.

$$A = \{1, \ 2, \ \ldots, \ 9, \ -45\} \quad \text{and} \quad B = \{-1, \ -2, \ \ldots, \ -9, \ 45\}.$$

For any number s from collection $A^{(5)}$, there are five numbers in B whose sum equals $-s$. Then the sum of the other five numbers in B equals s.

1991.21. Solution is absolutely analogous to solution of **1991.28**.

1991.22. Consider the line parallel to BC passing through point A. Select point Z such that $|AZ| = |BX|$. Now, it is easy to see that $|YZ| \geqslant |AC|$, and it follows from the triangle's inequality that $|XZ| + |XY| \geqslant |YZ| \geqslant |AC|$. Since triangles AXZ and BXY are congruent, we have $|XY| = |XZ|$ and therefore, $2|XY| \geqslant |AC|$.

1991.23. Let us prove that $x^2 + y^2 + z^2 = (x + y - z)^2$, where z is the shortest side of the triangle. That is equivalent to $xy = xz + yz$ or $\frac{1}{z} = \frac{1}{x} + \frac{1}{y}$. The last equality follows from $\frac{2S}{z} = \frac{2S}{x} + \frac{2S}{y}$, where S is the triangle's area.

1991.24. Note that this sequence can never "jump over" interval $(10^p - 9; 10^p - 1)$, where p is a natural number such that $10^{p-1} \leqslant a_1 < 10^p$.

1991.25. The circle inversion with respect to circumference ABC maps line BC to circumference OBC with diameter OX, and point P—to point P', the intersection of ray OP with circumference OBC. Therefore, OP' is perpendicular to $P'X$.

On the other hand, the same inversion maps line AP to the circumference with diameter AO. Hence, point P' also lies on this circumference, and it follows that $OP' \perp P'A$. Thus, broken line $AP'X$ is a segment and it must be perpendicular to OP.

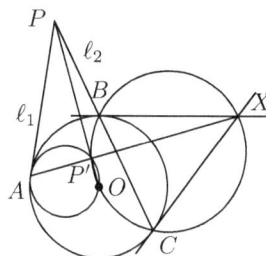

Note. For those who are familiar with the concepts of pole and polar line, there exists a somewhat "simpler" solution: line AP is a polar of point A, and line BC is a polar of point X. Thus, the pole for line AX can be found as the intersection of AP and BC, and that is point P. Therefore, $AX \perp OP$.

1991.26. Consider the acquaintance graph and delete the maximum possible number of edges from each of its components of connectedness while keeping them connected (e.g., for every such component, take its spanning tree and remove all the edges not included in it).

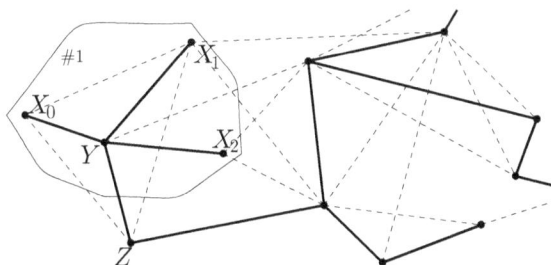

After that, find any pendant vertex X_0 adjacent to just one vertex Y, and then consider all vertices X_0, X_1, ..., X_k, adjacent **only** to vertex Y. These vertices plus Y form group #1. Among the remaining vertices, obviously, none is connected to all the others, otherwise it would be impossible to find one more vertex not connected to both Y and that vertex. At the same time, each of the remaining vertices is connected to at least one vertex not included in group #1. It is left to show that these vertices can be split into two groups so that each vertex is connected to at least one other in its group. This fact follows from the same reasoning we used at the beginning of this paragraph.

1991.27. Instead of the numbers, write and use only their remainders modulo 2. Obviously, the parity of the number of zeros and parity of the sum of all numbers would not change.

Then it is pretty obvious that if in the beginning there were an odd number of zeros while the sum of all numbers was even, then in the end we have only one number left, and that number is zero. Also, if in the beginning we had an odd number of zeros while the sum of all numbers was odd, then in the end we have two numbers, zero and one.

Now, consider the case when the original collection had an even number of zeros. Mark all the groups of consecutive zeros surrounded by ones such that the number of zeros in them is odd. Obviously, the number of such groups is even; let us denote it by $2n$. The rest is, therefore, split into $2n$ groups that we can denote by A_1, ..., A_{2n} (each group A_i contains a_i ones and is positioned between adjacent groups each consisting of odd number of zeros; enumeration is cyclical, we consider $A_{2n+1} \equiv A_1$). Also let B_i be the group that consists of odd number of zeros and separates A_i and A_{i+1}.

Examine the expression

$$S = |a_1 - a_2 + \cdots + a_{2n-1} - a_{2n}| \, .$$

Let us prove that S does not change under the permitted operations. That claim is obvious if the number of groups does not change. However, the number of groups can change only if one of the groups B_i contains exactly one zero, and the operation is performed with that zero. When the operation is carried out, we can see it as removing the surrounding digits 1 instead of removing zero and then adding the surrounding digits 1. So the only case when the number of groups changes is when (a) group B_i consists of exactly one digit 0, (b) this zero is the number on which the operation is performed, and (c) one of the groups A_i and A_{i+1} (or both) consists of exactly one digit 1.

If only group A_i consists of one 1, it disappears and groups B_i and B_{i+1} merge into one group with an even number of zeros. Thus, groups A_{i-1} and A_{i+1} (as well as B_i and B_{i+1} which separated them before) merge; therefore, the number of ones in this union is equal to $a_{i-1} + a_{i+1} - 1 = a_{i-1} + a_{i+1} - a_i$. Hence, the indexing of A_k shifts by 2, starting at some index, and it is easy to see that S stays the same.

If both groups A_i and A_{i+1} consist of just one digit 1, then they both disappear, groups B_{i-1}, B_i and B_{i+1} merge into one group with an odd number of zeros, separating groups A_{i-1} and A_{i+2}. Since in this case we, obviously, had $a_{i-1} = a_i = 1$, once again S does not change.

Now, consider the final collection. It has an even number of zeros, and therefore it consists of either two zeros, or several ones with no zeros. Using invariant S allows us to determine exactly what we actually obtain. If originally $S = 0$ and at the end we have at least one digit 1, then by definition of S we must have at least two ones and two zeros, and therefore one more operation can be carried out. Therefore, in this case the final collection consist of just two zeros.

1991.28. Assume the opposite. Let the maximum on the right be equal to $a^2 - a$. Then $a^2 - b < a^2 - a$, and $b > a$. Further, $b^2 - c < a^2 - a$, and therefore, $c > a$. Also $c^2 - d < a^2 - a$, and consequently $d > a$. In this case $d^2 - a > a^2 - a$, a contradiction.

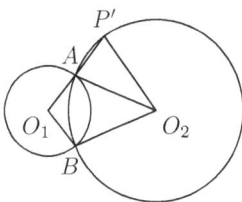

1991.29. There are several possible cases of how the points and circles can be arranged. For example, consider the case shown in the figure. Suppose that line $O_1 A$ crosses the second circumference at point P'. Then $\angle AP'O_2 = \angle P'AO_2 = 180° - \angle O_1 AO_2 = 180° - \angle O_1 BO_2$. Therefore, quadrilateral $O_1 BO_2 P'$ is inscribed, and it follows that points P and P' coincide.

1991.30. See solution to Question **1991.26**.

1991.31. Hint: See the figure below. It shows the graph of the given function $f(x)$, as well as the summands in the two expressions from the problem's statement represented as the collection of 18 thin rectangles. The area of the shaded square is $\frac{1}{100}$, and, as one can easily see, that square is not covered by these 18 rectangles.

1991.32. Answer: no.

Let us prove that any number divisible by 37 can be obtained only from another such number. For operation (a) that is obvious. If we have $(a + b) \vdots 37$, then $1000a+b$ is also divisible by 37 since $1000a+b-(a+b) = 37 \cdot 27a$. It is left to note that 703 is divisible by 37 and 604 is not.

1991.33. Point H is the foot of the perpendicular dropped from P onto line L. Let us index the lines through the sides of the polygon as $L_1, L_2, \ldots, L_n$ in the order of their intersections with L. Prove that pairs of lines (L_k, L_{k+1}) and (L_k, L_{n-k+1}) must form the neighboring sides of the polygon. Consider two cases: when points of intersection with L lie to one side of H, or otherwise.

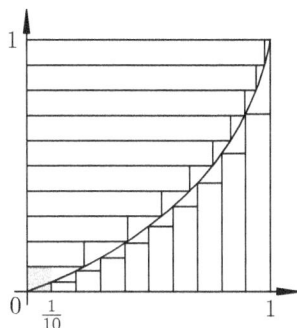

1991.34. Denote by B and G the numbers of blue and green squares, respectively. Obviously, $R \leqslant 3B$, $B \leqslant 3G$, $G \leqslant 3R$ (if $R > 3B$, then there must be four red squares—and consequently, no green ones—next to some blue square). Let us prove that $R \leqslant B+4G$, $B \leqslant G+4R$, $G \leqslant R+4B$. For every red square x_1, consider a chain of three squares—square x_1 itself, its blue neighbor x_2, and x_2's green neighbor x_3. If possible, choose them in such a way that their centers are not collinear. Now, in each chain mark one square: blue, if the chain is straight (collinear centers), and green if the chain is "crooked" (non-collinear centers). Then every blue square is marked no more than once, and every green square—no more than four times, proving the inequality $R \leqslant B + 4G$.

The required inequalities can be proved now as follows.

$$12n^2 = 12(R + B + G)$$
$$= 12R + 3 \cdot 3B + 3 \cdot (B + 4G) \geqslant 12R + 3R + 3R = 18R,$$
$$n^2 = R + B + G \leqslant R + (G + 4A) + G = 2G + 5R \leqslant 11R.$$

1991.36. Answer: no.

Assume the opposite—then the sums of the numbers in our three groups are $102x$, $203y$, and $304z$, respectively, where x, y, and z are some positive integers. Also we have $102x + 203y+304z = 1+2+\cdots+100 = 5050$. Therefore, $x+y+z = 50 \cdot 101 - 101(x+2y+3z) \vdots 101$, and it follows that $x+y+z \geqslant 101$. But then $5050 = 102x + 203y+304z \geqslant 102 \cdot 101 > 5050$.

1991.38. Note that this sequence can never "jump" over interval $[10^p - 9; 10^p - 1]$, where p is a natural number for which $10^{p-1} \leqslant a_1 < 10^p$. Also, if $10^{p-1} \leqslant a_1 \leqslant 4 \cdot 10^{p-1}$, then the sequence cannot jump over interval $[4 \cdot 10^{p-1} + 1; 4 \cdot 10^{p-1} + 8]$.

1991.39. Answer: no.

Multiplying the equation by $x - y$, we obtain $x^{11} - y^{11} = x - y$, that is, $x^{11} - x = y^{11} - y$. So it follows that for some real number C, polynomial $f(t) = t^{11} - t - C$ must have four different roots, which is impossible since its second derivative $f''(t)$ has only one root.

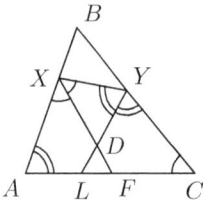

1991.40. The bisector of angle AXY intersects line AC at point F, the bisector of angle XYC intersects AC at point L, and point D is the intersection of these two angle bisectors. Then, triangles AXF, YLC, and YDX cover quadrilateral $AXYC$, and therefore, for their areas, we have the inequality

$$S_{AXYC} \leqslant S_{AXF} + S_{YLC} + S_{YDX}.$$

Note that all these triangles are similar to the original one with ratios $|AX|/|AC|$, $|YC|/|AC|$, and $|XY|/|AC|$, respectively. Therefore,

$$S_{AXF} = S_{ABC} \cdot \frac{|AX|^2}{|AC|^2}, \quad S_{YLC} = S_{ABC} \cdot \frac{|YC|^2}{|AC|^2}, \quad S_{YDX} = S_{ABC} \cdot \frac{|XY|^2}{|AC|^2},$$

and adding these up, we obtain the desired inequality.

1991.41. Hint: In the first 500 days, DoT needs to create a closed cycle of length 400 or more. Then, the number of the cities from which no more than 300 roads (not yet closed, of course) go into the cities of the cycle, does not exceed fifteen. Now, within the next thirty days, these cities must be connected by two roads ("there" and "back") with all other cities. After that, DoT acts according to the following strategy. Each of the remaining cities is to be connected by two roads ("there" and "back") with the cycle—for example, that can be done by connecting—on the specific day—the city with the minimum (at that particular moment) number of the open roads leading from it to the cities of the cycle.

1991.42. Let $a_1, a_2, \ldots, a_{70}$ be the given numbers. Consider the following collection of 210 numbers: $a_1, a_2, \ldots, a_{70}, a_1 + 4, a_2 + 4, \ldots, a_{70} + 4, a_1 + 9, a_2 + 9, \ldots, a_{70} + 9$. Since none of these numbers exceed 209, this collection must contain two equal numbers $a_i + x$ and $a_k + y$. Then numbers a_i and a_k differ either by 4, or by 5, or by 9.

1991.43. Answer: it is a circumference with diameter AB.

Denote by α the measure of arcs AB on these circumferences. Then $\angle AXY = \alpha/2 = \angle AYX$ (regardless of how the points B, X and Y are arranged). Therefore, point M is the midpoint of XY if and only if $\angle AMB = 90°$.

1991.44. Let us prove that by induction on the number of the teams. Write the given numbers in ascending order: $a_1 \leqslant a_2 \leqslant \cdots \leqslant a_n$. Note that $a_n \leqslant 2n - 2$, because the sum of all numbers equals $n(n-1)$ and $a_1 + a_2 + \cdots + a_{n-1} \geqslant (n-1)(n-2)$. Represent number $2n - 2 - a_n$ as a sum $y_1 + y_2 + \cdots + y_k$, where $y_1 = y_2 = \ldots = y_{k-1} = 2$, and y_k is equal to 1 or 2 depending on the parity of $2n - 2 - a_n$. It is easy to check that the collection

$$a_1, a_2, \ldots, a_{n-k-1}, a_{n-k} - y_k, a_{n-k+1} - y_{k-1}, \ldots, a_{n-1} - y_1,$$

which has $n-1$ numbers, satisfies the conditions of the problem. By the induction hypothesis, there exists a tournament in which the teams' results coincide with the numbers above. Now, we add a team which (a) defeated teams with a_i points, $i = 1, 2, \ldots, n - k - 1$; and (b) gained $2 - y_i$ points against teams with $a_{n-i} - y_i$ points, $i = 1, 2, \ldots, k$.

1991.45. Take, for instance, numbers $a_k = (2k + 1)^2$ for $k = 1, 2, \ldots, 8$.

1991.46. Answer: such a function does not exist.

Consider sequence defined by formulas $a_1 = 1$ and $a_n = f(a_{n-1})$. Note that this sequence covers all natural numbers because $a_{n+f(a_n)} = a_n + 1$. Moreover, if $a_k = a_n$, then $k + 1 = a_{k+f(a_k)} = a_{n+f(a_n)} = n + 1$. Therefore, all the terms in this sequence are different. It suffices now to note that number $n + 1$ must occur after n, and therefore, $a_n = n$ and $n + 1 = a_{n+1} = f(a_n) = f(n)$. But this function, obviously, does not satisfy the main recursive equation.

1991.47. Begin with a lemma that we leave as an exercise for the readers. **Hint:** Use induction.

Lemma. *Let S be a set of $k \leqslant 13$ integers not divisible by 13, and $A(S)$—the set of all remainders produced by dividing all possible sums of several numbers from S are 13. Then, $A(S)$ contains at least k elements.*

Let us split our 26-digit number into 13 two-digit numbers $\overline{a_1 b_1}$, $\overline{a_2 b_2}$, ..., $\overline{a_{13} b_{13}}$. Now, consider set S consisting of 13 numbers $9a_1, 9a_2, \ldots, 9a_{13}$. From the lemma, it follows that $A(S)$ consists of all 13 possible remainders modulo 13. In particular,

$$m \equiv 9a_{i_1} + 9a_{i_2} + \cdots + 9a_{i_p} \pmod{13},$$

where m is the sum of all thirteen two-digit numbers $\overline{a_i b_i}$. Now, split each of the two-digit numbers $\overline{a_{i_1} b_{i_1}}$, $\overline{a_{i_2} b_{i_2}}$, ..., $\overline{a_{i_p} b_{i_p}}$ into separate digits. The sum of the resulting one-digit numbers and the remaining two-digit numbers is equal to $m - 9a_{i_1} - 9a_{i_2} - \cdots - 9a_{i_p}$, which is divisible by 13.

1991.48. Draw the common bisector of angles AMB and CMD, and denote second points of its intersection with circumcircles of triangles AMB and CMD by K and E, respectively. Since $|KB| = |KA|$ and point M lies on arc AB of the circumference not containing K, we have the inequality $|KM| > |KB| = |KA|$. It follows then that $2|KM| > |KB| + |KA| > |AB|$. Similarly, we get $2|EM| > |CD|$. Adding these two inequalities we obtain $2|KE| > |AB| + |CD|$. It suffices now to note that $2|PQ| \geqslant |KE|$, because the projection onto line KE maps points P and Q to the midpoints of segments KM and ME.

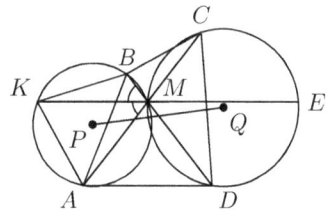

1991.49. Let us prove that even 48 shuffles are enough. Any shuffle can be represented as follows. The deck is split into several parts, each of the parts is turned over in its place, then the entire deck is turned over. We will call the first part of the shuffle (splitting and turning the parts over) a *flop*. Then index all the cards in the order that we need to obtain.

Lemma. *Let the deck contain no more than $2N$ cards, $2N \leqslant 3^m + 1$. Then with no more than $m + 1$ flops we can ensure that the cards 1, 2, ..., N lie atop the deck.*

The proof is done by induction on m. The basis for $m = 0$ is obvious.

Split the deck into two groups: group $A = \{1, \ldots, N\}$, and group B that consists of all other cards (thus, group B has no more than N cards).

Split the deck into the portions that consist of the cards of the same group. In the beginning, we have no more than N such portions per group. Let us represent the split by writing the group name instead of indexes of cards in each portion (left of the line corresponds to the top of the deck). Now, perform the flop

$$A|\underset{\leftrightarrow}{BA}|B|\underset{\leftrightarrow}{AB}|A|\underset{\leftrightarrow}{BA}|B|\ldots \; \to \; AA\,BBB\,AAA\,BB\ldots\,,$$

if the deck begins with portion A, and the flop

$$|\underset{\leftrightarrow}{BA}|B|\underset{\leftrightarrow}{AB}|A|\underset{\leftrightarrow}{BA}|B|\underset{\leftrightarrow}{AB}\ldots \; \to \; A\,BBB\,AAA\,BBB\,A\ldots\,,$$

if it begins with B.

It is easy to verify that in both cases we obtain the deck with no more than $[\frac{N}{3}+1] \leqslant 3^{m-1}+1$ consecutive portions marked with the same letter (that is, belonging to the same group). Replace every such portion by just one letter (the group's name). Now, by the induction hypothesis, m flops is enough to put all the portions from group A on the top of the deck. Therefore, in no more than $m+1$ flops we can have all cards from group A on the top of the deck. This completes the proof. $\square$

Now, let us return to the problem's solution. We will prove a much better upper bound—namely, 48—than the one claimed in the problem (56).

Begin with the estimate for the necessary number of operations. For convenience, we will work with a deck that has 1024 cards, simply adding 24 virtual, empty cards to the current deck. It follows from the lemma that using only 8 flops we can have the cards with indexes 1 through 512 in the top half of the deck; however, their precise order is unknown. After that, we deal separately with the upper and the lower halves. It is enough now to estimate the number of flops needed to reorder the top half of the deck; while we do that, we reorder the lower half in parallel, so the number of required flops remains the same.

It follows from the lemma that in 7 flops we can achieve the top half of the half-deck to consist of cards with indexes 1 through 256, and the lower half of the half-deck—to consist of cards indexed 257 through 512. Now, we deal with each quarter-deck in parallel, etc. It is easy to see that the entire reordering can be done with $48 = 8+7+7+6+5+5+4+3+2+1$ flops.

Finally, we recall that the shuffle differs from the flop by the full deck's turnover. However, since 48 is even, after 48 flops the cards are arranged in the desired order, not the reverse one (otherwise we would have to add one more shuffle).

1991.50. Answer: no.

When we place a checker on the board, connect the center of this square with the centers of adjacent free squares (if any exist). Thus, every two adjacent squares will eventually be connected by exactly one segment.

Since the total number of adjacent pairs is odd (remember, the upper left corner square was already occupied), at some point the just placed white checker has an odd number of adjacent free squares. That means that at the end, after all the following recolorings, this checker becomes black.

1991.51. Let point D be the midpoint of BB'. Since $\angle AMM' = \angle ABB'$, lines MM' and BB' intersect at point N' that lies on our circumference. Short computation shows that $|NM|/|MM'| = |NB|/|BD|$, and the equality of angles $\angle NMM'$ and $\angle NBD$ implies the similarity of triangles NMM' and NBD.

Therefore, $\angle MM'N = \angle BDN$; hence, quadrilateral $M'DN'N$ is inscribed, and so $\angle M'DN = \angle M'N'N = \angle MN'N = 90°$.

1991.52. Summing up the inequalities

$$4x_k^3 - 3x_k + 1 = (x+1)(2x-1)^2 \geqslant 0$$

for k from 1 through n, we get the required proof.

1991.53. Note that for every positive integer n there exists a number written only by ones and zeros which is divisible by n (because sequence 1, 11, 111, $\ldots$, contains two numbers with identical remainders modulo n, and their difference can serve as such a number). Thus, we can multiply n by some number and then remove zeros to obtain $11\ldots1$. Then multiply this result by 82 to get $911\ldots102$; remove zero and multiply by 9 to obtain $8200\ldots8$; remove zeros to get 828. Finally, $828 \to 20700 \to 27 \to 108 \to 18 \to 90 \to 9$.

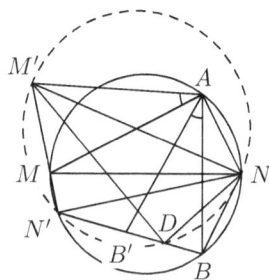

1991.54. Note that for every $t \in [0; 1]$, the inequality

$$\int_0^1 f(x)\,dx \geqslant \int_{1-t}^1 f(x)\,dx \geqslant (1-t)f(t) \geqslant f(t) - t$$

holds true. It suffices now to substitute $t = g(x)$ and integrate over x from 0 to 1.

1991.55. Consider $z \neq 1$, a complex root of unity of degree p, that is, a complex number such that $z^p = 1$. Since

$$0 = z^p - 1 = (z-1)(z^{p-1} + z^{p-2} + \cdots + z + 1),$$

it follows that $z^{p-1} + z^{p-2} + \cdots + z + 1 = 0$. Now, consider polynomial $f(t) = a_1 + a_2 t + a_3 t^2 + \cdots + a_n t^{n-1}$. If sequence (a_k) is p-balanced, then any pth root of unity, not equal to 1, is a root of polynomial f. Indeed,

$$\begin{aligned}
f(z) &= a_1 + a_2 z + \cdots + a_n z^{n-1} \\
&= (a_1 + a_{p+1}z^p + a_{2p+1}z^{2p} + \cdots) \\
&\quad + z(a_2 + a_{p+2}z^p + a_{2p+2}z^{2p} + \cdots) + \cdots \\
&= S(1 + z + z^2 + \cdots + z^{p-1}),
\end{aligned}$$

where $S = a_1 + a_{p+1} + a_{2p+1} + \cdots = a_2 + a_{p+2} + a_{2p+2} + \cdots$ and so on; therefore, $f(z) = 0$.

Since for every p there are $p-1$ complex roots of unity of degree p different from 1, polynomial $f(t)$ has $2 + 4 + 6 + 10 + 12 + 16 = 50$ different (!) roots. So we have a complex polynomial of degree 49 with at least 50 roots. We conclude that $f(t) \equiv 0$, and so all of its coefficients are zeros.

1991.56. Set $x = 512$, $y = 675$, $z = 720$. Then $2z^2 = 3xy$ and

$$\begin{aligned}
x^3 + y^3 + z^3 &= x^3 + y^3 - z^3 + 3xyz \\
&= (x+y-z)(x^2 + y^2 + z^2 - xy + xz + yz)
\end{aligned}$$

is divisible by $x + y - z = 467$.

Note. Actually, this number is equal to $229 \cdot 467 \cdot 7621$.

1991.57. Answer: no, that is not true.

Lemma. *For any $\varepsilon > 0$ there exists natural number m such that*

$$P = \prod_{k=1}^{m} \left(1 - \frac{1}{p_k}\right) < \varepsilon,$$

where p_k denotes the kth prime number.

To prove the lemma, note that

$$\left(1 - \frac{1}{p}\right)^{-1} = 1 + \frac{1}{p} + \frac{1}{p^2} + \frac{1}{p^3} + \cdots .$$

Thus,

$$P^{-1} = \prod_{k=1}^{m} \left(1 + \frac{1}{p_k} + \frac{1}{p_k^2} + \frac{1}{p_k^3} + \cdots\right) \geqslant \sum_{i=1}^{p_m} \frac{1}{i}. \qquad (*)$$

This inequality follows from the fact that every natural number not exceeding p_m can be uniquely represented as the product of the form $p_1^{\alpha_1} p_2^{\alpha_2} \ldots p_m^{\alpha_m}$, where $\alpha_1, \alpha_2, \ldots, \alpha_m$ are non-negative integers. It is well known that the sum $(*)$ can be arbitrarily large (that is, it tends to infinity as $m \to \infty$), and therefore, P^{-1} can also be arbitrarily large; consequently, P can be arbitrarily small. $\square$

Let us index the vertices of the polygon by numbers 1 through $2n$ in the clockwise order. Then a pattern is a set of n integers from interval $[1; 2n]$, and a rotation of a pattern simply means adding some fixed number modulo $2n$ to all members of the pattern.

Consider $\varepsilon = \frac{1}{2} \prod_{k=1}^{99} \left(1 - \frac{1}{p_k}\right)$. For some m we have

$$A = \prod_{k=100}^{m} \left(1 - \frac{1}{p_k}\right) < \frac{1}{2}.$$

If $n = p_{100}\, p_{101} \ldots p_m$, then the quantity of numbers from interval $[1; 2n]$, not divisible by any of the primes $p_{100}, p_{101}, \ldots, p_m$, equals $2nA < n$. Thus, there are at least n numbers less than $2n$ and divisible by at least one of the primes $p_{100}, p_{101}, \ldots, p_m$. Select n of them—this is the desired pattern.

Consider an arbitrary set of numbers $x_1, x_2, \ldots, x_{100}$, defining one hundred rotations of this pattern. For every $i = 100, 101, \ldots, m$, choose remainder r_i that is not equal to any of the remainders $x_j \pmod{p_i}$ (such remainder can be found because $p_i > 200$). Then it follows from Chinese remainder theorem that there exists number $a \leqslant n$ such that $a \equiv r_i \pmod{p_i}$ for all $i = 100, \ldots, m$. This number is not covered by any of the rotated patterns.

PART 4

Miscellany

Additional Problems

These problems were offered at some other[115] high-level mathematical contests held in Leningrad during 1961–1991. A few of them come from additional elimination rounds of LMO, some others— from math battles between high school teams,[116] science-and-math schools' olympiads, and other similar competitions.

On average, these problems are considerably more difficult than the ones from regular LMO problem sets, and quite a few of them were not solved by the contestants.

In parentheses, we show the year when the problem was originally offered (based on the available archival data). Solutions are not provided. However, some of them can be found in the archives of KVANT magazine as well as in the materials of Moscow Math Olympiads (see [15]), All-Russia Math Olympiads, and International Tournament of the Towns.

---

[115] That is, different from LMO.

[116] Math battle is a highly popular team contest, in which two or three teams compete in the art of math problem solving. Interested? See [25], p. 236 or run a web search for "math battle".

1. (1965) There are n large glasses and each one of them contains the same volume of water. It is permitted to pour from one glass to another as much water as is already present in the second glass. What are the values of n for which it is possible to collect all water in one glass using these operations?

2. (1965) Two people play the following game, making their moves in turn. The player's move consists of taking one, two, or three matches from the pile that originally contained 25 matches. The player who ends up with the even number of matches, wins. Who wins in the errorless game, the player who makes the first move or her opponent?

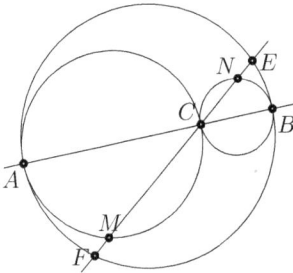

3. (1965) After an arbitrary point C is chosen on segment AB, three circumferences O_1, O_2, O_3 with diameters AB, AC, and BC, respectively, are drawn.

A straight line passes through point C and intersects circumference O_1 at points E and F, and circumferences O_2 and O_3 at points M and N, respectively. Prove that $|MF| = |EN|$.

4. (1965) Arbitrary point O is selected inside heptagon $A_1 A_2 \ldots A_7$ with sides of equal length. Let us denote by H_1, H_2, $\ldots$, H_7 projections of point O onto lines $A_1 A_2$, $A_2 A_3$, $\ldots$, $A_7 A_1$, respectively. Given that points H_1, H_2, $\ldots$, H_7 lie on the sides themselves, not on their extensions, prove that

$$|A_1 H_1| + |A_2 H_2| + \cdots + |A_7 H_7| = |H_1 A_2| + |H_2 A_3| + \cdots + |H_7 A_1|.$$

5. (1965) The scientific conference is attended by $2n$ scientists; each one of them is friends with at least n of the participants. Prove that we can select four scientists and sit them around the table so that everyone is sitting next to their friends.

6. (1965) Prove that any non-negative even integer can be uniquely represented as $(x + y)^2 + 3x + y$, where x and y are non-negative integers.

7. (1965) In triangle ABC, side BC equals half-sum of the two other sides. Prove that the bisector of angle A is perpendicular to the segment connecting the incenter and the circumcenter of ABC.

8. (1965) The city has twenty streets, ten of which are parallel to each other, and the other ten are perpendicular to the first ten. For a closed route that passes through all the intersections, what is the smallest possible number of turns?

9. (1965) Several vectors issue from point O on the plane. The sum of their lengths is 4. Prove that it is possible to select some of them so that the length of their sum is greater than 1.

10. (1965) Four pedestrians walk (each at his own constant speed) along four straight roads, no two of which are parallel, and no three intersect at one point. It is known that the first pedestrian will meet the second, the third, and the fourth, and the second will meet the third and the fourth. Prove that the third will meet the fourth.

11. (1965) Among $2n$ knights of the Round Table who have gathered at King Arthur's court, each one has no more than $n - 1$ enemies. Prove that Merlin[117] can sit the knights at the Round Table so that none of them would sit next to his enemy.

12. (1967) Sequence $a_1, a_2, \ldots, a_{n+1}$ consists of positive numbers. Denote by A_k the arithmetic mean of the first k terms of the sequence, and by G_k—their geometric mean. Prove that

$$n(A_n - G_n) \leqslant (n + 1)(A_{n+1} - G_{n+1}).$$

13. (1967) Fill the three-dimensional space with circumferences of non-zero radius so that there is exactly one circumference passing through every point of the space.

14. (1967) Ten nickels[118] are placed on the flat table, forming the closed chain. The eleventh nickel rolls, without slipping, around this configuration, touching, in turn, all ten coins of the chain. How many full rotations will this coin make upon returning to its original location?

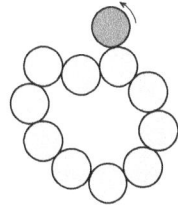

15. (1967) Prove that it is impossible to dissect the rectangle with dimensions $a \times b$ into squares if a/b is an irrational number.

16. (1967) Given seven points in general position on the plane, connect each two of them with a segment. What is the minimum and maximum possible number of intersection points between these segments?

17. (1968) Can numbers 27, 84, 110, 133, 144 satisfy equation $x^n + y^n + z^n + w^n = t^n$ for some natural n?

18. (1968) In the sequence of symbols $OXOXOX \ldots OXOX$ (n pairs OX) it is permitted to take any two neighboring symbols and move them as a contiguous block by writing them next to another (possibly empty) sequence. What is the minimum possible number of such operations which is necessary to obtain sequence $OO \ldots OOXX \ldots XX$?[119]

19. (1968) Prove the inequality

$$\sum_{i,j=1}^{\infty} \frac{a_i a_j}{i + j} \leqslant \pi \sum_{k=1}^{\infty} a_k^2.$$

20. (1968) Digits of four perfect squares were substituted by letters, with equal digits replaced by equal letters, to obtain words **MERRY**, **XMAS**, **TO**, and **ALL**. Determine the substitution rules.

---

[117] Legendary magician and King Arthur's chief councillor.
[118] Five cent coins.
[119] It is implied here that as an intermediate result, we are allowed to have a collection of several sequences.

21. (1968) Prove that on the sheet of graph paper with gridlines at unit distance, it is not possible to construct a segment of length $\sqrt{3}$, using only straightedge.

22. (1969) A circle of radius 10 cm is dissected into several parts by 32 straight lines. Prove that one of these parts contains a circle of radius 3 mm.

23. (1969) Find two integers greater than 1 whose product is equal to $235^2 + 972^2$.

24. (1970) It is time for the new presidential elections in the Republic of Anchuria. The incumbent president Miraflores is supported only by the army which constitutes 1% of the 20 million voters. Anchuria's elections are designed as follows. The electorate will be split into several **equal** groups, each one of them will also be split into several **equal** groups and so on.[120] Then the elections will proceed as follows. In each smallest group, the voters will choose the group's elector who will be then delegated to the meeting of the electors of the next level group where the delegates will again vote and choose the elector for the next level group and so on. Finally, the electors of the largest groups vote to elect the new president. At every level, in each group, the decision is made by simple majority, and in case of 50/50 split, the opposition wins by default. The president can choose the groups on all levels however he wants. Can he do so to guarantee his "democratic" re-election?

25. (1970) Every day a few students from the same school go to the ice cream parlor. However, every trip ends in a really bad argument[121] and no two kids from that group ever go for ice cream together. By the end of the school year, after $k > 1$ group trips, it turned out that each student could only go to the ice cream parlor by himself or herself. Prove that $k \geqslant n$, where n is the number of students in that school.

26. (1970) 2^{p-1} subsets are selected in the set with p elements. Prove that if any three of these subsets have a common element, then they all have a common element.

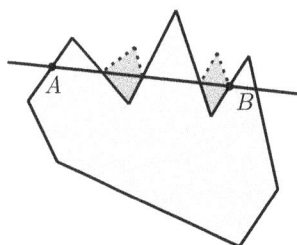

old plot
▨ new (additional) plot

27. (1970) A wily businessman bought a square plot of land. He built a fence along its border and obtained a special permission from the municipality which allows him to perform the following procedures. He can choose any two points A and B on the contour of the fence, draw the line through these points, take a portion of the fence between A and B which lies on one side of the line, demolish it, and then built the new portion of the fence which is symmetric to the old one with respect to line AB—after that, his plot will change to the area that lies inside the new fence. Could he, after a few such procedures, end up with a plot larger (by area) than his original one?

28. (1970) Prove that if $a + b + c = 0$, then
$$\frac{a^7 + b^7 + c^7}{7} = \left(\frac{a^5 + b^5 + c^5}{5}\right)\left(\frac{a^2 + b^2 + c^2}{2}\right).$$

---

[120] Number of subgroups can change from group to group; they only have to be equal within the larger group.
[121] Perhaps about which ice cream flavor is the best?

29. (1971) A number triangle is constructed in the following way. The upper row has k zeros and one 1. Every number in any lower row equals 0 or 1 based on whether the two numbers directly over it are equal or not (see example of such triangle for $k = 5$ in the figure). For which values of k the lowest number in the triangle does not depend on the position of the only 1 in the upper row?

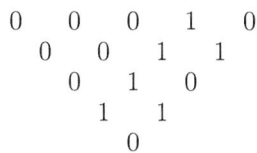

```
0   0   0   1   0
  0   0   1   1
    0   1   0
      1   1
        0
```

30. (1972) Two straight lines on the plane form angle φ. A flea sits at point M which belongs to one of these lines. The flea starts jumping from one line to the other, each jump exactly 1 cm long, so that it never jumps back to the point it has just visited on that line.[122] After several jumps, the flea has returned to M. Prove that fraction φ/π is a rational number.

31. (1972) Line segments AC and BD intersect at point E. Points K and M on segments AB and CD are such that KM passes through point E. Prove that the length of KM does not exceed the largest of lengths $|AC|$ and $|BD|$.

32. (1972) A fly sits in one of the cube's vertices, and two spiders—in the diametrically opposite vertex. All three can move along the cube's edges with the same speed. Can the spiders catch the fly?

33. (1972) Prove the equality

$$(a+b+n)^{n-1}\left(\frac{1}{a}+\frac{1}{b}\right) = \sum_{k=0}^{n}\binom{n}{k}(a+k)^{k-1}(b+n-k)^{n-k-1}.$$

34. (1972) Given n lines in general position on the plane, prove that at least $n-2$ of the parts into which the lines divide the plane are triangles.

$n=6$

35. (1973) There is an infinite sequence of light bulbs which are indexed by natural numbers, and an infinite sequence of switches which are also indexed by the natural numbers; each switch has a finite number of states. It is known that whatever the states of the switches are, at least one of the bulbs is lit. It is also known that the state of any bulb depends on the state of only finite number of the switches. Prove that there is a finite set of bulbs, at least one of which is always lit, regardless of how the switches are set.

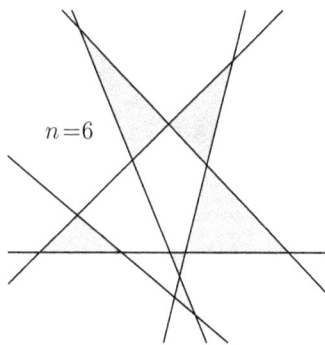

36. (1973) Let $f(x)$ be a continuous periodic function with period T defined on the entire real axis; $f = g + h$, where $g(x)$ and $h(x)$ are two continuous functions with period T_1 and T_2, respectively. Prove that periods T_1 and T_2 are commensurable with T.[123]

---

[122] Exactly two jumps ago.
[123] Two numbers a and b are called commensurable if their ratio a/b is rational.

37. (1974) Let us call four unit squares on the infinite sheet of graph paper, whose centers lie on the vertices of a rectangle with sides parallel to grid lines, a *quartet*. What is the maximum number of mutually disjoint quartets inside a 25×25 square?

38. (1974) A planar figure, obtained as the intersection of twenty circles, contains more than one point. Prove that the boundary of that figure can be represented as a union of 38 circular arcs.

39. (1974) Function $f(x)$ is defined on set $R = \mathbb{R} \setminus \{0, 1\}$. It is known that for every $x \in R$, the equality

$$f(x) + f\left(\frac{1}{1-x}\right) = x$$

holds true. Find all such functions.

40. (1974) Given convex polygon P with n sides, for every three consecutive vertices of P, we construct a circumference that passes through them. Prove that the largest of these circumferences contains entire polygon P.

41. (1975) Irreducible fraction x/y is called a *nice* approximation of number C if $|C - x/y| < 1/y^{100}$. Prove that any non-empty open interval contains a number which has infinitely many nice approximations.

42. (1975) On the plane, we have k points, as well as several rays which issue from these points (some of them possibly with the same endpoint) such that no two rays intersect. Prove that one can draw $k - 1$ segments connecting the given points so that no two of these segments have interior common points with each other or with any of the rays.

43. (1976) There is a reel-to-reel tape recorder; 25 reels with magnetic tape wound onto each one of them, and one empty reel. It is permitted to rewind the tape on any reel onto the empty one (after which, of course, the first reel will be empty, and the tape on the second reel will be wound in the opposite direction). Is it possible to get all the tapes on the same reels they were originally on, each one rewound in the opposite direction?

44. (1976) Several circles, each one with radius not exceeding 0.001, are placed inside the unit square. It is also known that the distance between any points of any two of these circles is different from 0.001. Prove that the area covered by all the circles does not exceed 0.34.

45. (1976) p is an odd prime number. We have $p - 1$ integers, none of them divisible by p. Prove that by reversing the sign for some of these numbers, we can obtain $p - 1$ numbers whose sum is divisible by p.

46. (1976) Squares of the infinite sheet of graph paper are filled with real numbers from $[0; 1]$ in such a way that every number is equal to the arithmetic mean of its four neighbors. Prove that all the numbers are equal to each other.

47. (1976) Prove that for any natural n, there exists an integer written by digits 1 and 2 only and divisible by 2^n.

48. (1977) Nine points are selected inside 2×2 square. Prove that the distance between some two of them does not exceed 1.

49. (1977) For any function $f \colon \mathbb{R} \to \mathbb{R}$ by $f^{[n]}$ we will denote function $f(f(...f(x)...))$ (n pairs of parentheses). If $P(t) = a_n t^n + a_{n-1} t^{n-1} + \cdots + a_0$ is some polynomial, then by $P(f)$ we will denote function

$$q(x) = a_n f^{[n]}(x) + a_{n-1} f^{[n-1]}(x) + \cdots + a_1 f(x) + a_0 x.$$

a) Let $P(t) = t^2 - t + 1$. Prove that if $P(f) = 0$, then $f^{[7]} = f$.
b) Let P, Q, R be polynomials such that $Q = P \cdot R$. Prove that if $P(f) = 0$, then $Q(f) = 0$.
c) Prove that for polynomial P from item (a), there exists a function f such that $P(f) = 0$.
d) Is the same true for any polynomial?
e) Let Q, P be polynomials such that $\{f \colon P(f) = 0\} \subset \{f \colon Q(f) = 0\}$. Prove that for some polynomial R, we have $Q = P \cdot R$. Start by proving that for $Q(t) = t^n - 1$.

50. (1978) Straight line L passes through the center of a regular polygon with $2n$ sides. Prove that the sum of the distances to L from the vertices situated on one side of L is the same as the sum of the distances to L from the vertices on the other side.

51. (1979) Some finite number of positive integers were named *lucky*. Let us denote by A_k the maximum amount of lucky numbers which can be written (with possible repeats) if we had k copies of every digit. Prove that for some value of k expression A_k/k reaches its maximum.

52. (1979) Set X on the coordinate plane satisfies the following conditions:

a) Intersection of X with any unit square whose vertices lie on the nodes of the grid consists of two parallel segments whose endpoints are the midpoints of the square's sides;
b) X maps into itself when translated by distance of 25 in any direction which is parallel to a coordinate axis.

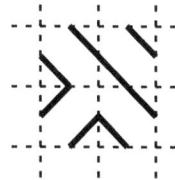

Prove that X contains an infinitely long broken line.

53. (1979) Fifty-five bricks are arranged into several stacks. Every minute we take one brick off the top of every stack and assemble these bricks into one new stack. Prove that at some point we will have ten stacks with 1, 2, 3, ..., 10 bricks in them.[124]

---

[124] An example of such an operation is shown in the figure.

$(13, 10, 10, 10, 6, 3, 3)$ $(12, 9, 9, 9, 7, 5, 2, 2)$

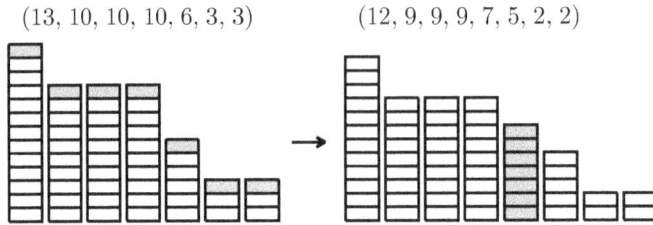

54. (1980) We will call two points (x_1, y_1) and (x_2, y_2) on the coordinate plane *dependent* if they satisfy the relation

$$(x_1 - x_2)^2 = (y_1 - y_2)(x_1 y_2 - x_2 y_1).$$

There are four points such that every two of them are dependent. Prove that they all lie on the same straight line.

55. (1980) $2n$ points on the plane are connected by curves of red, blue, and green color so that for every point there is exactly one curve of each color coming out of it. Let a, b, and c be the numbers of red-blue, red-green, and blue-green cycles, respectively. Prove that $n + a \geqslant b + c$.

56. (1981) There are n straight lines on the plane. On each one of them, the other $n - 1$ lines cut $n - 2$ segments of equal length. Prove that $n = 3$.

57. (1982) Strictly increasing sequence (a_k) of natural numbers is such that $a_2 = 2$, and for any co-prime m and n, we have $a_{mn} = a_m a_n$. Prove that $a_k = k$ for all k.

58. (1982) Real number is written into every square of the infinite sheet of graph paper in such a way that

   a) numbers in two squares located in the same row or in the same column at distance of 1982 are equal;

   b) each number is equal to the arithmetic mean of either its two neighbors in the row or its two neighbors in the column.

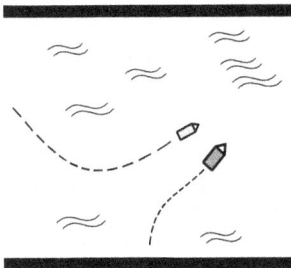

Prove that either in every column all the numbers are equal or in every row all the numbers are equal.

59. (1982) A ten kilometer wide strait is patrolled by a police boat whose maximum speed is seven times the maximum speed of a dingy used by the contraband runners. The patrol will see the dingy (and subsequently capture it) if the distance between them is less than or equal to 1 km. Can the patrol find a way to guarantee that the contraband runners never pass through the strait?[125]

---

[125] We should assume here that the patrol's trajectory is a closed curve, while the dinghy starts from somewhere sufficiently far from that trajectory.

60. (1982) Find real numbers a_1, a_2, a_3 and c_1, c_2, c_3 such that for any polynomial $P(x)$ of degree 5 we have

$$\int_0^1 P(x)dx = a_1 P(c_1) + a_2 P(c_2) + a_3 P(c_3).$$

61. (1982) Several positive numbers a_1, a_2, ..., a_n are arranged around the circle. It is known that if two numbers—a_k and a_ℓ—are neighbors, then expression $2/\left(\frac{1}{a_k} + \frac{1}{a_\ell}\right)$ coincides with one of the numbers a_i. Prove that all numbers a_i are equal.

62. (1983) Given a circumference S and point H inside S, prove that there exists circumference T such that for any triangle inscribed in S, which has point H as its orthocenter, the midpoints of triangle's sides lie on circumference T.

63. (1983) For any points A and B on the plane, define point $A * B$ as image of point B under rotation around point A by $60°$. Prove that for any points A, B, C, and D equality $(A * B) * (C * D) = (A * C) * (B * D)$ holds true.

64. (1983) Strictly monotonically increasing sequence of natural numbers (x_n) possesses the property that for any $n > 1982$ the equality

$$x_1^3 + x_2^3 + \cdots + x_n^3 = (x_1 + x_2 + \cdots + x_n)^2$$

holds true. Prove that $x_n = n$ for any n.

65. (1983) Find smallest number k such that there exists a polynomial of degree k in ten variables such that for any two different sets of arguments its values are different.

66. (1983) Two complex polynomials $P(z)$ and $Q(z)$ are such that one of them is not constant; it is known that the sets of their roots coincide as well as the sets of the roots of polynomials $P(z) - 1$ and $Q(z) - 1$. Prove that $P = Q$.

67. (1983) S is a set with n elements, and M_1, M_2, ..., M_{n+1} are its non-empty subsets. Prove that it is possible to find two different collections of these subsets such that the union of subsets in the first collection coincides with the union of subsets in the second collection.

68. (1984) A computer can find the sum and the difference of two numbers, can divide a number by any natural number, and can compute the tenth power of a number. Prove that using this machine it is possible to compute the product of ten given numbers.

69. (1984) In some graph, it is possible to reach any vertex from any other having visited no more than $n - 1$ additional vertices. It is also known that the shortest cyclical route has length $2n + 1$. Prove that all the vertex degrees[126] are the same.

70. (1984) A finite sequence of zeros and ones is given. Every minute, in the space between each pair of neighboring numbers, we write another number: one between two zeros, and zero for any other pair. Prove that based on the original sequence we can find number C such that however many times this operation is carried out the difference between the number of zeros and the doubled number of ones will never exceed C.

---

[126] The degree of a graph's vertex is the number of edges incident to the vertex.

71. (1984) In the Republic of Anchuria, due to the budget restrictions, it was decided to close several highways connecting the towns in such a way that the road system still remains connected, that is, one can always reach each town from any other. The Ministry of Transportation submitted a plan of closings with the smallest possible total length of the remaining roads. However, upon investigation, it was determined that the road maintenance expenses are proportional not to the length of the road but to the square of the length. Can the Ministry of Transportation claim with certainty that the original plan will also result in the minimal possible maintenance costs?

72. (1984) What is the maximum number of polygons on the plane such that the distance between any two of them is equal to 1?

73. (1984) There are p countries on the planet ($1000 < p < 2000$). Some trios of these countries form tripartite unions, and it is known that every two countries are members of exactly one such union. It is also known that if trios ABF, BCG, CAH are such unions, then FGH is such union as well. Find all p for which this is possible.

74. (1984) Continuous function f is defined on the set of all positive real numbers. It is known that for any $x > 0$ sequence $a_n = f(nx)$ converges. Is it true that limit $\lim_{x \to \infty} f(x)$ exists?

75. (1985) Several chess players played a round-robin tournament without draws. The referee has written all the results (ones and zeros) into the square table, and then she realized that she forgot how the players were indexed; only the final results (totals) for every participant are known.[127] Prove that the number of all possible ways to restore the indexing is odd.

76. (1986) Sum of k numbers is zero. M is the largest of the numbers, and m is the smallest of them. Prove that the sum of their squares does not exceed $(-kmM)$.

77. (1986) Let us denote by $F(A, B, C)$ the incenter of triangle ABC. Prove or disprove the identity

$$F(F(A, B, X), F(A, C, X), F(B, C, X)) = X.$$

78. (1986) Sequences (a_i) and (b_i) are defined as follows: a_0 and b_0 are positive, and

$$a_{k+1} = \min(a_k, b_k), \quad b_{k+1} = |a_k - b_k|.$$

Prove that

   a) sequence (a_i) tends to zero;
   b) sequence $\sum a_i^2$ has a limit, and find that limit as a formula in a_0 and b_0.

79. (1986) Among all convex quadrilaterals with the given diagonals and given angle between them, find the one with the minimum perimeter.

---

[127] This last condition was omitted; without it, however, the statement is obviously false.

80. (1986) A stone lies on each one of the lower 50 steps of the stairs with 101 steps. Sisyphus can take any stone and carry it up the stairs to deposit it onto the nearest free step. Hades can roll any stone onto the previous step if it is free. They make their "moves" in turn. Can Hades prevent Sisyphus from placing a stone on the topmost step of the stairs?

81. (1986) Let $a_1, a_2, \ldots, a_n$ be pairwise distinct positive numbers. Compute all possible sums of these numbers without repeating indexes. Find the smallest possible number of different numbers among these sums.

82. (1986) Circumferences centered at points A, B, C, and D are positioned on the plane in such a way that any two of them touch externally, with point D lying inside triangle ABC. Let P, Q, and R be the incenters of triangles ABD, BCD, and ACD, respectively. Prove that D is the incenter of triangle PQR.

83. (1986) Prove that polynomial

$$x^6 + 12x^5 + ax^4 + bx^3 + cx^2 + dx + 68$$

can never have exactly six real positive roots.

84. (1986) The tetrahedron has volume V and surface area S. The planes that contain its faces were pushed outside by the distance of h. Find the volume and the surface area of the resulting tetrahedron.

85. (1986) Consider all possible 100-digit numbers in which every two neighboring digits are different. Are there more even or odd numbers among them?

86. (1986) Function $f: [0; 1] \to \mathbb{R}$ is such that $f(0) = f(1)$, and for every two different numbers a and b from $[0; 1]$, we have

$$f\left(\frac{a+b}{2}\right) \leqslant f(a) + f(b).$$

Prove that this function has infinitely many zeros inside interval $[0; 1]$.

87. (1986) We will call two polygons positively homothetic if one of them can be obtained from the other by translation or by a homothety with positive coefficient. Prove that if two convex positively homothetic polygons have a common point, then one of them contains some vertex of the other.

88. (1987) Unit cubes that make up $n \times n \times n$ cube are colored into two colors. It is permitted to repaint the entire "line" $1 \times 1 \times n$ into the color that prevails in it. Is it always possible to perform several such operations to make all the unit cubes the same color?

89. (1987) Diagonals of convex quadrilateral $ABCD$ intersect at point O. Points K, L, M, and N are the incenters of triangles ABO, BCO, CDO, and DAO, respectively. Find maximum value of constant c such that the inequality

$$P(ABCD)P(KLMN) \geqslant c \cdot S(ABCD),$$

is always true. P and S denote perimeter and area functions, respectively.

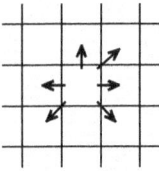

90. (1987) A "lame king" moves around the chessboard in the usual way except it is prohibited from making two types of move: "up-and-left" and "down". Two players move the king in turn, and it is not permitted to visit any square more than once. The player who cannot make a move loses. Who wins in the errorless game, the player who makes the first move or his opponent?

91. (1987) Coins in the Republic of Tiktakistan have value of 1, m, and n tiktaks, where m and n are coprime numbers greater than 1. What is the number of ways to pay the sum of Amn tiktaks using these coins?

92. (1987) Vertices of a convex polygon with n sides lie on the nodes of the grid. Prove that its perimeter is greater than $n\sqrt{n}/100$.

93. (1987) There are n light bulbs. What is the minimum number of tumbler switches, each one changing the state of some fixed set of the light bulbs, such that it would be possible to have any two light bulbs turned on with all the others off?

94. (1987) In convex quadrilateral $XYZT$, the extensions of sides XT and YZ intersect at point F, the extensions of sides XY and ZT—at point G, and diagonals XZ and YT—at point O, where point Y lies inside both segments FZ and XG. Straight line FO intersects sides XY and ZT at points A and B, and straight line GO intersects sides XT and YZ at points C and D, respectively. Prove that $|AB| + |CD| \leqslant |YT| + \frac{3}{2}|XZ|$.

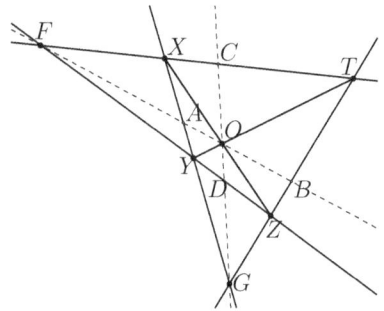

95. (1987) Two people play the following game, making their moves in turn. The first player, with his every move, places a point on the plane and connects it with some of the points already placed there by line segments so that no two segments on the plane intersect each other in interior points. The second player's move consists of coloring the new point into one of the ten fixed colors. Prove that the first player can play in such a way that at some moment, there will be a segment with equally colored endpoints.

96. (1987) Given a pentagon all of whose angles are equal, prove that the sum of the distances from a point inside the pentagon to all of its sides is constant.

97. (1987) Numbers 1 through n are written in order around the circle. It is permitted to replace any two neighboring numbers by their arithmetic mean. Prove that if $n > 4$, then it is impossible to make all the numbers equal.

98. (1987) A drunk librarian does the following every minute: he walks up to the shelf which houses all the volumes of British Encyclopedia, arbitrarily chooses one of the volumes which does not stand in its proper place (by index), and then puts it in its proper location. If at some point all the volumes are standing in their proper places, the librarian will join the Christian Temperance Union. Can the Union count on this librarian's eventual membership?

99. (1987) The Kingdom of Tardus has 16 towns. The king wishes to connect some of the towns with roads so that a traveler could reach any town from any other, while passing through no more than one other town on the way. He is also trying to save money, and therefore, wishes to do it so that there are no more than k roads going out of every town. What is the minimum possible value of k which will allow this plan to succeed?

100. (1987) Someone selected n secret numbers, then wrote all their pairwise sums on $\frac{1}{2}n(n-1)$ cards and thoroughly shuffled them. Is it possible to restore the original set of numbers from this deck?

a) Solve the problem for $n = 3, 4$.
b) Prove that, generally speaking, it cannot be done if n is a power of 2.
c) Prove that it can always be done if n is not a power of 2.
d) Is it possible to restore the original set from the randomly reordered collection of all possible triple sums (that is, the collection of $\binom{n}{3}$ sums of form $a_r + a_s + a_t$, where $\{a_1, a_2, \ldots, a_n\}$ are the unknown numbers, and $1 \leqslant r, s, t \leqslant n$ are any three different indexes)?
e) Is it possible if we have a randomly reordered collection of all possible k-sums of an n-number collection $\{a_1, a_2, \ldots, a_n\}$ (that is, the collection of $\binom{n}{k}$ sums of form $a_{i_1} + a_{i_2} + \cdots + a_{i_k}$, where $1 \leqslant i_1, i_2, \ldots, i_k \leqslant n$ are k different indexes)?

101. (1987) We have n circles with radii $r_1 < r_2 < \cdots < r_n$ and straight line L. How do we place these circles on the plane so that all of them touch L, with each circle, except the first one and the last one, touching both neighboring circles, so that the projection of this set of circles onto L has the maximum possible length?

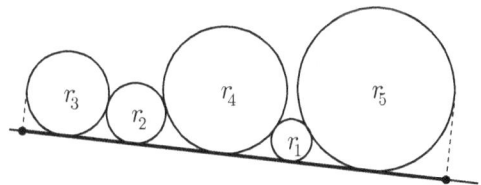

102. (1987) An automaton accepts two cards with numbers x and y and then prints out a card with their product xy (it also returns the input cards). Another automaton accepts one card with number x and produces a card with number x^2 (also returning the original input card).

At the beginning, we have one card with number a. Let us denote by $P(n)$ the smallest possible number of automata operations necessary to obtain a card with number a^n.

a) Find $P(n)$ for $n = 1, 2, \ldots, 16$.
b) Prove inequalities $\log_2 n \leqslant P(n) \leqslant 2 \log_2 n$.
c) Prove that $P(n) \leqslant \frac{3}{2} \log_2 n + C$.
d) Prove that $P(n) = \log_2 n + o(\log_2 n)$.

103. (1988) All values of a quadratic polynomial at points $-1, 0, 1$ lie inside interval $[0; 1]$.

Prove that any value of this polynomial inside interval $[-1; 1]$ does not exceed $\frac{9}{8}$.

104. (1988) Four of the cube's faces (all but the two opposite ones) are selected, and in each one, a diagonal is chosen in such a way that no two of them intersect. Prove that the four straight lines defined by these diagonals possess the following property: if some straight line L intersects three of them, then L either intersects the fourth line or is parallel to it.

105. (1988) Theo would like to split the set of all natural numbers greater than 1 into two subsets—"good" and "bad" numbers—in such a way that the product of any 239 different good numbers is bad, and the product of any 45 different bad numbers is good. Prove that he cannot do that.

106. (1988) $A_1 A_2 \ldots A_{2k}$ is a convex $2k$-gon with lengths of all of its sides equal to 1, where O is an arbitrary point on the plane. Prove that

$$\big||OA_1| - |OA_2| + \cdots + |OA_{2k-1}| - |OA_{2k}|\big| \leqslant \pi.$$

107. (1988) Consider equation

$$f(x) + f^{-1}(x) = g(x),\qquad\qquad (*)$$

where $f(x)$ is an unknown strictly monotonically increasing continuous function defined on $[0; \infty)$, and the given function $g(x)$ possesses the same properties while also being continuously differentiable. For which $g(x)$ does equation $(*)$ have solutions and how many?

a) Prove that if a solution exists, then for any non-negative x we have inequality

$$\int_0^x g(t)\,dt \geqslant x^2.\qquad\qquad (**)$$

b) Find all solutions of $(*)$ for $g(x) = cx$, c is a constant.

c) Prove that if $g'(0) > 2$ and for every positive x inequality $(**)$ holds true, then equation $(*)$ has no more than two solutions.

d) Prove that if $g'(x) \geqslant 2$ for all $x > 0$, and $g(x)$ is linear in some neighborhood of zero, then a solution exists.

e) Find an example of function $g(x)$ such that $g(x) > 10x$ for any positive x, but equation $(*)$ does not have any solutions.

108. (1988) Several circles on the plane are drawn in such a way that any two of them have a common point. Prove that there exist seven points on the plane such that every circle contains at least one of them.

109. (1988) Integers x, y, z are pairwise co-prime and such that $x^2 + y^2 = z^{2n}$, where $p = 4n - 1$ is a prime number. Prove that one of the numbers x and y is divisible by p.

110. (1988) Two people play the game, making their moves in turn. Originally, there is a pair of positive integers (a, b), and with each move, a player subtracts a multiple of the smaller number from the larger number so that the result is non-negative. The player who obtains a pair with zero, wins. Who wins in the errorless game? (The answer, of course, must be dependent on the numbers a and b.)

111. (1988) Every vertex of convex polyhedron M has exactly three incident edges. Prove that if the number of edges of M is $3n$, then any plane intersects no more than $n+2$ edges of M in interior points. Also show that this estimate is strict—there exists a convex polyhedron with the given properties and a plane which intersects exactly $n+2$ of its edges in their interior points.

112. (1989) Each one of the three cities has n inhabitants. It is known that each one of these $3n$ people has at least $2n$ acquaintances among them. Prove that we can select a person in each city so that these three people are all acquainted with each other.

113. (1989) Sequence of natural numbers (t_k) is defined as follows: $t_1 = 0$, $t_2 = 2$, $t_3 = 3$, and $t_n = t_{n-2} + t_{n-3}$ for any $n > 3$. Prove that for any prime p, number t_p is divisible by p.

114. (1989) Different real numbers a, b, c are such that $a + \frac{1}{b} = b + \frac{1}{c} = c + \frac{1}{a} = p$. Prove that $abc + p = 0$.

115. (1990) Let us glue to every face of unit cube $1 \times 1 \times 1$ a rectangular parallelepiped with dimensions $1 \times 1 \times 4$ (see the figure). Is it possible to tile the entire three-dimensional space with these "3d-crosses"?

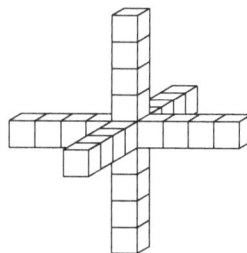

116. (1990) Let $M(p)$ be the number of the solutions of equation $x^2 + y^2 = p^2$ in integers. Prove that the sequence $(M(p)/p)$ has a limit as $p \to \infty$, and find that limit.

117. (1991) Given a set with 100 elements, how many ways are there to choose an ordered trio of its pairwise disjoint subsets?

118. (1991) Inscribed quadrilateral $ABCD$ has sides BC and CD of equal length. Prove that the area of this quadrilateral equals $\frac{1}{2}AC^2 \sin \angle DAB$.

119. (1991) For any two natural numbers a and b, define function

$$d(a, b) = \max \left(\frac{a}{\gcd(a, b)}, \frac{b}{\gcd(a, b)} \right).$$

Prove that $d(a, b)d(b, c) \geqslant d(a, c)$ for any three natural numbers a, b, and c.

120. (1991) One hundred vertices of a connected graph with 101 vertices are labeled with integers. We will call a *nudge* the following operation: adding 1 to or subtracting 1 from both numbers at the ends of any edge of the graph.

a) Prove that the one unlabeled vertex can be labeled by a natural number less than 100 so that it is possible to make all the numbers equal using nudge operations.

b) In the previous item, is it possible to replace 100 with a smaller number?

121. (1991) (a) Find three polynomials f, g, and h such that $f^3 + g^3 + h^3 = x^{1991}$. (b) Is it true that for any polynomial $p(x)$ one can find three polynomials with sum of their cubes equal to $p(x)$?

Problem Classification Guide

For the readers' convenience, we include here a problem classification guide which has all the questions from this collection assigned to certain general categories, themes, and methods. As you can easily see, some of these categories (such as "Polynomials" or "General Graph Theory") refer purely to the problem's statement, and some (e.g., "Mathematical Induction" or "Count the Angles")—to the problem's solution. As a consequence, most of the problems are listed at least twice.

Similarly, some of the categories cover the problems which have a specific mathematical "element" present in their statements, as well as the questions whose solution uses the corresponding method. Consider, for instance, category "Area", containing not only the questions about area computation or equations involving area but also the problems that use area as a helpful tool employed in their solutions.

1. NUMBERS AND FUNCTIONS

Arithmetic and Number Theory

Arithmetical Problems about Motion

1961.04, 1962.01, 1963.01, 1963.02, 1964.01, 1965.03, 1965.09, 1967.05, 1969.03, 1969.10, 1971.01, 1974.05, 1976.15, 1991.03.

Decimal Representation

1961.08, 1961.23, 1962.03, 1962.06, 1962.10, 1962.21, 1962.51, 1963.20, 1963.32, 1964.14, 1965.02, 1965.05, 1968.09, 1968.17, 1969.17, 1970.01, 1970.31, 1971.04, 1971.14, 1971.21, 1972.04, 1972.14, 1972.33, 1973.03, 1973.14, 1973.20, 1974.01, 1974.09, 1976.37, 1977.04, 1977.08, 1979.02, 1979.11, 1979.15, 1979.24, 1980.29, 1981.05, 1981.13, 1982.15, 1983.06, 1983.19, 1984.01, 1984.20, 1985.37, 1986.06, 1986.20, 1987.29, 1989.57, 1990.09, 1990.33, 1991.15.

Diophantine Equations and Inequalities

1962.14, 1962.36, 1963.11, 1963.25, 1963.37, 1964.13, 1964.26, 1965.01, 1965.18, 1965.25, 1966.19, 1968.01, 1969.02, 1970.23, 1971.03, 1971.37, 1973.01, 1974.14, 1979.05, 1979.32, 1980.02, 1980.13, 1981.06, 1982.40, 1984.10, 1984.27, 1985.04, 1985.09, 1986.15, 1986.22, 1986.31, 1987.04, 1987.20, 1988.25, 1989.10, 1990.07, 1990.16, 1991.08.

Divisibility, Modular Arithmetic

1961.22, 1962.06, 1962.10, 1962.52, 1963.03, 1963.12, 1963.17, 1963.18, 1963.35, 1965.05, 1967.14, 1967.18, 1967.24, 1970.03, 1972.06, 1974.36, 1975.17, 1976.01, 1976.07, 1978.15, 1979.17, 1981.01, 1981.07, 1981.16, 1982.04, 1982.09, 1983.05, 1985.19, 1987.02, 1987.11, 1987.35, 1988.04, 1988.09, 1988.39, 1990.14, 1990.21, 1990.43, 1991.02, 1991.36.

Divisibility Rules

1962.51, 1968.09, 1970.04, 1973.03, 1979.04, 1979.18, 1981.11, 1982.07, 1985.02, 1985.07.

Divisors, G.C.D., L.C.M., Co-prime Numbers

1961.02, 1961.22, 1962.02, 1963.06, 1964.11, 1964.18, 1966.30, 1976.14, 1978.05, 1979.22, 1979.26, 1989.22, 1989.37, 1989.49, 1990.59.

General Number Theory, Combinatorial Arithmetic

1964.28, 1964.30, 1965.07, 1965.23, 1968.36, 1969.23, 1969.28, 1970.34, 1972.18, 1972.27, 1980.25, 1981.36, 1982.44, 1983.07, 1983.41, 1984.44, 1988.05, 1989.50, 1989.60, 1989.64, 1991.47, 1991.57.

Prime and Composite Numbers, Prime Factorization

> 1962.21, 1963.08, 1963.40, 1964.08, 1964.21, 1966.08, 1966.15, 1966.21, 1966.32, 1971.17, 1972.16, 1972.35, 1973.10, 1974.38, 1975.08, 1975.14, 1975.25, 1975.33, 1976.02, 1976.18, 1976.37, 1977.27, 1978.03, 1979.06, 1981.37, 1983.17, 1985.51, 1986.03, 1987.42, 1989.12, 1991.56.

Algebra

Equations and Systems of Equations

> 1961.01, 1961.13, 1962.15, 1962.33, 1962.53, 1963.28, 1964.06, 1965.11, 1966.29, 1967.02, 1968.08, 1968.19, 1968.26, 1970.11, 1970.14, 1971.07, 1971.19, 1974.06, 1975.27, 1976.27, 1976.33, 1979.12, 1979.29, 1984.14, 1984.21, 1985.20, 1985.26, 1986.28, 1986.35, 1989.21, 1990.27, 1991.39.

Identities

> 1962.14, 1962.25, 1962.31, 1962.32, 1964.16, 1965.06, 1968.14, 1969.09, 1972.12, 1972.27, 1975.03, 1980.17, 1981.29, 1982.28, 1984.30, 1984.44, 1984.51, 1985.31, 1987.42, 1988.19, 1988.29, 1989.38, 1991.18, 1991.56.

Inequalities

> 1961.26, 1962.16, 1963.21, 1965.28, 1965.32, 1966.01, 1967.07, 1967.17, 1967.22, 1967.36, 1968.02, 1968.23, 1968.30, 1968.34, 1969.33, 1969.38, 1971.28, 1971.32, 1972.23, 1972.25, 1973.27, 1973.31, 1974.27, 1975.09, 1975.32, 1975.46, 1976.20, 1977.29, 1978.20, 1980.20, 1980.42, 1981.21, 1981.45, 1982.13, 1983.28, 1984.35, 1986.34, 1986.38, 1986.56, 1987.17, 1987.21, 1987.38, 1987.47, 1988.13, 1988.24, 1988.38, 1988.47, 1989.16, 1989.58, 1990.25, 1990.31, 1990.41, 1991.21, 1991.28, 1991.52.

Integer and Fractional Parts

> 1961.13, 1977.23, 1980.35.

Polynomials

> 1961.27, 1962.20, 1962.40, 1967.32, 1969.37, 1970.22, 1971.31, 1973.33, 1978.19, 1981.46, 1984.34, 1985.28, 1986.35, 1986.60, 1988.29, 1988.58, 1989.47, 1990.35, 1990.57.

Rational and Irrational Numbers

> 1961.12, 1961.27, 1962.24, 1963.41, 1969.21, 1973.26, 1981.40, 1982.29.

Quadratic Trinomials

> 1963.26, 1965.33, 1967.12, 1968.14, 1975.16, 1980.30, 1982.19, 1985.22, 1986.24, 1988.27.

Sequences

> 1963.15, 1963.30, 1966.17, 1967.20, 1967.36, 1969.18, 1969.23, 1969.28, 1970.32, 1971.29, 1972.24, 1972.31, 1973.26, 1974.30, 1977.05, 1978.32, 1981.28, 1982.42, 1983.31, 1983.43, 1985.14, 1985.29, 1985.42, 1985.43, 1985.44, 1987.17, 1987.48, 1987.56, 1988.50, 1989.61, 1989.71, 1990.23, 1991.24, 1991.38.

Trigonometry

> 1961.28, 1965.26, 1967.36, 1968.26, 1970.26, 1973.27, 1975.31, 1977.25, 1980.36, 1981.45, 1989.61, 1989.69.

Calculus

Derivative and Integral

> 1970.22, 1984.34, 1985.34, 1986.58, 1988.58, 1991.54.

Functions and their Graphs, Functional Equations and Inequalities

> 1961.26, 1969.10, 1970.29, 1975.42, 1976.26, 1977.23, 1986.55, 1987.54, 1988.34, 1988.54, 1989.40, 1990.65, 1991.31, 1991.46.

Real and Complex Numbers

> 1964.34, 1966.28, 1982.35, 1986.50, 1989.71, 1990.41, 1990.58, 1991.11, 1991.55.

2. Discrete mathematics

Combinatorics

Combinatorial Identities and Inequalities

> 1965.31, 1970.36, 1989.01.

Enumerative Combinatorics

> 1961.10, 1961.25, 1962.35, 1962.42, 1962.54, 1964.32, 1966.34, 1969.20, 1971.15, 1973.24, 1974.02, 1982.01, 1987.51, 1988.18, 1989.24, 1990.20.

General Combinatorics

> 1971.34, 1972.05, 1972.36, 1985.52, 1989.02.

Subsets, Permutations and Combinations

> 1961.03, 1966.23, 1967.35, 1987.40, 1987.60, 1990.50, 1990.56, 1990.66.

Graph Theory

Combinatorics in Graph Theory

> 1968.21, 1968.37, 1972.38, 1973.28, 1975.35, 1978.02, 1978.34, 1985.17, 1990.06, 1990.19.

General Graph Theory

1961.05, 1967.16, 1967.34, 1972.16, 1974.32, 1981.03, 1982.12, 1987.03, 1990.48, 1991.06, 1991.26, 1991.30.

Paths and Connectivity

1965.17, 1969.15, 1970.05, 1971.36, 1972.37, 1973.18, 1974.06, 1975.37, 1976.36, 1977.24, 1986.42, 1988.40, 1988.48, 1990.17, 1990.64, 1991.12, 1991.41.

Miscellaneous

Algorithms and Processes

1962.29, 1967.20, 1968.38, 1974.16, 1974.35, 1975.34, 1978.11, 1980.44, 1983.04, 1983.10, 1983.24, 1983.27, 1985.18, 1986.01, 1986.51, 1987.23, 1987.25, 1987.40, 1987.51, 1988.05, 1988.11, 1988.22, 1988.61, 1989.18, 1989.51, 1989.54, 1989.70, 1990.09, 1990.50, 1990.66, 1991.32, 1991.49, 1991.50, 1991.53.

Games and Strategies

1967.30, 1968.32, 1969.04, 1971.05, 1971.16, 1971.22, 1973.37, 1975.01, 1975.05, 1975.07, 1976.04, 1976.34, 1977.06, 1980.11, 1981.39, 1982.06, 1983.39, 1984.06, 1984.38, 1985.24, 1986.45, 1986.61, 1987.06, 1987.39, 1987.44, 1988.12, 1988.42, 1989.19, 1989.53, 1989.72, 1990.04, 1990.54, 1991.05.

Logic and Set Theory

1964.05, 1965.14, 1969.06, 1972.02, 1973.05, 1976.23, 1977.14, 1983.03, 1987.60, 1988.02, 1989.14, 1989.44, 1990.05, 1990.19, 1991.01.

Polyominoes

1961.21, 1963.13, 1964.02, 1971.12, 1980.23, 1984.19, 1991.13, 1991.19.

Problems on Graph Paper

1963.41, 1965.35, 1967.30, 1968.21, 1970.16, 1970.37, 1970.38, 1972.38, 1973.38, 1976.05, 1977.16, 1978.16, 1979.01, 1980.12, 1981.04, 1981.23, 1983.14, 1983.42, 1984.04, 1984.12, 1984.18, 1988.18, 1988.46, 1988.63, 1989.24, 1990.37, 1990.62, 1991.34.

Tables

1964.04, 1965.06, 1965.31, 1967.35, 1969.02, 1970.24, 1970.30, 1971.35, 1974.34, 1975.24, 1976.29, 1976.35, 1978.28, 1979.09, 1979.23, 1980.01, 1981.41, 1982.03, 1982.17, 1982.23, 1983.14, 1983.18, 1983.42, 1984.02, 1984.09, 1984.41, 1985.40, 1987.01, 1988.01, 1990.13.

Tournaments

> 1963.09, 1964.01, 1966.02, 1966.05, 1968.05, 1970.06, 1970.12, 1972.30, 1973.34, 1980.22, 1980.28, 1980.34, 1982.16, 1983.01, 1983.20, 1985.39, 1985.47, 1986.25, 1989.25, 1991.06, 1991.44.

Weighings and Similar Puzzles

> 1964.36, 1967.01, 1968.04, 1979.08, 1980.04, 1980.26, 1985.01, 1989.04, 1989.09, 1990.02, 1990.18.

3. GEOMETRY

Planar Geometry

Area

> 1962.49, 1963.07, 1964.33, 1965.08, 1965.34, 1966.18, 1966.31, 1967.10, 1967.21, 1968.25, 1969.08, 1972.07, 1973.11, 1973.21, 1973.36, 1973.40, 1976.17, 1977.03, 1977.12, 1978.27, 1980.24, 1981.20, 1983.21, 1983.22, 1983.38, 1985.15, 1985.38.

Circles and Inscribed Angles

> 1962.13, 1964.17, 1964.23, 1964.27, 1965.28, 1965.30, 1966.07, 1967.13, 1967.19, 1967.31, 1968.22, 1968.31, 1970.13, 1970.19, 1970.35, 1972.13, 1972.22, 1972.34, 1974.10, 1974.23, 1976.24, 1976.30, 1978.17, 1979.16, 1979.28, 1979.35, 1982.41, 1984.43, 1985.23, 1987.27, 1988.57, 1989.28, 1990.28, 1990.45, 1991.29, 1991.48.

Geometric Constructions

> 1961.07, 1961.11, 1961.17, 1962.07, 1963.22, 1964.09, 1964.15, 1964.29, 1967.06, 1968.20, 1969.14, 1987.18.

Geometric Inequalities

> 1961.09, 1962.01, 1962.04, 1962.17, 1962.19, 1962.27, 1962.30, 1962.47, 1964.12, 1964.25, 1964.31, 1964.35, 1965.19, 1965.22, 1966.31, 1968.22, 1968.25, 1969.11, 1969.19, 1970.09, 1970.17, 1971.02, 1971.08, 1971.18, 1971.30, 1972.29, 1972.32, 1973.13, 1973.36, 1973.39, 1974.29, 1975.12, 1975.15, 1975.21, 1976.28, 1978.06, 1979.19, 1980.43, 1980.45, 1981.23, 1981.38, 1984.22, 1984.33, 1985.41, 1987.41, 1987.57, 1988.49, 1988.60, 1989.34, 1989.52, 1990.47, 1990.55, 1991.40.

Inscribed and Circumscribed Polygons

> 1962.17, 1962.19, 1962.30, 1968.13, 1970.15, 1970.21, 1972.20, 1973.19, 1980.15, 1983.29, 1986.21, 1988.41, 1989.62, 1991.51.

Locus

> 1961.20, 1962.34, 1963.19, 1964.29, 1965.15, 1965.36, 1969.22, 1974.08, 1989.40, 1991.43.

Polygons

  1963.16, 1966.35, 1970.27, 1973.29, 1976.31, 1978.33, 1986.09, 1986.39,
  1989.08, 1990.61, 1991.33.

Quadrangles

  1962.12, 1966.16, 1967.08, 1968.07, 1969.13, 1969.36, 1972.09, 1972.15,
  1973.09, 1973.35, 1981.25, 1984.08, 1984.37, 1986.18, 1987.13, 1989.15,
  1989.23, 1990.10, 1990.24.

Similar Triangles

  1967.19, 1970.19, 1970.35, 1972.20, 1973.19, 1980.21, 1980.27, 1985.25.

The Triangle Inequality

  1963.05, 1964.07, 1968.03, 1972.01, 1975.04, 1976.03, 1976.10, 1976.25,
  1977.09, 1978.29, 1981.08, 1982.10, 1983.08, 1983.15, 1984.13, 1986.23,
  1987.30, 1988.10, 1989.19, 1991.05, 1991.10, 1991.22.

Triangles and their Elements

  1961.14, 1962.12, 1963.14, 1963.39, 1966.22, 1967.25, 1967.33, 1970.02,
  1970.07, 1971.23, 1971.33, 1973.13, 1974.12, 1975.13, 1976.19, 1977.13,
  1979.07, 1979.13, 1979.20, 1979.36, 1980.21, 1980.27, 1981.17, 1981.22,
  1982.20, 1983.25, 1984.15, 1984.23, 1985.08, 1986.16, 1987.08, 1987.22,
  1987.37, 1988.20, 1988.37, 1989.48, 1990.22, 1991.17, 1991.23.

Trigonometry in Planar Geometry

  1963.39, 1967.33, 1968.27, 1969.30, 1974.29, 1975.19, 1989.30.

Vectors

  1963.10, 1968.11, 1975.10, 1976.16, 1977.18, 1979.21, 1980.16, 1982.30,
  1982.35, 1982.38.

Combinatorial Planar Geometry

Convexity and Extremum Problems

  1962.38, 1962.47, 1963.34, 1967.23, 1970.33, 1971.11, 1971.38, 1973.16,
  1974.33, 1975.30, 1976.22, 1976.38, 1977.15, 1977.17, 1978.12, 1978.35,
  1979.27, 1980.06, 1980.40, 1980.48, 1982.30, 1987.43, 1988.53, 1988.60.

Coverings, Dissections and Tilings

  1962.48, 1963.23, 1963.38, 1964.25, 1965.04, 1965.10, 1965.13, 1965.21,
  1965.24, 1965.35, 1966.11, 1966.33, 1966.35, 1968.24, 1969.31, 1969.35,
  1971.06, 1972.17, 1973.06, 1973.08, 1973.38, 1974.15, 1975.38, 1976.06,
  1976.32, 1979.14, 1981.20, 1983.02, 1983.23, 1985.27, 1990.03, 1990.11,
  1990.32, 1990.44, 1990.49, 1990.60.

Planar Configurations of Points, Lines and other Figures

1961.16, 1961.19, 1962.11, 1962.28, 1962.38, 1963.31, 1964.20, 1965.12,
1966.04, 1966.09, 1966.20, 1967.21, 1967.29, 1968.12, 1968.33, 1969.11,
1969.32, 1969.34, 1970.18, 1972.03, 1973.32, 1974.17, 1974.37, 1975.23,
1978.24, 1978.31, 1979.25, 1980.03, 1982.24, 1982.43, 1983.12, 1983.44,
1986.44, 1986.54, 1986.59, 1988.44, 1989.03, 1989.26, 1989.55, 1989.63.

Solid Geometry

Combinatorial Geometry in Three Dimensions

1961.29, 1962.37, 1962.50, 1965.29, 1968.29, 1974.28, 1974.31, 1975.28,
1975.40, 1977.26, 1979.34, 1987.24, 1990.42.

General Solid Geometry

1963.27, 1963.29, 1968.35, 1969.25, 1970.25, 1971.26, 1971.27, 1972.26,
1977.28, 1980.39, 1981.34, 1982.34, 1983.34, 1983.35, 1985.33, 1986.37,
1988.30, 1988.36, 1989.41.

Volumes and Areas of Three-Dimensional Bodies and Surfaces

1961.24, 1963.33, 1963.36, 1974.26.

4. METHODS OF PROBLEM SOLVING

Upper/Lower Bound plus Example

1968.21, 1975.24, 1975.35, 1975.45, 1976.05, 1977.15, 1977.16, 1978.04,
1979.10, 1981.04, 1982.39, 1984.18, 1984.35, 1985.16, 1987.05, 1990.56.

Case-by-Case Analysis

1968.18, 1970.23, 1972.10, 1973.02, 1975.06, 1975.11, 1975.36, 1978.30,
1980.09, 1980.23, 1982.05, 1983.37, 1984.36, 1986.02, 1986.05, 1991.04.

Colorings

1961.21, 1963.04, 1963.13, 1964.02, 1968.15, 1969.01, 1971.12, 1980.12,
1981.24, 1981.27, 1981.35, 1984.12, 1984.42, 1986.11, 1986.53, 1987.16,
1987.19, 1988.06.

Constructions and Counterexamples

1962.05, 1969.05, 1969.24, 1969.29, 1971.24, 1972.14, 1973.04, 1973.06,
1974.04, 1974.13, 1974.25, 1976.09, 1977.02, 1978.16, 1979.01, 1980.05,
1980.41, 1981.03, 1981.05, 1981.19, 1981.42, 1981.44, 1982.17, 1982.37,
1986.04, 1986.22, 1986.31, 1987.01, 1987.03, 1987.12, 1988.03, 1988.06,
1988.21, 1988.44, 1988.62, 1989.20, 1989.31, 1990.08, 1990.50, 1991.06,
1991.07, 1991.20, 1991.45.

Count the Angles

1961.20, 1970.07, 1973.19, 1979.20, 1979.28, 1982.14, 1982.20, 1982.02, 1983.09, 1983.40, 1984.28, 1985.13, 1985.21, 1986.41, 1987.27, 1988.14, 1988.20, 1988.26, 1988.37, 1989.48, 1990.63.

Geometric Transformations

1962.12, 1966.22, 1971.09, 1971.13, 1974.22, 1976.38, 1978.10, 1986.29, 1986.57, 1991.25.

Invariants and Monovariants

1975.18, 1981.12, 1981.43, 1984.05, 1984.39, 1984.52, 1986.19, 1988.01, 1988.22, 1988.23, 1988.43, 1989.13, 1989.54, 1991.27.

Mathematical Induction

1961.16, 1962.05, 1962.35, 1962.42, 1963.30, 1964.28, 1965.17, 1965.28, 1966.23, 1966.32, 1967.14, 1967.29, 1967.36, 1968.33, 1969.05, 1969.06, 1970.05, 1971.32, 1971.36, 1972.37, 1974.24, 1974.30, 1975.03, 1975.23, 1976.35, 1976.36, 1977.29, 1980.33, 1980.38, 1980.46, 1982.42, 1985.44, 1987.23, 1987.26, 1987.48, 1987.56, 1988.40, 1989.18, 1989.49, 1989.60, 1990.46, 1990.48, 1991.44, 1991.49.

Moving Backwards

1969.18, 1982.21, 1986.43.

Parity

1961.15, 1961.21, 1961.30, 1963.04, 1963.13, 1964.02, 1966.03, 1968.15, 1971.12, 1972.04, 1972.12, 1973.15, 1974.03, 1975.02, 1976.06, 1976.13, 1977.01, 1978.01, 1978.07, 1979.23, 1980.01, 1981.02, 1982.02, 1982.11, 1983.26, 1984.03, 1984.16, 1985.11, 1986.40, 1988.35, 1989.11, 1990.01, 1991.09.

Pigeonhole Principle

1962.52, 1963.18, 1964.04, 1966.32, 1967.03, 1967.04, 1967.34, 1970.16, 1973.28, 1975.36, 1979.03, 1979.33, 1981.31, 1983.37, 1984.24, 1985.05, 1986.13, 1988.02, 1991.42.

The Extremum Principle (Infinite Descent)

1961.19, 1962.28, 1964.03, 1967.21, 1973.17, 1973.18, 1974.18, 1982.36, 1982.44, 1983.27, 1984.40, 1985.03, 1988.11.

List of Authors

Below we list the authors of nearly every problem offered at the Leningrad Mathematical Olympiads in 1980–1991. Some of the problems had more than one author. Unfortunately, there exists virtually no authorship data for the years before 1980.

E. V. ABAKUMOV	88.06, 88.46, 88.61, 91.07, 91.10, 91.26, 91.30
A. A. BERZIŇŠ	87.43
S. L. BERLOV	91.53
A. V. BOGOMOLNAYA	87.13, 87.35, 88.01, 88.04
A. YU. BURAGO	89.53
D. YU. BURAGO	85.42
V. N. DUBROVSKY	86.41
I. B. FESENKO	80.36, 80.39, 81.25

D. V. Fomin	82.24, 82.37, 83.18, 83.31, 83.38, 84.05, 84.19,
	84.24, 85.03, 85.07, 85.09, 85.19, 85.24, 86.42,
	86.44, 86.59, 87.13, 87.17, 87.18, 87.23, 87.27,
	87.30, 87.47, 87.51, 87.54, 88.02, 88.03, 88.09,
	88.11, 88.19, 88.21, 88.22, 88.24, 88.25, 88.26,
	88.35, 88.39, 88.47, 88.48, 88.60, 89.02, 89.03,
	89.04, 89.11, 89.13, 89.14, 89.18, 89.20, 89.22,
	89.23, 89.25, 89.26, 89.31, 89.34, 89.37, 89.38,
	89.40, 89.47, 89.49, 89.51, 89.52, 89.54, 89.55,
	89.58, 89.60, 89.61, 89.64, 89.69, 90.01, 90.02,
	90.03, 90.05, 90.07, 90.08, 90.10, 90.11, 90.14,
	90.16, 90.18, 90.20, 90.21, 90.22, 90.25, 90.27,
	90.28, 90.32, 90.41, 90.43, 90.44, 90.45, 90.47,
	90.48, 90.49, 90.55, 90.57, 90.58, 90.60, 90.61,
	90.62, 91.02, 91.03, 91.05, 91.06, 91.08, 91.09,
	91.11, 91.13, 91.15, 91.17, 91.18, 91.19, 91.20,
	91.22, 91.23, 91.25, 91.26, 91.27, 91.29, 91.31,
	91.33, 91.39, 91.43, 91.47, 91.49, 91.55, 91.56,
	91.57
S. V. Fomin	80.03, 80.05, 80.17, 80.26, 81.20, 81.38, 81.42,
	82.01, 82.02, 82.04, 82.06, 82.09, 82.15, 82.17,
	82.19, 82.36, 82.38, 82.40, 83.03, 83.27, 84.02,
	84.09, 84.10, 84.12, 84.34, 84.41, 84.51, 85.01,
	85.04, 85.05, 85.14, 85.18, 85.27, 86.03, 86.06,
	86.13, 86.22, 86.25, 86.31, 86.51, 87.02, 87.04,
	87.06, 87.12, 87.19, 87.42, 88.53, 89.21, 90.46
S. A. Genkin	81.22, 83.14, 84.01, 84.03, 84.06, 84.16, 84.23,
	84.42, 85.08, 85.17, 85.41, 85.44, 85.51, 86.01,
	86.02, 86.25, 86.45, 86.50, 86.55, 86.61, 87.01,
	87.05, 87.08, 87.39, 87.48, 88.05, 88.40, 88.54,
	88.58, 88.62, 89.01, 89.06, 89.12, 89.28, 89.44,
	89.63, 90.13, 90.17, 90.23, 90.24, 90.33, 90.35,
	90.56, 90.65, 91.12, 91.24, 91.38, 91.44
A. S. Golovanov	89.10
A. G. Goldberg	80.13, 80.21, 80.24, 80.27, 82.28, 83.40, 84.8, 84.14,
	85.26, 85.33, 86.56
V. A. Gordeyev	82.29
D. Yu. Grigoriev	80.40
M. N. Gusarov	81.24, 82.20, 83.44, 89.05, 89.41
O. T. Izhboldin	81.23, 81.38, 82.14, 82.44, 83.42, 83.43, 89.71,
	90.50, 90.66

Y. I. IONIN	83.28, 83.34, 84.13, 84.15, 84.27, 84.36, 84.46, 85.22, 85.39, 85.47
I. V. ITENBERG	84.4, 84.18, 88.12, 88.18, 88.40, 90.04, 90.07, 90.17, 90.24, 90.33, 90.37, 90.64, 91.06, 91.07, 91.12, 91.38
A. L. KIRICHENKO	90.54, 91.50
V. E. KOZYREV	86.43
A. L. KOLDOBSKY	83.24
N. N. KONSTANTINOV	87.40
K. P. KOKHAS	89.72, 90.59, 91.01, 91.32
L. D. KURLIANDCHIK	80.20, 81.29, 81.45, 85.28, 85.37, 86.15, 86.60, 87.38, 91.42, 91.45
A. N. LIVSHITS	83.41, 87.21, 87.26, 87.56
A. S. MERKURJEV	80.17, 80.22, 80.28, 80.29, 80.34, 81.36, 81.40, 81.44, 81.46, 82.21, 82.35, 82.42, 84.20, 84.39, 84.43, 84.44, 84.47, 84.52, 85.29, 85.52, 86.19, 86.28, 86.40, 87.41, 88.41, 88.43, 89.57, 89.62
F. L. NAZAROV	85.16, 87.10, 87.20, 87.22, 87.24, 87.57, 88.10, 88.13, 88.23, 88.29, 88.34, 88.38, 88.44, 88.49, 88.50, 88.63, 89.08, 89.15, 89.16, 89.19, 89.24, 89.50, 89.70, 90.06, 90.09, 90.19, 90.31, 91.04, 91.10, 91.21, 91.28, 91.34, 91.36, 91.40, 91.46, 91.48, 91.52, 91.54
Z. H. NASYROV	91.51
V. S. NEIMAN	80.33, 80.38
N. YU. NETSVETAEV	81.43, 85.43, 86.11, 86.38, 86.39, 86.50, 86.53, 86.54, 87.23, 87.44, 88.27, 89.30, 90.42
A. I. ORLOV	90.63
G. YA. PEREL'MAN	84.40
A. I. PLOTKIN	82.16, 85.34, 86.58
V. A. POGREBNYAK	87.37
V. YU. PROTASOV	88.37
I. S. RUBANOV	81.39, 88.42
S. E. RUKSHIN	80.23, 81.21, 81.31, 81.41, 82.12, 82.23, 83.21, 83.25, 83.29, 83.35, 83.37, 84.22, 84.28, 84.35, 84.38, 85.15, 86.24, 87.29, 88.20

Bibliography

[1] I. Chistyakov (**1934**), *Itogi leningradskoy matematicheskoy olimpiady (in Russian)*, MATEMATIKA I FIZIKA V SREDNEY SHKOLE, vol.4. pp. 134–136, Uchpedgiz, Moscow, USSR

[2] I. Chistyakov (**1935**), *Matematicheskaya olimpiada Leningradskogo gosudarstvennogo universiteta imeni A.S.Bubnova (in Russian)*, MATEMATICHESKOE PROSVESCHENIE, vol.3., pp. 59–65, ONTI NKTP, Moscow, USSR

[3] E. Beckenbach, R. Bellman (**1961**), *Introduction to Inequalities*, Random House, New York, NY

[4] A. Leman (**1965**), *Sbornik zadach Moskovskih matematicheskih olimpiad (in Russian)*, Prosveschenie, Moscow, USSR

[5] H.S.M. Coxeter, S. Greitzer (**1967**), *Geometry Revisited*, The Mathematical Association of America, New York, NY

[6] H.S.M. Coxeter (**1969**), *Introduction to Geometry*, John Wiley and Sons, New York, NY

[7] M. Aleksandrova (**1984**), *Pervaya matematicheskaya olimpiada (in Russian)*, Kvant, **9**

[8] S. Rukshin, N. Matveyev (**1984**), *50 let matematicheskih olympiad (in Russian)*, MATEMATIKA V SHKOLE, **4**, Moscow, USSR

[9] D. Pedoe (**1988**), *Geometry. A Comprehensive Course*, Dover Publications, Mineola, NY

[10] O. Ore (**1991**), *Graphs and their uses*, The Mathematical Association of America, New York, NY

[11] D. Fomin (**1994**), *Sankt-Peterburgskie matematicheskie olympiady (in Russian)*, Politekhnika, St. Petersburg, Russia

[12] D. Fomin, A. Kirichenko (**1994**), *Leningrad Mathematical Olympiads* 1987–1991, MathPro Press, Chelmsford, MA

[13] S. Golomb (**1994**), *Polyominoes*, Princeton University Press, Princeton, NJ

[14] F. Swetz (ed.) (**1994**), *From Five Fingers to Infinity. A Journey through the History of Mathematics*, Open Court Publishing, Chicago, IL

[15] D. Leites (ed.), G. Galperin, A. Tolpygo, et al. (**1997**), 60-*odd Years of Moscow Mathematical Olympiads*, preprint @ Library Genesis

[16] Sextus Empiricus, J. Annas, J. Barnes (eds.) (**2000**), *Outlines of Skepticism*, Cambridge University Press, Cambridge, UK

[17] S. Berlov, S. Ivanov, K. Kokhas (**2000**), *Peterburgskie matematicheskie olympiady (in Russian)*, Lan', St. Petersburg, Russia

[18] B. Yandell (**2002**), *The Honors Class. Hilbert's Problems and Their Solvers*, A K Peters, Natick, MA

[19] R. Guy (**2004**), *Unsolved Problems in Number Theory*, Springer-Verlag, New York, NY

[20] J. Michael Steele (**2004**), *The Cauchy-Schwarz Master Class. An Introduction to the Art of Mathematical Inequalities*, Cambridge University Press, New York, NY

[21] K. Kokhas, D. Fomin et al. (**2005**), *Peterburgskie matematicheskie olympiady, 1961–1993 (in Russian)*, Lan', St. Petersburg, Russia

[22] S. Vassar (**2006**), *Manifold Destiny*, New Yorker, New York, NY

[23] S. Berlov, S. Ivanov, D. Karpov, K. Kokhas et al. (**2006**), *Peterburgskie olympiady shkolnikov po matematike. 2000–2002 (in Russian)*, Nevsky Dialect (BHV Peterburg), St. Petersburg, Russia

[24] S. Berlov, S. Ivanov, D. Karpov, K. Kokhas et al. (**2007**), *Peterburgskie olympiady shkolnikov po matematike. 2003–2005 (in Russian)*, Nevsky Dialect (BHV Peterburg), St. Petersburg, Russia

[25] A. Karp, B. Vogeli (eds) (**2010**), *Russian Math Education, History and World Significance*, World Scientific, Singapore

[26] D. Epstein, Ya. Shapiro, S. Ivanov (eds.) (**2011**), *Math-Mekh LGU, shestidesyatye i ne tol'ko. Sbornik vospominaniy (in Russian)*, Kopi-R Grupp, St. Petersburg, Russia

[27] J. O'Rourke (**2011**), *How to Fold it. The Mathematics of Linkages, Origami and Polyhedra*, Cambridge University Press, New York,NY

[28] R.E. Raspe (**2012**), *The Travels and Surprising Adventures of Baron Munchausen*, Melville House Publishing, Brooklyn, NY

[29] Ya.N. Shapiro (ed.) (**2015**), *Math-Mekh LGU-SPbGU ot istokov do dney nedavnih. Dopolnitel'nye glavy. Sbornik materialov (in Russian)*, St. Petersburg, Russia

[30] O. Ore (**2017**), *Invitation to Number Theory*, The Mathematical Association of America, Washington, DC

[31] G.I. Sinkevich, A.I. Nazarov (eds.) (**2018**), *Matematicheskiy Peterburg. Istoriya, nauka, dostoprimechatel'nosti*, (in Russian), Obrazovatelnye Proekty, St. Petersburg, Russia

[32] D. Fomin, K. Kokhas (**2022**), *Leningradskie matematicheskie olympiady. 1961–1991 (in Russian)*, MCCME, Moscow, Russia

www.ingramcontent.com/pod-product-compliance
Lightning Source LLC
Chambersburg PA
CBHW081218220326
41598CB00037B/6819